T0181111

Advances in Intelligent Systems and Computing

Volume 515

Series editor

Janusz Kacprzyk, Polish Academy of Sciences, Warsaw, Poland
e-mail: kacprzyk@ibspan.waw.pl

About this Series

The series "Advances in Intelligent Systems and Computing" contains publications on theory, applications, and design methods of Intelligent Systems and Intelligent Computing. Virtually all disciplines such as engineering, natural sciences, computer and information science, ICT, economics, business, e-commerce, environment, healthcare, life science are covered. The list of topics spans all the areas of modern intelligent systems and computing.

The publications within "Advances in Intelligent Systems and Computing" are primarily textbooks and proceedings of important conferences, symposia and congresses. They cover significant recent developments in the field, both of a foundational and applicable character. An important characteristic feature of the series is the short publication time and world-wide distribution. This permits a rapid and broad dissemination of research results.

Advisory Board

Chairman

Nikhil R. Pal, Indian Statistical Institute, Kolkata, India
e-mail: nikhil@isical.ac.in

Members

Rafael Bello Perez, Universidad Central "Marta Abreu" de Las Villas, Santa Clara, Cuba
e-mail: rbellop@uclv.edu.cu

Emilio S. Corchado, University of Salamanca, Salamanca, Spain
e-mail: escorchado@usal.es

Hani Hagras, University of Essex, Colchester, UK
e-mail: hani@essex.ac.uk

László T. Kóczy, Széchenyi István University, Győr, Hungary
e-mail: koczy@sze.hu

Vladik Kreinovich, University of Texas at El Paso, El Paso, USA
e-mail: vladik@utep.edu

Chin-Teng Lin, National Chiao Tung University, Hsinchu, Taiwan
e-mail: ctlin@mail.nctu.edu.tw

Jie Lu, University of Technology, Sydney, Australia
e-mail: Jie.Lu@uts.edu.au

Patricia Melin, Tijuana Institute of Technology, Tijuana, Mexico
e-mail: epmelin@hafsamx.org

Nadia Nedjah, State University of Rio de Janeiro, Rio de Janeiro, Brazil
e-mail: nadia@eng.uerj.br

Ngoc Thanh Nguyen, Wroclaw University of Technology, Wroclaw, Poland
e-mail: Ngoc-Thanh.Nguyen@pwr.edu.pl

Jun Wang, The Chinese University of Hong Kong, Shatin, Hong Kong
e-mail: jwang@mae.cuhk.edu.hk

More information about this series at http://www.springer.com/series/11156

Suresh Chandra Satapathy · Vikrant Bhateja
Siba K. Udgata · Prasant Kumar Pattnaik
Editors

Proceedings of the 5th International Conference on Frontiers in Intelligent Computing: Theory and Applications

FICTA 2016, Volume 1

 Springer

Editors
Suresh Chandra Satapathy
Department of Computer Science
  and Engineering
ANITS
Visakhapatnam, Andhra Pradesh
India

Vikrant Bhateja
Department of ECE
Shri Ramswaroop Memorial Group
  of Professional Colleges
Lucknow, Uttar Pradesh
India

Siba K. Udgata
SCIS, University of Hyderabad
Hyderabad
India

Prasant Kumar Pattnaik
School of Computer Engineering
KIIT University
Bhubaneswar, Odisha
India

ISSN 2194-5357   ISSN 2194-5365   (electronic)
Advances in Intelligent Systems and Computing
ISBN 978-981-10-3152-6   ISBN 978-981-10-3153-3   (eBook)
DOI 10.1007/978-981-10-3153-3

Library of Congress Control Number: 2016957505

Printed on acid-free paper

This Springer imprint is published by Springer Nature
The registered company is Springer Nature Singapore Pte Ltd.
The registered company address is: 152 Beach Road, #21-01/04 Gateway East, Singapore 189721, Singapore

Preface

The volume is a collection of high-quality peer-reviewed research papers presented at the 5th International Conference on Frontiers in Intelligent Computing: Theory and Applications (FICTA 2016) held at School of Computer Engineering, KIIT University, Bhubaneswar, Odisha, India during 16–17 September 2016.

The idea of this conference series was conceived by few eminent professors and researchers from premier institutions of India. The first three editions of this conference: FICTA 2012, 2013 and 2014 were organized by Bhubaneswar Engineering College (BEC), Bhubaneswar, Odisha, India. Owing to its popularity and wider visibility in the entire country as well as abroad; the fourth edition of FICTA 2015 has been organized by the prestigious NIT, Durgapur, West Bengal, India. All papers of past FICTA editions are published by Springer AISC Series. Presently, FICTA 2016 is the fifth edition of this conference series which brings researchers, scientists, engineers, and practitioners on to a single platform to exchange and share their theories, methodologies, new ideas, experiences and applications in all areas of intelligent computing theories and applications to various engineering disciplines such as computer science, electronics, electrical, mechanical, biomedical engineering, etc.

FICTA 2016 had received a good number of submissions from the different areas relating to intelligent computing and its applications and after a rigorous peer-review process with the help of our program committee members and external reviewers (from the country as well as abroad) good quality papers were selected for publications. The review process has been very crucial with minimum two reviews each; and in many a cases three to five reviews along with due checks on similarity and overlaps as well. The number of papers received under FICTA 2016 touched 500 mark including the main track as well as special sessions. The conference featured eight special sessions in various cutting-edge technologies of specialized focus which were organized and chaired by eminent professors. The total number of papers received includes submissions from eight overseas countries. Out of this pool of received papers, only 150 papers were accepted yielding an acceptance ratio of 0.4. These papers have been arranged in two separate volumes: Volume 1: 79 chapters and Volume 2: 71 chapters.

The conference featured many distinguished keynote addresses by eminent speakers like Dr. João Manuel R S Tavares, Faculdade de Engenharia da Universidade do Porto Porto, Portugal on Computational Image Analysis: Methods and Applications; Dr. Swagatam Das, ISI Kolkata on Swarm Intelligence; Dr. Rajib Mall, Information Technology IIT, Kharagpur; Dr. V. Ravi, Centre of Excellence in Analytics Institute for Development and Research in Banking Technology (IDRBT), (Established by RBI, India) Hyderabad; Dr. Gautam Sanyal, Professor, Department of CSE, NIT Durgapur on Image Processing and Intelligent Computing; and a plenary session on 'How to write for and get published in scientific journals?' by Mr. Aninda Bose, Senior Publishing Editor, Springer India.

We thank the General Chairs: Prof. Samaresh Mishra and Prof. Jnyana Ranjan Mohanty at KIIT University Bhubaneswar, India, for providing valuable guidance and inspiration to overcome various difficulties in the process of organizing this conference. We extend our heartfelt thanks to the Honorary Chairs of this conference: Dr. B.K. Panigrahi, IIT Delhi and Dr. Swagatam Das, ISI Kolkota for being with us from the very beginning to the end of this conference, without their support this conference could never have been successful. Our heartfelt thanks are due to Prof. P.N. Suganthan, NTU Singapore for his valuable suggestions on enhancing editorial review process.

We would also like to thank School of Computer Engineering, KIIT University, Bhubaneswar, for coming forward to support us to organize the fifth edition of this conference series. We are amazed to note the enthusiasm of all faculty, staff and students of KIIT to organize the conference in a professional manner. Involvements of faculty coordinators and student volunteers are praiseworthy in every respect. We are confident that in future too we would like to organize many more international level conferences in this beautiful campus. We would also like to thank our sponsors for providing all the support and financial assistance.

We take this opportunity to thank authors of all submitted papers for their hard work, adherence to the deadlines and patience with the review process. The quality of a refereed volume depends mainly on the expertise and dedication of the reviewers. We are indebted to the program committee members and external reviewers who not only produced excellent reviews but also did these in short time frames.

We would also like to thank the participants of this conference, who have considered the conference above all hardships. Finally, we would like to thank all the volunteers who spent tireless efforts in meeting the deadlines and arranging every detail to make sure that the conference can run smoothly. Additionally, CSI Students Branch of ANITS and its team members have contributed a lot to this conference and deserve an appreciation for their contribution.

All efforts are worth and would please us all, if the readers of these proceedings and participants of this conference found the papers and conference inspiring and enjoyable. Our sincere thanks go to all in press, print and electronic media for their excellent coverage of this conference.

We take this opportunity to thank all keynote speakers, track and special session chairs for their excellent support to make FICTA 2016 a grand success.

Visakhapatnam, India   Suresh Chandra Satapathy
Lucknow, India   Vikrant Bhateja
Hyderabad, India   Siba K. Udgata
Bhubaneswar, India   Prasant Kumar Pattnaik

Organisation Committee

Chief Patron

Achyuta Samanta, KISS & KIIT, Bhubaneswar, India

Patron

P.P. Mathur, KIIT University, Bhubaneswar, India

Advisory Committee

Sasmita Samanta, KIIT University, Bhubaneswar, India
Ganga Bishnu Mund, KIIT University, Bhubaneswar, India
Madhabananda Das, KIIT University, Bhubaneswar, India
Prasant Kumar Pattnaik, KIIT University, Bhubaneswar, India
Sudhanshu Sekhar Singh, KIIT University, Bhubaneswar, India

General Chair

Samaresh Mishra, KIIT University, Bhubaneswar, India

General Co-chair

Jnyana Ranjan Mohanty, KIIT University, Bhubaneswar, India

Honorary Chairs

Dr. Swagatam Das, ISI Kolkota, India
Dr. B.K. Panigrahi, IIT Delhi, India

Convener

Chittaranjan Pradhan, KIIT University, Bhubaneswar, India

Organizing Chairs

Sachi Nandan Mohanty, KIIT University, Bhubaneswar, India
Sidharth Swarup Routaray, KIIT University, Bhubaneswar, India

Publication Chairs

Suresh Chandra Satapathy, ANITS, Visakhapatnam, India
Vikrant Bhateja, SRMGPC, Lucknow (U.P.), India

Steering Committee

Suresh Chandra Satapathy, ANITS, Visakhapatnam, India
Siba K. Udgata, UoH, Hyderabad, India
Manas Sanyal, University of Kalayani, India
Nilanjan Dey, TICT, Kolkota, India
B.N. Biswal, BEC, Bhubaneswar, India
Vikrant Bhateja, SRMGPC, Lucknow (U.P.), India

Editorial Board

Suresh Chandra Satapathy, ANITS, Visakhapatnam, India
Vikrant Bhateja, SRMGPC, Lucknow (U.P.), India
Siba K. Udgata, UoH, Hyderabad, India
Prasant Kumar Pattnaik, KIIT University, Bhubaneswar, India

Transport and Hospitality Chairs

Harish K. Pattnaik, KIIT University, Bhubaneswar, India
Manas K. Lenka, KIIT University, Bhubaneswar, India
Ramakant Parida, KIIT University, Bhubaneswar, India

Session Management Chairs

Manoj Kumar Mishra, KIIT University, Bhubaneswar, India
Suresh Ch. Moharana, KIIT University, Bhubaneswar, India
Sujoya Datta, KIIT University, Bhubaneswar, India

Registration Chairs

Himansu Das, KIIT University, Bhubaneswar, India
Manjusha Pandey, KIIT University, Bhubaneswar, India
Sharmistha Roy, KIIT University, Bhubaneswar, India
Sital Dash, KIIT University, Bhubaneswar, India

Publicity Chairs

J.K. Mandal, University of Kalyani, Kolkota, India
K. Srujan Raju, CMR Technical Campus, Hyderabad, India

Track Chairs

Machine Learning Applications: Steven L. Fernandez, SCEM, Mangalore, India.
Image Processing and Pattern Recognition: V.N. Manjunath Aradhya, SJCE, Mysore, India.
Big Data, Web Mining & IoT: Sireesha Rodda, Gitam University, Visakhapatnam, India.

Signals, Communication and Microelectronics: A.K. Pandey, MIET, Meerut (U.P.), India.

MANETs and Wireless Sensor Networks: Pritee Parwekar, ANITS, Visakhapatnam, India.

Network Security and Cryptography: R.T. Goswami, BITS Mesra, Kolkota Campus, India

Data Engineering: M. Ramakrishna Murty, ANITS, Visakhapatnam, India.

Data Mining: B. Janakiramaiah, DVR & Dr. HS MICCT, Kanchikacherla, India.

Special Session Chairs

SS01: Wireless Sensor Networks: Architecture, Applications, Data Management & Security: Pritee Parwekar, ANITS, Visakhapatnam, India and Sireesha Rodda, Gitam University, Visakhapatnam, India

SS02: Advances in Document Image Analysis:- V.N. Manjunath Aradhya, SJCE, Mysore, India

SS03: Computational Swarm Intelligence:- Sujata Dash, North Orissa University, Baripada, India and Atta Ur Rehman, (BIIT) PMAS Arid Agricultural University, Rawalpindi, Punjab, Pakistan

SS04: Wide-Area-Wireless Communication Systems-Microwave, 3G, 4G and WiMAX:- K. Meena (Jeyanthi), PSNACET, Dindigul, Tamil Nadu, India

SS05: Innovation in Engineering and Technology through Computational Intelligence Techniques:- Jitendra Agrawal, RGPV, M.P., India and Shikha Agrawal, RGPV, M.P., India

SS06: Optical Character Recognition and Natural Language Processing:- Nikisha B. Jariwala, Smt. Tanuben & Dr. Manubhai Trivedi CIS, Surat, Gujarat, India and Hardik A. Vyas, BMaIIT, Uka Tarsadia University, Bardoli, Gujarat, India

SS07: Recent Advancements in Information Security:- Musheer Ahmad Jamia Millia Islamia, New Delhi, India

SS08: Software Engineering in Multidisciplinary Domains:- Suma V, Dayananda Sagar College of Engineering, Bangalore, India

Technical Program Committee/International Reviewer Board

A. Govardhan, India
Aarti Singh, India
Almoataz Youssef Abdelaziz, Egypt
Amira A. Ashour, Egypt
Amulya Ratna Swain, India
Ankur Singh Bist, India
Athanasios V. Vasilakos, Athens
Banani Saha, India
Bhabani Shankar Prasad Mishra, India
B. Tirumala Rao, India
Carlos A. Coello, Mexico
Charan S.G., India

Chirag Arora, India
Chilukuri K. Mohan, USA
Chung Le, Vietnam
Dac-Nhuong Le, Vietnam
Delin Luo, China
Hai Bin Duan, China
Hai V. Pham, Vietnam
Heitor Silvério Lopes, Brazil
Igor Belykh, Russia
J.V.R. Murthy, India
K. Parsopoulos, Greece
Kamble Vaibhav Venkatrao, India
Kailash C. Patidar, South Africa
Koushik Majumder, India
Lalitha Bhaskari, India
Jeng-Shyang Pan, Taiwan
Juan Luis Fernández Martínez, California
Le Hoang Son, Vietnam
Leandro Dos Santos Coelho, Brazil
L. Perkin, USA
Lingfeng Wang, China
M.A. Abido, Saudi Arabia
Maurice Clerc, France
Meftah Boudjelal, Algeria
Monideepa Roy, India
Mukul Misra, India
Naeem Hanoon, Malysia
Nikhil Bhargava, India
Oscar Castillo, Mexcico
P.S. Avadhani, India
Rafael Stubs Parpinelli, Brazil
Ravi Subban, India
Roderich Gross, England
Saeid Nahavandi, Australia
Sankhadeep Chatterjee, India
Sanjay Sengupta, India
Santosh Kumar Swain, India
Saman Halgamuge, India
Sayan Chakraborty, India
Shabana Urooj, India
S.G. Ponnambalam, Malaysia
Srinivas Kota, Nebraska
Srinivas Sethi, India
Sumanth Yenduri, USA
Suberna Kumar, India

T.R. Dash, Cambodia
Vipin Tyagi, India
Vimal Mishra, India
Walid Barhoumi, Tunisia
X.Z. Gao, Finland
Ying Tan, China
Zong Woo Geem, USA
M. Ramakrishna Murthy, India
Tushar Mishra, India
And many more …

Contents

About the Editors

Dr. Suresh Chandra Satapathy is currently working as Professor and Head, Department of Computer Science and Engineering, Anil Neerukonda Institute of Technology and Sciences (ANITS), Visakhapatnam, Andhra Pradesh, India. He obtained his Ph.D. in Computer Science Engineering from JNTUH, Hyderabad and Master degree in Computer Science and Engineering from National Institute of Technology (NIT), Rourkela, Odisha. He has more than 27 years of teaching and research experience. His research interest includes machine learning, data mining, swarm intelligence studies and their applications to engineering. He has more than 98 publications to his credit in various reputed international journals and conference proceedings. He has edited many volumes from Springer AISC, LNEE, SIST and LNCS in past and he is also the editorial board member in few international journals. He is a senior member of IEEE and life member of Computer Society of India. Currently, he is the National Chairman of Division-V (Education and Research) of Computer Society of India.

Prof. Vikrant Bhateja is Associate Professor, Department of Electronics and Communication Engineering, Shri Ramswaroop Memorial Group of Professional Colleges (SRMGPC), Lucknow and also the Head (Academics and Quality Control) in the same college. His areas of research include digital image and video processing, computer vision, medical imaging, machine learning, pattern analysis and recognition. He has published 100 high-quality publications in various international journals and conference proceedings. Professr Vikrant has been on TPC and chaired various sessions from the above domain in international conferences of IEEE and Springer. He has been the track chair and served in the core-technical/editorial teams for international conferences: FICTA 2014, CSI 2014, INDIA 2015, ICICT 2015 and ICTIS 2015 under Springer-ASIC Series and INDIACom 2015, ICACCI 2015 under IEEE. He is associate editor of International Journal of Synthetic Emotions (IJSE) and International Journal of Ambient Computing and Intelligence (IJACI) under IGI Global. He also serves the editorial board of International Journal of Image

Mining (IJIM) and International Journal of Convergence Computing (IJConvC) under Inderscience Publishers. He has been editor of three published volumes with Springer and another two are in press.

Dr. Siba K. Udgata is Professor of School of Computer and Information Sciences, University of Hyderabad, India. He is presently heading Centre for Modelling, Simulation and Design (CMSD), a high-performance computing facility at University of Hyderabad. He received his Masters followed by Ph.D. in Computer Science (mobile computing and wireless communication). His main research interests include wireless communication, mobile computing, wireless sensor networks and intelligent algorithms. He was a United Nations Fellow and worked in the United Nations University/International Institute for Software Technology (UNU/IIST), Macau, as research fellow in the year 2001. Dr. Udgata is working as a principal investigator in many Government of India funded research projects mainly for development of wireless sensor network applications and application of swarm intelligence techniques. He has published extensively in refereed international journals and conferences in India as well as abroad. He has edited many volumes in Springer LNCS/LNAI and Springer AISC Proceedings.

Dr. Prasant Kumar Pattnaik Ph.D. (Computer Science), Fellow IETE, Senior Member IEEE, is Professor at the School of Computer Engineering, KIIT University, Bhubaneswar. He has more than a decade of teaching and research experience. Dr. Pattnaik has published a number of research papers in peer-reviewed International journals and conferences. His areas of specialization include mobile computing, cloud computing, brain computer interface and privacy preservation.

Human Action Recognition Using Trajectory-Based Spatiotemporal Descriptors

Chandni Dhamsania and Tushar Ratanpara

Abstract Human action recognition has gained popularity because of its wide applicability in automatic retrieval of videos of particular action using visual features. An approach is introduced for human action recognition using trajectory-based spatiotemporal descriptors. Trajectories of minimum Eigen feature points help to capture the important motion information of videos. Optical flow is used to track the feature points smoothly and to obtain robust trajectories. Descriptors are extracted around the trajectories to characterize appearance by Histogram of Oriented Gradient (HOG), motion by Motion Boundary Histogram (MBH). MBH computed from differential optical flow outperforms for videos with more camera motion. The encoding of feature vectors is performed by bag of visual features technique. SVM with nonlinear kernel is used for recognition of actions using classification. The performance of proposed approach is measured on various datasets of human action videos.

Keywords Human action · Recognition · Classification · Trajectory · Spatiotemporal · HOG · MBH · Bag of features · SVM

1 Introduction

Human action recognition is a way of retrieving videos using visual features. Enormous amount of videos are collected in a day, captured by video surveillance on roads, in hospitals, malls, YouTube, and many more places. It requires a lot of man power to analyze the videos manually to search for a particular action. According to the survey done by comScore [1] in March 2013, online consumption of videos has doubled during the 2 years, 2011–2013. 54 million viewers in India

C. Dhamsania (✉) · T. Ratanpara
Department of Computer Engineering, Dharmsinh Desai University, Nadiad, India
e-mail: chandni.dhamsania@gmail.com

T. Ratanpara
e-mail: tushar.ratanpara@gmail.com

© Springer Nature Singapore Pte Ltd. 2017
S.C. Satapathy et al. (eds.), *Proceedings of the 5th International Conference on Frontiers in Intelligent Computing: Theory and Applications*, Advances in Intelligent Systems and Computing 515, DOI 10.1007/978-981-10-3153-3_1

watched online videos in March 2013. Google Sites, driven by YouTube.com for viewing videos ranked as top most viewing source in India. The storage is also a severe problem. It can be solved by keeping the videos with interested actions by human action recognition from videos and eliminate others. Many people want to watch movie videos with particular action. To search for particular movie videos from enormous amount of movies is very time-consuming. In this era of technology where automation is very important, human action recognition is useful in video surveillance, entertainment, human–computer interface, etc.

2 Related Work

In recent years, much work is done in the field of automatic video classification. Video classification is performed based on the information such as text, audio, visual, and metadata associated with the videos. Here the focus is on video classification using visual information contained in it. Visual information can be of several types such as it can be based on object detection, motion information, human action recognition, etc. Human action recognition is a growing area of research for video classification. Jimenez et al. [2] also implemented work on human interaction recognition from TV videos. Histogram of Oriented Gradient (HOG) and Histogram of Optical Flow (HOF) are obtained for Spatiotemporal Interest Point (STIP). Bag-of-words model is used for encoding and Support Vector Machine (SVM) for classification of action videos. Slimani et al. [3] have based on two-dimensional co-occurrence matrix constructed by extracting 3D-XYT volume around two interacting persons. For classification, k-nearest neighbor and SVM classifier are used. Nguyen and Yoshitaka [4] introduced three-layer convolution network for action recognition from segmented and unsegmented videos. This network uses Independent Subspace Analysis algorithm. Dhamsania and Ratanpara [5] have presented a survey that shows the need and scope in the field of human action recognition from videos. Jiménez and Blanca [6] have introduced a method that detects shot from the video and then the bounding box around upper body of both actors is detected. Pyramid of accumulated histogram of optical flow (PaHOF) descriptor of successive frames is computed, bag of descriptor is used for video representation and bag-SVM is used for classification of action videos.

3 Proposed Approach

In the proposed system, human actions are recognized from real-time environment videos using visual features. Recognition of videos that consists of actions of two-person interaction such as hug, handshake, kiss, and kick is carried out. The abstract model of proposed approach is shown in Fig. 1.

The rest of the paper explains all the steps of proposed approach in detail.

Fig. 1 Abstract model of proposed approach

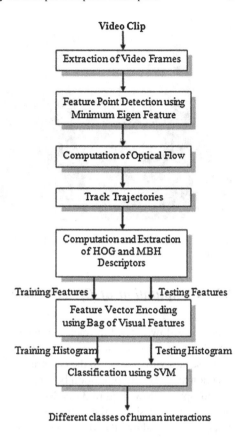

3.1 Extraction of Video Frames

Frames are extracted from videos. Here just for the purpose of understanding, we take a video consisting of action handshake from set1 of UT-interaction dataset. The numbers of frames present in video are 138, frame rate is 29.9997, and length of video is 4.7 s.

3.2 Feature Points Detection Using Minimum Eigen Feature

Feature points are the pixels with interest points in an image. Min Eigen feature corner detection technique is used to find corner points [7]. Figure 2 shows the corner points detected by the minimum Eigen feature detection technique from the frames of handshake video clip.

(a) (b)

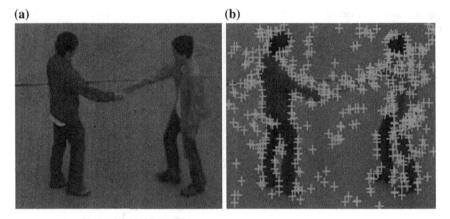

Fig. 2 **a** Original frame of handshake video and **b** feature points detected in a video frame of handshake video

3.3 Computation of Optical Flow

The optical flow is used to track the feature points obtained in previous section till next 14 frames to obtain trajectories. Optical flow shows the shape of motion of visual scenes with respect to the camera or observer. The optical flow represents the direction and velocity of motion between frames. Horn–Schunck [8] method is used to find optical flow. Figure 3a and b shows the consecutive frames of handshake video. Figure 4c shows the optical flow of all the pixels between two successive frames, arrows show the direction of motion and size of the arrows show the velocity of motion.

(a)   (b)   (c)

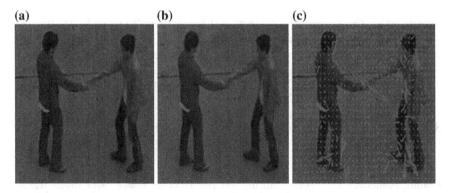

Fig. 3 **a, b** are consecutive frames of handshake video and **c** optical flow between the consecutive frames of video

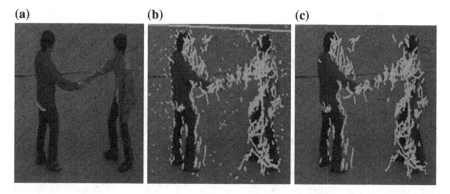

Fig. 4 a Original frame of handshake video, **b** video frame without removing unwanted trajectories and **c** video frame after removing unwanted trajectories

3.4   Track Trajectories

Trajectory is the path of motion, in other words it shows the pattern of motion of any person, body parts, animals, or any moving object. Each point $P_t(x, y)$ is tracked to next frame $t + 1$ by median filtering in optical flow field. The formula for obtaining point in next frame of trajectory [9] can be given as

$$P_{t+1}(x, y) = P_t(x, y) + (M * \omega)|_{(\bar{x}, \bar{y})}, \qquad (1)$$

where $(\bar{x}, \bar{y})$ is the rounded position of (x, y), M is median filter, and ω is the optical field (u_t, v_t). Points of N = 14 successive frames are concatenated to form trajectory that consists of points $(P_t, P_{t+1}, P_{t+2}, \ldots, P_{t+N})$. Trajectories that possess sudden drift and are static need to be removed. The variance of trajectories that is more than the maximum variance equal to 50 and less than the minimum variance equal to $\sqrt{3}$ is removed. Feature is extracted around the remaining trajectories. Figure 4a shows the original frame of video, (b) shows the trajectories obtained by tracking all feature points of a frame. 43483 numbers of trajectories are obtained by tracking feature points of all the frames of video, and (c) shows the trajectories remained after removing static trajectories, trajectories with sudden drift and the trajectories that moves out of the frame. 14980 numbers of trajectories are left after removing unwanted trajectories from whole video.

3.5   Computation and Extraction of HOG and MBH Descriptors

The descriptors HOG and MBH are extracted in spatiotemporal grid to capture spatial and temporal information. Spatiotemporal grid of size $n_\sigma \times n_\sigma \times n_\tau$ is obtained around the trajectory. Grid consists of 32 × 32 pixels patch around trajectory located spatially and such three temporal patches are considered. Each patch is of four cells. In each cell, gradient vector is calculated and angle of gradient vector is obtained. The histogram of nine bins is obtained using full orientation. The histograms of all cells of a grid are concatenated at last. Histogram is then normalized.

3.5.1 Histogram of Oriented Gradient (HOG)

HOG [10, 11] is used to show the static appearance of an object, person, or any other thing. HOG feature is calculated around the trajectory in a spatiotemporal grid. HOG feature is computed around trajectories for all the frames from which the trajectories are successfully tracked. At last histograms are concatenated to form feature vector of size 108. For the handshake video, the number of features obtained are 14881 × 108.

3.5.2 Motion Boundary Histogram (MBH)

Motion Boundary Histogram (MBH) [12] is calculated for human action recognition by the gradient of optical flow. The gradient of optical flow removes the smooth camera motion and the changes in motion boundaries are kept. From the optical flow between two successive frames horizontal MBH_X and vertical MBH_Y components are separated. Now MBH_X and MBH_Y are considered as separate images. The size of MBH_X and MBH_Y features is same as that of HOG feature, 14881 × 108 for handshake video.

3.6 Feature Vector Encoding Using Bag of Visual Features

Bag of visual words [13] is vector consisting of counts of occurrence of visual features. For this a codebook of visual words need to be generated by clustering. 100002 features are selected randomly that are divided equally among all videos of training for each descriptor HOG, MBH_X, and MBH_Y. For Bollywood movie video dataset, 85428 features are randomly selected. Then these 100002 HOG, MBH_X, and MBH_Y features are clustered to 800 visual words using k-means clustering.

3.7 Classification Using SVM

Nonlinear SVM is used with RBF-X2 kernel [11]. Histograms of size N × 800 are formed by bag of visual words techniques of N training videos for each descriptor separately. Then the histograms of all the descriptors are concatenated to form training model of size $N \times 2400$ and $M \times 2400$ for M testing videos will be formed. Then based on similarity measure classification of videos are done into different actions of human interaction.

4 Experimental Results

In this section, the proposed approach is validated by testing human action recognition algorithms on three datasets. Sevenfold cross-validation is used for the evaluation of accuracy. 70 and 30% videos are used for training and testing, respectively.

First, dataset consists of video clips from Hollywood movies with 60 videos that consist of handshake, kiss, and hug actions. Second, dataset consists of video clips from Bollywood movies with 30 videos that consist of handshake, kiss, and hug classes. As videos are taken from movies, they consist of high variations, camera motion, and illumination changes. Results of Hollywood and Bollywood video dataset are shown in Table 1.

Third, dataset used is UT-interaction dataset [14]. It consists of set1 and set2 with 30 videos in each set. Figure 5 shows the performance through bar graph for set1, set2 and for combined set1 and set2 of UT-interaction dataset videos. For set1, the accuracy of handshake, kick, and hug obtained is 84.32, 88.91, and 76.18%, respectively. Combined set1 and set2 gives accuracy for handshake, kick, and hug classes around 78.8, 80.4, and 73.5%, respectively. For set2, accuracy obtained for handshake, kick, and hug class is 72.72, 81.37, and 65.3%. Hence the overall

Table 1 Results for Hollywood and Bollywood movie video dataset

Datasets	Actions (%)			
	Handshake	Kiss	Hug	Accuracy
Hollywood movie video clips	61.88	70.22	55.05	62.38
Bollywood movie video clips	65.87	79.16	68.52	71.2

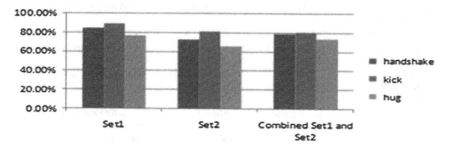

Fig. 5 Accuracy of each class of UT-interaction dataset

accuracy obtained on set1, set2 and on combined set1 and set2 is 83.13, 73.13, and 77.56%, respectively.

5 Conclusion and Future Work

In this proposed work, an approach is introduced for human action recognition using trajectory-based spatiotemporal descriptors. The combination of HOG and MBH descriptors gives better discriminative power for the recognition of human actions. Experiments are performed on different datasets consisting of videos from Hollywood, Bollywood movies, and also on UT-interaction dataset. On Bollywood movie video dataset, the overall recognition rate is higher as compared to Hollywood movie video dataset. The experiment is performed on UT-interaction dataset, accuracy obtained on set1 is highest.

In the future work, many challenges like high computation cost, change in appearance due to clothes, change in illumination, and low recognition rate need to be solved.

References

1. Online Video Consumption in India has Doubled in the Past 2 Years, https://www.comscore.com/Insights/Press-Releases/2013/5/Online-Video-Consumption-in-India-May-2013
2. Jiménez, M., Yeguas, E., Blanca, N.: Exploring STIP-Based Models for Recognizing Human interactions in TV videos. Pattern Recognition Letters, Vol. 34. Elsevier (2013) 1819–1828
3. Slimani, K., Benezeth, Y., Souami, F.: Human Interaction Recognition Based on the Co-occurrence of Visual Words. Computer Vision and Pattern Recognition Workshops, IEEE (2014)
4. Nguyen, N., Yoshitaka, A.: Human Interaction Recognition using Independent Subspace Analysis Algorithm. International Symposium on Multimedia (ISM), IEEE (2014) 40–46
5. Dhamsania, C., Ratanpara, T.: A Survey on Human Action Recognition from Videos. International Conference on Innovations in Information Embedded and Communication Systems, IEEE (2016)
6. Jiménez, M., Blanca, N.: Human Interaction Recognition by Motion Decoupling. Pattern Recognition and Image Analysis, Springer Berlin Heidelberg (2013) 374–381
7. Shi, J., Tomasi, C.: Good Features to Track. Computer Vision and Pattern Recognition, IEEE (1994) 593–600
8. Horn, B., Schunck, B.: Determining Optical Flow. Artificial Intelligence, Vol. 17, (1981)
9. Wang, H., Kläser, A., Schmid, C., Liu, C.: Action Recognition by Dense Trajectories. Computer Vision and Pattern Recognition, IEEE (2011) 3169–3176
10. Dalal, N., Triggs, B.: Histograms of Oriented Gradients for Human Detection. Computer Society Conference on Computer Vision and Pattern Recognition, Vol. 1, IEEE (2005) 886–893
11. Laptev, I., Marszałek, M., Schmid, C., Rozenfeld, B.: Learning Realistic Human Actions from Movies. Computer Vision and Pattern Recognition, IEEE (2008) 1–8
12. Dalal, N., Triggs, B., Schmid, C.: Human Detection using Oriented Histograms of Flow and Appearance. Computer Vision–ECCV, Springer Berlin Heidelberg (2006) 428–441

13. Yang, J., Jiang, Y., Hauptmann, A., Ngo, C.: Evaluating Bag-of-Visual-Words Representations in Scene Classification. IN: ACM SIGMM Int'l Workshop on Multimedia Information Retrieval (MIR 2007), Augsburg, Germany (2007)
14. UT-Interaction Dataset, ICPR Contest on Semantic Description of Human Activities (SDHA), http://cvrc.ece.utexas.edu/SDHA2010/Human_Interaction.html

Mathematical Modeling of Specific Fuel Consumption Using Response Surface Methodology for CI Engine Fueled with Tyre Pyrolysis Oil and Diesel Blend

Saumil C. Patel and Pragnesh K. Brahmbhatt

Abstract In this study, response surface methodology (RSM)-based prediction model was prepared for specific fuel consumption (SFC) as a response. A regression model was designed to predict SFC using RSM with central composite rotatable design (CCRD). In the development of regression models, injection timing, compression ratio, injection pressure, and engine load were considered as controlled variables. Injection pressure and compression ratio were observed as the most influencing variables for the SFC. The predicted SFC values and the succeeding verification experiments under the optimal conditions established the validity of the regression model.

Keywords Specific fuel consumption (SFC) · Compression ignition engine (CI) · Tyre pyrolysis oil (TPO) · Response surface methodology (RSM)

1 Introduction

The diesel engines are generally used for transportation, engineering industrial, and agricultural machinery due to its better fuel efficiency. The increasing cost of fuel has made nation to depend on diesel-based engines. Due to reduction and high cost of petroleum-based fuels investigators around the globe look for alternate fuel and trying to discover the best alternative fuel.

S.C. Patel (✉)
Mechanical Engineering Department, PAHER University, Udaipur, Rajasthan, India
e-mail: saumil_patel278@yahoo.com

P.K. Brahmbhatt
Government Engineering College-Dahod, Gujarat, India

© Springer Nature Singapore Pte Ltd. 2017
S.C. Satapathy et al. (eds.), *Proceedings of the 5th International Conference on Frontiers in Intelligent Computing: Theory and Applications*, Advances in Intelligent Systems and Computing 515, DOI 10.1007/978-981-10-3153-3_2

2 Literature Review

Atmanlı et al. investigated optimal blend ratios for compression ignition (CI) engine applications for diesel, *n*-butanol, and cotton seed oil by using RSM. Experiments were performed at rated load and steady speed to found engine performance and emission parameters. Results showed that optimization was done by using RSM to recognize the optimum blend. According to performance tests BT, BP, Brake thermal efficiency, and Brake mean effective pressure of DnBC diminished as SFC increased relative to diesel [1]. Researcher was investigating the performance and exhaust parameters of CIDI engine for the result of injection factors. The included parameters were IP, nozzle tip protrusion, and IT. The biodiesel fuel used in experiment was derived from pongamia seeds. RSM and DoE methods were used to design experiments. RSM results were helpful to predict BSFC, BTHE, CO, HC, NOx, burn opacity and additional to recognize the important relations between the contribution parameters on the output parameters [2]. One of the researchers was optimizing performance parameters related to BP and economy of fuel during experimental investigation and RSM. CR, IP, and IT were used as performance parameters and BTE and BSFC was the response parameters. A single cylinder engine was used in experimentation. RSM was used to design experiments. The results showed that at most favorable factors, the values of the Brake thermal efficiency and Brake specific fuel consumption were found to be 29.76%, respectively [3, 4]. Silva et al. investigated the parameters which affect the transesterification process. The RSM and factorial design method were used to optimize the biodiesel production process. RSM method was used to finding out the combined effects of various factors. The optimum conditions were catalyst concentration (1.3 M), mild temperature (56.7 °C), molar ratio (9:1), and reaction time (80 min) [5]. Rashid et al. investigated optimum condition for the transesterification of Moringa oleifera oil using RSM with CCRD. In optimization various parameters were used [6].

Theme of present work is to examine the influence of performance variables of CI engine utilizing tyre pyrolysis oil (TPO) as fuel using RSM-based experimental design. The other objective is to determine the optimal values parameters which would be resulting in improved.

3 Experimental Details

In this study, TPO and diesel blend were used as a working fluid in CI engine. In experiments, four parameters (injection timing, compression ratio, injection pressure, and engine load) with five levels for each were used to find out optimization. The TPO and diesel blend were taken as D85T15 for better performance [7]. During experiments, vibrometer was used for homogeneous mixing of blend. An engine

Fig. 1 The engine test rig

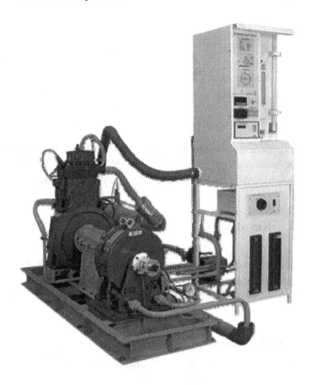

test rig with computer based data acquisition system was used. Figure 1 shows the engine test rig. Table 1 shows the engine specifications.

Investigation has been carried out using developed empirical model of RSM in order to check the effect of variables (injection timing, compression ratio, injection pressure, and engine load) on the response (SFC). Variables are varied up to five different levels and each design position for the proposed methodology is simulated multiple times in order to eliminate the chance of error. Table 2 shows the

Table 1 Engine Specifications

Sr no.	Item	Specification
1	Model	TV1
2	Make	Kirlosker oil engines
3	Type	Four stroke, water cooled, diesel
4	No. of cylinder	One
5	Bore	87.5 mm
6	Stroke	110 mm
7	Compression ratio	12–18
8	Power rating	7.5 HP
9	Injection timing	$\leq 25^0$ BTDC

Table 2 Variables with range

Variables/range	Lowest	Lower	Centre	Higher	Highest
Injection timing	21	22	23	24	25
Compression ratio	14	15	16	17	18
Injection pressure	140	160	180	200	220
Engine load	1	4	7	10	13

experiments conditions. RSM and DFA have been employed simultaneously for the purpose of optimizing the variables and conformation experiments have been conducted in order to check the validity of prediction model.

4 Results

Table 3 indicates 21 sets of experiment based on the central composite rotatable design. Some of the replications are having same set of parameters in order to obtain a more precise result and to estimate the experimental error. Hence, it is not

Table 3 Experimental results

Experiment no.	Injection timing (°BTDC)	Compression ratio	Injection pressure (bar)	Engine load (kg)	Specific fuel consumption (kg/kwh)
1	21	16	180	7	1.259988
2	23	16	180	1	1.938195
3	24	15	200	4	0.383044
4	23	16	180	7	0.096909
5	23	16	180	7	0.096909
6	22	15	200	10	1.033158
7	23	16	180	7	0.096909
8	24	17	200	4	0.821836
9	23	16	220	7	0.0557
10	23	16	180	13	0.425037
11	23	16	180	7	0.096909
12	23	16	180	7	0.096909
13	24	17	160	4	1.010105
14	22	17	160	4	0.671339
15	24	15	160	4	0.115779
16	23	16	140	7	0.934611
17	23	18	180	7	0.819101
18	22	17	200	10	0.386156
19	25	16	180	7	1.033265
20	22	15	160	4	1.757947
21	23	14	180	7	0.01955

obligatory to obtain the same results with same set of parameter due to uncontrollable conditions/error/variables.

The regression equation in terms of actual factors for the SFC as a function of four input process variables was developed using experimental information. The insignificant coefficients identified from ANOVA of some terms of the quadratic equation have been omitted. Equation (1) represents the prediction equation for SFC in terms of actual variables.

$$
\begin{aligned}
SFC = {}& 169.16187 - 20.14749 * IT + 10.08648 * C - 0.25033 * IP + 4.51159 * \\
& LOAD + 0.38334 * IT * CR + 6.80605E - 003 * IT * IP + 0.044350 * IT * LOAD - \\
& 9.95550E - 005 * CR * IP - 0.35156 * CR * LOAD - 1.72260E - 003\,8\,IP * LOAD + \\
& 0.27004 * IT^2 - 0.54779 * CR^2 + 2.67929E - 004 * IP^2 + 0.011424 * LOAD^2
\end{aligned}
$$

$$(1)$$

The ANOVA table for quadratic model for the SFC is given in Table 4. The value of p for the term of model is less than 0.05 indicates the significance of the term or model, as the confidence level of experiment is set at 95% for the proposed

Table 4 Analysis of variance table

Source	Sum of squares	DF	Mean square	F value	p-value prob > F	
Model	6.38	14	0.46	14.97	0.0016	**Significant**
A-injection timing	0.026	1	0.026	0.84	0.3938	
B-compression ratio	0.41	1	0.41	13.50	0.0104	
C-injection pressure	0.39	1	0.39	12.68	0.0119	
D-load	0.15	1	0.15	5.07	0.0652	
AB	1.18	1	1.18	38.59	0.0008	
AC	0.15	1	0.15	4.87	0.0695	
AD	0.071	1	0.071	2.32	0.1782	
BC	3.172E-005	1	3.172E-005	1.041E-003	0.9753	
BD	0.36	1	0.36	11.80	0.0139	
CD	0.043	1	0.043	1.40	0.2811	
A2	1.69	1	1.69	55.60	0.0003	
B2	0.40	1	0.40	13.12	0.0111	
C2	0.27	1	0.27	8.76	0.0253	
D2	0.011	1	0.011	0.37	0.5659	
Residual	0.18	6	0.030			
Lack of fit	0.18	2	0.091			
Pure error	0.000	4	0.000			
Cor total	6.57	20				

study. This value showed that the quadratic model fits well to the experimental results.

The value of p for the term of model is less than 0.05 indicates that it is considered to be statistically significant. This value showed that the quadratic model fits well to the experimental results. ANOVA results in Table 4 indicate compression ratio and injection pressure are the most significant variables. It has been also observed that linear effects of injection timing and engine load are insignificant parameters even though their quadratic effects are significant.

Figure 2 shows the RSM predictions against the experimental results. Figure 3 shows that the low specific fuel consumption can be achieved at smaller value of injection pressure and compression ratio. The iterative effect of compression ratio and injection pressure on the specific fuel consumption is shown in Fig. 4. Consider a measure of the model's overall performance referred to as the coefficient of determination and denoted by R^2. In the model, R^2 is obtained equal to 97.22%. The R^2 value indicates that the burnishing parameters explain 97.22% of variance in SFC.

The validating experiments was done using regression model on four sets of parameters, and results show that the numerical optimization technique confirm the effectiveness of RSM. Table 5 shows that percentage error of validating experiments.

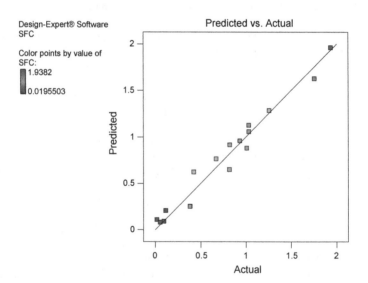

Fig. 2 RSM predictions against the experimental results

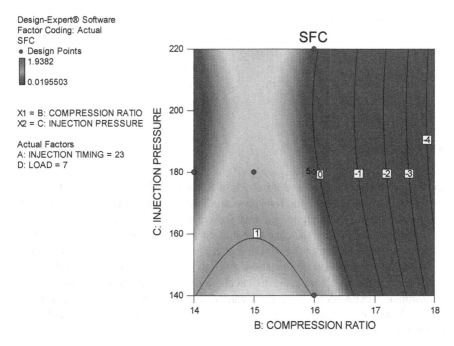

Fig. 3 Desirability function for IP versus CR

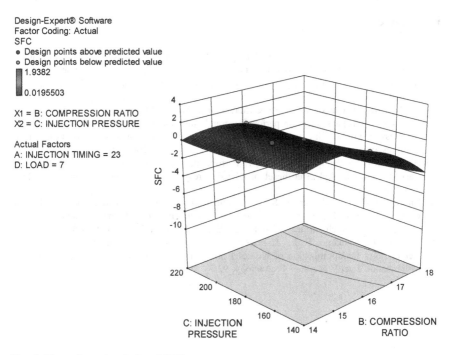

Fig. 4 Three dimensional plot of SFC

Table 5 Percentage error of validating experiments

Experiment no.	Injection timing (°BTDC)	Compression ratio	Injection pressure (bar)	Engine load (kg)	% Error
1	22	16	220	13	0.025
2	21	18	200	10	0.019
3	23	17	180	13	0.028

5   Conclusion

In this present work, combined approach of RSM and DFA has been implemented to investigate the effect of the four process variable engine load, injection timing, compression ratio, and injection pressure. It is clearly indicated from the results that the quadratic model in all the characteristics does not demonstrate a significant lack of fit; hence the adequacy of quadratic model is confirmed at 95% confidence level.

Injection pressure and compression ratio are observed as the most influencing variables for the SFC. In order to check the validity of the developed model, conformation experiments have been carried out. RSM was found to be a useful approach, and it should be recommended that this methodology be adapted to all optimization studies.

References

1. Atmanlı, A., Yüksel, B., İleri, E., & Karaoglan, A. D. (2015). Response surface methodology based optimization of diesel–n-butanol–cotton oil ternary blend ratios to improve engine performance and exhaust emission characteristics. *Energy Conversion and Management, 90,* 383–394.
2. Pandian, M., Sivapirakasam, S. P., & Udayakumar, M. (2011). Investigation on the effect of injection system parameters on performance and emission characteristics of a twin cylinder compression ignition direct injection engine fuelled with pongamia biodiesel–diesel blend using response surface methodology. Applied Energy, 88(8), 2663–2676.
3. Hirkude, J. B., & Padalkar, A. S. (2014). Performance optimization of CI engine fuelled with waste fried oil methyl ester-diesel blend using response surface methodology. Fuel, 119, 266–273.
4. Hirkude, J. B., & Padalkar, A. S. & Vedartham D. (2014), Investigations on the effect of waste fried oil methyl ester blends and load on performance and smoke opacity of diesel engine using research surface methodology. Energy Procedia 54 (2014) 606–614.
5. Silva, G. F., Camargo, F. L., & Ferreira, A. L. (2011). Application of response surface methodology for optimization of biodiesel production by transesterification of soybean oil with ethanol. Fuel Processing Technology, 92(3), 407–413.

6. Rashid, U., Anwar, F., Ashraf, M., Saleem, M., & Yusup, S. (2011). Application of response surface methodology for optimizing transesterification of Moringa oleifera oil: Biodiesel production. Energy Conversion and Management, 52(8), 3034–3042.
7. Patel, M. H. M., & Patel, T. M. (2012, June). Performance analysis of single cylinder diesel engine fuelled with Pyrolysis oil-diesel and its blend with Ethanol. In *International Journal of Engineering Research and Technology* (Vol. 1, No. 4 (June-2012)). ESRSA Publications.

Ensemble Learning for Identifying Muscular Dystrophy Diseases Using Codon Bias Pattern

K. Sathyavikasini and M.S. Vijaya

Abstract Hereditary traits are anticipated by the mutations in the gene sequences. Identifying a disease based on mutations is an essential and challenging task in the determination of genetic disorders such as Muscular dystrophy. Silent mutation is a single nucleotide variant does not result in changes in the encoded protein but appear in the variation of codon usage pattern that results in disease. A new ensemble learning-based computational model is proposed using the synonymous codon usage for identifying the muscular dystrophy disease. The feature vector is designed by calculating the Relative Synonymous Codon Usage (RSCU) values from the mutated gene sequences and a model is built by adopting codon usage bias pattern. This paper addresses the problem by formulating it as multi-classification trained with feature vectors of fifty-nine RSCU frequency values from the mutated gene sequences. Finally, a model is built based on ensemble learning LibD3C algorithm to recognize muscular dystrophy disease classification. Experiments showed that the accuracy of the classifier shows 90%, which proves that ensemble-based learning, is effective for predicting muscular dystrophy disease.

Keywords Codon · Codon usage bias · LibD3C · Positional cloning · RSCU

1 Introduction

A muscular dystrophy is a collection of consecutive muscle disorders induced by mutations in genes that encode for proteins that are essential for regular muscle function [1]. At present, there is no remedy for muscular dystrophy and the disease affected patients will have the chance of losing their life at earlier stages. Various

K. Sathyavikasini (✉) · M.S. Vijaya
PSGR Krishnammal College for Women, Coimbatore, India
e-mail: Mail2sathyavikashini@gmail.com

M.S. Vijaya
e-mail: msvijaya@psgrkc.com

© Springer Nature Singapore Pte Ltd. 2017
S.C. Satapathy et al. (eds.), *Proceedings of the 5th International Conference on Frontiers in Intelligent Computing: Theory and Applications*, Advances in Intelligent Systems and Computing 515, DOI 10.1007/978-981-10-3153-3_3

types of muscular dystrophy are Duchenne, Becker, Emery–dreifuss, Limb-Girdle muscular dystrophy, Facioscapulohumeral, Myotonic and Congenital Muscular Dystrophy.

Amend in the genetic code that causes a permanent change in the DNA sequence is termed as mutation. DNA mutations perceptibly root to genetic diseases. Single character change in a gene makes an impact on the gene which in turn changes the function of the gene.

Missense mutations are the substitution in a codon that encodes a different amino acid and cause a small change in the protein [2]. Nonsense mutations are those where the protein attains to stop codon when a change occurs in the DNA sequence. Deletions are the mutations when a base or an exon is deleted from a sequence the mutations [3, 4].

Silent mutations are changes in codon that encodes for the same amino acid and therefore the protein is not altered [5]. The information from the genes transfers the nucleic acid to proteins in the form of codons. During the process of translation, the synonymous codons have different frequencies [6] which are referred as codon usage bias that is dynamic. The functionality of the gene depends on the codon usage bias [7]. Therefore, the codon usage is important in mutation studies and in molecular evolution.

2 Synonymous Mutation

The silent mutation is a kind of point mutation which changes the codon usage pattern. The translated protein in the amino acid sequence is not modified with the synonymous codon changes. Only very few research has been carried out on gene sequences based on RSCU (Relative Synonymous Codon Usage) to predict or classify either type of gene or virus or diseases. The authors in [7] proposed a model to classify the types of Human Leukocyte Antigen (HLA) gene into different functional groups by choosing the codon usage bias as input. In their work, they converted the gene sequence into 59 vector elements by calculating the RSCU values for the gene sequence.

The authors C.M. Nisha, Bhasker Pant, and K.R. Pardasani proposed a new approach based on codon usage pattern to classify the type of Hepatitis C virus (HCV) that are the primary reason for the liver infection. To classify the subclass of its genotype, a model was created using codon usage bias as input to multi class SVM [8].

The authors in [9] employed a machine learning approach based on ensemble classifier LibD3C to predict the cytokines. The analysis was made on the physicochemical properties and the distribution of whole amino acids.

The above literature survey motivates that the classification of disease can also be carried out by modeling silent mutations with RSCU values through ensemble-based learning. Hence, it is proposed in this paper to demonstrate the application of RSCU in disease gene sequences to model the silent mutations for predicting the type of muscular dystrophy diseases.

3   Muscular Dystrophy Disease Identification

The gene sequences and its pattern vary in every human. Also the pattern gets altered when mutations occur in the chromosome. The principal focus of this research is to provide an efficient machine learning solution for predicting the type of muscular dystrophy disease with the silent mutations. The mutational gene sequences are generated as the synonymous mutated gene sequences are not explicitly available for this complicated disease. Five types of muscular dystrophy namely DMD, BMD, EMD, LGMD, and CMT have been considered for building the disease prediction model.

3.1   Disease Identification Model

The development of Muscular dystrophy disease Identification model comprises of five phases such as mutational gene sequence generation, feature extraction, RSCU calculation, building the model and classification. The framework of the proposed model is illustrated in Fig. 1.

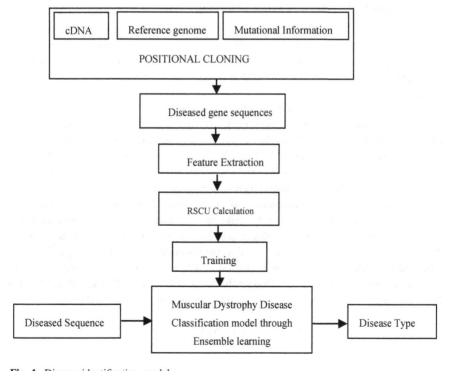

Fig. 1 Disease identification model

3.2 Mutational Gene Sequences Formation

Mutated gene sequences are generated through positional cloning based on the mutation and its location on the chromosome. The information on the position of mutations in the gene sequences is available in HGMD (Human Gene Mutation Database) [10] is a core collection of data on germline mutations in genes coupled with the human-inherited disease which are grasped from various literatures.

The positional change of the nucleotide is done in cDNA sequence against the reference gene sequence and the new mutated gene sequences for muscular dystrophy are generated through R script. Using the functions from the library, the required position to be altered is identified and replaced with the nucleotide specified in the nucleotide change column of HGMD database. Using the traditional positional cloning approach, the mutated sequences are generated and stored as FASTA files.

There are about 55 genes evacuated with five types of muscular dystrophy. For each phenotype, 60 mutated gene sequences are generated and dataset comprises of 300 mutated gene sequences combining all forms of muscular dystrophy is developed.

3.3 Feature Extraction and Training

A codon is the triplet of nucleotides that code for a specific amino acid. Many-to-one relationship occurs between the codon and amino acid. Many amino acids are coded by more than one codon because of the degeneracy of the genetic codes.

A total number of codons in a DNA sequence count to 64. Since methionine (ATG) and tryptophan (TGG) have only one corresponding codon, they are not counted and are eliminated from the analysis as their RSCU values are always equal to 1. The three stop codons (TGA, TAA, TAG) are also not included. Accordingly, the number of codons considered is 59. Therefore, irrespective of the size, the DNA sequence is converted to a feature vector of 59 elements.

The differences in the frequency of occurrence of synonymous codons are referred as codon usage bias. The formula for calculating RSCU can be explained as, the number of times a particular codon is observed, relative to the number of times that the codon would be observed in the absence of any codon usage bias [8]. The RSCU carries the value 1.00 if the codon usage bias of that particular codon is absent. If the codon is used less frequently than expected, the RSCU values tend to have the negative values. Following formula is used to calculate RSCU.

$$RSCU = X_{ij} / \left(1/n_i * S\{X_{ij}; j = 1, n_i\} \right)$$

Table 1 RSCU values for 59 codons

Codon	Value	Codon	Value	Codon	Value	Codon	Value
AAA	1.05	CCC	0.97	GGC	0.92	CAA	0.87
AAC	0.812	CCG	0.12	GGG	0.75	CAC	0.86
AAG	0.948	CCT	1.64	GGT	0.64	CAG	1.13
AAT	1.18	CGA	0.87	GTA	0.81	CAT	1.14
ACA	1.52	CGC	0.54	GTC	0.93	CCA	1.25
ACC	0.76	CGG	0.66	GTG	1.40	GCA	1.23
ACG	0.24	CGT	0.63	GTT	0.85	GCC	1.18
ACT	1.48	CTA	0.73	TAC	0.61	GCG	0.15
AGA	1.84	CTC	0.87	TAT	1.38	GCT	1.42
AGC	0.99	CTG	1.41	TCA	1.23	GGA	1.67
AGG	1.42	CTT	1.03	TCC	0.91	TTA	0.71
AGT	1.36	GAA	1.22	TCG	0.14	TTC	0.64
ATA	0.52	GAC	0.81	TCT	1.33	TTG	1.23
ATC	1.10	GAG	0.77	TGC	1.16	TTT	1.63
ATT	1.36	GAT	1.18	TGT	0.833		

where X_{ij} is the number of occurrences of the jth codon for the ith amino acid, and n_i is the number of alternative codons for the ith amino acid. If the synonymous codons of an amino acid are used with equal frequencies, then their RSCU values are 1 (Table 1).

In this manner, the RSCU values are derived for 59 codons from each mutated gene sequence which forms a feature vector for classification task.

3.4 Ensemble Learning

In machine learning, the hybrid approach has been an ongoing research area for gaining best performance for classification or prediction problems over a single learning approach. Hybridization is based on combining two different machine learning techniques. The motivation of the hybrid model is that a hybrid classification model can be composed of one unsupervised learner to preprocess the training data and one supervised learner to learn the clustering result [11].

The ensemble learning methods can yield robustness to multi-label classification. The familiar ensemble learning methods for classification problems are bagging, boosting, and random forests. To reduce the information redundancy within multi-label learning, a model shared subspace boosting algorithm was constructed [12] that automatically finds shared subspace models, where every model was made to learn from the random feature subspace and bootstrap data and combined a number of base models through multiple labels [13].

The prediction ability of an ensemble classifier is excellent to that of a single classifier because the former can address the differences produced by the latter more efficiently when challenged with different problems [14].

LibD3C is a kind of Ensemble classifiers with a clustering and dynamic selection strategy. A method that blends two types of discriminating ensemble techniques called as dynamic selection and circulating combination-based clustering (D3C). LibD3C employs two types of selective ensemble techniques, such as, ensemble pruning based on k-means clustering and dynamic selection and circulating combination [15].

In this paper, mutated gene sequences are taken as input to the classifier and a model is build to identify the type of muscular dystrophy from synonymous mutations. Hence, the ensemble-based classifier LibD3C is employed to predict the type of disease.

4 Experiments and Results

Muscular dystrophy disease identification model for silent mutations is developed using WEKA, a software environment for solving the pattern recognition problems. The Weka, Open Source, Portable, a GUI-based workbench is a collection of machine learning algorithms and data preprocessing tools [16]. In this experiment, the mutated gene sequences of muscular dystrophy are generated through positional cloning. The training data set with instances related to five categories of muscular dystrophy has been developed as described in Sect. 3.

The LibD3C package was installed through the package manager in WEKA 3.7. The parameters were set as default. Evaluating the generalization power of the classifiers and to estimate their predictive capabilities for unknown samples, a standard 10-fold cross-validation technique is used to split the data randomly and repeatedly into training and test sets. The prediction accuracy is defined as the ratio of the number of correctly classified instances in the test dataset and the total number of test cases. In this work, LIBD3C algorithm shows an accuracy of 90%. The results of the experiments are summarized in Table 2. The performance of the

Table 2 Predictive performance of the LibD3C classifier

Performance criteria	LibD3C classifier
Kappa statistic	0.85
Mean absolute error	0.09
Root mean squared error	0.202
Relative absolute error	29.844
Root relative square error	50.801
Time taken to build the model (in sec)	7.36
Correctly classified Instance	202
Incorrectly classified instance	28
Prediction accuracy	90%

Fig. 2 Prediction accuracy of LibD3C classifier

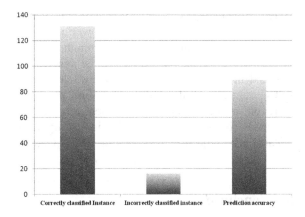

Fig. 3 Comparision of mean absolute error and root mean squared error

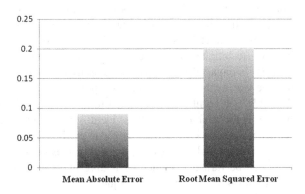

classifier is depicted in Fig. 2. The Comparison of the Mean Absolute Error and Root Mean Squared Error is shown in Fig. 3.

The experiment is compared with the standard classification algorithms such as Naïve bayes, Decision Tree, Artificial neural network, and Support vector machine and LibD3c. Table 3 shows the comparative analysis of the all the standard algorithms with the ensemble learning algorithm.

From the above results, it is perceived that LibD3C is suitable for predicting the disease from the mutated gene sequences as desirable accuracy is attained from the ensemble learning methodology. The algorithm attains high kappa statistic and prediction accuracy. The mean absolute error is minimized so the reliability of the

Table 3 Comparative analysis of classifiers in terms of accuracy

Classifiers	Accuracy (%)
Naïve Bayes	85.2
Decision tree	85.8
ANN	84.3
SVM	87.3
LibD3C	90

system is increased. LibD3C is a hybrid approach that is a combination of unsupervised over supervised learning, a powerful approach for predicting the instance. The time taken to build the model is minimal of 3.36 s.

Identifying a genetic disease from the mutated gene sequences is a challenging task and in this work it is achieved by selecting the features by calculating the RSCU values from the codon usage bias pattern. This experiment concentrates more on predicting the disease effectively and so the features are vigilantly designed. The proposed model aids in classifying type of muscular dystrophy in mutated gene sequences by capturing RSCU features of silent mutations.

5 Conclusion

This research work demonstrates the development of muscular dystrophy disease prediction model using mutated gene sequences and codon usage bias. The RSCU values are extracted as features and a model is built by employing the machine learning technique through ensemble learning method. The Ensemble classifier LibD3C is engaged to perform the classification. The performance of the learning method was evaluated based on their predictive accuracy. The results indicate that the LibD3C algorithm is best suited for predicting the type of muscular dystrophy. In the field of bioinformatics, identifying genetic disease through the gene sequences plays a very important role in predicting the genetic disorder. This model can also be applied in investigating the changes in protein folding and function. The work can be further extended by adding more sequences and repeating the experiment with other ensemble techniques.

References

1. Lenka Fajkusova, ZdeneIk LukasIb, Miroslava Tvrdoakova a, Viera Kuhrova a, Jirioa Haajekb, Jirioa Fajkusc, Novel dystrophin mutations revealed by analysis of dystrophin mRNA: alternative splicing suppresses the phenotypic effect of a nonsense mutation, Neuromuscular Disorders Vol 11, (2001)
2. Kann, M.G., Advances in translational bioinformatics: computational approaches for the hunting of disease genes, Briefings in Bioinformatics 11, 96–110 (2009)
3. Tranchevent, L.-C., et al., A guide to web tools to prioritize candidate genes, Briefings in Bioinformatics, 12, 22–32 (2010)
4. KN North and KJ Jones, Diagnosing childhood muscular dystrophies, Journal of Paediatrics and Child Health
5. Koenig M, Hoffman EP, Bertelson CJ, Monaco AP, Feener C, Kunkel LM., Complete cloning of the Duchenne muscular dystrophy (DMD) cDNA and preliminary genomic organization of the DMD gene innormal and affected individuals, Cell 1987;50:509 ± 517
6. Charif D, Thioulouse J, Lobry J. R and Perrière G, Online synonymous codon usage analyses with the ade4 and seqinR packages, Bioinformatics Oxford Journal,2005,21(4):545–547

7. Jianmin Ma, Minh N. Nguyen, Gavyn W.L. Pang, and Jagath C. Rajapakse, Gene Classification using Codon Usage and SVMs, IEEE, 2005
8. C.M. Nisha, Bhasker Pant, and K. R. Pardasani, SVM model for classification of genotypes of HCV using Relative Synonymous Codon Usage Journal of Advanced Bioinformatics Applications and Research ISSN 0976-2604. Online ISSN 2278 – 6007 Vol 3, Issue 3, 2012, pp 357–363
9. Quan Zou, et al., An approach for identifying cytokines based on a novel Ensemble classifier, Hindawi Publishing Corporation BioMed Research International, 2013
10. Peter D. Stenson, Matthew Mort, Edward V. Ball, Katy Shaw, Andrew D. Phillips, David N. Cooper, The Human Gene Mutation Database: building a comprehensive mutation repository for clinical and molecular genetics, diagnostic testing and personalized genomic medicine, July 2013
11. Chen, et al., LibD3C: Ensemble classifiers with a clustering and a dynamic strategy, Elseiver's Neurocomputing 123 (2014) pp 424–435
12. Gulisong, et al., A Triple-Random Ensemble Classification Method for Mining Multi-label Data, IEEE 2010, 978-0-7695-4257-7/10
13. R. Yan, et al., Model-shared subspace boosting for multi-label classification, in: Proceedings of the 13th ACM SIGKDD International Conference on Knowledge Discovery and DataMining, ACM, 2007, pp. 834–843
14. Z.-H. Zhou, J. Wu, andW. Tang, Ensembling neural networks: many could be better than all, Artificial Intelligence, vol. 137, no. 1–2, pp. 239–263, 2002.
15. Electronic Supplementary Material (ESI) for Molecular BioSystems. The Royal Society of Chemistry 2015
16. Ian H. Witten, Eibe Frank, Len Trigg, Mark Hall, Geoffrey Holmes, Sally Jo Cunningham. Weka: Practical Machine Learning Tools and Techniques with Java Implementations

Basic Arithmetic Coding Based Approach to Compress a Character String

Ipsita Mondal and Subhra J. Sarkar

Abstract Data compression plays an important role for storing and transmitting text or multimedia information. This paper refers to a lossless data algorithm is developed in C-platform to compress character string based on Basic Arithmetic Coding. At the preliminary stage, this algorithm was tested for the character array comprising of vowels only and the probability distribution is assumed arbitrarily. The result being obtained is encouraging with compression ratio far beyond unity. Though the algorithm was tested for vowels only but the work can be extended for any character array with probability of distribution as obtained from the survey of few randomly selected articles.

Keywords Data compression technique · Basic arithmetic coding · Probability distribution · Encoding–Decoding · Compression ratio

1 Introduction

In the present age of digitization, data compression becomes extremely important for reducing the bit size of the data. With reduced number of bits, there will reduced memory requirement thereby eliminating the memory constraints of the system. In context of data communication, reduced number of bits implies lesser energy requirement thereby leading toward energy efficiency. Data compression not only reduces the data size but it also has the inherent capability of data encryption thereby ensuring data security. A typical data compression algorithm can be represented by the block diagram, as given in Fig. 1 [1–3].

I. Mondal (✉)
Department of CSE, Techno India Batanagar, Kolkata, India
e-mail: ipsita.mondal@yahoo.com

S.J. Sarkar
Department of EE, Techno India Batanagar, Kolkata, India
e-mail: subhro89@gmail.com

© Springer Nature Singapore Pte Ltd. 2017
S.C. Satapathy et al. (eds.), *Proceedings of the 5th International Conference on Frontiers in Intelligent Computing: Theory and Applications*, Advances in Intelligent Systems and Computing 515, DOI 10.1007/978-981-10-3153-3_4

31

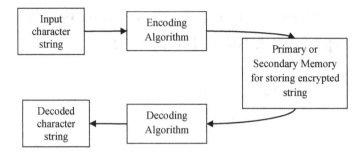

Fig. 1 Block diagram of the proposed system

There are numerous methods of data compression. Broadly, the compression can be classified as lossy or lossless compression. In lossy compression, there is some removal of some unimportant data values present in the file while performing these algorithms. Some of its examples include transform coding, Karhunen–Loeve Transform (KLT) coding, wavelet-based coding, etc. Real-time applications of these compression algorithms are in compression of multimedia files like audio, video, images, etc. [1]. On the other hand, there is no loss of data information in lossless data compression techniques like Shannon–Fano algorithm, Huffman algorithm, arithmetic Coding, etc. [1, 4]. Lossless data compression is more popular for compressing text documents, images of higher importance like image of cancerous tissues, etc. [4].

The application of the work done in [1] was confined to the compression of data string for power system applications only. If the algorithm can be extended to compress any character string, it can be used for the applications like compression of files present in any office or compression of the contents of books present in the library, etc. As the actual data obtained after data compression algorithm is encrypted, it becomes impossible for any external agency to decode the data. So, this method can allow only the authenticated users to access and use the data [5, 6]. The proposed algorithm is developed in C-language and the tested offline to obtain the results.

2 Arithmetic Coding

Basic arithmetic coding is a lossless data compression technique where a data or character string is encoded in form of a fractional single number, n where $0.0 \leq n < 1.0$. In this method of data compression, the probability distribution is applied of the content of source message to narrow the interval successively. Considering some source message comprising of symbol set S {'1', '2', '3', '4', '5', '6'} with probability of occurrence as given in Table 1 [1, 6–9].

Table 1 Probability distribution table for symbol set, S

Sl. no.	Character	Probability	Cumulative probability	Range [r_low, r_hi)
1	1	0.3	0.3	[0, 0.3)
2	2	0.2	0.5	[0.3, 0.5)
3	3	0.1	0.6	[0.5, 0.6)
4	4	0.2	0.8	[0.6, 0.8)
5	5	0.1	0.9	[0.8, 0.9)
6	6	0.1	1.0	[0.9, 1.0)

Table 2 Shrinking of range in arithmetic coding for different iterations

Iteration no.	Character (x)	Min	Max	r
1	1	0	0.3	0.3
2	1	0	0.09	0.09
3	2	0.027	0.045	0.018
4	5	0.0414	0.0432	0.0018
5	6	0.04302	0.0432	0.00018

The algorithm for basic arithmetic coding in order to compress a string comprising of the characters given in Table 2, is given in the subsequent section [1, 9].

STEP 1: Obtain the string, s and calculate its length (l).
STEP II: Initialize variables min = 0, max = 1, and r = 1.
STEP III: Set a counter i = 1.
STEP IV: Repeat steps V–IX until i != (l + 1).
STEP V: x = s (i).
STEP VI: Obtain r_low and r_hi corresponds to x.
STEP VII: Update min = min + r * r_low and max = min + r * r_hi.
STEP VIII: r = max − min.
STEP IX: i = i + 1.
STEP X: End of the loop.
STEP XI: Obtain a number num with minimum binary string length such that min < num < max.
STEP XII: End.

Considering a string {'1', '1', '2', '5', '6'} with 5 characters upon which arithmetic coding algorithm is going to be implemented. The process of execution is illustrated below [1].
Initialization: min = 0, max = 1and r = 1.
l = 5 → No. of iterations = 5.

Table 3 Extracting the encoded string elements for different

Iteration no.	Value	x	l	h	r
1	0.0430908203125	1	0	0.3	0.3
2	0.143636067	1	0	0.3	0.3
3	0.478786892	2	0.3	0.5	0.2
4	0.893934461	5	0.8	0.9	0.1
5	0.939344618	6	0.9	1.0	0.1

Output: num = 0.0430908203125 having binary string (.) 0000101100001 which lies between 0.04302 and 0.0432 having minimum binary string length (13 in this case).

This compressed binary string can be then be utilized for communication or storage purpose, as required. While decoding the actual string, the output num of the previous algorithm is compared with Table 3 continuously to extract the actual string. In this present algorithm, it is also required to give the number of characters in the actual string as the probability of occurrence of string termination is not considered. The algorithm for decoding the actual string is given below where the input is encoded binary array and the actual string size [1, 9].

STEP I: Obtain the binary string and length of encoded string (l).
STEP II: Determine the float number (num) corresponding to the binary string.
STEP III: Set a counter i = 1 and define a null array arr.
STEP IV: Repeat steps until i != (l + 1).
STEP V: Find the range between which num lies and the character x corresponds to it.
STEP VI: arr (i) = x, l = r_low (x), h = r_hi (x) and r = h − l.
STEP VII: num = (num − l)/r.
STEP VIII: i = i + 1.
STEP IX: End of the loop.
STEP X: Array arr is the encoded string.
STEP XI: End.

The execution of the algorithm for the binary string 000010110001 is illustrated below [1].

Since, l = 5 → No. of iterations = 5.
Output: arr = {1, 1, 2, 5, 6}

3 Proposed Algorithm

Encoding algorithm

STEP 1: Start

STEP 2: Input the string i.e. str.

STEP 3: Count the length of the string.

STEP 4: Initialize pini = 0.0, pfin = 1.0 and r = pfin-pini.

STEP 5: i = 0

STEP 6: Repeat steps 7 and 8 while i < length do

STEP 7: Fetch the r_min[i] and r_max[i] values from Table No. 4 and detemine the corresponding stringthat lies between the ranges.

STEP 8: i = i+1

      [End of loop]

STEP 9: i1 = 0

STEP 10: Repeat STEP XI and XII while i1< length do

STEP 11: pini = pini + r * r_max[i1] and pfin = pini + r * r_max[i1]

STEP 12: i1 <- i1+1

      [End of loop]

STEP 13: Select a value i.e. val which lies between pini and pfin i.e. pini < val < pfin

STEP 14: Convert val to binary and store it in str1

STEP 15: Check if str1 % 7 != 0 do

           t = str1%7

           t1 = 7 − t

           add t1 number of 0's at the start bits of the string

      [End of if]

STEP 16: Count the length of str1 i.e. l

STEP 17: l1 = l / 7

STEP 18: i2 = 0

STEP 19: Repeat steps 20 to 25 while i2 < l1

STEP 20: j = 0

STEP 21: Repeat while j < 7

STEP 22: Store str1[j] in a new array

STEP 23: Convert it to decimal equivalent value

STEP 24: j = j+1

      [End of inner loop]

STEP 25: i2 = i2+1

      [End of outer loop]

STEP 26: Print the decimal string which is the compressed string.

Table 4 Minimum and maximum values for different strings

Sl. no.	String	r_min	r_max	Range
1	a	0.0	0.3	0.3
2	e	0.3	0.55	0.25
3	i	0.55	0.75	0.2
4	o	0.75	0.9	0.15
5	u	0.9	1.0	0.1

Decoding algorithmtpb 2

See Table 4.

STEP 1: Read the number of zero added   i.e. t1
STEP 2: i = 0
STEP 3: Repeat steps 4 and 5 while i < 11 do
STEP 4: Read the 7 bit decimal number and convert it in equivalent binary
STEP 5: i = i + 1
          [End of loop]
STEP 6: Concatenate all the strings and delete the t1 number of 0's from the string
STEP 7: i1 = t1
STEP 8: Repeat steps 9 and 10 while i1 < length do
STEP 9: Store the elements in an array
STEP 10: i1 = i1 + 1
          [End of loop]
STEP 11: Determine the decimal equivalent of the string
STEP 12: Check the range of pini and pfin where the decimal value lies in between i.e. pini < deci < pfin
STEP 13: i2 <- 0
STEP 14: Repeat while i < length do
STEP 15: Check the r_min[i2] and r_max[i2] values from table 4 and print the correspondingstring.
STEP 16: i2 = i2 + 1
          [End of loop]
STEP 17: The string is the required output that is the input string
STEP 18: End

4   Results and Analysis

The proposed algorithm is tested with input of various string length and corresponding output size is obtained. Compression ratio is an important parameter for any data compression algorithm which gives the effectiveness of the compression.

Table 5 Variation of output string size with string length for best, intermediate, and worst cases

Sl. no.	String length	Compression ratio		
		Best case	Intermediate case	Worst case
1	5	2.5	2.5	2.5
2	15	3.75	3.75	3.75
3	25	5	4.167	3.571
4	35	7	5	4.375

The value of compression ratio being obtained by the proposed algorithm is much beyond unity. The compression ratio being obtained for different string length is given in Table 5. The input string can have three possible combinations, i.e., string containing characters with highest probability only (best case), string containing characters with lowest probability only (worst case), and any random combination of characters (intermediate case). It is obvious that compression ratio for the best case will have highest possible value than that obtained for intermediate or worst case. From Table 5, it is also clear that compression ratio increases with the input string for all three cases and thereby can be used for compressing large strings quite effectively.

5   Conclusions

From Table 5, it is clear that the compression ratio being obtained is pretty impressive for longer strings. In this paper, only vowel characters, i.e., a, e, i, o, u are considered with arbitrary probability to test the algorithm. The algorithm can extended to be implemented for compressing the character string containing all the characters including special characters. But it is obvious that the compression ratio will not be as high as obtained in this case. Accurate determination of the probability of occurrence of the characters is required to improve the compression ratio. This is possible either by following the character probability pattern of previous available data or by employing adaptive algorithm. But the adaptive algorithm has its own limitations due to the requirement of probability distribution table for decoding purpose. The variation of actual string and encrypted data size for the three possible cases with the length of input string is provided in Fig. 2. From the graph given in Fig. 2, it is clear that though the encrypted data size for all the three cases are same for lower string length, but for larger string, there is a significant variation of encrypted data size between the best and worst case.

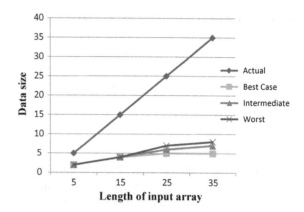

Fig. 2 Variation of data size with the length of input array

References

1. Sarkar, S. J., Das, B., Dutta, T., Dey, P., Mukherjee, A.: An Alternative Voltage and Frequency Monitoring Scheme for SCADA based Communication in Power System using Data Compression. In: International Conference and Workshop on Computing and Communication (IEMCON), pp. 1–7 (2015)
2. Takahashi, Y., Matsui, S., Nakata, Y., Kondo, T.: Communication Method with Data Compression & Encryption for Mobile Computing Environment, https://www.isoc.org/inet96/proceedings/a6/a6_2.html
3. Liu, H.-S., Chuang, C.-C., Lin, C.-C., Chang, R.-I, Wang, C.-H., Hsieh, C.-C.: Data Compression for Energy Efficient Communication on Ubiquitous Sensor Network. In: Tamkang Journal of Science and Engineering, Vol. 14, No. 3, pp. 345–354 (2011)
4. Kodituwakku, S. R., Amarasinghe, U. S.: Comparisons of Lossless Data Compression Algorithms for Text Data. In: Indian Journal of Computer Science and Engineering, Vol. 1, No. 4, pp. 406–425
5. Brar, R. S. and Singh, B.,: A survey on different compression techniques and bit reduction algorithm for compression of text data. In: International Journal of Advanced Research in Computer Science and Software Engineering (IJARCSSE) Volume 3, Issue 3 (March 2013)
6. Theory of Data Compression, http://www.data-compression.com/theory.shtml
7. Porwal, S., Chaudhary, Y., Joshi, J., and Jain, M.: Data Compression Methodologies for Lossless Data and Comparison between Algorithms. In:International Journal of Engineering Science and Innovative Technology (IJESIT) Volume 2, Issue 2 (March 2013)
8. Shanmugasundaram, S., and Lourdusamy, R.: A Comparative Study of Text Compression Algorithms. In:International Journal of Wisdom Based Computing, Vol. 1 (3) (December 2011)
9. Li, Z.-N., Drew, Mark S., Liu, J.: Fundamentals of Multimedia, 2nd Edition, Springer (2014)

Comparative Analysis of Different Feature Ranking Techniques in Data Mining-Based Android Malware Detection

Abhishek Bhattacharya and Radha Tamal Goswami

Abstract Malwares have been rising in drastic extent as Android operating system enabled smart phones and tablets getting popularity around the world in last couple of years. For efficient detection of Android malwares, different static and dynamic malware detection methods have been proposed. One of the popular methods of static detection technique is permission/feature-based detection of malwares through AndroidManifest.xml file using machine learning classifiers. But ignoring important feature or keeping irrelevant features may specifically cause mystification for classification algorithms. So to reduce classification time and improvement of accuracy different feature reduction tools have been used in different literature. In this work, we have proposed a framework that extracts the permission features of manifest files, generates feature vectors and uses six different feature ranking tools to create separate feature reducts. On those feature reducts different machine learning classifiers of Data Mining Tool, Weka have been used to classify android applications. We have evaluated our method on a set of total 734 applications (504 benign, 231 malwares) and results show that highest TPR rate observed is 98.01% while accuracy is up to 87.99% and highest F1 score is 0.9189.

Keywords Android · Malwares · Static detection · Classification · Feature ranking · Feature reduction

A. Bhattacharya (✉)
Department of Computer Science & Engineering,
Institute of Engineering & Management, Kolkata, India
e-mail: abhishek.bhattacharya@iemcal.com

R.T. Goswami
Department of Computer Science & Engineering,
Birla Institute of Technology, Mesra, India
e-mail: rtgoswami@bitmesra.ac.in

S.C. Satapathy et al. (eds.), *Proceedings of the 5th International Conference on Frontiers in Intelligent Computing: Theory and Applications*, Advances in Intelligent Systems and Computing 515, DOI 10.1007/978-981-10-3153-3_5

1 Introduction

The Android operating system has become soft target for attackers and Android applications are easy targets for reverse engineering, which is a specific characteristic of Java applications. It is often abused by malicious attackers, who attempt to embed malicious program into benign applications, hence creating subspecies of existing malware. Moreover, Android permits the users to download apps from untrusted third-party markets. To protect mobile users from rigorous threats of malwares, various approaches have been proposed. Though some static analysis approaches have been successful, different obfuscation techniques have also evolved against it. In this paper, we have proposed a data mining based malware detection framework for Android devices. Summering our main contributions in this paper are:

1. We have described the process of extracting permissions from .apk files.
2. We have applied six featured reduction tools to rank extracted features and created data sets using those selected features.
3. We have performed empirical validation using machine learning classifiers and showed the comparison of performances of different Weka based machine learning classifiers on different reducts.

We have collected 734 samples of diverse categories of apks from different Android markets. The main categories of benign apps are system tools, entertainment, news, music, audio, games, sports, etc. Malware apks are downloaded from Contagio malware dump. The reminder of this paper is organized as follows: Sect. 2 details the related works. Section 3 describes feature engineering, feature reduction, and detection methods. Section 4 shows the empirical validations. Finally, Sect. 5 shows the ways to future works.

2 Related Work

Burguera et al. [1] proposed Crowdroid-an approach that analyzed the behavior of the applications through clustering techniques in Malware detection and they considered two categories of Android applications: tools and business. Permlyzer [2] discussed the usage of permission through both static and dynamic analysis. In [3], apk's permissions were extracted from manifest file and especially <uses-permission> tag was used for this purpose. Here Euclidian, Cosine, and Manhattan distances were considered. Average accuracy obtained was 85% through Manhattan distance and using Euclidian, Cosine distances they obtained 87.57% and 90% accuracy, respectively. In [4], authors proposed a static detection method where they used binormal separation (BNS) and mutual information (MI) as feature selection tool that resulted 81.56% accuracy with 15 selected features through MI. Aswini [5], used five feature selection tools like BNS, MI, RS, KO, and KL. They obtained highest 92.51%

accuracy with 30 top features in BNS. In [6], authors used Manhattan, Euclidean, and Cosine distance as feature selector and obtained AUC 0.88 using Manhattan distance with average accuracy 85%. In [7], authors used chi-square, Fisher score and information gain as feature selection tool and they obtained average accuracy 90.77% with average AUC 0.931.

3 Feature Engineering

In this work, we have proposed a framework for detecting Android malwares and compared the performances of different machine learning algorithms.

3.1 Feature Extraction

We have first extracted the permissions using Androguard tool [8]. The structure for declaring permission in AndroidManifest.xml is shown in below.
      < uses-permission android: name ="String" />
These permissions are generally requested by any application during installation in mobile devices. After processing all 504 benign and 231 malwares we have created binary feature vector of permissions (0 means presence and 1 means absence).

3.2 Preprocessing

In our work, we have carried out feature elimination to remove irrelevant features. It is actually carried out to select a subset of significant installation-time features by eliminating features which have no predictive information. Ignoring important attribute or keeping irrelevant attributes may cause mystification for mining algorithms. The main goal of this phase is to find a minimum set of features such that the resulting probability distribution of data class is as close as possible to the original distribution created with all features. Information Gain (IG), Pearson Coefficient, Gain Ratio, Chi-Square, One R, and Relief are applied to our model as a feature ranking tools to our combined feature set and we have selected top 10 features (according to ranks) from total 83 combined and common features (both malware and benign) according to their feature ranking scores. IG and chi-square are having a specific problem that is if two features are having high ranks they will be included in the reduct but if they are highly dependent of each other then addition of both of them will not add any extra information in the reduct. Moreover, IG is biased to attribute having larger number of values. So gain ratio is used here to overcome those limitations. Finally, we have prepared six feature sets (Feature set #1, #2, #3,

Table 1 Summary of top 10 selected features through different feature selectors

Pearson Coefficient	Information Gain	Gain Ratio	Chi-Square	One R	Relief
ACCESS_NETWORK_STATE	ACCESS_NETWORK_STATE	ACCESS_CACHE_FILESYSTEM	ACCESS_NETWORK_STATE	DELETE_PACKAGES	ACCESS_COARSE_LOCATION
DELETE_PACKAGES	DELETE_PACKAGES	DELETE_CACHE_FILES	DELETE_PACKAGES	DEVICE_POWER	ACCESS_FINE_LOCATION
INSTALL_PACKAGES	INSTALL_PACKAGES	DELETE_PACKAGES	INSTALL_PACKAGES	INSTALL_PACKAGES	ACCESS_NETWORK_STATE
MODIFY_PHONE_STATE	MODIFY_PHONE_STATE	HARDWARE_TEST	MODIFY_PHONE_STATE	MODIFY_PHONE_STATE	DELETE_PACKAGES
READ_LOGS	READ_LOGS	INSTALL_PACKAGES	READ_LOGS	READ_LOGS	INSTALL_PACKAGES
READ_PHONE_STATE	READ_PHONE_STATE	MODIFY_PHONE_STATE	READ_PHONE_STATE	READ_PHONE_STATE	READ_PHONE_STATE
RECEIVE_SMS	RECEIVE_SMS	READ_OWNER_DATA	RECEIVE_SMS	RECEIVE_SMS	SEND_SMS
SEND_SMS	SEND_SMS	READ_SECURE_SETTINGS	SEND_SMS	SEND_SMS	VIBRATE
VIBRATE	VIBRATE	SEND_SMS	VIBRATE	WRITE_APN_SETTINGS	WAKE_LOCK
WRITE_EXTERNAL_STORAGE	WRITE_EXTERNAL_STORAGE	WRITE_APN_SETTINGS	WRITE_EXTERNAL_STORAGE	WRITE_SMS	WRITE_EXTERNAL_STORAGE

Table 2 Different feature sets with feature selectors

Feature set name	No. of features	Tools used for feature ranking
Feature set #1	10	Pearson coefficient
Feature set #2	10	Information Gain
Feature set #3	10	Gain Ratio
Feature set #4	10	Chi-Square
Feature set #5	10	One R
Feature set #6	10	Relief

Table 3 Result comparison with existing works

Existing works	No. of samples: MAL/BEN	TPR %	Accuracy	AUC
Yerima et al. [7]	1000/1000	90.6	91.8	0.974
PUMA [10]	249/357	92	85.8	0.920
Yerima et al. [11]	1000/1000	90.9	93.1	0.977
Yerima et al. [4] with top 20 features with RF classifier	2925/3938	90.2	91.2	0.933
Our framework	**231/504**	**98.6**	**87.99**	**0.910**

#4, #5 and #6) having top 10 features which are selected using those six feature selection tools (Pearson Coefficient, information gain (IG), gain ratio, chi-square, One R, and relief), respectively. Table 2 shows different feature sets with feature selectors. Table 1 shows summary of top 10 selected features which are selected by different feature selectors.

4 Empirical Validation

To evaluate our method, we have used the datasets described in Sect. 3 which are composed of 734 sample Android applications. Weka tool is used to analyses the evaluation of proposed model.

4.1 Evaluation Metrics

We have evaluated the model by measuring the following parameters.

1. **True Positive Ratio (TPR)**: TPR = TP/(TP + FN) where TP represents the quantity of benign applications properly classified as benign and FN represents the quantity of benign applications incorrectly classified as malware.

2. **Accuracy**: It determines the percentage of the total quantity of predictions is correct.
3. **F1-Measure**: It computes the harmonic mean of precision (P) and true positive Ratio (TPR). F1 = (2*P*TPR)/(P + TPR). F1 score 1 indicates that good performance in classification.
4. **Area under ROC Curve (AUC)**: It sets up the relation between TPR against FPR. It shows the predictive capacity of classifier.

4.2 Classifiers

One of the most engaging features of machine learning algorithms is that they improve their ability to discriminate normal behavior from anomalous behavior with experience. K-fold cross-validation has been used for evaluating the results of a numerical analysis, which generates an independent dataset. Using K = 10 folds in cross-validation means 90% of data is utilized for training purpose and 10% for testing in each fold test. To visualize the classification performance of the models, we have constructed the receiver operating characteristic (ROC). AUC values closer to 1 denotes better classifier. Figure 2 shows that for Feature set #2 highest AUC of 0.91 in SMO classifier is seemed to be the most predictive of all. The second best AUC is obtained with random forest classifier (0.879). Figure 3 shows the AUC results for six different Feature sets. Figure 1 shows that random tree classifier provides the best TPR % (98.6) for Feature sets #1, #2, and #4. Decision table classifier also provides same TPR % for Feature set #6. Figure 4 shows that average true positive rate (TPR) is highest (97.8%) in Feature set #1 followed by Feature set #2 and Feature set #4. In [9] TPR% of top 10 selected features using ID3, J48, KNN, and SVM was nearly 81% but in our framework average TPR % for all six Feature set is 97.44, which is significantly better. In our framework, Naïve Bayes

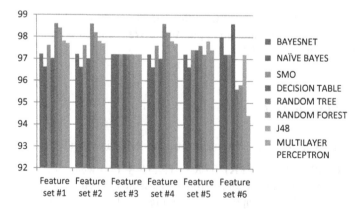

Fig. 1 Comparison of TPR %

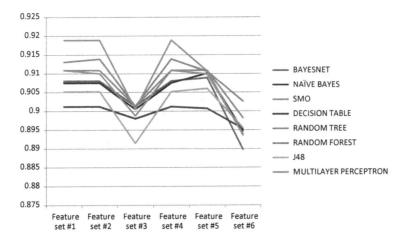

Fig. 2 Comparison of ROC

and J48 have generated average TPR 96.81% and 97.61%, respectively, which is significantly better than [9]. In [6], J48 generated highest TPR % (90.7) in dataset 1 and RF generated highest TPR % (91.8) in dataset 2 but in our framework, J48 generates highest TPR as 97.81% and RF generates highest TPR as 96.9% which are significantly better than that of [6].

Figure 5 demonstrates the comparison of different F1 score with all six feature sets using different Weka classifiers. It shows that Feature set #1, #2, and #4 obtain significantly higher F1 score (0.912) using multilayer perceptron classifier. Figure 6 shows that average F1 score is highest (0.9095) in Feature set #1, #2, and #4. Figure 7 reveals that though individually multilayer perceptron generates highest Accuracy % (88) with Feature set #1, #2 and #4. Figure 8 shows that average Accuracy is highest (86.66) in Feature set #2 and #4. Table 3 shows how the results

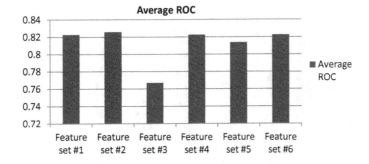

Fig. 3 Comparison of average ROC

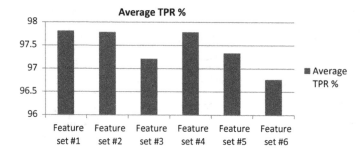

Fig. 4 Comparison of average TPR %

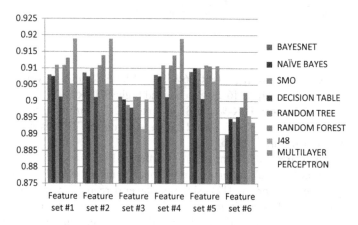

Fig. 5 Comparison of F1 values

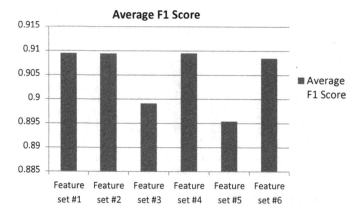

Fig. 6 Comparison of average F1 score

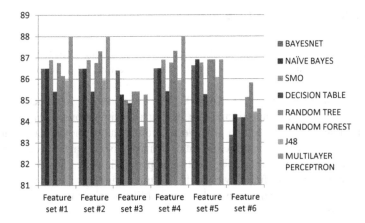

Fig. 7 Comparison of accuracy %

Fig. 8 Comparison of average accuracy %

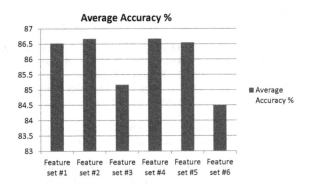

Table 4 Result comparison with existing work

Framework	Feature selector name	TPR
Wei Wang	Information gain	0.9228
	Correlation coefficient	0.9232
Our framework	Information gain	**0.9778**
	Pearson coefficient	**0.9780**

of our framework measures against the best results of [4, 7, 10, 11], respectively. This is highlighted in the bottom of the table where the result of our proposed approach shows highest TPR %. Table 4 shows the comparison of our framework with Wang's [12] framework.

5 Conclusion and Future Works

In this paper, we have made the comparative studies of six different feature reducts, created by IG, Pearson coefficient, gain ratio, chi-square, One R, and relief as feature ranking techniques and measured the performances of machine learning algorithms in terms of ROC curve (AUC), TPR, F1 scores and Accuracy. Machine learning-based detection approaches are having two limitations: they have high false alarm rates [13]. Through proper investigation, it can be shown that because of sparseness of feature vector, detection ratio of may be sometimes inaccurate [12]. Moreover, a number of malwares request same permissions that are also requested by benign applications In future, we would like train classifiers with larger datasets to get more accurate classification and we would like to use clustering concepts in feature reduction.

References

1. I. Burquera, U. Zurutuza, and S. Nadjm-Tehrani: Crowdroid: behavior-based malware detection system for Android. In: 1st ACM workshop on Security and privacy in smartphones and mobile devices, 2011, pp. 15–26, 2011.
2. W. Xu, F. Zhang, S. Zhu: Permlyzer: Analyzing permission usage in Android applications. In: IEEE International Symposium on Software reliability Engineering (ISSRE), pp. 400–410 (2013).
3. B. Sanz, I. Santos, X. U. Pedrero, C. Laorden, J. Nieves, P. Garcia Bringas: Instance-based Anomaly Method for Android Malware Detection. SECRYPT, SciTePress, pp. 387–394 (2013).
4. S.Y. Yerima, S Sezer, G. McWilliams: A new android malware detection using Bayesian classification. In: 27th IEEE International Conference on Advanced Information Networking and Applications (AINA), pp. 121–128 (2013).
5. A. M. Aswini, P. Vinod: Android Malware Analysis Using Ensemble Features. Security, Privacy, and Applied Cryptography Engineering Lecture Notes in Computer Science, vol. 8804, pp. 303–318 (2014).
6. A. M. Aswini, P. Vinod: Droid Permission Miner: Mining Prominent Permissions for Android Malware Analysis. In: 5th International Conference on the Applications of the Digital Information and Web Technologies (ICADIWAT), pp. 81–86 (2014).
7. S.Y. Yerima, S Sezer, G. McWilliams, I. Muttik: Analysis of Bayesian classification-based approaches for Android malware detection. IET Information Security, vol. 8, issue 1, pp. 25–36 (2014).
8. Androguard Project in Google Code Archive, https://code.google.com/p/androguard.
9. Y. Aafer, W. Du, H. Yin: DroidAPIMiner: Mining API-Level Features for Robust Malware Detection in Android. Lecture Notes on Security and Privacy in Communication Networks, vol. 127, pp. 86–103, Institute for Computer Sciences, Social Informatics and Telecommunications Engineering (2013).
10. Z. Aung, W. Zaw: Permission-Based Android Malware Detection. International Journal Of Scientific & Technology Research, vol. 2, issue 3, pp. 228–234 (2013).
11. S.Y. Yerima, S Sezer, G. McWilliams, I. Muttik: High Accuracy Android malware detection Using Ensemble Learning. IET Information Security, vol. 9, issue 6, pp. 313–320 (2015).

12. W. Wang, X. Wang, D. Feng, J. Liu, Z. Han, X. Zhang: Exploring Permission-Induced Risk in Android Applications for Malicious Application Detection. IEEE Transactions on Information Forensics and Security, vol. 9, issue 11, pp. 1869–1882 (2014).
13. K. Allix, T. F. D. A. Bissyande, J. Klein, and Y. Le Traon: Machine Learning-Based Malware Detection for Android Applications: History Matters!. Technical Report, University of Luxembourg, pp. 1–17 (2014).

Feature Optimality-Based Semi-supervised Face Recognition Approach

Taqdir and Renu dhir

Abstract In this paper, a novel approach is proposed that cope with challenges such as illuminations, expressions, poses, and occlusions. The proposed methodology is a non-domination-based optimization technique with a semi-supervised classifier for recognizing a known and unknown face based on different scenarios. The classification is a robust method attaining aptness at different stages resulting in identification of proper training set with actual face image. Different datasets Yale Face Database, Extended Yale Face Database B, ORL database has been considered for our experiments. The performance of the proposed method has been evaluated on several grounds. Results show that the proposed method attains a better performance than the statistical methods.

Keywords Face recognition · Discrete wavelet transform · Principle component analysis · Linear discriminant analysis · Non-Domination optimality

1 Introduction

Face recognition has emerged as an area of interest over past few years. In the broader area of application, such as surveillance systems, identification system, human computer interaction and many more led the researchers to focus in this domain [1]. A human face is a dynamic object having variation in appearance, pose, and orientation that makes the face identification problem more difficult. The face recognition system comprises of three major steps, such as face detection, feature

Taqdir (✉)
Computer Engineering Department, Guru Nanak Dev University Regional Campus,
Gurdaspur, India
e-mail: taqdir_8@rediffmail.com

Renu dhir
Computer Engineering Department, National Institute of Technology (NIT),
Jalandhar, India
e-mail: dhir@nitj.ac.in

© Springer Nature Singapore Pte Ltd. 2017
S.C. Satapathy et al. (eds.), *Proceedings of the 5th International Conference on Frontiers in Intelligent Computing: Theory and Applications*, Advances in Intelligent Systems and Computing 515, DOI 10.1007/978-981-10-3153-3_6

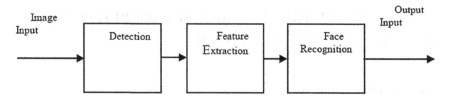

Fig. 1 Basic phases of face recognition

extraction, and face recognition [1]. The first step is to find the existence of a human face in the image and to locate them. The preprocessing is performed with scaling and orientation of the image. In the second step, image is converted into vector representing set of fiducial points and their location on the face. Finally, face recognition is performed in two step process. The initial step matches the face template with the face image known as face verification and the later step is matching the face image against all the template images in database known as face identification. The test image features are compared to the feature of the images in the database and similarity is determined for finding the match. The basic steps for the face recognition can be depicted as in Fig. 1 [2].

Much of work has been carried out with geometric-based features and holistic-based approaches. Most of the geometric-based approaches, such as elastic bunch graph matching (EBGM) [3] and active appearance graph model (AAM) [4] are dependent on the selection of the local facial features that makes the system semi-automatic. Whereas with the holistic methods the entire pixel of the face image vector is considered that increases the computational complexity of the system. These approaches are highly expensive and require a high degree of correlation between the training and the test images. Moreover, the dimensionality becomes another major factor for the detection process as the feature set in high-dimensional data needs to be reduced for the selection optimizing the detection. As these methods perform the classification by majority schemes based on training examples, the annotator labeling [5] is dependent on the facial expression that transitively depends on correlation among expression, reduces the performance due to same labeling in some cases.

In this work we have proposed a feature optimality-based semi-supervised face recognition system. The proposed method addresses the problem of high-dimensionality images by optimizing the feature vectors. A non-dominating Pareto-based optimization technique is implemented for selecting the relevant features. These optimized set of features obtained from the training and test set results in reducing the computation complexity for classification. We formulate a local neighbor structure (LNS) which is a transformation of training set for compressed representation. A dissimilarity matrix is computed for identifying the most correlated face image with the obtained LNS. Finally, the most recognized face is obtained based on higher similarity value. Figure 2 shows the block diagram of the proposed method.

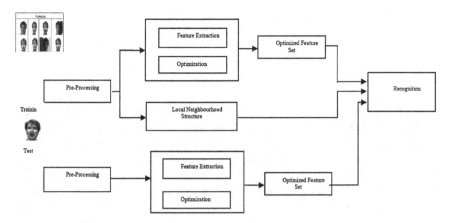

Fig. 2 Block diagram of proposed work

We have extensively evaluated the performance of our method with different face images under variability, such as lightening conditions, orientations, different facial expression, and occlusions. Experiments have been done for comparing the performance of the proposed method with existing methods based on various measures, such as precision, recall, confusion matrix, and recognition rate. Different datasets namely, Yale Face Database, Extended Yale Face Database, ORL database, has been considered for our experiments. Results show that our method performs significantly well with a better recognition accuracy.

The remaining part of the paper is organized as follows. Section 2 presents a review state-of-the-art literature. In Sect. 3 the proposed approach is detailed with algorithmic steps and flowchart. Section 4 deals with the experimental results and discusses the outcomes considering the performance of the proposed method and Sect. 5 draws the conclusion.

2 Overview of Existing Methods

Faces have high variability and are dynamic in nature. A number of conditions such as orientation, lighting, scale, and expressions [6] variate over period of time under circumstances become problematic in some cases. Face recognition is done in two approaches holistic approach and feature-based approach. Features obtained for holistic approach are global features from faces whereas in feature-based approach the features are obtained from the local features from the face. Among the different dimensionality reduction methods from the holistic approach principal component analysis and singular value decomposition has been massively used. As compared to the holistic methods the feature-based methods have more robust with scale, size, and location of the face but the methods are susceptible to occlusion caused by glasses, facial expressions, and head orientation [7].

The discrete wavelet transform (DWT) [8] method is one of the prominent holistic methods and has advantages of simultaneous localization in time and frequency domain that decomposes the component wavelet making the computation faster. With smaller number of coefficients a better approximation can be done that reveals breakdown points, discontinuities in higher derivatives, and self-similarity for analyzing the data. Another most effective face recognition approach is principal component analysis (PCA) [17], the method reduces the dimensionality by computing maximum scatter and correlations of the projected vectors. The test image is projected onto the training set of eigenfaces outcomes the recognition of face. The process involves computation of the covariance matrix and invariance that becomes difficult when the training data is unavailable. As PCA encodes only the orthogonal linear space information, another linear combination features separating two or more classes is linear discriminant analysis (LDA) [9]. The features of class are obtained through eigenvector analysis of scatter matrix maximizing between-class variation and minimizing the within-class variations. Another face recognition method based on Karl Pearson's coefficient is coefficient of correlation (CoC) [10]. From 1999 onwards the research toward the face recognition gained a pace. A huge number of research works has been done in this field [11].

In 2009, Latha et al. [12], presented a neural-based algorithm in which the dimensionality of the image is reduced with PCA followed by the back propagation neural network (BPNN) for recognition. The algorithm performs by obtains the eigenfaces, that is, the variation of the between the faces. The face is computed by reconstructing weighted combination of eigenfaces. The new face is combined by projecting the image into eigenface vectors. The random weights and threshold of the network are initialized. These values are updated each iteration thereby minimizing the total squared error of target and output vector. Finally, the input image is classified as face image or other object.

In 2010, Aishwarya and Marcus [13] proposed a method to compute the image subspace known as face space to project a face image for computing the Euclidean distance between the vectors. The image subspace is constructed by computing the eigenvector of the covariance matrix obtained from set of images. A number of eigenvectors with highest eigen values is selected forming the face space. The acceptance of the face image is determined with a threshold value. But sometimes the estimation of the threshold value becomes difficult as facial expression, lightening conditions may variates.

In 2011 Sparse Representation gained a lot of attention for representing the training samples with sparse linear combination. The method performs $l1$-minimization that makes the sparsity-based classification methods more expensive. For improving the robustness and effectiveness of sparse representation a robust sparse coding (RSC) is proposed by Yang et al. [14]. The method relies on MLE-estimator that robustly regress the signal with sparse regression coefficients. The minimization problem is transformed into reweighted sparse coding such that the outliers are compressed with low weight values. The method is robust with different types of outliers and performs better with images with variation of illumination, expression and occlusions.

The identification of face in controlled environment with large number of sample of face is performed and results in better accuracy. However, the identification of face under uncontrolled environment with lack of training data is yet a problem as the lightening and facial expression changes that need to be addressed [15]. For addressing such a problem Schwartz et al. [1] performed a combination of feature weighting by partial least square (PLS) that handles high-dimensional data with less number of samples. This PLS is converted to a tree-structure for reducing the computational complexity. The classification of one-against-all classifier is done for handling unbalanced class distribution.

In 2013, an advance descriptor-based face recognition technique [16] is designed with an over-complete representation. The comparison of the actual image with the pair is signature based, where the signature are stored in order of hundred floating point numbers. In the over-complete representation the LBP that computes over-lapping blocks with horizontal and vertical overlaps. A within-class covariance normalization (WCCN) reduces the features in low-dimensional subspace and produces score applying cosine similarity. In 2014 Gaidhane et al. [17] proposed a simple method for face recognition. The method illuminates the computation of eigenvalues and eigenvectors which reduces the complexity. For set of training images, a polynomial characteristic is computed that obtains the features for each image. These features of training image are stored in a matrix form known as companion matrix. Similar procedure is repeated for a test image. A sequence of vectors for both the test image matrix and training image matrix is computed. The determinant of the symmetric matrix is calculated for finding the singularity. If the matrix is singular the test image is similar to training set else no common features are detected. The maximum value of nullity of matrix shows recognition of unknown face. The method reduces the computation complexity and the problem of high dimensionality of data is resolved but requires more number of steps for computations.

Mehta et al. [18] proposed an optimal directional face (ODF) recognition method that extracts the directional information using the image directive. The method performs image derivative with local polynomial approximation (LPA) which is a directional filter at multiple scales. As the directional features captures increases the dimensionality, hence the intersection of confidence interval (ICI) is applied for selection of scale at each pixel. The textural features are extracted by applying modified local binary pattern followed by partitioning the feature image and finally the histogram of each feature values are computed and concatenated together. After portioning the ODF at different level, the LDA is applied for reducing in dimensionality; finally the SVM is applied for classification.

Seo and Park [19] developed a method that is robust to partial variations by extracting the features through scale invariant feature transform (SIFT). The method is test on different bench mark datasets with facial image expression and occlusions. Facial image is represented with local features using SIFT that uses scale-space difference of Gaussian to detect key points in the image. The SIFT is applied on set of training image and test image and estimation of probability density function is performed. With the density function weight of each descriptor smaller weight value are used to represent occluded area. Once the weight for feature values for test image are computed, the K-NN classifier is applied for finding the similarity and assign the test image to a class. However, the method when applied for occluded images and partial variations results in high-dimensionality and improper results with local variations.

The problem of overlapping features in spontaneous expression with geometric and appearance features became a major problem for the classifiers to separate the boundaries. In 2014, Wan and Aggarwal [20] proposed method utilizes specificity and sensitivity model that contains mislabeled expression existing in the training data with different annotators. The expectation–maximization (EM) procedure is deployed to estimate the true label of expression by computing the likelihood with a metric. An adaptive online metric learning is proposed to solve optimal positive semi-definite metric with the steepest descent method. The method performs the classification by simple means of voting process based on training examples, however, the annotator labeling is dependent on the facial expression that transitively depends on correlation among expression that reduces the performance due to same labeling in some cases.

Luan et al. [21] considers the problem of human face recognition with variations, such as illumination and occlusions and proposed method deal with them. The method can be directly applied on face despite of any feature selection procedure. Two descriptors namely sparsity descriptor and smoothness descriptor are proposed followed by the recognition process. The sparsity descriptor computes the sparsity of image matrix to find the degradation of the image with noise and error. With the error image, the image of an individual looks smooth in some non-occluded regions. The sparsity descriptor is computed in pixel domain whereas the smoothness descriptor assumes to in gradient domain. These two descriptors jointly determine the true identity. In 2015, a method to overcome the drawbacks of conventional filters resulting in blur edges was addressed with the concept of anisotropic diffusion (AD) [22]. The method performed features extraction with Gabor filter followed by the binary particle swarm optimization (BPSO) for optimal feature selection. The BPSO is applied results in reduces number of features by optimizing the extracted features form after the application of the Gabor filter. This decreases the classification testing time thereby improving the performance of recognition system.

3 Proposed Work

The proposed methodology is an address to the subjects that optimizes the features set resulting in reduction of the dimensionality and performs the recognition in an unsupervised manner thereby reducing the computation time. The method initiates with set of training images that are preprocessed with median filter as the boundaries remains definite and are less affected with presence of outliers. For each training set the feature selection is performed by searching peaks in the scale-space from a difference of Gaussian function [23]. The optimization is performed on the obtained selected features based on the Non-Dominance-based optimality [24, 25], thereby reducing the dimensionality and feature vector. From the optimized set of training images and test image the decision is performed. We have categorized the decision step into three stages:

(i)   Non-Domination-based Dissimilarity (NDD)
(ii) Sparse Dissimilarity (SD)
(iii) Decisive-based Feature Similarity (DFS)

The Non-Dominance optimization is a multi-objective optimization (MO) that sorts the population into different non-domination levels. The procedure begins with initialization of the population, sorted to each front. The sorting procedure assigns a rank to individual in each front it belongs. Each solution in the population set is assigned two values, domination count, and the number of solutions that dominate the solution and the set of solution that dominates [22]. Initially, all the solution have zero as their domination value. As each solution is visited the count is reduced by one and the solution is put in a separate list that belongs to second non-dominated front. This process continues until all fronts are identified resulting in a Pareto-front optimality. The solutions are sorted from the higher to lower ranking based on average distance with density estimation. A fitness value is computed for each individual that measures the closeness of individual to its neighbors known as co-related metric (CM). The selection of parents is based on the rank and CM. The CM is calculated as the sum of individual distance value corresponding to each objective function as given in Eq. 1. These selected parents undergo the process of crossover, mutation for generating off springs. The flowchart of the preprocessing stages is shown in Fig. 3.

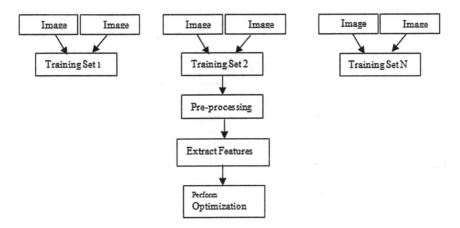

Fig. 3 Preprocessing and optimization

For each individual $k = 2$ to $(n - 1)$

$$I(\text{dist}_k) = I(\text{dist}_k) + \frac{I(k+1) \cdot m - I(k-1) \cdot m}{f_m^{\max} - f_m^{\min}}, \tag{1}$$

where $I(k) \cdot m$ is the value of mth objective function of the kth individual in I.

In the decision step, from the obtained optimized feature vector, NDD is computed which is a pixel-based approach for finding less-dissimilar faces thereby reducing the number of comparisons. It computes the L^2 norm between the test images and obtained LNS of each training set as given in Eq. 2. The LNS of the training set selects the relevant face images, increasing the chances of obtaining the exact match.

$$d^2 E = \sum_{i,j=1}^{MN} g_{ij}\left(x^i - y^i\right)\left(x^j - y^j\right), \tag{2}$$

where M and N are dimensions of images and g_{ij} is a metric matrix with orientation.

Here we have a sparse dissimilarity (SD) computation that groups the similar test images based on the distance value from the LNS of each training set as shown in Fig. 4. This grouping is one for identifying similar persons face image that validates the decision of known or unknown face. The dissimilarity is formulates as given in Eq. 2.

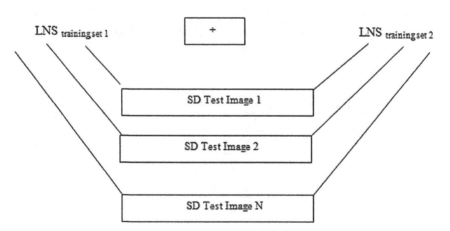

Fig. 4 Sparse dissimilarity for 2 training set

$$\text{Sparse Dissimlarity } T_m = \sum_{i=1}^{m} \sum_{j=1}^{n} \text{dist}(T_i, \text{LNS}_j), \tag{3}$$

where m = number of test images and n = training sets.

After obtaining the SD value of test images a normalized max–min sparse dissimilarity is computed for that normalizes the value to a range for finding the correlated face images. The correlation is minimum difference value in the comparison matrix of the test image dissimilarity value. However, the difference value is threshold certain conditions where the test image is an unknown face. The method of SD has a major contribution in grouping and identifying the faces in correct training set, correct training set with same expression given the test image, correct training set with invalid expression given the test image and incorrect training set with invalid expression. The L^2 norm value of training set LNS and test image abet identification of the training set for respective test image. The lower is the L^2 value of test image, higher is belongingness to the training set. The final recognition phase is the decisive feature-based similarity (DFS), the obtained optimum features of the test set is correlated with the obtained optimum features of the training set corresponding to the respective LNS as given in Fig. 5. The recognition result identifies the exact image of the person with valid expression given the expression exists in training set otherwise it outputs a nearby-value face. The method also segregates an unknown face when the person does not exist in the dataset.

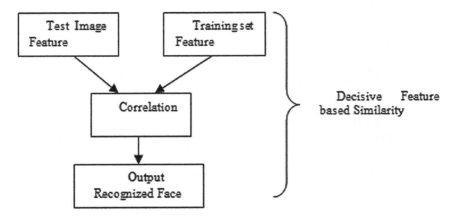

Fig. 5 Decisive Feature-based Similarity

Algorithm

Input: Set of training images T_m
        Test image T_i
Output: Recognition rate R

Input the T_m and T_i
Perform pre-processing of the images
Initialize the population P
Set T be number of iterations
Generate N solutions (*Sol*) to form initial population P
Divide the P into m objectives.
Evaluate the fitness function of each solution

For each $p, q \in P$
  If (p <q)
 Add q to Sol set
    Else
     Select the next solution
     Each solution *Dcount (P=0)*
  Add to Front (P)
Rank the non-dominated fronts Front (P).
Density (Sol) be averaging the distance points along objectives
$Dist_j = 0$
  For each individual k=2 to (n-1)

$$I(dist_k) = I(dist_k) + \frac{I(k+1).m - I(k-1).m}{f_m^{\max} - f_m^{\min}}$$

Perform crossover and mutation to create offspring P_{t+1} of size N
Increment the next iteration
LNS Tr = Avg (T_m)
  For each T_i
$NDD_{vec} = L^2 (T_i, LNS_{Tr})$

$$SD = \sum_{i=1}^{m} \sum_{j=1}^{n} dist(T_i, LNS_j)$$

Obtain the group in T_i
Find the training set $SELT_r$
$DFS_{vec} = (Ti\ (feature\ set), SELTr)$
Check T_i for different cases
 Output the recognized face with recognition
rate R

4 Results and Analysis

4.1 Dataset Description

For analyzing the recognition accuracy and efficiency, the proposed method has been experimented on a number of publicly available face databases, such as the Yale Database, Extended Yale Face Database, and the ORL database.

The ORL database [26] contains 10 different images of 40 distinct individuals. For some of the individuals, the images we recaptured at the different intervals of time, varying the poses, lighting, facial expressions (smiling/non-smiling, open or closed eyes) along with or without glasses. All the images were taken against a dark homogeneous background with the individual upright frontal position. The Yale database contains 165 gray scale images of 15 different persons. There are 11 images of each person having different facial expression under various conditions (centre-light, left-light, right-light, with glasses, without glasses, normal, happy, sad, sleepy, surprised, and wink). The Extended Yale database [27] contains 5760 single light source images of 10 persons, 9 different poses, and 64 illumination conditions of each person.

The proposed approach has been tested on different datasets ORL Database, the Yale Database, the Extended Yale Database, and evaluated with measures precision, recall, confusion matrix, and recognition accuracy on different criterion. The method is compared with the conventional holistic method, such as PCA, DWT, LDA, and correlation methods, SSIM and CoC. A new robust classifier incorporated enhances the recognition rate by identifying exact match of test image person with actual expression person in the training set. All the experiments have been carried out on Intel Core I5 processor with 2.6 GHz frequency and 1 TB RAM. All the programs have implemented and tested on MATLAB platform of version R2014b.

4.2 Analyzing Yale Database

The Yale database is a standard face recognition database. For this experiment we have considered images of three persons with nine different expressions as training set as shown in Fig. 6. The test set comprises of six different person images excluding the training set from Yale database as shown in Fig. 7. The experiment initially finds the feature set of matrix dimension 59×2 for training set and 69×2 for test set. The method optimizes the selected set of features that reduces the dimensionality, increasing the accuracy rate thereby reducing the computation cost. The optimization method ranks the solution obtained in the non-domination front which includes those selected features that are prominent for decision. The optimized feature set reduces its dimensionality to 20×2 and 26×2 for training set and test set, respectively. The classification step performs LNS that computes

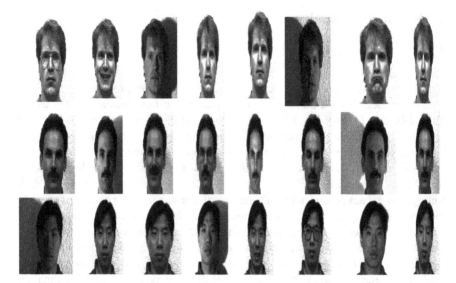

Fig. 6 Sample training set of Yale database

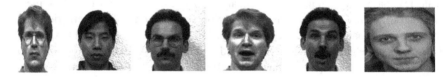

Fig. 7 Test set Images

the Sparse Dissimilarity for given test image against the training set validating known or unknown face image. The minimum dissimilarity value selects the training set for given test image. The optimized features set of the test image 26 2 is computed against the LNS of the training set. The DFS outputs the image matrix with recognition value showing the true positive, false positive and true negative results

- The true positive (TP) illustrates the person exists in the database with valid expression present in the training set.
- The false positive (FP) illustrates person with invalid expression with respect to the person present in the training set
- The true negative (TN) illustrates invalid expression of unknown person not present in the entire database

Figure 8a–c shows TP for recognition the actual image of the person, Fig. 8d, e shows FP recognizing valid person with different expression and Fig. 8f, g shows the TN, recognition no image for invalid expression of invalid person.

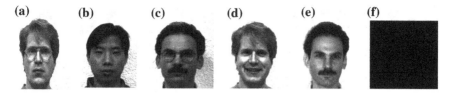

Fig. 8 **a–c** Correctly Recognized faces, **d**, **e** are correct person with different expressions, **f** image of invalid person

4.3 Analyzing ORL Database

The experiment on the ORL database comprises the training set of four persons with different expressions and various poses as shown in Fig. 9. The test set comprises of nine images of six different persons as shown in Fig. 10. The algorithm selects the set of features that are optimized for finding optimal feature set in the training as well as test set. The LNS for training set is computed for finding the grouping among the test set images and finding the relevant training set for given test image. The DFS is computed for finding recognition rate from the selected training set and the optimum features in test image. Figure 11a–d shows TP for recognition the actual image of the person, Fig. 11e–g shows FP recognizing valid person with different expression and Fig. 11h, i shows the TN, recognition no image for invalid expression of invalid person.

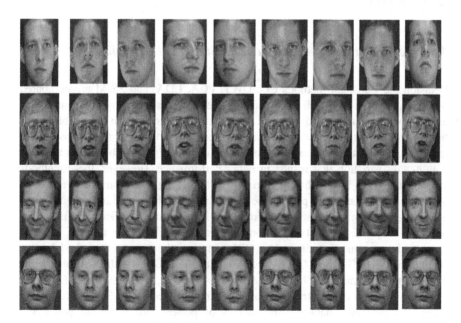

Fig. 9 Sample training set of ORL database

Fig. 10 Test set Images

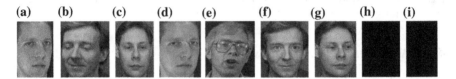

Fig. 11 a–d Correctly Recognized faces, **e–g** are correct person with different expressions, **h, i** image of invalid person

4.4 Analyzing Extended Yale Database

In the Extended Yale database, the training set comprises of images of nine different expressions of a person under various lightening conditions. The test set comprises of seven images of three different persons. The results shows three images of the person that are correctly recognized represented as TP. The 2 images represent FP with different expression and lightening condition whereas the TN is represented by the null images.

4.5 Performance Evaluation with Statistical Methods

Decision the statistical methods CC, DWT, PCA, LDA, and SSIM are compared with the proposed approach for evaluating the performance recognition rates. The experiments are performed on ORL, Yale and Extended Yale B databases for finding the results. The statistical methods have the advantage of dimensionality reduction, wavelet decomposition for faster computation and between-class and within-class variation that results in better performance. However, these methods are incapable with images having variation in lightening conditions, occurrences of occlusions and changes in poses. The proposed methodology address the problem by finding the optimal feature set from the large number of feature set thereby reducing the dimensionality and computation of the LNS for training set also reduces the computation time. The ORL database the proposed method has an accuracy of 98.97%. The CC, DWT, DWT with LDA, DWT with PCA, and SSIM has recognition rate of 92.2317, 93.7571, 94.7322, 96.7339, and 69.383%. The SSIM is similarity metric that computes the distance for finding the correlated faces. For the ORL database the images of persons are varied in poses and expressions that detoriate the performance with SSIM. With the Yale database the

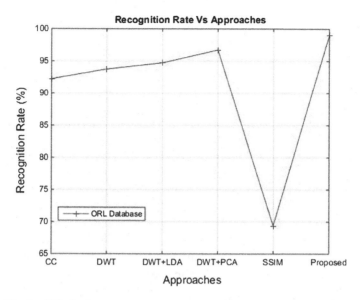

Fig. 12 Plot for ORL database

proposed method performs outstanding with the exact identification with the accuracy of 99.00% and 86.4335, 94.9819, 95.9583, 96.9598, 68.8798% with different statistical approaches. Experiments show that with the Extended Yale Database B the proposed method attains the accuracy of 98.98%. The plot of recognition rate for ORL, Yale and Extended Yale Face Database B is shown in Figs. 12, 13 and 14.

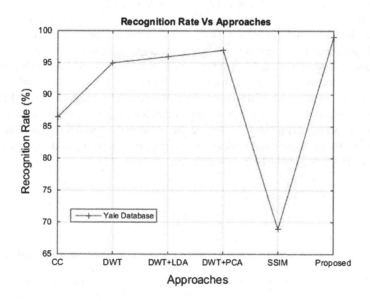

Fig. 13 Plot for Yale database

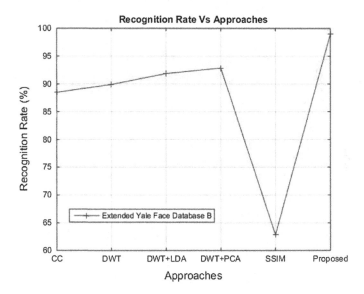

Fig. 14 Plot for extended Yale face database B

5 Conclusions

In this paper we have studied a number of conventional methods for face recognition. The statistical methods having advantage in terms of dimensionality reduction and computation cost, thereby results in poor performance when applied on datasets having various facial expressions, poses and lightening conditions. The proposed method is a robust approach for recognizing the actual face and identifying the known or unknown face. The method optimizes the feature set for finding the grouping among the test images and extracting the correct training set for the test image by computing the LNS of the training set. Recognized face is obtained by the computation of the selected training set features and the optimized test set features. The performance of the proposed method is evaluated with the TP, FP, and TN classification terms. The correct face identification, correct face with invalid expression and invalid person face image are some major grounds for performance evaluation. Result shows the performance of the proposed methods with the statistical methods is outstanding with different datasets.

References

1. W. R. Schwartz, H. Guo, J. Choi, L.S. Davis, "Face Identification Using Large Feature Sets", IEEE Trans. On Image Processing, vol. 21(4), pp. 2245–2255, 2012.
2. K. Roy, P. Bhattacharya, C.Y. Suen, "Iris recognition using shape guided approach and game theory", Pattern Anal. Applications, vol. 14(4), pp. 329–348, 2011.

3. S. Shin, S.-D. Kim, H.-C. Choi, "Generalized elastic graph matching face recognition", Pattern Recognition Letters, vol. 28 (9), pp. 1077–1082, 2007.
4. H. Zhang, Q.M.J. Wu, T.W.S Chow, M. Zhao, "A two-dimensional neighborhood preserving projection for appearance-based face recognition", Pattern Recognition, vol. 45 (5), pp. 1866–1876, 2012.
5. S. Wan, J.K. Aggarwal, A scalable metric learning-based voting method for expression recognition, in: 10th IEEE International Conference on Automatic Face and Gesture Recognition, pp. 1–8, 2013.
6. X. Luan, B. Fang, L. Liu, W. Yang, J. Qian, "Extracting sparse error of robust PCA for face recognition in presence of varying illumination and occlusion", Pattern Recognition vol. 47, pp. 495–508, 2014.
7. R. Mehta, J. Yuan, K. Egiazarian, "Face recognition using scale-adaptive directional and textural features", Pattern Recognition vol. 47, pp. 1846–185, 2014.
8. Wenkai Xu and Eung-Joo Lee, "Face Recognition Using Wavelets Transform and 2D PCA by SVM Classifier", International Journal of Multimedia and Ubiquitous Engineering, vol. 9 (3), pp. 281–290, 2014.
9. J.L. Tirupathamma, M.K. Rao, K.V. Swamy, "LDA based face recognition using DCT and hybrid DWT", International Journal of Electronics Signals and Systems, vol. 1 (3), pp. 7–10, 2012.
10. A. Kaur, L. Kaur, S. Gupta, "Image Recognition using Coefficient of Correlation and Structural Similarity Index in Uncontrolled Environment", vol. 59(5), pp. 32–39, 2012.
11. K. S. Kinage and S. G. Bhirud, "Face Recognition based on Two- Dimensional PCA on Wavelet Subband", International Journal of Recent Trends in Engineering, vol. 2 (2), pp. 51–54, 2009.
12. P. Latha, L. Ganesan, S. Annadurai, "Face Recognition using Neural Networks", *Signal Processing: An International Journal (SPIJ)*, vol. 3, pp. 153–160, 2009.
13. P. Aishwarya and K. Marcus, "Face recognition using multiple eigenface subspaces", *Journal of Engineering and Technology Research*, vol. 2(8), pp. 139–143, 2010.
14. M. Yang, L. Zhang, J. Yang and D. Zhang, "Robust sparse coding for face recognition", International Conference in Computer Vision and Pattern Recognition, pp. 625–632, 2014.
15. W. R. Schwartz, H. Guo, J. Choi, L. S. Davis, "Face Identification Using Large Feature Sets", vol. 21 (4), pp. 2245–2255, 2012.
16. O. Barkan, J. Weill, L. Wolf, Lior H. Aronowitz, "Fast High Dimensional Vector Multiplication Face Recognition", The IEEE International Conference on Computer Vision (ICCV), pp. 1960–1967, 2013.
17. V.H. Gaidhane, Y. V. Hote, V. Singh, "An efficient approach for face recognition based on common eigen values", Pattern Recognition, vol. 47, pp. 1869–1879, 2014.
18. R. Mehta, J. Yuan, K. Egiazarian, "Face recognition using scale-adaptive directional and textural features", Pattern Recognition, vol. 47, pp. 1846–1858, 2014.
19. J. Seo, H. Park, "Robust recognition of face with partial variations using local features and statistical learning", Neurocomputing, vol. 129, pp. 41–49, 2014.
20. S. Wan, J.K. Aggarwal, "Spontaneous facial expression recognition: A robust metric learning approach", Pattern Recognition, vol. 42, pp. 1859–1868, 2014.
21. X. Luan, B. Fang, L. Liu, W. Yang, J. Qian, "Extracting sparse error of robust PCA for face recognition in the presence of varying illumination and occlusion", Pattern Recognition, vol. 47, pp. 495–508, 2014.
22. T. M. Abhishree, J. Latha, K. Manikantan, S. Ramachandran, "Face recognition using Gabor filter based feature extraction with Anisotropic Diffusion as a Pre-processing Technique", vol. 45, pp. 312–321, 2015.
23. Y.Y. Wang, Z. M. Li, L. Wang, M. Wang, "A Scale Invariant Feature Transform Based Method", Journal of Information Hiding and Multimedia Signal Processing, vol. 4 (2), pp. 73–89, 2013.

24. K. Deb, A. Pratap, S. Aggarwal, T. Meyarivan, "A fast and Elitist Multiobjective Genetic Algorithm: NSGA II", IEEE Transactions on Evolutionary Computation, vol. 6 (2), pp. 182–197, 2002.

25. S. F. Ghribi, A. Y. Jammoussi, D.S. Masmoudi, "A Multi Objective Genetic Algorithm based Optimization of Wavelet Transform Implementation for Face Recognition Applications", World of Academy Science, Engineering and Technology, 56, pp. 1590–1593, 2011.

26. The ORL Database of Faces, AT&T Research Laboratories, Cambridge, [Online]. Available: <http://www.cam.ac.uk/Research/DTG/attarchive/pub/data/attfaces.tar.Z>.

27. The Yale Face Database B [Online]. Available http://cvc.yallefacesB.html.

Fuzzy-Based Algorithm for Resource Allocation

Gurpreet Singh Saini, Sanjay Kumar Dubey and Sunil Kumar Bharti

Abstract The algorithm presented in this paper deals with use of soft computing technique of Fuzzy logic applied with dynamic graph theory to create graphs which can be efficient in resource allocation process in varied environments, i.e., software project management, operating systems, construction models, etc. The algorithm implies one unique factor of dynamicity which makes graph of resource allocation evolving even after primary design due to chaotic nature of the afore mentioned nature of environments. The use of Fuzzy imparts a logical inference mechanism which rules out non-monotonous reasoning perspective of this dynamicity. The algorithm is robust and adaptive to varied environments. The proposed algorithm will be beneficial for more accurate Engineering in terms of reducing the failures and being more specific in answering the allocation of the resources and how the work has to be undertaken using those resources. It will also emphasize on devising a model which can be adhered to with the proper follow ups such that it could be referred to at the time of chaos or failures. "The development of the Algorithm will be much more product centric and will stick to developer's view of development along with customer's view of required functionalities."

Keywords Dynamic graph theory · Resource allocation · Fuzzy logic · Chaos theory · Soft computing

G.S. Saini (✉) · S.K. Dubey
Amity University Uttar Pradesh, Noida, India
e-mail: g.saini4888@live.com

S.K. Dubey
e-mail: skdubey1@amity.edu

S.K. Bharti
Central University of Haryana, Mahendergarh, India
e-mail: sunilbharti@cuh.ac.in

© Springer Nature Singapore Pte Ltd. 2017   69
S.C. Satapathy et al. (eds.), *Proceedings of the 5th International Conference on Frontiers in Intelligent Computing: Theory and Applications*, Advances in Intelligent Systems and Computing 515, DOI 10.1007/978-981-10-3153-3_7

1 Introduction

Dynamic graph theory [1, 2] is quite new to computing environment and relatively well known in field of Engineering [3]. Through review of publications in renowned journals and few case studies available over the research channels the idea of implementation of dynamic graph theory in the domain of resource allocation struck.

In simple language dynamic graph [4] word is relative term in field of Mathematics and is applied to a situation, where occurs a lot of paths (which could be termed into anything ranging from man power, time, skills, etc.) but, most of them lead to failure. In such situations, a human being driven by his natural conscience works for minimizing the damages based upon his/her evaluation of current situation. But, if a person is unable to evaluate upon his/her current status of work; he/she can never determine or plan the series of events to be undertaken in future to minimize the damages. This very phenomenon gave rise to Chaos theory in Physics [5] which stated "When the present determines the future, but the approximate present does not approximately determine the future." However, with the increase in knowledge the Chaos theory saw evolution to many levels and the latest turning out to be Dynamic Graph Theory "facilitating construction of networks with multifactor-based path discovery from links within the network" [6].

There has been lot of proposals of using dynamic graph in the various fields [7, 8] and few of computer science specifically the field of software engineering is still not tapped to its full potential and still limited to testing domains [3, 6, 9, 10]. Software engineering is already predominated with few successful models of resource allocation Engineering which have been widely accepted in term of Development and rating Quality. But, with new innovations chaos has been successfully implied in development of dynamic graph theory [11]. Practically all the models are totally developer oriented in terms of their approaches and most of the times it leads to, non-satisfaction of the end user in terms of quality they expect out of the developed system. The Software quality is generally defined as "quality is a perceptual, conditional and somewhat subjective attribute and may be understood differently by different people" and is one such factor which determines the successful nature of the system developed through extreme hardwork of developer; but is totally a factor reliant over the end user or the vendor for which it is developed. The idea of dynamic graph-oriented resource allocation is quite simple in terms of understanding.

As per the reports by Richard Schmidt, Sirrush Corporation in October 2012 [12], around Two-third of the software projects developed in 2009 either "failed" or were "critically challenged" leading to their extensions in timeline by many years and hence incurring huge losses. To this very point, we present an efficient algorithm for resource allocations to be conducted.

The idea is simple and it states "Simply take the requirements providing very critical and uttermost required functionalities based on decisions taken through Dynamics network generated through the Stable marriage resource allocation based on stated requirements during earliest phase of software cycle, and build a system around it. The leftover requirements should be imparted to the system as an update."

1.1 Literature Background

As mentioned earlier dynamic graph theory [11] was developed out of Chaos Strategy in view to answer the context relative to future events which are going to occur, based upon the current future. The official statement of Chaos theory states "When the present determines the future, but the approximate present does not approximately determine the future." This theory was highly used in answering the future states in operations of nonlinear, complex systems wherein uncertainty and predictability play a vital role.

Similarly, the field of engineering relies totally on uncertainty and prediction to drive upon and reach the goals stated during the initial phase of project development life cycle. Initial literature review lead to discovery of simple strategy of dividing the problem state into substantial sized individual and nonconflicting smaller problems each having its own state definition. The division of the problem is as follows [13–15]:

(a) A complex Module is an incomplete task with lot of dependencies.
(b) The most important fact is to divide modules on the basis of large, timely, and dependent issues.

   1. Large issues provide a functionality to a user and are basis of this development over which they will be imparted to the project.
   2. Timely issues are the ones which require immediate assistance or else they will delay the other work.
   3. Dependent modules are already tested components which are being reused as per the requisite of the project.

(c) The output is presented when the problem has been addressed completely and a state of stability is achieved.

However, so close the "Chaos Model and Graph model" came in answering many prominent questions related to "what is to be done?," it failed in answering most of the questions relative to "How it is to be done?." To this very point fuzzy algorithm solution using stable marriage resource allocation will be applicable. The Models had major backdrops in form of answers to the following:

(a) What requirement or functionality has to be designed at a given time and what resources, time, testing strategy (Graph Model had this feature) has to be employed over it?

(b) In defining, how one could break down the system and produce an early working model?

(c) Finally, to what extent we can minimize the risk of "failures"?

2   Algorithm

The work aims at implementation of soft computing techniques using dynamic graph for resource allocation strategy in prioritizing the requirements using specific weights allocated to them based upon their nature. The weights will be based upon probabilistic data which will be refined during the course of complete fuzzy process. Then, finally after finishing the of fuzzy process, the priorities will be given to requirements keeping into mind applied algorithm for predicting future steps based upon the current status. The complete process will take into account eradicating the drawbacks faced in chaos model of engineering such that the team gets an idea of putting appropriate resources on the given module/requirement(s) such that it does not leads to a failure. After each module completion an iteration will be made to early working model created at the beginning step. Hence, also giving this approach an evolutionary feature. The key steps involved in this approach are:

(a) The major problems should be broken down into smaller ones and a search must be conducted to convert each sub-problem into a question.

(b) Conduct an exhaustive search through various modus operandi into relevant text for answering the questions generated.

(c) Conduct tests of validity over all the answerable questions for finding impact of each question over the project and it is domain of applicability.

(d) Convert the findings into a practical evidence to support decisions and revise them as per the customer's values.

(e) Conduct performance evaluations over the set parameters and do evolutionary improvements as per the integrated modules into the project.

2.1   Advantages of Fuzzy Algorithm

There have been numerous constraints that evolved in past as well as while designing this algorithm. There were quite a few models designed and they did answer few of the backdrops but in the process, they moved away from the basic fact of dynamicity in development and hence, the requisite of efficient resource allocation came into existence which was totally based upon predictive nature of future given the current statuses of the system. Each one of the models were clear in their approach but lacked in answering following few questions:

(a) Defining "what requirement or functionality has to be designed at a given time and what resources, time, testing strategy has to be employed over it."

(b) Defining "how one could break down the system and produce an early working model"; hence providing more productivity to the end-user at a very early phase and cutting down losses at both developer and end user site.

(c) Finally, "minimizing the risk of "failures" by minimizing the chances of facing "critical situations" leading to permanent shutdowns."

In general situations of Dynamicity, the software development team implies damage recovery procedures based upon any of the above available methods or some classical software development approaches.

As the problem is answering all nonpredictive models of development, the dynamic strategy works in chaotic manner while meeting deadlines. The project managers aim at finishing the tasks given the list of problems. This is usually done through a process of prioritizing them through an experienced professional of the same domain and accepting it as the only valid way of fixing all the issues.

This approach is implied at every Software Development Cycle and to overcome these fuzzy algorithm have been designed to answer all the above question which adds an advantage to the project planning team [16–18]. The algorithm has been successfully implied in answering the major other backdrops which are mostly human perspective based. The simplest possible perspective-based reason which could be quoted out based upon literature and observation could be stated as "How one can say that "someone" who prioritizes the tasks has done that efficiently as the software quality states *"quality is a perceptual, conditional and somewhat subjective.""*

This definition lays down the approach for Improvements. The last part of sentence marks that each individual has its own expectations. At the stage of "critically challenged," an end user cannot expect a standard working system, however, he does expects a system fulfilling the basic needs of its business. And that could only be provided in stipulated time if the prioritization of work has been done in efficient manner and is subjective to the functionalities expected out of the system.

2.2 Design

The idea is simple and it states *"Simply take the requirements providing very critical and uttermost required functionalities, based on decisions taken through Dynamic network. One must take into account that those functionalities are generated through the applied algorithm on stated requirements during earliest phase of Engineering cycle. Once identified start building a system around it. The leftover requirements should be imparted to the system as an update".*

The approach is as simple as its statement. It can be simply understood as following steps.

(a) Retrieve all the requirements. Divide them on the basis of priorities not set by an individual but through the collective weighing by end users and an experienced Team.
(b) Prepare a heuristic to determine weights or priorities of modules.
(c) Apply stable marriage approach for resource allocation at each of the requirements and start predicting the beginning to end prioritization of the work. The initial module or the iterations of module to be processed could be prioritized using fuzzy logic techniques [17–19].
(d) Start working on the modules and prepare working models implemented in live environment, and add more and more functionalities with new Iterations or updates to the older module.

Identify the Membership Function over which requirements could be fuzzified
Creation of weights for each Membership Function and preparing rules to prioritize the requirements based upon them
Refining the weights while Fuzzy Inference Engine is ON
Repeating the process for each individual which is part of prioritization
Accommodating each requirement dynamically by creating a union between dynamic graph strategy and fuzzy output
Theoretical verification through Fractal Identification for future projects
Repeating the complete process at each new iteration to be introduced into the first evolution

This strategy can eradicate following Problems:

(a) Only an individual taking up the task of prioritization.
(b) Transparency in Development cycle for the end user.
(c) Early working system.
(d) Expanded timeline for complete Development of system.
(e) Independence to explore new methods for solution.
(f) Independent modules, resulting into reusability in terms of software or civil constructions.

2.3 Backdrops

The process of algorithm design, however reflected few important points which are hard to neglect and may not be reformed to some extent:

(a) *History is not always the best support to build future upon*: Historical data can teach us how to proceed but can never inform of efficient use of resources. This is the dynamic nature of projects. This dynamic aspect voids all the constants and results into new complex equations.

(b) *The issue of nonpredictable values*: When one discusses and plans one complex system, one cannot ensure all the constraints have been defined and all the variable shave been assigned some values. There will always be few unpredicted values which may pop up due to dynamic nature of the problem.

(c) *Self-defeat of algorithm*: After a practice has been generalized into mode of operations and people have understood its protocol or algorithm; they may use it as per their personal modifications and this beats its essence.

2.4 Algorithmic Outcomes

Along with above-mentioned important outcome of literature survey, research will also take into consideration many factors:

(a) Time period of every iteration
(b) The man power required
(c) Testing required
(d) Future functionality considerations
(e) Coupling of new updates into older versions
(f) Training of end users

The strategy can be implemented at the beginning itself or could be used whenever during any phase of project cycle the development team faces a Challenged situation.

The algorithm has answered almost all the backdrops with efficient deployment of Fuzzy for computing the relevant no. of resources and the allocation process is efficiently done using the stable marriage algorithm. The fusion will result in efficient planning of projects and the deployments may be done even outside the IT projects with few customizations relevant to the domain of implementation.

3 Conclusion

The proposed research work will be beneficial for more accurate Engineering in terms of reducing the failures and being more specific in answering the allocation of the resources and how the work has to be undertaken using those resources. The results of the algorithm will also emphasize on devising a model which can be adhered to with the proper follow ups such that it could be referred to at the time of chaos or failures. "*The development of the model will be much more product centric and will stick to developer's view of development along with customer's view of required functionalities.*"

4 Future Scope

The design of algorithm started with developing a generalized framework which could be employed in varied environments and hence a few backdrops did evolve as stated in sections above. The future actions may comprise undertaking this generalized framework and turning it to environment specific framework which could turn out to be a real-time project management solution to the specific nature of the problem.

References

1. Konig, M.D., Battiston, S., Napoletano, M. And Schweitzer, F., "On Algebraic Graph Theory and the Dynamics of Innovation Networks", Networks and Heterogeneous Media, Volume 3, Number 2, June (2008)
2. Shai, O., Preiss, K., "Graph theory representations of engineering systems and their embedded knowledge", Artificial Intelligence in Engineering, Elsevier, Vol 13(1999)
3. Toffetti, G., Pezze, M., "Graph transformations and software engineering: Success stories and lost chances", Journal of Visual Languages and Computing, Elsevier, Vol 24(2013)
4. Konig, M.D., "Dynamic R&D Networks - The Efficiency and Evolution of Interfirm Collaboration Networks", Dissertation – 18182, ETH ZURICH (2010)
5. Boccaletti, S., Grebogi, C., Lai, Y.C., Mancini, H., Maza, D., "The Control of Chaos: Theory and Applications", S. Boccaletti et al. / Physics Reports 329, Elsevier (2000)
6. Chong, C.Y., Lee, S.P., "Analyzing maintainability and reliability of object oriented software using weighted complex network", The Journal of Systems and Software, Elsevier, Vol 110 (2015)
7. Attri, R., Grover, S., Dev, N., "A graph theoretic approach to evaluate the intensity of barriers in the implementation of total productive maintenance (TPM)", International Journal of Production Research, Taylor and Francis (2013)
8. Schweitzer, F., Fagiolo, G., Sornette, D., Redondo, F.V., White, D.R., "Economic Networks: What do we know and what do we need to know?", ACS - Advances in Complex Systems, Vol 12, Number 4(2009)
9. Robinson, H., "Graph Theory Techniques in Model-Based Testing", International Conference on Testing Computer Software (1999)
10. Lane, P.C.R., Gobet, F., "A theory-driven testing methodology for developing scientific software", Journal of Experimental & Theoretical Artificial Intelligence, Taylor & Francis (2012)
11. Saini, D. K., Ahmad, M., "Software Failures and Chaos Theory", WCE -London, Vol II (2012)
12. Schmidt, R., "Software engineering: Architecture-driven Development", NDIA 15th Annual Systems Engineering Conference, October (2012)
13. Jifeng, H., Li, X., Liu, Z., "Component-Based Software Engineering-The Need to Link Methods and their Theories", 973 project 2002CB312001 of the Ministry of Science and Technology of China
14. Boehm, B., "Value-Based Software Engineering: Overview and Agenda", USC-CSE-2005-504, February (2005)
15. Dybå, T., Kitchenham, B.A., Jørgensen, M., "Evidence-Based Software Engineering for Practitioners", IEEE Software, Published by the IEEE Computer Society, January-February (2005)

16. Kelly, D., "Scientific software development viewed as knowledge acquisition: Towards understanding the development of risk-averse scientific software" The Journal of Systems and Software, Elsevier, Vol 109(2015)
17. Kumar, G., Bhatia, P.K., "Neuro-Fuzzy Model to Estimate & Optimize Quality and Performance of Component Based Software Engineering", ACM SIGSOFT Software Engineering Notes, Vol 40(2015)
18. Mishra, S., Sharma, A., "Maintainability Prediction of Object Oriented Software by using Adaptive Network based Fuzzy System Technique", International Journal of Computer Applications, Vol 119(2015)
19. Tyagi, K., Sharma, A., "A rule-based approach for estimating the reliability of component-based systems", Advances in Engineering Software, Elsevier, Vol 54 (2012)

Agreement-Based Interference-Aware Dynamic Channel Allocation in Cognitive Radio Network (CRN)

Diksha and Poonam Saini

Abstract Cognitive Radio Networks (CRNs) is an intelligent wireless communication network that senses its environment to adjust the transmitter parameters in order to exploit the unused portions of available spectrum. The objective here is to ensure reliable communication with minimum intereference to Primary Users (PUs) and efficient spectrum utilization. The spectrum assigned to licensed users is underutilized and the growing demand causes starvation to the unlicensed users. Thus, CRN senses the available spectrum to find the most appropriate spectrum for allocation. Further, to maximize the efficient use of available spectrum, agreement (consensus) may be used wherein all users agree on a common decision value. In the paper, we discuss various techniques of spectrum allocation in CRN. Lastly, we propose an interference-aware protocol that achieves load balancing, high throughput and less number of reallocations to maximize spectrum utilization. Also, the paper validates the proposed algorithm using the simulation results.

Keywords Cognitive Radio Network (CRN) · Primary user (PU) · Secondary user (SU) · Spectrum assignment (SA) · Spectrum sharing

1 Introduction

The wireless networks are driven by fixed spectrum allocation policy, i.e., spectrum is allocated to licensed users for a long term by the government agencies. However, the usage of the allocated spectrum by the licensed users is very sparse and underutilized. Hence, with an increase in the demand for the spectrum, the unlicensed users may starve of the spectrum. Therefore, there is a need for optimal utilization of the available spectrum. The spectrum allocation process must be

Diksha (✉) · P. Saini
PEC University of Technology, Sector-12, Chandigarh 160012, India
e-mail: dikugarg42@gmail.com

P. Saini
e-mail: nit.sainipoonam@gmail.com

© Springer Nature Singapore Pte Ltd. 2017
S.C. Satapathy et al. (eds.), *Proceedings of the 5th International Conference on Frontiers in Intelligent Computing: Theory and Applications*, Advances in Intelligent Systems and Computing 515, DOI 10.1007/978-981-10-3153-3_8

dynamic for its efficient management. It has been observed in the existing literature that agreement is an inherent approach used for dynamic allocation of channels with minimum interference to the Primary Users (PUs). Dynamic allocation of channel allows the cognitive radio to operate in the best available channel.

Cognitive Radio Network (CRN) is a radio network that changes its transmitter parameters based on the interaction with its surrounding environment in order to exploit the unused portions of spectrum in an opportunistic manner [1]. CRN provides the dynamic allocation of the spectrum by allocating the spectrum holes to the unlicensed users keeping interference to PUs minimum. A spectrum hole is the portion of the spectrum that is not used by a licensed network. The cognitive network has following major functions:

(a) *Spectrum sensing*: It detects the unused spectrum that can be shared without any interference to other users.
(b) *Spectrum management*: It selects the best available channel for sharing.
(c) *Spectrum mobility*: The secondary user vacates the channel required by primary user and uses some other channel for communication.
(d) *Spectrum sharing*: The channel is allocated and shared among the users in an efficient manner.

2 Spectrum Assignment (SA) in Cognitive Radio Networks

Spectrum assignment (SA) is a basic functionality in CRN. The SA is performed with minimum interference to the licensed users. Also, the SA helps in optimal utilization of the available spectrum. The cognitive radio reuses the available spectrum by exploiting the spectrum holes present in the spectrum. The utilization of spectrum holes results in interference to the other CR users as well as to the primary users. In order to avoid the interference, efficient spectrum assignment (SA) should be performed. The SA assigns the most appropriate channel to the secondary user according to predefined criteria that reduces the interference to other users. The approaches to solve the SA problem [2] are centralized, distributed, segment-based, and cluster-based.

3 Related Work

In the existing literature, the Spectrum Assignment (SA) problem was solved using different algorithms. The SA algorithm [3] has multiple objectives. The multi-objective problem is translated into single-objective problem using modified game theory (MGT). The SA algorithm [4] allocates the spectrum dynamically by evaluating the impact of multi-cell and multi-operator interference on the radio resources. The SA algorithm [5] provides distributed coordination in dynamic spectrum allocation networks. The users organize themselves into groups and select the coordination channel using voting procedure. The distributed SA algorithm [6] divides the CR nodes into multiple clusters. The algorithm allocates the channel in order to maximize the spectral efficiency, to minimize the transmission power, and to maximize the data rate.

The segment-based channel assignment approach [7] divides the nodes into different segments and same channel is assigned to all the nodes in a particular segment. The SA algorithm [8] assigns the spectrum to the most hungry user first. The algorithm performs minimum reallocations to achieve maximum utilization and load balancing. The SA algorithm [9] is a fault-tolerant cluster-based algorithm used to allocate the channel in an efficient manner. The algorithm handles the failure of SUs carefully.

The existing consensus protocols may be used to handle the problem of channel allocation. The consensus protocol [10] consists of two-layer hierarchy. The protocol is executed in two phases, namely, message exchange phase and decision-making phase. Also, local majority is applied to reduce the message complexity. [11] is a two-layer hierarchical consensus protocol. The protocol performs four tasks to reach the consensus. In [12], the performance of the protocol [11] is evaluated and analyzed on metrics like number of rounds, execution time, and the number of messages exchanged.

4 Proposed System Model

The section describes the assumptions, data structures, and messages used in the proposed system model. Further, the detailed description of proposed protocol is given.

4.1 Assumptions

(i)   Assume the network has Z users with $(U_1, U_2, U_3 \ldots \ldots U_{Z-1}, U_Z)$.
(ii) Users are of two types, namely, Primary Users (PUs) and Secondary Users (SUs).
(iii) There is M number of channels in the network, say, $(1, 2, 3 \ldots M)$.
(iv) Message Passing System is used for decision-making.
(v)   Cluster-based SA approach and cooperative spectrum access techniques are used.
(vi) Underlay Control Channels (UCCs) are used by SUs for communication. UCCs are used because they are always available for communication.

4.2 Types of Messages

(i) *CHANNEL_UPDATE (SU$_i$, AC$_i$)*: The message contains the list of channels AC$_i$ sensed by the user SU$_i$. The list is sent to the respective CHs in two cases:

- When the user senses the environment for the first time.
- When there is change in the list sensed by the user.

(ii) *DECISION (SU$_i$, j)*: The message is sent by CHs to SUs to deliver the decision of channel allocation. The decision is to allocate channel *j* to secondary user SU$_i$.

(iii) *UTILIZE (SU$_i$, δ)*: The message is sent by CH to the SU in order to allow the SU$_i$ to initiate the communication on the allocated channel for the δ period of time.

4.3 Types of Data Structures and Definitions

At each Secondary User:

(i) *Channel ID (Ch_id$_j$)*: The *Ch_id$_j$* indicates the id of the channel j which is equal to j.

(ii) *Interference (I$_j$)*: I$_i$ refers to the sum of interference and noise incurred in the channel j due to other spectrums present in the network.

(iii) *Bandwidth (W$_j$)*: The W$_j$ refers to the bandwidth of the channel j sensed by the SU.

(iv) *Timestamp (ts$_i$)*: The timestamp refers to the time at which SU$_i$ sends the available channel list. The value of timestamp is initialized to 0 and is incremented each time a message is sent by the SU$_i$.

(v) *Avail_channel (AC$_i$)*: The *AC$_i$* is the list prepared by the SU$_i$ after sensing its environment. The list has four parameters, namely, channel id (Ch_id$_j$), interference (I$_j$), bandwidth (Wj) of the channel j, and the timestamp (ts$_i$) of the message.

At each Cluster Head:

(i) *Available Channel Table (ACT)*: The CHs prepare the *ACT* table to store the channels sensed by the SUs. Two copies of the ACT table are maintained, one is kept as original O_ACT table and the other is used for modification.

(ii) *Channel Allocation Vector (CAV)*: The *CAV* is a vector used to store the list of SUs allocated to a channel.

(iii) *Channel Allocation Table (CAT)*: The *CAT* table consists of CAV for every channel.

(iv) *Channel_demand*: The *channel_demand* is an array used to store the value of occurrence of channels in ACT table.

4.4 Proposed Algorithm

Consider N number of SUs from $(SU_1, SU_2, SU_3 \ldots\ldots SU_N)$ and $(Z\text{-}N)$ number of PUs from $(PU_1, PU_2, PU_3 \ldots\ldots PU_{(Z\text{-}N)})$ enter into the network. The proposed algorithm consists of two phases, namely, Cluster Formation Phase and Channel Allocation Phase.

Cluster Formation Phase: The proposed cluster formation phase consists of following steps:

1. The network is divided into *K* number of clusters.
2. Each cluster has an associated Cluster Head (CH).
3. Every CH broadcasts the signal in the network.
4. The SUs receive signal from the CHs and compare the strength of the signal.
5. The SU join the cluster with maximum signal strength.

Channel Allocation Phase: In the proposed channel allocation phase, the channel is allocated to the SUs by respective CHs. The SUs sense their environment and prepare a list of available channels along with the bandwidth (*W*) and the interference (*I*) level of the channel sensed using the control channels. Afterwards, the list is piggybacked with the timestamp of the message and sent to respective CH by the SU. Every CH merges the received messages into one message and sends the same to other CHs. The channels are allocated to the users depending on the timestamp of the message, (*I/W*) ratio, and the size of the list. Finally, the channel is allocated to every user and the decision is sent to the users by their respective CHs. In case the channel is required by the PU, the SUs need to vacate the channel and the channel allocation process restarts. The proposed channel allocation phase has arrangement rules and procedure described below.

Arrangement Rules: The arrangement rules are the rules used to order the SUs and the channels in the ACT table. The rules used in the proposed algorithm are given below.

Rule 1: The SU with single channel in the available channel list is placed at the top of ACT table. Then, the SUs are arranged in the order of decreasing timestamp *ts*. In case of conflict, order the users according to their list size.
Rule 2: The channel with 1 value in *channel_demand* array is put at front in the list. In case of conflict, the channels are ordered in increasing value of *I/ W*.
Rule 3: The channels with value other than 1 in *channel_demand* array are arranged in increasing order of value *I/W*.

Procedure: The proposed channel allocation phase consists of the intra-cluster and inter-cluster execution of the algorithm as given below.

   A. Intra-cluster execution:

1. The SUs sense the environment and prepare a list of available channels *avail_channel* with the following 4 parameters $<Ch\_id_j, I_j, W_j, ts_i>$.
2. Further, the SUs send the *avail_channel* list to their respective CHs through CHANNEL_UPDATE message.
3. The CH waits until it receives CHANNEL_UPDATE from all the SUs in its cluster.
4. After receiving the lists from the users, CH merges the received lists into single list.

   B. Inter-cluster execution:

1. Every CH sends the merged list to other CHs in the network.
2. Each CH waits until it gets the list from all other CHs.
3. The CHs prepare the *ACT* table according to arrangement rule 1.
4. The CHs calculate the value of occurrence of each channel in the *ACT* table and store it in an array *channel_demand*.
5. Now, the channels in each row of the *ACT* table are rearranged according to arrangement rules 2 and 3.
6. The CHs execute the CHANNEL_ALLOCATION () procedure (Appendix A).
7. Then each CH sends the DECISION (SU_i, j) message to the SUs in its cluster.
8. The SUs utilize the allocated channel in a periodic manner with other SUs having same allocated channel. Thus, the communication on the allocated channel is started by the user on receiving the message UTILIZE (SU_i,δ) for δ period of time.

5   Example Execution

Consider the network containing 5 SUs, 4 channels, and 2 clusters. Each cluster has an associated Cluster Head (CH), namely, CH_1 and CH_2. The secondary users SU_1, SU_2 belong to cluster 1 and SU_3, SU_4, and SU_5 belong to cluster 2. The SUs sense their environment and prepare *avail_channel* (*AC*) list as given in Table 1.

Table 1 AC list of SUs

At CH_1	$AC_1 = \{\{2, I_2, W_2, 1\}, \{4, I_4, W_4, 1\}\}$
	$AC_2 = \{\{2, I_2, W_2, 1\}, \{1, I_1, W_1, 1\}\}$
At CH_2	$AC_3 = \{\{1, I_1, W_1, 2\}, \{3, I_3, W_3, 2\}, \{2, I_2, W_2, 2\}\}$
	$AC_4 = \{\{2, I_2, W_2, 1\}\}$
	$AC_5 = \{\{1, I_1, W_1, 2\}, \{3, I_3, W_3, 2\}\}$

Afterwards, the SUs send the list to their respective CHs. Now, each CH shares the list with each other. Further, the CHs prepare *channel_demand* array, the *ACT* table using the arrangement rules, and the *CAT* table (shown in Table 2). Assume the interference to bandwidth ratio (I/W) of the channels has order: $I_3/W_3 > I_4/W_4 > I_2/W_2 > I_1/W_1$.

Afterwards, the CH starts the channel allocation process by allocating the channel to the first SU in the *ACT* table. The corresponding updated *channel_demand* array, *ACT* and *CAT* table are shown in Table 3. Here, the SU with allocated channel is removed from *ACT* table and its entry is made in *CAT* table. The process continues till all the channels are allocated to at least one SU as shown in Tables 4, 5, and 6.

Table 2 Channel_demand array, ACT table with 5 SUs, and CAT with 4 channels and no allocations

3	4	2	1	SU$_4$	2			CAV$_1$	CAV$_2$	CAV$_3$	CAV$_4$
CH$_1$	CH$_2$	CH$_3$	CH$_4$	SU$_5$	1	3					
				SU$_3$	1	2	3				
				SU$_1$	4	2					
				SU$_2$	1	2					

Table 3 Channel_demand array, ACT table with 4 SUs, and CAT with 4 channels and 1 allocation

3	0	2	1	SU$_1$	4			CAV$_1$	CAV$_2$	CAV$_3$	CAV$_4$
CH$_1$	CH$_2$	CH$_3$	CH$_4$	SU$_2$	1				SU$_4$		
				SU$_5$	1	3					
				SU$_3$	1	3					

Table 4 Channel_demand array, ACT table with 3 SUs, and CAT with 4 channels and 2 allocations

3	0	2	0	SU$_2$	1			CAV$_1$	CAV$_2$	CAV$_3$	CAV$_4$
CH$_1$	CH$_2$	CH$_3$	CH$_4$	SU$_5$	1	3			SU$_4$		SU$_1$
				SU$_3$	1	3					

Table 5 Channel_demand array, ACT table with 2 SUs, and CAT with 4 channels and 3 allocations

0	0	2	0	SU$_5$	3			CAV$_1$	CAV$_2$	CAV$_3$	CAV$_4$
CH$_1$	CH$_2$	CH$_3$	CH$_4$	SU$_3$	3			SU$_2$	SU$_4$		SU$_1$

Table 6 Channel_demand array, ACT table with 1 SU, and CAT with 4 channels and 4 allocations

0	0	0	0	SU$_3$				CAV$_1$	CAV$_2$	CAV$_3$	CAV$_4$
CH$_1$	CH$_2$	CH$_3$	CH$_4$					SU$_2$	SU$_4$	SU$_5$	SU$_1$

Table 7 Channel_demand array, ACT table with 2 SUs, and CAT with 4 channels and 4 allocations

1	1	1	0	SU₃	2	1	3	CAV₁	CAV₂	CAV₃	CAV₄
CH₁	CH₂	CH₃	CH₄					SU₂	SU₄	SU₅	SU₁

Table 8 Channel_demand array, ACT table with 0 SU, and CAT with 4 channels and 5 allocations

0	0	0	0					CAV₁	CAV₂	CAV₃	CAV₄
CH₁	CH₂	CH₃	CH₄					SU₂	SU₄	SU₅	SU₁
								SU₃			

Now, the *ACT* table copies the entries for the remaining SUs from *O_ACT* table and corresponding updated tables and array are shown in Table 7. The channel allocation procedure repeats till the channel is allocated to all SUs (shown in Table 8).

The above approach successfully allocates the available channel as proved by the example execution.

6   Simulation Details

The CRCN patched network simulator is used to validate the proposed algorithm. The nodes are assumed to be randomly distributed in an area of 500×500 m². The nodes move with a speed of 0–50 m/s. Fig 1 shows the % of SUs served and % of SUs reallocated versus evaluation time in case number of SUs are 13 and number of PUs are 7. Further, Fig. 1a shows that with increase in the evaluation time, % of SUs served becomes constant and Fig. 1b shows that with increase in evaluation time, % of SUs reallocated increases. In future simulation, proposed algorithm is compared with existing algorithms on metrics like throughput.

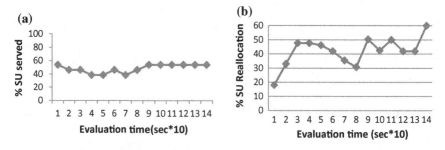

Fig. 1 **a** Evaluation Time versus % SU served **b** Evaluation Time versus % SU Reallocation

The channels are allocated to the users in the order of (I/W) ratio. The channel with least value of (I/W) is allocated prior to other channels. Thus, the channel with higher transmission power due to low value of (I/W) is allocated to more number of users in comparison to other channels. Thus, the throughput of the network is high in the proposed protocol as compared to existing channel allocation algorithms which do not consider the interference factor. Also, the proposed algorithm achieves load balancing, allocation of channels is mutually exclusive, and spectrum utilization is maximum.

7   Conclusion

Cognitive Radio Networks (CRNs) senses the environment to exploit the unused portions of available spectrum. The aim is to utilize the available spectrum optimally along with minimum interference to the Primary Users (PUs). The paper presents the Spectrum Assignment (SA) problem and the existing algorithms to solve the problem. In the last, an *interference-aware* protocol is proposed for dynamic spectrum allocation based on agreement. The design of proposed protocol provides better spectrum utilization, load balancing, high throughput, and requires low number of reallocations. The static analysis is presented with the help of an example execution scenario. Also, the proposed algorithm is validated using simulation results. The future work focuses on the comparison of the proposed algorithm with existing algorithms on parameters like throughput, number of reallocations, and number of SUs served.

A. Appendix

Procedure: CHANNEL_ALLOCATION ()

    i. Allocate the first channel to the first user in the *ACT* table and make the entry of SU in *CAT* table.

    ii. Remove the channel allocated in step i and the user with allocated channel from the *ACT* table.

    iii. Update *channel_demand* array.

    iv. If *channel_demand* array contains value 1 for any channel, rearrange the channels in the *ACT* table according to arrangement rules. Else, do nothing.

    v. Repeat steps from i to iv until all the channels are allocated to some users.

vi. If *ACT* table is not empty, then copy the rows from table *O_ACT* to
ACT table corresponding to the users that are not allocated any
channels yet.

vii. Update the *channel_demand* array and rearrange the channels in the
ACT table according to arrangement rules.

viii. Repeat the above steps until the *ACT* table becomes empty.

References

1. Akyildiz, I., Lee, W., Vuran, M., Mohanty, S.: NeXt generation/dynamic spectrum access/cognitive radio wireless networks: A survey. Computer Networks. 50, 2127–2159 (2006).
2. Tragos, E., Zeadally, S., Fragkiadakis, A., Siris, V.: Spectrum Assignment in Cognitive Radio Networks: A Comprehensive Survey. IEEE Communications Surveys & Tutorials. 15, 1108–1135 (2013).
3. Byun, S., Balasingham, I., Liang, X.: Dynamic spectrum allocation in wireless cognitive sensor networks: Improving fairness and energy efficiency. 68th IEEE Vehicular Technology Conference. pp. 1–5. IEEE (2008).
4. Alnwaimi, G., Arshad, K., Moessner, K.: Dynamic Spectrum Allocation Algorithm with Interference Management in Co-Existing Networks. IEEE Communications Letters. 15, 932–934 (2011).
5. Zhao, J., Zheng, H., Yang, G.: Distributed coordination in dynamic spectrum allocation networks. 1st IEEE International Symposium on New Frontiers in Dynamic Spectrum Access Networks (DySPAN). pp. 259–268. IEEE (2005).
6. Li, X., Zekavat, S.: Distributed Channel Assignment in Cognitive Radio Networks. International Conference on Wireless Communications and Mobile Computing: Connecting the World Wirelessly (IWCMC'09). pp. 989–993. ACM (2009).
7. Bian, K., Park, J.: Segment-Based Channel Assignment in Cognitive Radio Ad Hoc Networks. 2nd International Conference on Cognitive Radio Oriented Wireless Networks and Communications (CrownCom). pp. 327–335. IEEE (2007).
8. Pareek, H., Singh, A.: An Adaptive Spectrum assignment Algorithm in Cognitive Radio Network. International Conference on Recent Trends in Information, Telecommunication and Computing, ITC. pp. 408–418. ACEEE (2014).
9. Pareek, H., Singh, A.: Fault Tolerant Spectrum Assignment in Cognitive Radio Networks. International Conference on Information and Communication Technologies, ICICT. pp. 1188–1195. Elsevier (2014).
10. Wang, S., Yan, K., Wang, S.: An optimal solution for byzantine agreement under a hierarchical cluster-oriented mobile ad hoc network. Computers & Electrical Engineering. 36, 100–113 (2010).
11. Wu, W., Cao, J., Yang, J., Raynal, M.: A Hierarchical Consensus Protocol for Mobile ad Hoc Networks. 14th Euromicro International Conference on Parallel, Distributed, and Network-Based Processing (PDP). IEEE (2006).
12. Wu, W., Cao, J., Yang, J., Raynal, M.: Design and Performance Evaluation of Efficient Consensus Protocols for Mobile Ad Hoc Networks. IEEE Transactions on Computers. 56, 1055–1070 (2007).

Energy Efficient Resource Allocation for Heterogeneous Workload in Cloud Computing

Surbhi Malik, Poonam Saini and Sudesh Rani

Abstract Cloud computing is an internet based technology that provisions the resources automatically on the pay per use basis. With the development of cloud computing, the amount of customers and requirement of resources increases exponentially. In order to balance the load, the tasks must be equally distributed among multiple computing servers thereby, fulfilling Quality of Service (QoS) with maximum profit to cloud service providers. In addition, cloud servers consume huge amount of electrical energy leading to increased expenditure and environment degradation. Therefore, certain solutions are needed that results in efficient resource utilization while minimizing the environmental influence. In the paper, we present a survey of load balancing algorithms along with their limitations and propose a framework for an energy efficient resource allocation and load balancing for heterogeneous workload in cloud computing along with the validation of the framework using CloudSim toolkit.

Keywords Cloud computing · Energy consumption · Fault tolerant · Load balancing · Resource allocation

1 Introduction

Cloud computing is defined by NIST [1, 2] as a model to enable ubiquitous, convenient, on-demand network access to a shared pool of computing resources each of which being configurable which can be rapidly delivered and released without minimal management intervention. Cloud computing is a rising internet

S. Malik (✉) · P. Saini · S. Rani
PEC University of Technology, Sector-12, Chandigarh 160012, India
e-mail: surbhimalik22@gmail.com

P. Saini
e-mail: poonamsaini@pec.ac.in

S. Rani
e-mail: sudeshrani@pec.ac.in

© Springer Nature Singapore Pte Ltd. 2017   89
S.C. Satapathy et al. (eds.), *Proceedings of the 5th International Conference on Frontiers in Intelligent Computing: Theory and Applications*, Advances in Intelligent Systems and Computing 515, DOI 10.1007/978-981-10-3153-3_9

based distributed computing technology that provides shared computing resources and other devices on request and on pay-per-usage model [3–5]. The various services available on cloud can be classified into different service layers as *Infrastructure as a Service* (IaaS) [6] that allows to provision hardware, software, servers, storage and other infrastructure components to the end user, *Platform as a Service* (PaaS) that provides toolkit, application programming interfaces and standards for development allowing users to develop, run and manage applications without the complexity of hosting the infrastructure associated with the development of application, *Software as a Service* (SaaS) [4, 7, 8] that delivers domain-specific applications or services that are developed on a cloud platform and hosted in a cloud infrastructure. Cloud computing is based on virtualization that allows creation of several virtual machine instances to provide concurrent processing of various tasks over a shared hardware platform. Hypervisor or Virtual machine manager provisions the amount of resources required to each operating system and provides abstraction of the hardware to the virtual machines.

1.1 Load Balancing

The remarkable growth of cloud computing increases the number of clients and in addition increases the demand of resources. Consequently, this leads to heavy workload on the servers which further degrade the overall performance. Though virtualization [8] balances the load dynamically, the resources may be over-utilized causing performance degradation or under-utilized resulting in increased power consumption. Therefore, efficient utilization of resources and balance among servers is essential. Load balancing is the mechanism to divide the load of processing over a number of separate systems for an overall performance increase. It represents the ability to transfer some amount of processing to another system that will execute the request. The task of load balancing is divided into two subtasks [9]:

- Allocation of virtual machine instances on host for new request [10] (Scheduling)
- Reallocation or migration of VMs (Load balancing)

The first subtask can be solved by various load balancing algorithms. The second subtask involves critical decisions like which VM to migrate, when to migrate and where to migrate.

The rest of the paper is organized as follows: Sect. 2 presents the related work. In Sect. 3, the proposed system model is discussed. Section 4 contains the experimental setup along with the simulation results followed by conclusion mentioned in Sect. 5.

2 Related Work

There have been a variety of efforts for allocation of resources and to balance the load among cloud datacenters. This section gives a brief review of the various load balancing algorithms in order to distribute load among various machines efficiently.

Chen et al. [11] proposed two improved min-min algorithms i.e., load-balanced min-min scheduling and user-priority guided min-min scheduling on grounds of work of Yu et al. [11, 12]. In these algorithms, initially the tasks are assigned to the available resources in accordance with the min-min algorithm. Further, to balance the load, the completion time for the smallest size task on the overloaded server is considered and the task is migrated to the resource that gives minimum completion time for that job. With the purpose of considering different types of users and their priority user-priority aware load balance improved min-min scheduling algorithm is proposed that focuses on dividing the set of tasks into two groups, one for higher priority tasks i.e., VIP user tasks and the other i.e., ordinary user tasks. The algorithm results in efficient resource utilization and reduces the overall completion time.

Soni et al. [3] proposed a centralized algorithm for balancing the load named as *Central load balancer*. The central load balancer creates a table mapping the information about the virtual machine ID (VMid), VM current allocation state (Busy/available), VM priority and the allocation is done according to the resource availability and state of the VM. The algorithm balances the load among virtual machines by proper resource utilization according to the computing capacity.

Reddy et al. [13] suggested a local optimized approach to distribute the load between the existing resources. The proposed algorithm optimizes the basic *Throttled Load Balancing* algorithm [14–16]. The proposed system consists of a load balancer i.e. *Modified Throttled load balancer* that maintains an index table of VMs along with their states (BUSY/IDLE). On arrival of a new request the table is scanned from the first entry and an idle VM is selected for allocation. When next request comes, the VM entry after the one that is already assigned is selected on the basis of its VM state.

Beloglazov et al. [17] introduced a high level architecture consisting of consumers/brokers, green service allocator, virtual machines and physical machines. In this approach the energy consumption can be minimized by logically resizing the VMs and mergence of VMs into minimum amount of physical servers. The new request is handled by allocation of VM to the host that results in minimum rise in power consumption. The optimization of current VM allocation is done by selection of the VMs to be migrated and afterward placement of those VMs on other hosts. The discussed protocol results in substantial decrease in power consumption of cloud data centers.

Lee et al. [18] discussed two energy conscious task consolidation heuristics (ECTC and MaxUtil) that aims to efficiently utilize the resources while considering active and idle energy consumption. The heuristics assign each task to a resource which results in minimum power consumption without degrading the performance.

In both algorithms, the energy consumption of a resource is computed by a cost function and the task is assigned to the resource with minimum cost function value.

Shu et al. [19] introduced a clonal optimization algorithm on account of makespan and energy consumption model. On arrival of a new request, the system runs an *immune clonal selection algorithm* (ICSA) to optimize the resource allocation in which the resources and tasks are mapped to create the initial population. Afterwards, an affinity function is designed on base of energy consumption and makespan and the ones with higher value of affinity are selected for the next generation and a new population is further created by performing mutation operation on the present population.

Garg et al. [20] focused on allocating resources to different types of applications mainly non-interactive and transactional applications. In the allocation process, the VM scheduler checks for the available servers to run the application. Firstly, the web applications are allocated and thereafter the non-interactive jobs are hosted on the VM in a way that it causes minimum penalty of the SLA between the user and provider.

3 Proposed System Model

The proposed system assumes a cloud data center with N different physical machines each with R different types of resources (e.g. CPU, RAM, bandwidth etc). The cloud users submit their workloads according to their heterogeneous resource demands and select the VM that satisfies their demands with minimum resource wastage which further results in reduction in energy consumption of resources. The model assumes different types of workload that can be provisioned to a single machine.

- *Parallel batch jobs*—The jobs depend upon CPU and memory and mainly focus on performance and completion time along with deadline constraint.
- *Web applications*—The jobs depend upon network bandwidth and latency and focuses on response time and throughput.
- *Big data applications*—The jobs consist of large amount of heterogeneous data each containing two subtasks, namely, *Map jobs* and *Reduce jobs*.
- *Scientific Workload*—The application, as a whole, is represented using Directed Acyclic Graph (DAG) denoting the dependency between tasks which is considered while scheduling the application along with the deadline constraint.

The data centers comprises of machines with different processor architecture, memory and disk capacities and energy consumption rates. In cloud, the tasks are resource specific and at times do not utilize the same share of different resource types. Hence, the user's demand for a particular resource may soon exhaust that resource while other resources may still be available. Therefore, to ensure efficient resource consumption, the number of active physical machines should be calculated efficiently and skewness among resources should be minimized.

3.1 Performance Metrics

(i) *Skewness factor*—It measures the degree of variation in usage among multiple resource types (inner node) as well as between multiple physical machines (inter node).

$$s_n = \frac{Mean\ difference\ of\ utilization\ of\ R\ resource\ types}{Average\ Utilization\ of\ all\ resource\ types} = \frac{\sum\limits_{i,j=1}^{R} |u_i - u_j|}{(R-1) * \sum\limits_{i=1}^{R} u_i} \tag{1}$$

$R =$ Number of resource type,
$u_i =$ Utilization of a particular resource

(ii) *Number of physical Machines*—It measures the amount of energy consumed by various resources like CPU, memory, secondary storage and network communication.

(iii) *SLA Violation Percentage*—It is the percentage of SLA violations relative to the total number of processed time frames.

3.2 Power Model

The power model has been devised on the fact that resource utilization has a linear relationship with the energy consumption. The power consumption is determined by the CPU, memory, disk storage and network interfaces. Power can be expressed as:

$$P(u) = k \cdot P_{max} + (1-k) \cdot P_{max} \cdot u \tag{2}$$

$P_{max} =$ Power consumed by fully utilized server
$k =$ Fraction of power consumed by the idle server
$u =$ CPU utilization

3.3   Working of the Algorithm

The working of proposed energy efficient algorithm for resource allocation and load balancing among available machines is mentioned in Fig. 1. The requests from the user enter into the scheduling queue and are scheduled if the resources for the request are available. Next, the workload profiler distributes the requests into different clusters depending upon the performance characteristics and resource demands. Further, the amount of resources required by cloudlets is calculated and appropriate machines based on the configuration are selected. Afterwards, for the selected machines, the energy consumption is calculated using the power equation. A set of three machines is chosen based on minimum energy consumption. Further, the skewness factor for the selected machines is calculated. The machine with minimum skewness among resources is finally chosen to be allocated for the request. This ensures maximum resource utilization. The pseudo code of the algorithm is mentioned in Algorithm 1.

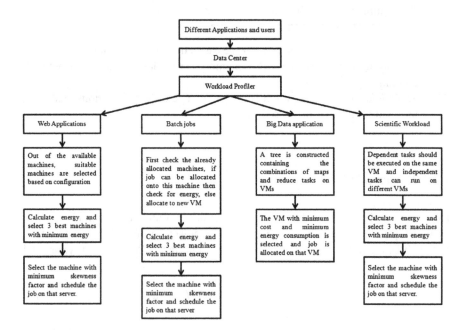

Fig. 1 Detailed description of the proposed algorithm

Algorithm 1 for allocation of VM on the data center resources
Input-hostlist, vmlist, cloudlet list   Output-Allocation of VMs

```
1.  for each VM in the scheduling queue do
2.     firstmin,secondmin,thirdmin ←MAX
3.     allocatedHost,FirstHost,SecondHost,ThirdHost←NULL
4.     for each host in the host list do
5.        if host is suitable for VM then
6.           powerDiff=powerafterallocation-power(host)
7.           if(powerDiff<first)
8.              firstmin=powerDiff
9.              firstHost=host
10.          else if(powerDiff<second)
11.             secondmin=powerDiff
12.             secondHost=host
13.          else if(powerDiff<third)
14.             thirdmin=powerDiff
15.             thirdHost=host
16.          determine value of skewness factor for FirstHost,SecondHost,ThirdHost and select the
             host with minimum skewness
17.       if allocatedhost ≠NULL then
18.          allocate vm to allocated host
19.    Return allocation
```

As discussed in system model, there can be different categories of workload viz., batch, web-based, big data and scientific. The parallel batch and web-based workload follow the above mentioned steps to allocate the resources. To assign the resource in Big data application, a tree is constructed that contains the combination of Map tasks and Reduce tasks on VMs and a combination of VM with minimum cost and energy consumption is chosen. In case of a scientific workflow, the dependency among the tasks is represented using Directed Acyclic Graph (DAG) and the dependent tasks must be allocated to the same VM. The process to allocate the dependant task to same VM is carried in accordance with the algorithm.

4 Simulation Details

The proposed framework is based on virtual machines and focuses on Infrastructure as a service (IaaS). However it is not practically as well as economically feasible to perform various experiments on actual infrastructure. CloudSim Toolkit is chosen as simulation platform for performing simulations for cloud computing environment. In the proposed algorithm a datacenter of containing different combinations of physical machines and virtual machines is created. The physical machine consists of CPU cores with performance of 2000, 2500, 3000 or 3500 MIPS and 2, 2.5, 3, 3.5 GB of RAM and 1 TB of storage and bandwidth of 10 GB/s. The power consumed by each host can be determined by the power equation described in Sect. 3.2. According to this model, a host consumes from 175 W with 0% CPU utilization, up to 250 W with 100% CPU utilization. Every incoming cloudlet creates a virtual machine instance that requires one CPU core with 100, 200, 250,

Fig. 2 Comparison of the utilization of the proposed algorithm

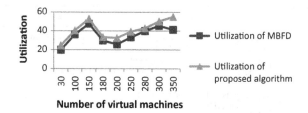

325 MIPS and 256, 400, 500 520 MB of RAM and bandwidth of 100, 200, 250, 325 KB/s. The users submit their requests for provisioning heterogeneous VMs that efficiently utilize the simulated data center. Each VM runs a different kind of application with variable workload that generates the CPU utilization according to a uniformly distributed random variable. The results are compared with the Modified best fit decreasing algorithm (MBFD) that is used for allocation of virtual machines on physical host. The results show that the proposed algorithm provides better utilization results as compared to the MBFD algorithm as shown in Fig. 2. In future the simulation for different types of workloads will be done and the results will be compared by the existing algorithms for different performance parameters.

5 Conclusion

Load balancing divides the workload between the available physical machines at the same time to provide in-time services to the end-users. The problem is to allocate the available resources and balance the load for optimized resource utilization. The paper presents a brief survey of various techniques used for load balancing in cloud computing systems. In addition, we present an *energy-efficient* load balancing and resource allocation algorithm to handle heterogeneous workload that optimizes the resource utilization with maximum throughput along with reduction in energy consumption of the datacenters. The future work focuses on the optimization of the proposed allocation algorithm and migration of virtual machines resulting in efficient resource utilization by VM consolidation and switching off idle physical machines.

References

1. Mell P, Grance T: The NIST Definition of cloud computing. NIST (2012).
2. Hameed, A., Khoshkbarforoushha, A., Ranjan, R., Jayaraman, P., Kolodziej, J., Balaji, P., Zeadally, S., Malluhi, Q., Tziritas, N., Vishnu, A., Khan, S., Zomaya, A.: A survey and taxonomy on energy efficient resource allocation techniques for cloud computing systems. Computing (2014).

3. Soni, G., Kalra, M.: A novel approach for load balancing in cloud data center. Advance Computing Conference (IACC), 2014 IEEE International. pp. 807–812. IEEE (2014).
4. Rodriguez, M., Buyya, R.: Deadline Based Resource Provisioning and Scheduling Algorithm for Scientific Workflows on Clouds. IEEE Transactions on Cloud Computing. 2, 222–235 (2014).
5. Alrokayan, M., Dastjerdi, A., Buyya, R.: SLA-Aware Provisioning and Scheduling of Cloud Resources for Big Data Analytics. 2014 IEEE International Conference on Cloud Computing in Emerging Markets (CCEM). pp. 1–8. IEEE (2014).
6. Vecchiola, C., Calheiros, R., Karunamoorthy, D., Buyya, R.: Deadline-driven provisioning of resources for scientific applications in hybrid clouds with Aneka. Future Generation Computer Systems. 28, 58–65 (2012).
7. Jennings, B., Stadler, R.: Resource Management in Clouds: Survey and Research Challenges. J Netw Syst Manage. 23, 567–619 (2014).
8. Manvi, S., Krishna Shyam, G.: Resource management for Infrastructure as a Service (IaaS) in cloud computing: A survey. Journal of Network and Computer Applications. 41, 424–440 (2014).
9. Shaw, S., Singh, A.: A survey on scheduling and load balancing techniques in cloud computing environment. Computer and Communication Technology (ICCCT), 2014 International Conference on. pp. 87–95. IEEE (2014).
10. Wei, L., Foh, C., He, B., Cai, J.: Towards Efficient Resource Allocation for Heterogeneous Workloads in IaaS Clouds. IEEE Transactions on Cloud Computing. 1–1 (2015).
11. Chen, H., Wang, F., Helian, N., Akanmu, G.: User-priority guided Min-Min scheduling algorithm for load balancing in cloud computing. 2013 National Conference on Parallel Computing Technologies (PARCOMPTECH). pp. 1–8. IEEE (2013).
12. Yu, X., Yu, X.: A New Grid Computation-Based Min-Min Algorithm. Sixth International Conference on Fuzzy Systems and Knowledge Discovery, 2009. FSKD'09. pp. 443–45. IEEE (2009).
13. Nuaimi, K., Mohamed, N., Nuaimi, M., Al-Jaroodi, J.: A Survey of Load Balancing in Cloud Computing: Challenges and Algorithms. 2012 Second Symposium on Network Cloud Computing and Applications (NCCA). pp. 137–142. IEEE (2012).
14. Wickremasinghe B: CloudAnalyst: A CloudSim-based Tool for Modelling and Analysis of Large Scale Cloud Computing Environments (2010).
15. Wickremasinghe, B., Calheiros, R., Buyya, R.: A CloudSim-Based Visual Modeller for Analyzing Cloud Computing Environments and Applications. 2010 24th IEEE International Conference on Advanced Information Networking and Applications (AINA). pp. 446–452. IEEE (2010).
16. Domanal, S., Reddy, G.: Load Balancing in Cloud Computing using Modified Throttled Algorithm. 2013 IEEE International Conference on Cloud Computing in Emerging Markets (CCEM). pp. 1–5. IEEE (2013).
17. Beloglazov, A., Abawajy, J., Buyya, R.: Energy-aware resource allocation heuristics for efficient management of data centers for Cloud computing. Future Generation Computer Systems. 28, 755–768 (2012).
18. Lee, Y., Zomaya, A.: Energy efficient utilization of resources in cloud computing systems. J Supercomput. 60, 268–280 (2010).
19. Shu, W., Wang, W., Wang, Y.: A novel energy-efficient resource allocation algorithm based on immune clonal optimization for green cloud computing. EURASIP J Wirel Commun Netw. 2014, 64 (2014).
20. Garg, S., Toosi, A., Gopalaiyengar, S., Buyya, R.: SLA-based virtual machine management for heterogeneous workloads in a cloud datacenter. Journal of Network and Computer Applications. 45, 108–120 (2014).

Accent Recognition System Using Deep Belief Networks for Telugu Speech Signals

Kasiprasad Mannepalli, Panyam Narahari Sastry and Maloji Suman

Abstract Accent and Emotion recognition for speech has become most important research area because of the increased demand of speech processing systems in handheld devices. Most of the research in speech processing is done for the English language only. In this paper, we present accent recognition system for Telugu speeches. Three important accents of Telugu were chosen and text-dependent speeches of Coastal Andhra, Rayalaseema, and Telangana accents were collected. Features like tonal power ratio, spectral flux, pitch chroma, and MFCC were extracted from these speeches. deep belief networks are used for the classification purpose. The recognition accuracy obtained in this work is 93%.

Keywords Accent recognition · Speech recognition · Deep belief networks

1 Introduction

The speech signal consists of the information about the accent of the speech within the words. A deviation of speech signal from its neutral behavior makes it difficult to recognize properly by speech the recognition systems. The speech recognition system's efficiency can be increased, if the accent of speech is detected before the speech recognition task [1]. In any language accents will be formed due to the geographical and ecological state of the area of the speakers, individuals have their

K. Mannepalli (✉) · M. Suman
K L University, KLEF, Vijayawada, Guntur, AP, India
e-mail: mkasiprasad@gmail.com

M. Suman
e-mail: suman.maloji@gmail.com

P.N. Sastry
CBIT, Hyderabad, Telangana, India
e-mail: ananditahari@cbit.ac.in

© Springer Nature Singapore Pte Ltd. 2017   99
S.C. Satapathy et al. (eds.), *Proceedings of the 5th International Conference on Frontiers in Intelligent Computing: Theory and Applications*, Advances in Intelligent Systems and Computing 515, DOI 10.1007/978-981-10-3153-3_10

own style of speaking, dialect, and accent and as well as their social background. Telugu language has mainly three different accents called as Coastal Andhra (CA), Rayalaseema (RS), Telangana (TG). These three accents are used in this work.

2 Related Work

Eriksson et al. [2] have studied the selection of features for speaker recognition from the information theory view. They have reported that the classification error probability to the mutual information across the speaker's identity and features are closely related. Qualitative statements about feature selection can be made using information theory. Features like different LPC parameterizations and mel-warped cepstrum coefficients were studied.

Ortega-Garcia et al. [3] have proposed a method on speaker recognition in the area of security applications through speech input. Nevertheless, variability of speech degrades the performance of speaker recognition. The external variability and intra-speaker variability sources produces mismatch across training and testing phases. The channel and inter-session variability would be explored for accomplishment of real automatic speech systems for the commercial as well as forensic speaker recognition. The experiments have shown that combination of score normalization and CMN techniques reduce ERR significantly.

Saeed and Kheir [4] explained a speech-and-speaker (SAS) identification system based on recognition of spoken Arabic digits. The speech samples of numbers from zero to ten collected in Arabic language. The conventional and the neural network based classifications were used. The successful recognition of the speaker identifying system touched about 98.8% in few cases. The efficiency of successful recognition is about 97.45% in recognition of the uttered word and identifying its speaker. For a three-digit password, the percentage of accuracy obtained is 92.5%.

Aronowitz and Burstein [5] presented the efficient techniques for speaker recognition. These techniques involve approximated cross entropy (ACE) for approximation of the Gaussian mixture modeling (GMM) likelihood scoring. The training and testing session were represented by Gaussian mixture modeling. The algorithm was efficient in performing speaker recognition with very less degradation when compared to classical Gaussian mixture model algorithm.

Stolcke et al. have described [6] a novel approach for speaker recognition. They have used the maximum likelihood linear regression (MLLR) adaptation transforms as features and support vector machine (SVM) for classification. They have shown that how MLLR–SVM approach can be enhanced by the combination of transforms relative to multiple reference models. A comparison is made between two techniques for compensating for intersession variability (WCCN and NAP) as applied to MLLR–SVM system. The results obtained have shown that the NAP is high sensitive to the choice of data for procuring covariance statistics, and the WCCN is very much influenced by the choice of background set.

Parthasarathi et al. presented a paper on "Privacy-Sensitive Audio Features for Speech/Non-speech Detection" [7]. The objective of the work was to examine the features for speech and non-speech detection (SND) having low linguistic information. Three different approaches were examined for privacy-sensitive features. Methods for Instantaneous feature extraction, excitation source information, and feature obfuscation such as local (less than 130 ms) temporal averaging. Randomization is applied on the information of excitation source. The application of obfuscation methods on the excitation features resulted low phoneme efficiencies in concurrence with SND performance comparable to that of MFPLP.

Satya Dharanipragada et al. have presented a robust technique of feature extraction for continuous speech recognition [8]. The important aspect of the technique is the minimum variance distortion-less response (MVDR) method of spectrum estimation. They have incorporated perceptual information in two approaches: (1) after the MVDR power spectrum is computed and (2) directly during the MVDR spectrum estimation. The technique used was MVDR Spectral Envelope Estimation. The proposed feature extraction method gave a lower WER compared to MFCC (Mel Frequency Cepstral coefficients) and PLP (perceptual linear prediction) feature extraction in number of cases. The technique is most robust to noise though the performance in clean conditions is not degraded.

3 Features

Tonal power ratio: Tonal power ratio [9] can be obtained by taking the ratio of the tonal power of the spectrum components to the overall power.

Spectral flux: the squared difference between the normalized magnitudes of consecutive spectral distributions that correspond to consecutive signal frames gives the spectral flux of the input speech signal.

MFCC: Consecutive applications of Fourier transform and logarithm to a speech signal, some coefficients are obtained. For the coefficients obtained, if the inverse Fourier transform is applied and the result is called cepstrum. There can be a complex, real, power, or phase cepstrum and the power cepstrum is widely used for analysis of human speech applications.

"MEL" frequency cepstral coefficients are obtained by applying logarithm with base 10 to the amplitude spectrum and DCT coefficients. There is a change in slope when there is a transition between one frame to another. These change in slope features are considered as important features of the speech.

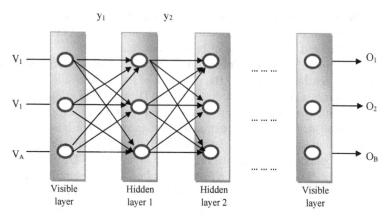

Fig. 1 Architecture of DBN

4 Deep Belief Networks

The deep belief networks formed with the connection of Restricted Boltzmann machines (RBM). It consists of multiple layers. Each layer consists of a set of binary or real-valued units and every adjacent unit have connected with every node but there is no connection within the layer. Figure 1 shows the architecture of the deep belief network. Let v be stochastic visible variable and p be stochastic hidden variable of the deep belief model. The energy function of the model is shown in the Eq. (1).

$$J(v, p, s) = - \sum_{i=1}^{A} \sum_{j=1}^{B} y_{ij} v_i p_j - \sum_{i=1}^{A} b_i p_i - \sum_{j=1}^{B} a_j p_j \qquad (1)$$

where y_{ij} represents the weight values between hidden units i and j, a_i and b_j represents the bias terms.

5 Methodology

The primary phase in the development of the Telugu accent recognition system is the collection accented speech data from Telugu speakers from different regions. The accents of the regions Telangana, Coastal Andhra, and Rayalaseema considered. The sentence selected for testing is text-dependent "ఎవరో అన్నం తిన్నారు నేను ఎవరిని చూడలేదు" (evaroo annam tinnaaru nenu evarini chuudaledu). The feature like spectral flux, Tonal power ratio, Mel frequency cepstral coefficients were extracted for each speech sample. 70% of the samples were used to train the network. The system is tested with all the speech samples. The methodology is shown in the Fig. 2.

Fig. 2 Methodology of
Telugu accent recognition
system using DBN

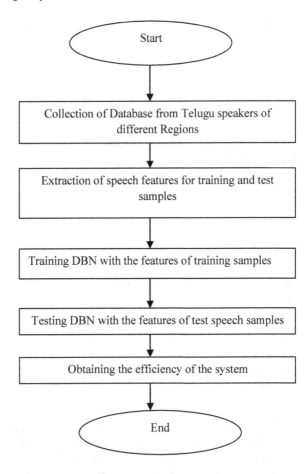

6   Results and Discussions

The Recognition accuracy for "Coastal Andhra accent" test samples is 94%. Further
Coastal Andhra accent was wrongly recognized as Rayalaseema accent for 2% of
the test samples. Also 4% of the test samples were wrongly recognized as Telangana accent speeches. For the Rayalaseema test samples, 92% of the samples are
correctly recognized, and 8% of them are wrongly recognized as Telangana samples. For the Telangana accent test samples, 94% of the samples are correctly
recognized, and 6% of them are wrongly recognized as Coastal Andhra. The overall
recognition accuracy found is 93% in this proposed system. The following Table 1
shows confusion matrix of the accent recognition system using DBN.

The efficiency of accent recognition system using MFCC features and GMM
modeling [10] is 91%. The proposed system has shown improvement in recognition
accuracy of the accent recognition.

Table 1 Result of Telugu accent recognition system using DBN

	CA	RS	TG
CA	47	1	2
RS	0	46	4
TG	3	0	47

Table 2 Comparison of accent recognition systems using DBN (Proposed) and MFCC-GMM (Published) methods

	DBN (Proposed)	MFCC-GMM (Published)
Features set	Tonal power ratio, spectral flux, MFCC	MFCC features
Classification/modeling	DBN	GMM
Data base used	13 speaker data base	13 speaker data base
Percentage efficiency of Coastal Andhra accent recognition	94	88
Percentage efficiency of Rayalaseema accent recognition	92	92
Percentage efficiency of Telangana accent recognition	94	92
Percentage efficiency of overall Telugu accent recognition	93	91

The comparison between the published and the proposed methods are shown in the Table 2. The published method has used MFCC features and GMM as the classifier. However in the proposed method, different features and also deep belief networks classifier could achieve recognition accuracy of 93%.

7   Conclusions

1. Telugu speeches of Costal Andhra, Rayalseema, and Telangana were collected to develop database for testing and training, as there is no standard data base for Indian languages as per the literature survey.
2. Feature sets consisting of MFCC, spectral flux, pitch chroma, and tonal power ratio were extracted successfully for both training and testing speech accents.
3. The efficiency of accent recognition system using the above features and deep belief networks (DBN) as classifier is found to be 93%.

References

1. Mingkuan Liu et al "Mandarin accent adaptation based on context independent/ Context-dependent pronunciation modeling". In Proceedings of the Acoustics, Speech, and Signal Processing, ICASSP '00, pages: II1025–II1028, Washington, DC, USA. IEEE Computer Society.
2. Thomas, Eriksson et al "An Information-Theoretic Perspective on Feature Selection in Speaker Recognition". IEEE signal processing letters, vol. 12, no. 7, july 2005.
3. J. Ortega-Garcia et al "Speech Variability in Automatic Speaker Recognition Systems for Commercial and Forensic Purposes", IEEE AES systems magazine, november 2000.
4. Khalid Saeed and Mohammad Kheir Nammous, "A Speech and Speaker Identification System: Feature Extraction, Description, and Classification of Speech-Signal Image", IEEE Transactions on Industrial Electronics Vol. 54, No.2, April 2007, pages: 887–897.
5. Hagai Aronowitz and David Burshtein, "Efficient Speaker Recognition Using Approximated Cross Entropy (ACE)", IEEE Transactions on Audio, Speech and Language Processing, Volume number 15, No. 7, September 2007, pp. 2033–2043.
6. Speaker Recognition with Session Variability Normalization Based on MLLR Adaptation Transforms. Andreas Stolcke, Senior Member, IEEE, Sachin S. Kajarekar, Luciana Ferrer, and Elizabeth Shrinberg. IEEE transactions on audio, speech, and language processing, vol. 15, no. 7, september 2007.
7. Sree Hari Krishnan Parthasarathi et al "Privacy-Sensitive Audio Features for Speech/Non-speech Detection", IEEE Transactions on Audio, Speech and Language Processing, Vol. 19, No. 8, November 2011, pp. 2538–2551.
8. Satya Dharanipragada et al "Robust Feature Extraction for Continuous Speech Recognition Using the MVDR Spectrum Estimation Method", IEEE Transactions on Audio, Speech and Language Processing, Vol. 15, No. 1, January 2007, pp. 224–234.
9. Alexander Lerch, "An Introduction to Audio Content Analysis: Applications in Signal Processing and Music Informatics", 272 pages, Wiley-IEEE Press, July 2012.
10. kasiprasad. Mannepalli, P. Narahari Sastry, Suman. Maloji "MFCC-GMM BASED ACCENTRE COGNITION SYSTEM FOR TELUGU SPEECH SIGNALS", International Journal Speech Technology, Vol 19. doi: 10.1007/s10772-015-9328-y.

Text Document Classification with PCA and One-Class SVM

B. Shravan Kumar and Vadlamani Ravi

Abstract We propose a document classifier based on principal component analysis (PCA) and one-class support vector machine (OCSVM), where PCA helps achieve dimensionality reduction and OCSVM performs classification. Initially, PCA is invoked on the document-term matrix resulting in choosing the top few principal components. Later, OCSVM is trained on the records of the matrix corresponding to the negative class. Then, we tested the trained OCSVM with the records of the matrix corresponding to the positive class. The effectiveness of the proposed model is demonstrated on the popular datasets, viz., 20NG, malware, Syskill, & Webert, and customer feedbacks of a Bank. We observed that the hybrid yielded very high accuracies in all datasets.

Keywords Text mining · Dimensionality reduction · Document classification · Principal component analysis · One-class support vector machine

1 Introduction

This text document classification is defined as the task of assigning text documents to predefined classes. Statistical and machine learning techniques cannot analyze text documents since text data is in an unstructured format. Therefore, the unstructured data must be converted into a structured form before any classifier is

B. Shravan Kumar · V. Ravi (✉)
Centre of Excellence in Analytics, Institute for Development and Research
in Banking Technology, Castle Hills Road No. 1, Masab Tank,
Hyderabad 500057, India
e-mail: padmarav@gmail.com

B. Shravan Kumar
e-mail: shravan.springer@yahoo.com

B. Shravan Kumar
School of of Computer & Information Sciences, University of Hyderabad,
Hyderabad 500046, India

© Springer Nature Singapore Pte Ltd. 2017   107
S.C. Satapathy et al. (eds.), *Proceedings of the 5th International Conference on Frontiers in Intelligent Computing: Theory and Applications*, Advances in Intelligent Systems and Computing 515, DOI 10.1007/978-981-10-3153-3_11

invoked. Text classification is fraught with challenges, including high dimensionality of the feature space, where each unique word represents a feature [1]. Sometimes it is also essential to reduce the input (document) space dimension, documents can be sparse with respect to the features when mapped into a structured format. In this paper, our objective is to reduce the feature space dimension, without compromising the performance of a classifier. According to Dorre et al. [2], text mining extracts the implicit knowledge from text documents. First step in text mining is to transform the text corpus into a document-term matrix. This requires preprocessing of text including the steps of tokenization, stop words removal, and stemming [3]. Once the document-term matrix is formed, data mining techniques are applied on the matrix to solve the underlying problem. Given the high dimensionality of the data, feature selection and/or dimensionality reduction is performed before invoking classifiers. Our research proposes a new method for document classification by performing dimensionality reduction with PCA followed by classifying the resultant matrix with OCSVM.

The structure of the rest of the paper is as follows. Section 2 reviews the works related to text mining, one-class SVM, and PCA. Section 3 overviews of the methods applied in this work. Section 4 presents the proposed methodology. Section 5 describes about the proposed methodology and Section 6 presents the results of our analysis. Finally, we summarize our work in Section 7.

2 Literature Review

In this section, we discuss the past works on text mining, One-Class SVM, and PCA. Text categorization was pioneered by Maron [4] in 1961. By looking at the occurrences of selected terms, the classification task was performed. Masand et al. [5] proposed a method to classify the news stories with the help of memory-based reasoning. They trained the model almost with 50 K stories from Dow Jones Press Release News and reported that recall of 80% and precision of 70%.

Manevitz and Yousef [6] have classified standard Reuters dataset with one-class SVM and analyzed different versions of term weighting schemes of document-term matrix including TF–IDF, bit vector as well as with different kernels. Yu et al. [7] performed web page classification using support vector machine (SVM). They introduced positive example-based learning in their work. Vert and Vert [8] discussed the convergence criterion of one-class SVMs. Metsis et al. [9] discussed the Spam e-mail classification. They tested with the five different versions of Naive Bayes (NB) classifiers, like multivariate Bernoulli, multinomial NB with term frequency, multinomial NB with binary attributes, multivariate Gauss NB and flexible Bayes. They experimented with 500, 1000, and 3000 features. The overall performance of the classifier was high with 3000 features. Chinta and Murthy [10] analyzed the various feature subset selection methods for categorizing the text documents. Feature subsets were selected by Information gain, Fisher score, Chi-square statistic, Mutual information, and Document frequency.

Pandey and Ravi [11] performed phishing and spam detection using text and data mining. They extracted 17 features from the source code of the URL. Sensitivity and accuracy values reported are higher than that of the previous studies. In this study, they built various models with Genetic Programming (GP), Logistic Regression (LR), Probabilistic Neural Network (PNN), Multilayer Perceptron (MLP), Classification And Regression Tree (CART), GP + CART, and reported higher sensitivity values. They reported rules for legitimate and phishing detection. Jun et al. [12] initiated a model to overcome the sparsity problem for document clustering. They combined dimension reduction with K-means clustering algorithm and experimented with patent documents which were retrieved from the United States patent office. Sundarkumar and Ravi [13] recently worked with OCSVM for data imbalance problem by employing k-Reverse Nearest Neighborhood (k-RNN).

We now present various works involving one-class classification problems. The essential task of web mining is web page classification. It identifies user open web pages, and requires a lot of effort including the preprocessing of all pages, etc. The collection of non-interested one is the complicated process when compared to the interested one. To solve this problem Yu et al. [7] introduced a new framework called positive example-based learning (PEBL) for negative page collection. They presented mapping convergence criteria and with this, they reported higher accuracy compared to the binary SVM classifier on the DMOZ and WebKB datasets. Denis et al. [14] trained the Naive Bayes with positive class documents. They compared the results with the regular approach. To evaluate the proposed model WebKB dataset was used. Use of linear functions to train on positive and unlabeled data was reported by Lee and Liu [15]. They proposed a model using logistic regression with weighted samples and performance index. To evaluate the efficacy of the model they tested it on the 20NG dataset. Manevitz and Yousef [16] trained a neural network with positive class data, and observed that the new approach is better than other standard methods on Reuters 21578 collection. Elkan and Noto [17] proposed a model which trains on the positive samples. They applied this concept to the biological database.

3 Overview of Methods Applied

3.1 *Principal Component Analysis (PCA)*

PCA is one of the most useful multivariate statistical techniques. Applications of PCA can be found in many branches of engineering and sciences for reducing the feature space dimensionality and thereby removing the multi-collinearity in datasets. The hallmark of PCA is that each principal component (PC) is a linear combination of the original features and that the first PC consists of maximum variance (information), then the second PC explains second highest variance (information) in the original data and so on. Therefore, if one selects first few PCs, then he/she is assured of accounting for maximum information. For further details refer the works of Anderson [18], Jolliffe [19], Burges [20], Ferre [21], and Lian [22].

3.2 One-Class Support Vector Machine

One-Class Support Vector Machine (OCSVM) is similar to SVM except that it deals with a training data consisting of only one class. It builds a boundary space from other class examples. It works well in high-dimensional settings when other methods (e.g., density estimation) fail. It has the same disadvantages, viz., choice of kernels and inability to perform multiclass classification as that of SVM. OCSVM is using in several applications including retrieval of images [23], anomaly detection in videos [24], document categorization [6], and sound recognition system [25].

4 Proposed Methodology

In the proposed method, our goal is to build a novel one-class classifier. Binary classifiers assume the availability of data of both classes. However, in practical scenarios, data of the only one class (i.e., negative class) is available. It means that we do not have prior knowledge of positive class pattern behaviors. Therefore, one is forced to build a model based on the negative class dataset and perform one-class classification. Consequently, whenever a data sample of the positive class (fraud, phishing, spam, malware, the churn of customers, etc.) is encountered, the one-class classifier built from the negative class would correctly identify the positive class sample, albeit, with some errors depending on how well the classifier was built.

Our proposed methodology comprises three phases: preprocessing, dimensionality reduction phase, training and test phase as depicted in Fig. 1. First, we collected all the samples of the given dataset and performed text preprocessing. Then, we mapped this text into a document-term matrix with term frequency (Term Occurrences) approach. Then, in the second phase, we applied PCA for dimensionality reduction. In the final phase, we trained the OCSVM classifier by feeding it with the chosen PCs corresponding to negative samples. Finally, OCSVM is tested with the selected PCs corresponding to positive samples.

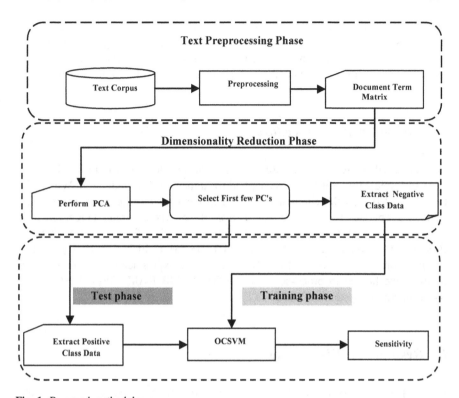

Fig. 1 Proposed methodology

5 Experimental Methodology

5.1 Datasets Description

To validate our proposed method, we conducted novel experiments with the following datasets, mostly available in the web. 20NG (20 News Groups) [26] is the collection of nearly 20 K newsgroup documents, separated evenly across 20 different groups. Here, we considered a total of 1997 documents, which is a subset of two groups. Malware dataset [27] contains a selection of Windows application program interface (API)/system-call trace files, for testing on classifiers treating with sequences. It contains two parts: 388 logs out of which there are 68 benign software traces labeled as '0' and 320 malware traces labeled as '1'. Syskill & Webert (SW) [28] is the dataset consisting of the web pages of user interest on a particular topic. It consists of four groups, namely band, sheep, goat, and biomedical. We combined the sheep and goat into one group and biomedical as another group by ignoring the band group for the binary classification task. Imperial Bank [29] dataset collection took from IBM SPSS modeler tool. We extracted the feedbacks of the 786 customers. Among these, 132 are satisfied customers, 148 are very satisfied, 166 are unsatisfied, 172 are very unsatisfied, and 168 are neutral customers. We combined the satisfied and very satisfied into one group; unsatisfied and very unsatisfied into another group. We ignored the neutral comments. The, group-1 consists of 280 documents and group-2 consists of 338 documents.

5.2 Experimental Procedure

We converted the text from unstructured format into document-term matrix using the tool Rapid Miner [30]. PCA is applied for dimensionality reduction. We used a threshold of 0.5 on the cumulative sum of λ_i's, where i is the index of the PCs. Later, we separated the negative records from the matrix and trained OCSVM with negative class examples. Finally, we performed the classification task with these PCs. For this, we used the library called LIBSVM available in Java [31]. For computation of PCA, we used MATLAB [32]. All experiments are conducted on the machine having i5 processor with 2.6 GHz, 8 GB RAM, 500 GB HDD, and 64-bit operating system of Windows 8.

6 Results and Discussions

In this paper, we evaluated the performance of the proposed system with the following metric, which is defined as follows:

$$\text{Sensitivity} = \frac{TP}{TP + FN}$$

where TP = True Positive, TN = True Negative, FP = False Positive, FN = False Negative.

The number of features extracted from the various datasets is as follows: 313 features from malware dataset, 216 features from Imperial Bank, 2038 features from SW dataset and finally, 16237 features for 20NG dataset. After feature extraction, we applied PCA and chose top PCs which have more than 50% of the contribution of variance of the original data. The number of principal components ($\lambda_i \geq 0.50$) computed from various datasets is as follows: 33 from 20NG, one from malware, two from SW, and ten from Imperial Bank.

The principal component distribution is as follows: Primarily, we will start discussing from malware dataset the first principal component having weightage of more than 70%, later 20% occupied by the second component. The first two components are holding more than 91%, whereas in SW dataset the first component is occupying 43, 17% by second component, and 6% by the third component. Top three components themselves are of more than 65% weighted. For the 20NG dataset, the first component occupies 16, 8% by the second component, and 3% by

Table 1 Performance of the proposed model

Dataset	Sensitivity
20NG	100
Malware	100
Imperial Bank	99.81
SW	99.62

third component. Similarly, first component of the Imperial Bank dataset has a weightage of 9%, the second component by 7%, and third and fourth components by 6% respectively. From the above assertion, it is clearly observed that the first principal component of malware occupies more weight than other principal components. Similarly for SW dataset, first, two principal components have higher contribution than remaining, whereas the cumulative sum of Imperial Bank was distributed among ten components. For the 20NG dataset, distribution of components is occupied by top 33 components.

There are different parameter settings available in OCSVMs. The best parameter combination always gives the lowest error rate. We experimented with four kernels on all datasets, and they are compared with the performance of four kernels: Linear, polynomial, RBF, and sigmoid available in OCSVM. The types of the kernels and their performance are discussed in next paragraph. However, we reported best values irrespective of the choice of kernels in Table 1.

Initially, we discuss from 20NG results. The sensitivity values in percentages for various kernels are as follows: 99.9 (linear), 99.9 (polynomial), 100 (RBF), and 99.9 (sigmoid). Similarly, for malware dataset, the sensitivity values obtained are 95.94 (linear), 98.12 (polynomial), 100 (RBF), and 100 (sigmoid). For Imperial Bank dataset these values are 95.27 (linear), 99.81 (polynomial), 94.86 (RBF), and 98.53 (sigmoid). For SW dataset sensitivity values are 57.36 (linear), 58.24 (polynomial), 99.62 (RBF), and 98.53 (sigmoid). All kernels performed the almost similarly on 20NG dataset with very slight variation. We reported the highest sensitivity with RBF kernel. On the Imperial Bank data, the polynomial kernel performed the best followed by sigmoid, whereas on malware dataset, RBF and Sigmoid Kernels performed well and the values obtained from these kernels were identical. For SW dataset, RBF kernel performed well, followed by sigmoid. In this work, we performed one-class classification task with sensitivity as the performance metric. So, we did not report any other metrics such as specificity, area under the curve (AUC), etc. Similarly, we did not compare the proposed classifier results with any other binary classifiers.

7 Conclusion

In this paper, we proposed a novel model for document classification using PCA and OCSVM. The experimental results indicated that the proposed approach yielded higher performance. To assess the effectiveness of the hybrid, experiments are conducted on 20NG, Imperial Bank, SW, and malware datasets. We conclude that the proposed method yielded the highest sensitivity values.

References

1. Joachims, T.: Text categorization with Support Vector Machines: Learning with many relevant features. In: Machine Learning: ECML 98, LNCS, Vol. 1398, pp. 137–142 (1998).
2. Dorre, J., Gerstl, P., Seiffert, R.: Text mining: Finding nuggets in mountains of textual data. In: KDD 99, San Diego, CA, USA, pp. 398–401 (1999).
3. Salton, G., and McGill, M. J.: Introduction to Modern Information Retrieval. McGraw-Hill, Inc. New York, NY, USA (1986).
4. Maron, M. E.: Automatic Indexing: An Experimental Inquiry, Journal of the ACM 8 (3), 404–417 (1961).
5. Masand, B., Linoff, G., Waltz, D.: Classifying news stories using memory based reasoning. In: 15th ACM International Conference on Research and Development in Information Retrieval (SIGIR 92), Copenhagen, Denmark, pp. 59–65 (1992).
6. Manevitz, L. M., Yousef, M.: One-Class SVMs for document classification. Journal of Machine Learning Research, 139–154 (2001).
7. Yu, H., Han, J., Chang, K. C-C.: PEBL: Positive Example Based Learning for web page classification using SVM. In: KDD 02, Edmonton, Alberta, Canada, pp. 239–248 (2002).
8. Vert, R., and Vert, J-P.: Consistency and Convergence Rates of One-Class SVMs and Related Algorithms. Journal of Machine Learning Research, 817–854 (2006).
9. Metsis, V., Androutsopolos, I., Paliouras, G.: Spam filtering with Naive Bayes - Which Naive Bayes?. In: 3rd Conference on Email and AntiSpam (CEAS 06), Mountain view, California, pp. 28–69 (2006).
10. Murthy, P. M., and Murthy, M. N.: Discriminative Feature Analysis and Selection for Document Classification. ICONIP, Part I, LNCS 7663, pp. 366–374 (2012).
11. Pandey, M., Ravi, V.: Text and data mining to detect phishing websites and spam emails. In: Swarm, Evolutionary, and Memetic Computing Conference (SEMCCO), Part II, LNCS 8298, pp. 559–573 (2013).
12. Jun, S., Park, S. S., Jang, D. S.: Document clustering method using dimension reduction and support vector clustering to overcome sparseness. Expert Systems with Applications 41 (7), pp. 3204–3212 (2014).
13. Sundarkumar, G. G., Ravi, V.: A novel hybrid undersampling method for mining unbalanced datasets in banking and insurance. Engineering Applications of Artificial intelligence 37, 368–377 (2015).
14. Denis, F., Gilleron, R., and Tommasi, M.: Text classification from positive and unlabeled examples. In: 9th International Conference on Information Processing and Management of Uncertainty in Knowledge-Based Systems (IPMU), Annecy, France, pp. 1927–1934 (2002).
15. Lee, W. S., and Liu, B.: Learning with positive and unlabeled examples using weighted Logistic Regression. In: 12th ICML' 03, Washington, DC, pp. 448–455 (2003).
16. Manevitz, L. M., Yousef, M.: Document classification on neural networks using only positive examples. In: 23rd ACM International Conference on Research and Development in Information Retrieval (SIGIR 00), Athens, Greece, pp. 304–306 (2000).
17. Elkan, C., and Noto, K.: Learning classifiers from only positive and unlabeled data. In: KDD 08, August 24–27, Las Vegas, Nevada, USA, pp. 213-220 (2008).
18. Anderson, T. W.: Asymptotic theory for principal component analysis. Annals of Mathematical Statistics 34 (1), 122–148 (1963).
19. Jolliffe, I.T.: Principal Component Analysis. Springer Verlag (1986).

20. Burges, C. J. C.: Dimension reduction: A guided tour. Foundations and trends in Machine Learning 2 (4), 275–365 (2009).
21. Ferre, L.: Selection of components in principal component analysis: A comparison of methods. Computational statistics and data analytics 19 (6), 669–682 (1995).
22. Lian, H.: On feature selection with principal component analysis for one-class SVM. Pattern Recognition Letters 33 (9), 1027–1031 (2012).
23. Chen, Y., Zhou, X., and Huang, T. S.: One-class SVM for learning in image retrieval. In: International Conference on Image Processing (ICIP), Thessaloniki, Greece, pp. 34–37 (2001).
24. Liu, C., Wang, G., and Ning, W., Lin, X., Li, L., Liu, Z.: Anomaly detection in surveillance video using motion direction statistics. In: 17th International Conference on Image Processing, Hong Kong, pp. 717–720 (2010).
25. Wan, C., Mita, A.: An automatic pipeline monitoring system based on PCA and SVM. International Journal of Mathematical, Computational, Natural and Physical Engineering 2 (9), 90–96 (2008).
26. Newsgroups, http://qwone.com/~jason/20Newsgroups.
27. Csmining group. http://www.csmining.org/index.php/malicious-software-datasets-.html.
28. Syskill & Webert web page ratings, https://archive.ics.uci.edu/ml/machine-learning-databases/SyskillWebert-mld/.
29. IBM SPSS, http://www-01.ibm.com/software/in/analytics/spss/products/data-collection/.
30. Rapid Miner, https://rapidminer.com (2013).
31. LIBSVM, http://www.csie.ntu.edu.tw/~cjlin/libsvm/#download.
32. MATLAB, www.mathworks.com. (2012).

Data Mining Approach to Predict and Analyze the Cardiovascular Disease

Anurag Bhatt, Sanjay Kumar Dubey, Ashutosh Kumar Bhatt and Manish Joshi

Abstract This paper presents the experimental analysis of data provided by UCI machine learning repository. Weka open source machine learning tool provided by Waikato University reveals the hidden fact behind the datasets on applying supervised mathematical proven algorithm, i.e., J48 and Naïve Bayes algorithm. J48 is an extension of ID3 algorithm having additional features like continuous attribute value ranges and derivation of rules. The data sets were analyzed using two approaches, i.e., first taken with selected attributes and taken with all attributes. The performance of both the algorithm reveals the accuracy of algorithm and predicting the various reasons behind this increasing problem of cardiovascular diseases.

Keywords Cardiovascular disease · J48 · Naïve Bayes · Data mining · Weka

1 Introduction

Cardiovascular disease is the leading cause of death in both young and old age group people. There are certain reasons behind this increasing problem in all the age group people, i.e., changing lifestyle, unbalanced diet, improper nutrition, hyper tension, high cholesterol, physical inactivity, stress level, etc. We can identify these

A. Bhatt (✉) · S.K. Dubey
Amity University Uttar Pradesh, Noida, India
e-mail: anurag15bhatt@gmail.com

S.K. Dubey
e-mail: skdubey1@amity.edu

A.K. Bhatt
Birla Institute of Applied Sciences, Bhimtal, India
e-mail: ashutoshbhatt123@gmail.com

M. Joshi
Amity University, Lucknow Campus, Lucknow, UP, India
e-mail: manishjoshi0903@gmail.com

© Springer Nature Singapore Pte Ltd. 2017
S.C. Satapathy et al. (eds.), *Proceedings of the 5th International Conference on Frontiers in Intelligent Computing: Theory and Applications*, Advances in Intelligent Systems and Computing 515, DOI 10.1007/978-981-10-3153-3_12

risk factors in terms of modifiable and non-modifiable risk factors. Modifiable factors can be cured or in other terms, and they can be suppressed through proper medication which will result less probability of having cardiovascular diseases, while on the other hand, non-modifiable risk factors like family history of cardiovascular diseases and first degree blood relation increase the risk factor of having attack.

Knowledge extraction plays an important role here by allowing us providing a novel approach to mine knowledge out of the bulky datasets available at healthcare research centers, medical informatics department. In recent years, research trends are inclined to medical field analysis where huge data sets are required to be analyzed through various data mining and machine learning algorithms. Naive Bayes is a conditional probability model: given a problem instance to be classified, represented by a vector $X = (x_1, \ldots x_n)$ representing some n features (independent variables), it assigns to this instance probabilities

$$p(C_k | x_1, \ldots x_n)$$

for each of K possible outcomes or *classes*. [1]

2 Related Work

Researcher [2] has demonstrated an experiment on data set taken by UCI machine learning repository and three different supervised algorithms are used, i.e., Decision tree, Bayesian classifier, and neural network. These experiments are conducted through Weka tool. Two approaches are used to collect results and their comparisons, i.e., with all attributes and with selected attributes. Researcher [3] has focused on predictive analysis of diabetes treatment using regression-based data mining techniques in Saudi Arabia. Oracle data miner (ODM) is taken as the software mining tools for predicting modes of treating diabetes. Datasets for analysis have been taken from WHO (World Health Organization) and these data mining techniques have been applied to identify effectiveness of different treatments types for different age groups. Five age groups are mainly consolidated into two age groups, i.e., p(y) and p(o) where p(y) stands for young people and p(o) stands for old people. In the paper [4], genetic algorithm approach has been used for getting high accuracy in prediction of disease. Researchers [5] analyzed heart disease symptoms and predicted the results on applying K-means clustering producing cluster relevant data and then applying mafia (maximum frequent itemset algorithm) that helped to recognize frequent patterns and then classification of that patterns using C4.5 algorithm. K-means based on mafia algorithm with ID3 and C4.5 showed accuracy of 92%. Researcher [6] has proposed types of heart disease including its symptoms and seriousness showing effect in human body. Researchers in [7, 8, 9] have proposed a system that can predict code blue (an indicator for sudden cardiac arrest) and can help hospital administration to predict upcoming attacks on the basis of analysis of previous heart patient ECG records and it also reveals that SVM (Support vector machine) works much better than other algorithms. Researcher [10] has used neural network approach to classify medical database obtained from Cleveland database and it has been seen that satisfactory results are derived from single-layer and multilayer implementation of neural network. Researcher [11] has presented a case report on sudden cardiac death in young adults due to HCM (Hypertrophic Cardiomyopathy). The paper [12] presents the medical investigation of reasons behind the sudden cardiac arrest in adults. Researcher [6] has given the theory in study of heart disease prediction using data mining techniques and he has introduced some of the most common and effective techniques like neural networks, decision trees, Naïve Bayes, etc. to apply in medical datasets. In paper [13], associative classification and genetic algorithm are introduced as the data mining techniques and in order to improve the accuracy of associative classification, informative attribute entered rule generation, and hypothesis testing Z-statistics for heart disease prediction is proposed. Researchers in paper [14] have given a review concluding with the performance on mathematically proven algorithms used in heart disease prediction system. Researchers [5] analyzed heart disease symptoms and predicted the results on applying K-means clustering producing cluster relevant data and then applying mafia (maximum frequent item set algorithm) that helped to recognize frequent patterns and then classification of that patterns using C4.5 algorithm. K-means based on mafia algorithm with ID3 and C4.5 showed accuracy of 92%.

3 Experimentation Results

In this paper, we have done experiments on the data provided by UCI machine learning repository and these data sets are available online. Hungarian database of heart disease is taken and echocardiograph raw data is analyzed to predict the sudden cardiac death ratio among heart disease patients. The risk of getting sudden cardiac arrest increases with the patients having medical history of heart disease that causes arrhythmia as well as rhythmic unbalance of heart. Various attributes are taken from the database that are collectively affecting the experiment results as well as it shows the efficiency or accuracy of particular algorithm applied in the data set. These two selected datasets are run with J48 pruned and Naïve Bayes algorithm showing various results which will be analyzed later and we are going to provide confusion and performance matrix from the results generated by data mining software "Weka." We have also proposed a framework for the software implementation of this system through which we can analyze and predict the data in single click.

Attribute description (Hungarian Database)

1. Age Numeric value of age in years.
2. Sex-Binary representation, 1 = male, 0 = female.

3. Cp Type—Chest pain types are as follows:
   Value 1—Typical Burning Sensation in heart
   Value 2—Acute stabbing like pain
   Value 3—Burning Sensation
   Value 4—Acute Crushing Pain in heart
4. Restbps—resting blood pressure values (numeric)
5. Chol—Cholesterol level in patient (numeric)
6. Fbs Value—fasting blood sugar level where 1 = true, 0 = false
7. Restecg—Rested electrocardiographic report indication
   Value 0: normal
   Value 1: ST-T wave Abnormality
   Value 2: Ventricular Abnormality
8. Thalach Value—It shows maximum heart rate achieved
9. Exang—Anigna caused by exercising (1 = yes, 0 = no)
10. Oldpeak—ST wave depression induced by exercising relative to rest.
11. Slope—the slope of the peak exercise ST segment.
    Value 1: slope in up direction.
    Value 2: flat or no slope.
    Value 3: downward slope.
12. Ca—number of major vessels.
13. Thal Value—It stands for Thalassemia value
    Value '3' = Normal.
    Value '6' = Fixed defect.
    Value '7' = Reversible defect.
14. Num: Number of Patients with heart disease.
    Value 0: No heart Disease
    Value 1: Heart Disease Level-1
    Value 2: Heart Disease Level-2
    Value 3: Heart Disease Level-3
    Value 4: Heart Disease Level-4

3.1 Hungarian Database (J48 Pruned Algorithm)

3.1.1 Using All Attributes

See Table 1.

3.1.2 Using Selected Attributes

See Table 2.

Table 1 Confusion matrix

A	B	C	D	E	Classification
13	0	0	0	2	A = 0
2	17	0	0	7	B = 1
3	3	13	0	9	C = 2
1	4	0	14	18	D = 3
2	0	1	0	185	E = 4

Number of leaves in model: 156
Size of the tree: 158

Table 2 Confusion matrix

A	B	C	D	E	Classification
0	0	12	0	3	A = 0
0	1	15	0	10	B = 1
0	0	23	0	5	C = 2
0	0	22	0	15	D = 3
0	0	19	0	169	E = 4

Number of leaves: 4
Size of the tree: 5

3.1.3 J48 Pruned with All Attributes

See Table 3.

3.1.4 J48 Pruned with Selected Attributes

See Table 4.

In Table 1, confusion matrix and detailed performance of experiment 1 is listed below showing results on 14 attributes and 294 instances. Whole data is classified into five classes as 0, 1, 2, 3, and 4. "0" class indicates "*No Heart Disease*" where "1" indicates "*Level-1 Heart disease,*" "2" indicates "*Level-2 Heart disease,*" "3" indicates "*Level-3 Heart Disease,*" and "4" indicates "*Level-4 heart Disease.*" J48 pruned with all attributes generated a tree with a size of 158 and 156, while J48 pruned with selected attributes generated tree of size 5 with four leaves. Experiment

Table 3 Performance matrix

Accuracy (%)	TP rate	Precision	F-measure	ROC area
82.3	0.823	0.844	0.804	0.849

Table 4 Performance matrix

Accuracy (%)	TP rate	Precision	F-measure	ROC area
65.64	0.656	0.647	0.598	0.773

1 achieved accuracy of "82.3%" with all attributes and "65.64%" with selected attributes. In the results shown in Table 1, "21" instances are classified as A, "24" are classified as B, "14" goes with both C and D, and "221" as E. Table 2 has taken selected attributes and provided the results which vary with all attribute datasets, i.e., "0" classified as A, "1" classified as B, "91" classified as C, and "202" classified as E with accuracy of 65.64%.

Attribute Description (Echocardiogram database)

1. Survival period (in months)—Number of patients has survived of attack.
2. Still-alive value—Binary value representation, 0 = dead within a period, 1 = Still alive.
3. Age during heart attack—Age of patient during heart attack.
4. Pericardial effusion liquid—Denoted by binary value. It is a liquid present around the heart muscles where 0 = no fluid, 1 = fluid.
5. Fractional contraction—Value showing abnormality in heart muscle contraction.
6. Epss value—Value used to measure the degree of contraction.
7. Lvdd Value—Stands for "Left ventricular end diastolic value" and it is diastolic dimension of left ventricle.
8. Wall motion value.
9. Wall motion index value.
10. Mult value.
11. Alive at 1 year: "0" indicates patient death within 1 year by SCD and "1" indicates patient survived for 1 year.

The above attribute information is gathered from again UCI machine learning repository under Echocardiogram dataset. This dataset again will help to get results disclosing patients having chronic heart disease using Naïve Bayes algorithm in data mining.

3.2 Echocardiogram Dataset (Naïve Bayes Algorithm)

3.2.1 Using All Attributes

See Table 5.

Table 5 Confusion matrix

A	B	Classification
49	1	A = 0
0	24	B = 1

Number of leaves in the model: 2
Size of the tree: 3
Time taken to build model: 0.02 s

Table 6 Confusion matrix

A	B	Classification
45	5	A = 0
0	24	B = 1

Table 7 Performance matrix

Accuracy	TP rate	Precision	F-Measure	ROC area
98.64	0.986	0.987	0.987	0.946

Table 8 Performance matrix

Accuracy	TP rate	Precision	F-Measure	ROC area
93.24	0.932	0.944	0.934	0.877

3.2.2 Using Selected Attributes

See Table 6.

3.2.3 Naïve Bayes with All Attributes

See Table 7.

3.2.4 Naïve Bayes with Selected Attributes

See Table 8.

The above experiment produced good results on applying Naïve Bayes algorithm to Echocardiogram database. Naïve Bayes with all attributes showed highest accuracy of 98.64% while with selected attributes, accuracy was 93.24%.

4 Analysis and Prediction

In this paper, we have come across various experimentation results and data that produce different results using data mining algorithms. Our mathematical results show that data mining algorithms applied to the data sets have performed well with maximum accuracy. Through these results we can predict that prediction of cardiovascular disease can be understood from the following mathematical expression:

$Y = F(X1, X2, X3, X4, X5, X6 \ldots\ldots Xn)$ where Y = Cardiovascular disease $X1, X2, X3, X4, X5, X6 \ldots\ldots Xn$ = Factors affecting cardiovascular Disease

The previous results of the above experimentation show different analytical results for different datasets, i.e., in case of Hungarian database we analyzed that large part of the subjects were found with level-4 heart disease or final stage heart patients. These patient's medical records clearly indicate that large number of their diagnostic parameters are found more abnormal than their standard values. A few subjects out of total were diagnosed as in early stage of heart disease and can be cured with proper medication and awareness. Levels of the heart disease are indication of the status of the heart malfunctioning and proper consultation and medical attention can be prescribed according to it. Experiment is carried out in two modes: with selected attributes and all attributes. Talking about the selected attributes in Hungarian database, it has been analyzed that there is contradiction in the number of healthy patients when experimented with selected attributes close to the heart diagnosis standard. In this way, we can understand that these mathematically proven algorithms work differently with different datasets, keeping attribute selection in focus.

Echocardiograph dataset with Naïve Bayes algorithms shows that patients with no heart disease are large in number as compared to patients having other heart problems with highest accuracy as compared to J48 pruned algorithm. In this paper, above analysis is limited to the attribute subset taken from the UCI machine learning repository, i.e., society responsible for donating a large variety of data sets to the researchers. Taking an important consideration, considering missing attribute and their values that are not used in the experimentation is handled and learned using Bernoulli mixture model and expectation maximization techniques. Cardiovascular disease is age independent because hyper tension, high cholesterol level, and physical inactivity are some of the most highlighted reasons of this disease. Throughout this paper, our results have shown us the analytical and predictive background of the datasets. In this way, we can analyze the data through its accuracy and we can predict that heart diseases are one of the leading causes of death. There are various factors available like changing lifestyles, eating habits, etc., which cause improper functioning of heart. These data sets, i.e., echocardiogram and Hungarian data, are mostly donated from health care facilities and clinical research institutions and according to a survey it has been found that urban people and their changing lifestyles is one of the major reasons of having cardiovascular diseases, which leads to sudden cardiac arrest.

5 Conclusion

The prevalence of cardiovascular diseases is increasing in the world as the records provided by the medical institutions reveal that after going through the process of knowledge extraction. This paper is focused towards the analysis of data sets provided from the open source. This paper is limited to the extent of using open source donated dataset by machine learning repository, but future scope of this paper can be marked with the fact that, in future research works, comparative analysis and prediction will be performed on clinical outcomes using real patient data acquired from hospitals and medical research institutions. Another advancement in future research work will be the inclusion of detailed comparative study of various machine learning algorithms for more accuracy and effectiveness. Through the mathematically proven algorithms run on the data sets and their respective outputs, we can predict that a large part of the world is getting into this problem due to its speedy lifestyle and lack of proper availability of healthcare. In nutshell, we can predict that CVD is one of the most fatal diseases for both young and old patients. Proper diagnosis and treatment of this disease is required along with balanced lifestyle for both social and physical wellbeing.

References

1. Patil, R.R.: Heart Disease Prediction system using Naïve Bayes and Jelinek mercer smoothing, International Journal of Advance Research in Computer and Communication Engineering, ISSN: 2278-1021, Volume 3, Issue 5, pp. 6787–6792, May (2014).
2. Taneja, A.: Heart Disease Prediction System Using Data Mining Techniques, Oriental Journal of Computer Science and technology, Vol. 6, No. 4, pp. 457–466, December (2013).
3. Aljumah, A.A., Ahamad, M.G., Siddiqui, M.K.: Application of data mining: Diabetes health care in young and old patients, Journal of King Saud University – Computer and Information Sciences, pp. 127–136, (2013).
4. Pandey, A.K., Pandey, P., Jaiswal, K.L., Sen, A.K.: Data Mining Clustering Techniques in the Prediction of Heart Disease using Attribute selection Method, International Journal of Science, Engineering and Technology Research (IJSETR), ISSN:2278-7798, Volume 2, Issue 10, pp. 2003–2008, October (2013).
5. Manikantan, V., Latha, S.: Predicting the Analysis of Heart Disease Symptoms Using Medical Data Mining methods, International Journal on Advanced Computer Theory and Engineering (IJACTE), ISSN: 2319-2526, Volume-2, Issue-2, pp. 5–10, (2013).
6. Sudhakar, K., Manimekalai, M.: Study of Heart Disease Prediction using Data Mining, International Journal of Advance Research in Computer Science and Software Engineering, Vol. 4, Issue 1, ISSN:2277 128X, pp. 1157–1160, January (2014).
7. Somanchi, S., Adhikari, S., Lin, A., Eneva, E., Ghani, R.: Early Prediction of Cardiac Arrest (Code Blue) using Electronic Medical Records, ACM, ISBN:978-1-4503-3664, pp. 2119–2126, August 11–14, (2015).
8. Sathawane, N.K.S., Kshirsagar, P.: Prediction and Analysis of ECG Signal Behaviour using Soft Computing, International Journal of Research in Engineering & Technology, Vol. 2, Issue 5, pp. 199, May (2014).
9. Palaniappan, S., Awang, R.: Intelligent Heart Disease Prediction System Using Data Mining Techniques, IEEE, pp. 108–115, (2008).
10. Rani, K.U.: Analysis of Heart Disease Dataset using Neural Network Approach, International Journal of Data Mining & Knowledge Management Process (IJDKP), Vol.1, No.5, pp. 1–8, September (2011).
11. Murty, N., M., Devi, S., V.: Pattern Recognition: An Algorithmic Approach, *ISBN 0857294946* (2011).
12. Nayak, S.R., Dash, B.K., Mishra, S., Bhutada, T., Jena, M.K.: Sudden Cardiac death in Young Adults, J Indian Acad. forensic Med., ISSN: 0971-0973, Volume 37, No. 4, pp. 438–440, October–December (2015).

13. Jabbar, M.A., Chandra, P., Deekshatulu, B.L.: Heart Disease Prediction System using Associative Classification and Genetic Algorithm, International Conference on Emerging Trends in Electrical, Electronics and Communication technologies-ICECIT, (2012).
14. Kaur, B., Singh, W.: Review on Heart Disease Prediction System using Data Mining Techniques, International Journal on Recent and Innovation trends in computing and Communication, ISSN: 2321-8169, Volume 2, Issue 10, pp. 3003–3008, October (2014).
15. Karthiga, G., Preethi, C., Devi, R.D.H.: Heart Disease Analysis System using Data Mining Techniques, International Journal of Innovative research in Science, Engineering and Technology, ISSN: 2319-8753, Volume 3, Special Issue 3, pp. 3101–3105, March (2014).
16. Chaurasia, V., Pal, S.: Early Prediction of Heart disease using Data Mining Techniques, Caribbean Journal of Science and Technology, ISSN: 0799-3757, Volume 1, pp. 208–217, (2013).
17. Srinivas, K., Rao, G.R., Govardhan, A.: Analysis of Attribute Association in Heart Disease using Data Mining Techniques, International Journal of Engineering Research and Applications (IJERA), ISSN:2248-9622, Volume 2, Issue 4, pp. 1680-1683, July–August (2012).
18. Masethe, H.D., Masethe, M.A.: Prediction of Heart Disease using Classification Algorithms, Proceedings of the World Congress on Engineering and Computer Science, ISSN: 2078-0966, San Francisco, USA, WCECS 2014, pp. 22–24 October (2014).
19. Ebrahimzadeh, E., Pooyan, M.: Early Detection of Sudden Cardiac Death by using Classical Linear techniques and Time-Frequency methods on Electrocardiogram signals, Journal of Biomedical science and Engineering (JBISE), pp. 699–706, (2011).
20. Sundar, N.A., Latha, P.P., Chandra, M.R.: Performance Analysis of Classification Data Mining Techniques over Heart Disease Database, International Journal of Engineering Science and Advanced Technology, ISSN: 2250-3676, Vol-2, Issue-3, pp. 470–478, May–June (2012).
21. Chitra, R., Seenivasagan, V.:Review of Heart Disease prediction system using Data Mining and Hybrid Intelligent techniques, ICTACT Journal on Soft Computing, ISSN:2229-6956, Vol.03, Issue-04, pp. 605–609, July (2013).
22. Aljumah, A.A., Ahamad, M.G., Siddiqui, M.K.: Application of Data Mining: Diabetes Health Care in Young and Old Patients, Journal of King Saud University-Computer and Information Sciences, pp. 127–136, (2013).
23. Soni, J., Ansari, U., Sharma, D.,Soni, S.: Predictive Data Mining for Medical Diagnosis: An Overview of Heart Disease Prediction, International Journal of Computer Applications (0975-8887), Volume 17-No.8, pp. 43–48, March (2011).

A Hybrid Genetic Algorithm for Cell Formation Problems Using Operational Time

Barnali Chaudhuri, R.K. Jana and P.K. Dan

Abstract This paper presents a two-stage approach consisting of a real-coded genetic algorithm and goal programming to obtain improved cell formation. In the first stage, the minimum value of each objective is determined using a single-objective genetic algorithm. In the second stage, goal programming is incorporated and the final objective is constructed as the minimization of sum of deviational variables of corresponding objectives. The proposed technique is implemented as a software toolkit using C Sharp.net programming language. Modified grouping efficiency is used as the performance measure to test the efficiency of the proposed technique. Five problems with different sizes have been considered from the literature to show the potentials of the proposed technique.

Keywords Genetic algorithm · Goal programming · Cellular manufacturing system · Modified grouping efficiency · Cell formation

1 Introduction

Cell formation (CF) has been accepted as one of the most effective techniques to organize manufacturing systems in an efficient manner to minimize the setup cost, flow time, material handling, and overall production cost. The production industries employ the advantages of group technology (GT) [1] to get benefit of both mass

B. Chaudhuri (✉) · R.K. Jana
Indian Institute of Social Welfare & Business Management, Kolkata, India
e-mail: cbarnali@yahoo.com

R.K. Jana
e-mail: rkjana1@gmail.com

P.K. Dan
Indian Institute of Technology Kharagpur, Kharagpur 721302 WB, India
e-mail: pkdan@see.iitkgp.ernet.in

© Springer Nature Singapore Pte Ltd. 2017   127
S.C. Satapathy et al. (eds.), *Proceedings of the 5th International Conference on Frontiers in Intelligent Computing: Theory and Applications*, Advances in Intelligent Systems and Computing 515, DOI 10.1007/978-981-10-3153-3_13

production and cost reduction. CF technique organizes the machines into groups, termed as cells, and each cell is dedicated to produce similar parts. The similar parts are identified based on either by their shape and size or by processing sequences, and collection of such similar parts is identified as part families. Cells should be designed in such a way that a single cell can completely process a single part family.

Several techniques have been proposed to solve CF problems [2, 3]. But majority of these techniques deal with single-objective problems [4], but CF problems are inherently multi-objective in nature. Therefore, techniques have been proposed in the literature [5, 6] to solve multi-objective CF (MOCF) problems. Weighted sum approach is the most common approach which considers all the objectives simultaneously to form the final objective. Weights are assigned to individual objectives according to the decision maker's preference.

To overcome the limitations of these techniques, different hybrid genetic algorithm (GA) approaches have been proposed in the cellular manufacturing system (CMS) domain. Using GA in fuzzy environment [7], an approach was proposed to minimize exceptional elements and maximize number of parts in cells. Group efficacy was used as the performance measure for cell formation. Aiming to minimize cell load variation and exception elements, a weighted sum approach [8] was proposed. Operational time was considered instead of 0–1 machine-part matrix to calculate modified grouping efficiency (MGE). It was shown that the obtained result is better than K-means and C-means clustering.

GP is a multi-criteria decision-making tool used to handle multiple and conflicting objectives with a given target. Unwanted deviations from the target values are minimized in GP. It has been applied to many problem domains. In CMS, GP was first applied with traveling salesman problem and p-median to obtain minimum set up time [9]. A linear programming embedded GA technique [10] was used to minimize the production cost and quality-related cost to show the impact of lot size on product quality.

In this paper, a GP-based hybrid GA approach is introduced to avoid the difficulties of handling the large number of binary variables in CF problem by incorporating the benefits of GA. Since no weight is assigned to individual objectives, all the objectives are equally prioritized. A user-friendly graphical user interface (GUI) is developed to run the proposed technique. Five small and medium size benchmarked problems are considered to illustrate the effectiveness of the proposed technique.

The paper is organized as follows: Sect. 2 provides the details of model development. Section 3 presents the proposed hybrid GA. Section 4 presents the results and analysis. Section 5 presents concluding remarks.

2 Model Development

This section explains the notations, decision variables, constraints, and the final objective function of the CF problem.

2.1 Indices and Notations

c: index for the cell, $c \in (1, 2, …, C)$.
m: index for the machine, $m \in (1, 2 ,…, M)$.
p: index for the parts processed in machines, $p \in (1, 2, …, P)$.

2.2 Decision Variable and Coefficients

x_{mc}: machine and cell incidence matrix.

   $= \,'1'$, if machine m is in cell c.

   $= \,'0'$, otherwise.

w_{mp}: work load on machine m induced by parts.

m_{cp}: matrix of cell and parts which indicates the average cell load,

$$\text{where } m_{cp} = (\sum_{m=1}^{M} x_{mc}*w_{mp})/ \sum_{m=1}^{M} x_{mc}$$

a_{mp}: element of zero-one machine-part incidence matrix.

y_{pc}: part and cell incidence matrix.

   $= \,'1'$, if part p is processed in cell c.

   $= \,'0'$, otherwise.

p_1: positive deviational variable for cell load variation.

p_2: positive deviational variable for exceptional element.

Z_1^*: minimum value of cell load variation.

Z_2^*: minimum value of exceptional element.

F_t: fitness value of t-th chromosome.

Obj_{max}: maximum objective value in a generation.

Obj_t: objective value of t-th chromosome in the same generation.

T_{ptm}: total workload inside the cells.

T_{ptc}: total processing time inside the cell c.

T_{pto}: total processing time outside the diagonal blocks.

N_{vc}: number of voids in cell c.

N_{ec}: number of elements in cell c.

2.3 Objective Functions of the Model

There are two objectives in the model. One is cell load variation and the other is exceptional elements.

Cell load variation: The minimum variation of cell load can be written as [8]

$$\min: Z_1 = \sum_{m=1}^{M} \sum_{c=1}^{C} x_{mc} \sum_{p=1}^{P} (w_{mp} - m_{cp})^2 \tag{1}$$

Exceptional element: Minimization of exceptional element indicates the reduction of intercellular movements and corresponding objective function can be written as [8]

$$\min: Z_2 = 0.5*(\sum_{c=1}^{C} \sum_{p=1}^{P} \sum_{m=1}^{M} |x_{mc} - y_{pc}| *a_{mp}) \tag{2}$$

2.4 Constraints of the Model

The model constraints can be formulated as follows:
Machine constraints:

$$\sum_{c=1}^{C} x_{mc} = 1, \forall_c \tag{3}$$

$$\sum_{m=1}^{M} x_{mc} \geq 1, \forall_m \tag{4}$$

Part constraints:

$$\sum_{c=1}^{C} y_{pc} = 1, \forall_c \tag{5}$$

$$\sum_{p=1}^{P} y_{pc} \geq 1, \forall_p \tag{6}$$

Goal constraints:

$$\sum_{m=1}^{M} \sum_{c=1}^{C} x_{mc} \sum_{p=1}^{P} (w_{mp} - m_{cp})^2 - p_1 = Z_1^*; \tag{7}$$

$$0.5*(\sum_{c=1}^{C} \sum_{p=1}^{P} \sum_{m=1}^{M} |x_{mc} - y_{pc}| *a_{mp}) - p_2 = Z_2^*; \tag{8}$$

$$p_1, p_2 \geq 0. \tag{9}$$

3   The Proposed Hybrid GA

In this study, we have implemented a real-coded GA using C sharp.net platform. A sequence of genes is represented by chromosomes. The length of the chromosome is considered as the summation of number of machine and parts. Next-generation populations are selected on the basis of fitness function. Single-point crossover and inversion mutation operations are used. Due to space limitation, detailed description of the GA is not provided here. However, in the following table the fitness function, crossover probability, mutation probability, size of population, and generation numbers are mentioned.

The final objective is determined as follows:

$$\min: p_1 + p_2, \tag{10}$$

where p_1 represents the amount of variation of cell load and p_2 represents the amount of deviation of number of exceptional elements for overall operations.

Steps of the proposed approach

Initially, the individual objectives are minimized using a single-objective GA considering the machine and part constraints. In the next stage, two positive deviational variables are introduced for corresponding objectives and the final objective is constructed as the minimization of sum of these two deviational variables by satisfying all the constraints using GP.

The steps of the proposed hybrid GA approach are presented as follows:

Step 1:   Formulate the CF problem (1)–(6)
Step 2:   Run the GA using the designed GUI to obtain the minimum value of each objective. At this stage, the fitness function of individual objective functions is considered as mentioned in Table 1
Step 3:   Formulate the deviational variables for GP as (7)–(9)
Step 4:   Construct the final objective as defined in (10)
Step 5:   Run the GA again with the same parameter values
Step 6:   Select the chromosome corresponding to the machine-cell orientation having the minimum objective value as the solution

The computational complexity of the proposed GA can be calculated as O (generation number × population size × chromosome length), where 'chromosome length' is the sum of the number of machines and parts.

Table 1 Different genetic operators used in the proposed technique

Fitness function	Crossover	Mutation	Population size	Stopping criterion
$F_t = (Obj_{max} - Obj_t)$	Single point with probability = 0.25	Inversion mutation with probability = 0.05	20	Number of generation = 100

4 Results and Analysis

Five benchmarked problems with different sizes are considered and their corresponding MGE is compared (Table 2) with the results of [8]. Figure 1 shows convergence curve of problem size (7 × 11) for an instance. It has been observed that as the number of iteration increases, the objective value decreases. Figure 2 represents graphical comparison of the improvement of the proposed technique over the others. The values of the deviational variables are shown in Table 2. It shows that as the total deviation from the target value increases the MGE decreases. Figure 3 represents a screenshot of the software toolkit for problem size (5 × 8) as an instance.

Table 2 Comparison of MGE (%) for different techniques and deviation variables for GP–GA

Reference number	Size	K-means clustering	C-link clustering	GA	GP–GA	p_1	p_2
[14]	5 × 8	100	100	100	100	0.56	0
[15]	7 × 11	63.42	63.42	63.42	73.88	1.33	6
[16]	8 × 20	59.74	59.74	59.74	89.13	0.36	2
[17]	8 × 20	72.11	72.11	72.11	100	1.09	0
[18]	10 × 15	72.19	72.19	72.19	84.58	0.49	2

Fig. 1 Convergence curve for problem size (7 × 11)

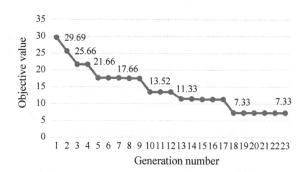

Fig. 2 Graphical comparison of MGE

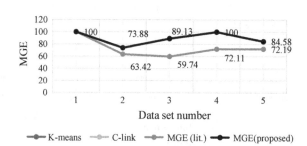

Fig. 3 Screenshot of the software toolkit for problem size (5 × 8)

Here, we consider MGE to overcome the limitations of group efficiency [11], efficacy [12], and generalized grouping efficacy [13]. The MGE is calculated using the following formula:

$$\text{MGE} = \frac{T_{ptm}}{\text{T}_{\text{pto}} + \sum_{c=1}^{C} T_{\text{ptc}} + \sum_{c=1}^{C} T_{\text{ptc}} * W_{\text{v}}}$$

where $W_{\text{v}} = N_{\text{vc}} / N_{\text{ec}}$.

MGE does not handle all the operations equally like grouping efficiency.

5 Conclusions

This study attempts to improve the cell formation using the GA in combination with the GP approach. The proposed technique is tested on small and medium size problems taken from the literature. Minimization of sum of deviations of cell load variation and exceptional elements are considered as objectives of this study. MGE is considered as the performance measure and the proposed technique performs better than the other techniques as mentioned in the results and analysis section. Since reduction of production cost is one of the primary objectives of cell formation in manufacturing systems, the proposed technique generates the better results compared to the other techniques as mentioned in this study. The proposed technique can further be tested for other problems with large size considering different objectives and GA parameter values. Different group performance measures, including grouping efficiency, may be compared using this proposed technique. The industrial data set can also be used for the purpose of testing.

References

1. Burbidge, J. L.: The introduction of group technology. Heinemann Press, London (1975).
2. Dimopoulos, C., Zalzala, A. M.S.: Evolutionary Computation Approaches to Cell Optimization. Adaptive Computing in Design and Manufacture, Parmee, I. C. (Ed.), pp. 69–83. Springer-Verlag, London (1998).
3. Mak, K. L., Wong, Y.S.: Genetic design of cellular manufacturing systems. Human Factors and Ergonomics in Manufacturing, 10(2), 177–192 (2000).
4. Shanker, R., Vrat, P.: Post design modeling for cellular manufacturing system with cost uncertainty. International Journal of Production Economics, 55, 97–109 (1998).
5. Chi, S.C., Yan, M.C. : A fuzzy genetic algorithm for high-tech cellular manufacturing system design. IEEE Annual Meeting of the Fuzzy Information, 2, 907–912 (2004).
6. Gupta, Y., Gupta, M., Kumar, A., Sundaram, C.: A genetic algorithm-based approach to cell composition and layout design problems. International Journal of Production Research, 34(2), 447–482 (1996).
7. Pai, P.F., Chang, P.T., Lee, S.H.: Part-machine family formation using genetic algorithms in a fuzzy environment. International Journal Advanced Manufacturing Technology, 25(11–12), 1175–1179 (2005).
8. Mahapatra, S.S., Pandian, R.S.: Genetic cell formation using ratio level data in cellular manufacturing systems. The International Journal of Advanced Manufacturing Technology, 38(5), 630–640 (2008).
9. Shafer, S.M., Rogers, D.F.: A goal programming approach to the cell formation problem. Journal of Operations Management, 10(1), 28–43 (1991).
10. Defersha, F.M., Chen, M.: A linear programming embedded genetic algorithm for an integrated cell formation and lot sizing considering product quality. European Journal of Operational Research, 187, 46–69 (2008).
11. Chandrasekharan, M.P., Rajagopalan, R.: An ideal seed non-hierarchical clustering algorithm for cellular manufacturing. International Journal of Production Research, 24(2), 451–464 (1986a).
12. Kumar, C.S., Chandrasekharan, M.P.: Grouping Efficacy: A quantitative criterion for goodness of block diagonal forms of binary matrices in group technology. International Journal of Production Research, 28, 233–243 (1990).
13. Zolfaghari, S., Liang, M.: A new genetic algorithm for the machine/part grouping problem involving processing times and lot sizes. Computers and Industrial Engineering, 45, 713–731 (2003).
14. Venugopal, V., Narendran, T.T.: Cell formation in manufacturing systems through simulated annealing. European Journal of Operations Research, 63, 409–422 (1992a).
15. Venugopal, V., Narendran, T.T.: A Genetic algorithm approach to the machine component and grouping problem with multiple objectives. Computers and Industrial Engineering, 224, 469–480 (1992b).
16. Venugopal, V., Narendran, T.T.: Neural network model for design retrieval in manufacturing systems. Computers in Industry, 20, 11–23(1992c).
17. Srinivasan, G., Narendran, T.T.: GRAFICS: a non-hierarchical clustering algorithm for group technology. International Journal of Production Research, 29 (3), 463–478 (1991).
18. Kusiak, A.: The generalized group technology concept. International Journal of Production Research, 25(4), 561–569 (1987).

Efficient and Parallel Framework for Analyzing the Sentiment

Ankur Sharma and Gopal Krishna Nayak

Abstract With the advent of Web 2.0, user-generated content is led to an explosion of data on the Internet. Several platforms such as social networking, microblogging, and picture sharing exist that allow users to express their views on almost any topic. The user views express their emotions and sentiments on products, services, any action by governments, etc. Sentiment analysis allows quantifying popular mood on any product, service or an idea. Twitter is popular microblogging platform, which permits users to express their views in a very concise manner. In this paper, a new framework is crafted which carried out the entire chain of tasks starting with extraction of tweets to presenting the results in multiple formats using an ETL (Extract, Transform, and Load) big data tool called Talend. The framework includes a technique to quantify sentiment in a Twitter stream by normalizing the text and judge the polarity of textual data as positive, negative, or neutral. The technique addresses peculiarities of Twitter communication to enhance accuracy. The technique gives an accuracy of above 84% on standard datasets.

Keywords openNLP (NER tagger) · Sentiment analysis · Sentiwordnet · Talend · Twitter

1 Introduction

In 2004, O'Reilley [1] espoused the advent of Web 2.0, which is characterized by user-generated content, scalability, and interoperability. The most important characteristic of Web 2.0 is that users generated content. This feature allows creating content with ease, modify existing content, comment on existing content, spread existing content, etc. Blog sites, wiki sites, comment sections on regular pages, and

A. Sharma (✉) · G.K. Nayak
International Institute of Information Technology, Bhubaneswar, Odisha, India
e-mail: ankursharma6592@gmail.com

G.K. Nayak
e-mail: gopal@iiit-bh.ac.in

© Springer Nature Singapore Pte Ltd. 2017   135
S.C. Satapathy et al. (eds.), *Proceedings of the 5th International Conference on Frontiers in Intelligent Computing: Theory and Applications*, Advances in Intelligent Systems and Computing 515, DOI 10.1007/978-981-10-3153-3_14

retweeting/sharing are examples of the above-mentioned feature. Social media have provided an important platform for user-created content. Social media such as Facebook, Twitter, and Google plus are extremely popular among the user across the world. The social media are characterized by the following: a. participation, b. openness, c. conversation, d. community, and e. connectedness.

One very popular social media service is Twitterm, which is a microblogging service. Twitter users can post the text of 140 characters called tweets to share their current thoughts, emotions, sentiments, etc. Around 320 million MAU people are using Twitter, 1.3 billion estimated total number of registered users and around 100 million daily active users [2], which are increasing every year. User-generated content is growing rapidly. According to a report, every second around 7,059 Tweets are posted, 481 Instagram photos are uploaded, 1,421 Tumblr are posted, 1,974 Skype calls made, 31,895 GB of data moves in Internet traffic, 113,863 YouTube videos viewed, and 2,453,467 Emails sent on the internet [3]. The large amounts of data created through social media offer opportunities for the study of sentiments. For example, on October 3, 2012, in the University of Denver in Denver, the first presidential 90 min debate between President Barack Obama and former Massachusetts Governor Mitt Romney on the topic of domestic issues triggered the 10 million tweets within few hours [4] presenting an opportunity to perform sentiment analysis on the debate.

The objective sentiment analysis is to identify the emotions, attitude, or opinion of an author from the textual information penned by him. The measurement of sentiment is an important feedback for companies, institutions, political parties, and others such organizations regarding opinion people carry regarding their products, services, images, and actions. While traditionally sentiments were measured through a survey using a Likert scale, today social media offer an alternate to measure sentiments in an automated manner from the live data available from the media. In this paper, we propose a framework to measure sentiments from social media and other sources to improve the accuracy of measurement, decrease neutrality and improve the speed of measurement.

We have used an open source tool Talend Open Studio for Big Data [5] that provides big data, Cloud, master data management, data quality, data integration, data management, and various other services that provide flexibility so as to solve many integration challenges in an easier manner. It is also an ETL tool capable of extraction, transformation, and loading, while Talend provides a GUI (Graphical user interface) in which we can drag, drop, and configure the components which will automatically generate the code, it also allows users to write custom code to extend and modify functionality.

The remaining part of the paper is crafted as follows. Section 2 deals with the review of literature on sentiment analysis; in Sect. 3 details of the framework is presented. Section 4 explains the preprocessing steps of the framework. Sections 5 and 6 show the algorithm of the framework and discusses the experimental results with comparisons to show the effectiveness of our scheme. Section 7 presents conclusion and gives directions for future work.

2 Literature Review

Sentiment analysis is the area of study that evaluates whether a phrase of either writer or speaker is positive, negative, or neutral. It is also known as opinion mining, deriving the opinions, sentiments, appraisals, emotions, and attitude of a writer or speaker.

Several approaches have been applied in sentiment analysis for the improvement of the results. Turney and Bo Pang were the pioneers in this field. Turney [6] uses the simple unsupervised learning algorithm for classifying the reviews. About 410 reviews are categorized into four domains Automobiles, Banks, Movies, and Travel Destinations and achieve the accuracy 84%, 80%, 65.83%, and 70.53%, respectively, and the average accuracy of all four domains is 74.39%. Pang et al. [7] tested on the movies review data using the maximum entropy classification, Naive Bayes, and support vector machines and achieved the accuracy on unigrams feature of 80.4%, 81%, and 82.9%, respectively. They report that support vector machines have the best performance as compared to the other machine learning algorithms.

Alec Go et al. [9] performed sentiment analysis on twitter dataset of the total of 1,600,000 tweets using three machine learning algorithm Naive Bayes, maximum entropy, and support vector machines and achieved the accuracy above 80%. Bifet and Frank [8] used the sliding window Kappa statistics on Twitter time-changing data streams. Using different machine learning algorithm stochastic gradient descent, multinomial naive Bayes, and Hoeffding tree on two different data stream statistics analyze the result. Agarwal et al. [10] examined the sentiment analysis using the POS specific prior polarity features and uses two new preprocessing resources, i.e., emoticon dictionary and acronym dictionary on 11,875 manually annotated Twitter data from a commercial source and got the accuracy of 73.4% with F-measure for positive and negative sentiment is 71.13% and 71.50%, respectively, on unigrams.

Mudinas et al. [11] conducted experiment on software reviews and movie reviews using a pSenti (hybrid) approach, which is the combination of lexicon-based and learning-based approaches and achieved the accuracy more than 78% in software reviews and 82.3% in movies reviews approaches. Soo-Guan Khoo et al. [12] applied the appraisal theory on some of the political news articles and try to identify a various aspect of sentiments such as attitude type and emotion in sentiment expression, bias of the appraisers and the author.

Mane et al. [13] performed sentiment analysis on Twitter using Hadoop on 1465 tweets and achieved the accuracy 72.27%. Hopper and Uriyo [14] applied the time-to-next-complaint analytical methods to review the feedback of the patient comments, so that the hospital managers can view the sentiments and able to derive the valuable information for satisfying the patients. Hridoy et al. [15] design the framework using SNLP, NamSor Gender API, and SentiWordNet dictionary. This framework is tested on the iPhone 6 tweets, which are extracted using the Twitter API and find the sentiment of the product and also conclude the analysis of male- female specific feature.

The article reviewed has focused on deploying algorithms to identify the sentiment. The underlying data being derived from social media is changing constantly. Hence, there is a need to develop a framework that can work for all data format such as JSON, CSV, and flat text files that are derived from a database, big data systems such as Hadoop or a social media live stream, which can perform the end-to-end tasks from receiving the data stream to publishing the sentiments. This article proposes the framework using open source tools to achieve this. The article also proposes an improved algorithm to obtain a sentiment score.

3   Details of the Framework for Analyzing Sentiments

Our parallel framework is implemented in Talend Open Studio for Big Data version 6.1.1 in such a way that it will give you the end result in Excel sheet and graph format on a single click within a minimum time span. The flowchart for the same can be found in Fig. 1.

- Step-1: Receive the live stream of user views from a source like Twitter using Twitter API [16] or an RDBMS or flat text file.
- Step-2: Preprocess the user views to remove noise, hashtags, etc., and expand the acronyms and smileys.
- Step-3: Sentiment classification using SentiWordNet [17] into positive, negative, or neutral sentiments.
- Step-4: For neutral user view, reclassification by breaking into tokens and using Opinion Lexicon dictionary [18, 19] and AFFIN polarity [20] dictionary to determine the polarity.
  Steps 2–4 are explained in subsequent sections.
- Step-5: Presents the data in the tabular and visual form.

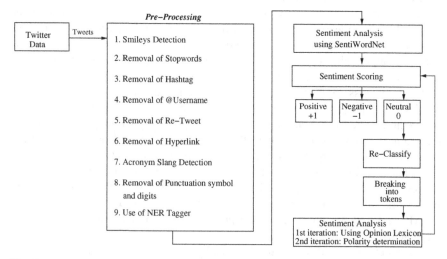

Fig. 1 Flow chart of our framework

Table 1 Step-by-step process of preprocessing

Tweet	Theo Walcott is still shit, **D:** @rafa u watch him #tag https://link
Smiley	Theo Walcott is still shit, disgust @rafa **u** watch him #tag https://link
Acronym (slang)	Theo Walcott is still shit, disgust @rafa you watch him **#tag** https://link
Hashtag	Theo Walcott is still shit, disgust **@rafa** you watch him https://link
@ username and RT	Theo Walcott is still shit, disgust you watch him **https://link**
Hyperlink	Theo Walcott **is** still shit, disgust **you** watch **him**
Stop words	Theo Walcott still shit, disgust watch
Punctuation symbol and digits	**Theo Walcott** still shit disgust watch
NER tagger	still shit disgust watch

4 Preprocessing of Tweets

The user views expressed through social media like Twitter and in other media are quite noisy and colloquial. Hence, there is a need to process user views before analyzing the sentiment. The following steps are used to preprocess the data as shown as bold in the Table 1.

(1) Smiley Detection: We prepare the emoticon dictionary by labeling the 170 smileys with their emoticons name. For example, :) is labeled as a smile.

(2) Hashtags, @Username, Retweets, Hyperlink, Stop Words, Punctuation symbol, and digits: Since these words do not have any sentiment value, they are filtered.

(3) Acronym (Slang) Detection: Acronyms are replaced with full abbreviations. For example, rofl is translated to rolling on the floor laughing.

(4) NER Tagger: Stanford Named Entity Recognizer [21] labels the text as a person, organization, date, location, money, money, percent, and time. Since these tags are not used for sentiment analysis, they are filtered.

5 Algorithm for Classification of Sentiments

The proposed algorithm uses SentiWordNet to classify a user view into positive, negative, or neutral sentiment. Neutral views are split into tokens (words) and the sentiment on each word is determined using the opinion lexicon. The sum of sentiments expressed in each relevant word is the sentiment of the user view. User views that still carry neutral sentiment, AFFIN polarity dictionary is used to determine the polarity for each word and aggregated polarity determines the sentiment of user view. The detailed algorithm is given below in Algorithm 1.

Algorithm 1: Algorithm for Analyzing Sentiment

Let,

S: Sentiment

T: Tweet

Score of SWN: Function of SentiWordNet which calculates the sentiment score for eachcomplete tweet.

T_i : The word number i of the tweet.

S(T): Sentiment score of the tweet T.

$S(T_i)$:Sentimentscore of word number i of the tweet.

$D(P_j)$: Data list of positive word number j of Opinion lexicon word list.

$D(N_k)$: Data list of negative word number k of Opinion lexicon word list.

$P(W_i)$: AFINN polarity word number i.

$P(S_i)$: AFINN polarity score of P (W_i).

1. **foreach *T* do**
2. Pre-process T
3.   S(T) = Score of SWN(T) /* *For neutral tweets, re classify using Opinion Lexicon.* */
4. **if** (*S (T)* == *0*) **then**
5.   Wordcount = 0
6. **foreach**T_i *(tweet)***do**
7. wordcount ++
8. **if** (*T_i in D(P_j)*) **then**/* *Checking positive lexicon.* */
9. $S(T_i) = S(T_i) + 1$
10. **end**
11. **if** (*T_i in D(N_k)*) **then**/* *Checking negative lexicon.* */
12. $S(T_i) = S(T_i) - 1$
13. **end**
14. **end**
    /* *Aggregating the sentiment score for each Tweet.* */
15. Sum = 0
16. **for** (*i = 0 to wordcount*) **do**
17. Sum = Sum + $S(T_i)$
18. **end**
19. **end**
    /* *For neutral tweets after using Opinion Lexicon, reclassify using AFFIN Polarity Dictionary* */
20. **if** (*S (T)* == *0*) **then**
21.   wordcount= 0
22. **foreach** T_i *(tweet)* **do**
23. wordcount ++
24. **if** (*T_i inP (W_i)*) **then**
25. $S(T_i) = P(V_i)$)

26. **end**
27. **end**
    /* Aggregating the sentiment score for each Tweet. */
28.   Sum = 0
29. **for** (*i* = *0 to wordcount*) **do**
30.   Sum = Sum + S(T$_i$)
31. **end**
32. **end**
33. **end**

It may be noted that the complexity of the algorithm proposed is O(n log n).

6 Analysis of the Results

The framework and the algorithm proposed were implemented using Talend Open Studio for Big Data version 6.1.1. To test the accuracy and efficiency of the framework, four data sets were used. Three of them are standard data sets used by many articles to test accuracy [13], STS-Gold Tweet [22], STS-Test [22]. The results of our proposed algorithm were compared with those reported in the literature to compare the accuracy and efficiency of the algorithm. The fourth dataset is a live twitter feed for some issues being debated at present. The framework was tested on a commodity hardware using Core i5 processor, 8 GB RAM, and running Windows 10 64bit OS.

The accuracy of our framework is measured by the percentage identification of the correct sentiment in the standard datasets. The efficiency is measured by the total time required to process the entire workflow starting with the acquisition of the data and ending with the presentation of the results.

The result of our framework after running on the above datasets can be found in Table 2.

The average accuracy of the dataset 1, 2, and 3 is 84.57% and it is efficiently better than the evaluated accuracy as 72.27% in [13]. The efficiency is partly derived from the fact that the algorithm proposed is a parallel one using this framework performed on the ETL tool Talend.

Table 2 Dataset result

Dataset	Sentiment	Dataset count	Correctly identified	Accuracy
1.	Positive	732	678	
	Negative	730	566	
	Neutral	4	1	
	Total	1466	1245	84.92%
2.	Positive	632	594	
	Negative	1402	1120	
	Neutral	0	0	
	Total	2034	1245	84.27%
3.	Positive	182	168	
	Negative	177	142	
	Neutral	139	111	
	Total	498	421	84.54%

Table 3 The result of the different Twitter hashtag

Sentiment	Count for #jnu	Count for #jatreservation
Positive	865	808
Negative	1111	1100
Neutral	24	92
Total	2000	2000

Average time required to process a user view (tweet) is 0.98 s.

- Time taken for Preprocessing Steps from 1 to 3 per user view: 0.03 s.
- Time taken for Preprocessing Step 4 per user view: 0.8 s.
- Time taken for sentiment identification per user view: 0.15 s.

It may be noted that the NER Tagger step in preprocessing takes about 0.8 s per user view, which is significantly higher than the time taken by other steps.

The result of our framework on a data stream obtained from Twitter using hashtags #JNU and #JATRESERVATION tested on date 25 February, 2016 can be found in Table 3. One of the sample bar chart is shown below in Fig. 2.

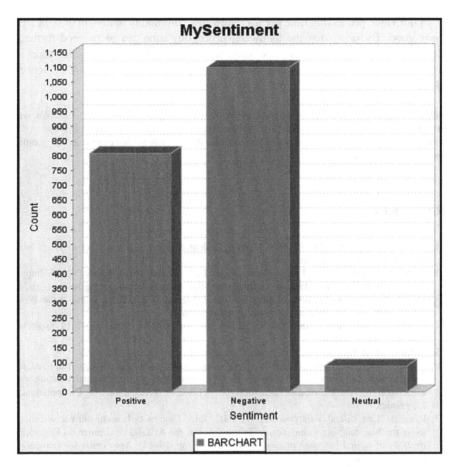

Fig. 2 An example of a Bar chart on #jatreservation

7 Conclusion and Future Work

In this paper, a framework is proposed to measure the sentiments in data sets obtained from social media, websites, or common databases. We also propose an algorithm to identify sentiment expressed by a user on a particular conversation. The framework is implemented using an open source tool called Talend open studio for big data. The accuracy obtained is a step improvement in the accuracy reported in the literature. However, there is still a significant scope for improving the accuracy. Our approach uses unigrams to assigning sentiment polarity. N-gram approached perhaps will provide better accuracy to identifying sentiments.

In our view, processing time per user view in a commodity hardware of 0.98 s is fairly good. By improving the hardware, processing time can be reduced further. However, on substep of the preprocessing namely named entity recognizer takes a significant amount to process. Optimizing this step will lead to significant reduction of the processing time.

Acknowledgements This research was also supported by Tiger Analytics Pvt. Ltd. We are thankful to them for providing insight and expertise that greatly assisted our research.We are also thankful to Prachi Khokhar for her assistance in editing the research and her comments that greatly improved the manuscript.

References

1. O'Reilly, T. and Battelle, J., 2004. Opening welcome: State of the internet industry. San Francisco, California, October, 5.
2. C. Smith, "170 Amazing Twitter Statistics and Facts: Social Media Article," 2015. http://expandedramblings.com/index.php/march-2013-by-the-numbers-a-few-amazing-twitter-stats/.
3. Google Search Statistics Internet Live Stats in 1 Second: The Official World Wide Web Anniversary. http://www.internetlivestats.com/one-second/.
4. A. Sharp, Dispatch from the Denver Debate, 2012. https://blog.twitter.com/2012/dispatch-from-the-denver-debate.
5. Talend: About Talend. https://www.talend.com/about-us.
6. Turney, P.D., 2002, July. Thumbs up or thumbs down?: semantic orientation applied to unsupervised classification of reviews. In Proceedings of the 40th annual meeting on association for computational linguistics (pp. 417–424). Association for Computational Linguistics.
7. Pang, B., Lee, L. and Vaithyanathan, S., 2002, July. Thumbs up?: sentiment classification using machine learning techniques. In Proceedings of the ACL-02 conference on Empirical methods in natural language processing-Volume 10 (pp. 79–86). Association for Computational Linguistics.
8. Bifet, A. and Frank, E., 2010, October. Sentiment knowledge discovery in twitter streaming data. In Discovery Science (pp. 1–15). Springer Berlin Heidelberg.
9. Go, A., Bhayani, R. and Huang, L., 2009. Twitter sentiment classification using distant supervision. CS224 N Project Report, Stanford, 1, p. 12.
10. Agarwal, A., Xie, B., Vovsha, I., Rambow, O. and Passonneau, R., 2011, June. Sentiment analysis of twitter data. In Proceedings of the workshop on languages in social media (pp. 30–38). Association for Computational Linguistics.
11. Mudinas, A., Zhang, D. and Levene, M., 2012, August. Combining lexicon and learning based approaches for concept-level sentiment analysis. In Proceedings of the First International Workshop on Issues of Sentiment Discovery and Opinion Mining (p. 5). ACM.
12. Soo-Guan Khoo, C., Nourbakhsh, A. and Na, J.C., 2012. Sentiment analysis of online news text: a case study of appraisal theory. Online Information Review, 36(6), pp. 858–878.
13. Mane, S.B., Sawant, Y., Kazi, S. and Shinde, V., 2014. Real Time Sentiment Analysis of Twitter Data Using Hadoop. IJCSIT) International Journal of Computer Science and Information Technologies, 5(3), pp. 3098–3100.
14. Hopper, A.M. and Uriyo, M., 2015. Using sentiment analysis to review patient satisfaction data located on the internet. Journal of health organization and management, 29(2), pp. 221–233.

15. Hridoy, S.A.A., Ekram, M.T., Islam, M.S., Ahmed, F. and Rahman, R.M., 2015. Localized twitter opinion mining using sentiment analysis. Decision Analytics, 2(1), pp. 1–19.
16. Twitter Application Management. https://apps.twitter.com/.
17. Baccianella, S., Esuli, A. and Sebastiani, F., 2010, May. SentiWordNet 3.0: An Enhanced Lexical Resource for Sentiment Analysis and Opinion Mining. In LREC (Vol. 10, pp. 2200–2204).
18. Hu, M. and Liu, B., 2004, August. Mining and summarizing customer reviews. In Proceedings of the tenth ACM SIGKDD international conference on Knowledge discovery and data mining (pp. 168–177). ACM.
19. Liu, B., Hu, M. and Cheng, J., 2005, May. Opinion observer: analyzing and comparing opinions on the web. In Proceedings of the 14th international conference on World Wide Web (pp. 342–351). ACM.
20. Hansen, L.K., Arvidsson, A., Nielsen, F.Å., Colleoni, E. and Etter, M., 2011. Good friends, bad news-affect and virality in twitter. In Future information technology (pp. 34–43). Springer Berlin Heidelberg.
21. Finkel, J.R., Grenager, T. and Manning, C., 2005, June. Incorporating non-local information into information extraction systems by gibbs sampling. In Proceedings of the 43rd Annual Meeting on Association for Computational Linguistics (pp. 363–370). Association for Computational Linguistics.
22. Saif, H., Fernandez, M., He, Y. and Alani, H., 2013. Evaluation datasets for Twitter sentiment analysis: a survey and a new dataset, the STS-Gold.

Content-Aware Video Retargeting by Seam Carving

Shrinivas D. Desai, Mahalaxmi Bhille and Namrata D. Hiremath

Abstract Due to the rapid growth of digital gadgets with various screen sizes, resolutions and hardware processing capabilities, robust video retargeting is of increasing relevance. An efficient retargeting algorithm should not only retain semantic content, but also maintain spatiotemporal resolution of video data. In this paper, the effective seam carving technique for content-aware video retargeting is discussed. Retargeting video is of immense importance as it is frequently played on several gadgets such as television, mobile, tablet, and notebook. The proposed method considers each video frame as an independent image entity and tries to resize it. Our main contribution is a formulation of seam carving using graph cut method. Convention cut techniques fail to defend a meaningful seam. Single monotonic well connected by pixel to pixel is most desirable property in seam carving process. The traditional seam carving method is designed to work based on the minimum energy concept, while ignoring the energy that has been introduced by the operator. To address this issue, we propose a new design criterion in which least amount of energy is introduced in retargeted video.

Keywords Seam carving · Video resizing · Pixel · Output devices

S.D. Desai (✉) · M. Bhille · N.D. Hiremath
Department of Information Science and Engineering,
B.V.B. College of Engineering and Technology, Hubballi, India
e-mail: sd_desai@bvb.edu

M. Bhille
e-mail: mahalaxmi.bhille@bvb.edu

N.D. Hiremath
e-mail: namrata_hiremath@bvb.edu

© Springer Nature Singapore Pte Ltd. 2017
S.C. Satapathy et al. (eds.), *Proceedings of the 5th International Conference on Frontiers in Intelligent Computing: Theory and Applications*, Advances in Intelligent Systems and Computing 515, DOI 10.1007/978-981-10-3153-3_15

1 Introduction

Transferring the data plays a very important role in the modern era. Huge amount of data is transferred every day which needs compression techniques so that the data size should not tamper the data transfer. Here the present techniques are discussed in brief. The existing system mainly focuses on cropping method, which is inappropriate if there is a mismatch in the resolution of prior and after retargeting of video. Here there is a probability of loosing signal of interest. Advanced cropping methods could be considered, such as pan-and-scan methods, which usually require user intervention to identify the region of interest in the video frame, especially to improve the performance [1]. These approaches are of limited capability when the region of interest does not fall within cropping area. These approaches are best suited for specific applications.

Scaling is a well-known method for video retargeting [2]. A better video retargeting is observed when linear or higher order interpolation methods are adopted. However, distortions are observed when aspect ratio gets mismatched. Sometimes scaling results in an unacceptable aspect ratio presenting distorted video content. The proposed method will allow the user to enjoy resized video with spatiotemporal coherence of video sequences, so that the video is compatible to be displayed on any electronic gadgets. It also overcomes the disadvantages of the present system, i.e., Cropping method.

Recently a method combining improved seam carving and frequency domain analysis is proposed. The study is limited to a specific video [3]. Another method based on flow-guided seam carving is proposed [4]. But author proposes combining this method with genetic algorithm could be better. Another method based on modeling for seam carving is proposed [5] but is applicable for security system.

2 Methodology

By finding the path of minimum energy cost from one end of the image to another seam is computed. In this process for computing seam, each pixel in the row is processed for minimum cost by considering upper three neighboring pixels.

2.1 Proposed System

The proposed system shown in Fig. 1 gives the brief details about the method being developed.

It explains the detailed design model and to describe the software structure, composition, interfaces, and data necessary for the implementation phase of video resizing system.

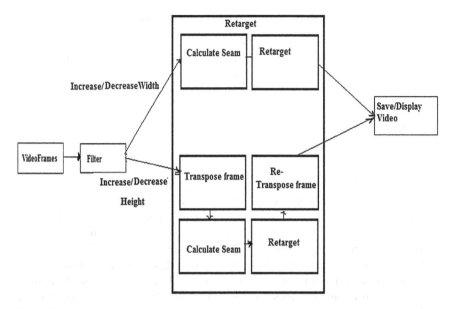

Fig. 1 Proposed system

Figure 2 explains about the Architecture of the system, including description of the various modules such as discontinuous _seam carving, transpose, re-transpose, filter and retarget of the system, and the details of the rationale behind adopting one of the various architectural alternatives. It explains how the system deals with data structures used, the user interface, the reports generated.

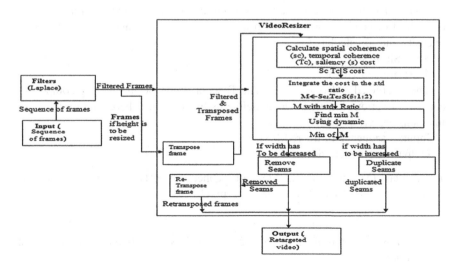

Fig. 2 System architecture

Following gives the procedure followed for the proposed system.

1. Input Video: Select a video of any format, e.g., .mp4, .3gp, .flv etc., as an input.
2. Filters: Some unwanted noise present in the image should be removed for better clarity of video. In order to achieve this, we use three kinds of filters and provide an option for the user to choose the required among them.
3. Discontinuous seam carving

   Sc: Spatial coherence.
   Tc: Temporal coherence.
   S: Saliency cost.
   Temporal coherence

Replicating the seams in the frames which vary smoothly over time is the sufficient condition to achieve temporal coherence, but through experiment we demonstrate that the similarity between retargeted video frame and the one obtained by applying optimal temporal coherent system is same.

A seam from the video frame $(M \times N)$ is removed, so that the resulting $(M - 1) \times N$ frame Ri would be similar to the most temporally coherent one Rc. Here Rc is obtained by reusing the previous seam $Si - 1$ and applying it to the current frame Fi. The meaning of the terms is as follows

$M \times N$: Dimension of the current frame.

$M - 1 \times N$: Dimension of the current frame after a seam is removed from it.

Ri: New frame obtained after seam computation and removal.

Rc: New frame obtained by reusing of previous frame $Si - 1$ and applying it to current frame Fi to find new seam Si and then eliminating the seam. Now to calculate Tc $x - 1$ m $- 1Tc(x, y) = \Sigma \|Fik, y - Rc k, y\|2 + \Sigma\|Fik, y - Rc k - 1, y\|2 K = 0 k = x + 1$

Spatial coherence

To estimate the extent of spatial error introduction after removal of a seam, the concept of piecewise seam is used. It is measured by the change in gradient not he intensity and to calculate Sc

Sc = Sh + Sv

4. Retargeted video:

The retargeted video is saved. The Discontinuous Seam Carving component consists of various functions which define its functionality, the algorithm for these functions are given below.

   Input: a sequence of filtered frames
   Output: Retargeted video

1 Calculate (M) the spatial and temporal coherence costs (Sc and Tc) as well as the saliency (S) cost of removing that pixel.

2 With a weight ratio of Sc: Tc: S of 5:1:2, combine linearly the above three cost measures.

3 Calculate the minimum cost seams (N) w.r.t. M for that frame using dynamic programming concept

4 Remove N seams

5 Resize height by repeating above procedure on Transposed frame

6 Display retargeted video.

3 Results and Discussions

We demonstrate our results for video retargeting in two ways, reducing the size based on discontinuous seam carving algorithm and increase the size based on interpolation method which applies to all the extracted frames. We have also used Laplacian filter, which is used to remove the noise present in the frames in order to get a good quality video as output. In the following: Table 3.1 the original video of resolution 640 × 360 is considered. It is retargeted to different resolutions by proposed method and standard available method. After retargeting, their video quality is compared with respect to three parameters such as PSNR, MSE, and VQM. For comparative study, we used open-source MSU video quality measurement tool. Table 1 indicates the values of abovesaid parameter over a range of frames. The datasets are taken from Weizmann action recognition dataset, All-free-download.com, Eesy.inc

Figure 3a, b, and c are graphs for images with 176 × 144 for the different formats like MSE1, PSNR1,and VQM1 respectively. Figure 4a, b, and c are graphs for images with resolution 320 × 240 for the different formats like MSE2, PSNR2, and VQM2 respectively (Table 2).

Figure 5a, b, and c are graphs for images with resolution 480 × 360 for the different formats such as MSE3, PSNR3, and VQM3 respectively. Figure 6a, b, and c are graphs for images with resolution 320 × 240 for the different formats such as MSE4, PSNR4, and VQM4 respectively (Table 3).

Table 1 Very slow motion video

Duration	Input video resolution	Comparison parameter	Retargeted video resolution (176 × 144) and its average values	Retargeted video resolution (320 × 240) and its average values	Retargeted video resolution (480 × 360) and its average values
40 s	640 × 360	MSE (R-RGB)	14003.19043	14220.4050	14977.18262
		PSNR (R-RGB)	6.44724	6.38976	6.37646
		VQM (Y-YUV)	21.27193	20.08118	19.0792

Fig. 3 176 × 144 images

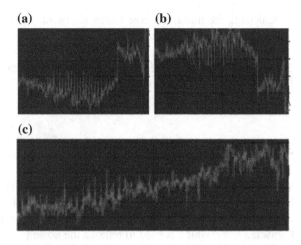

Fig. 4 320 × 240 images

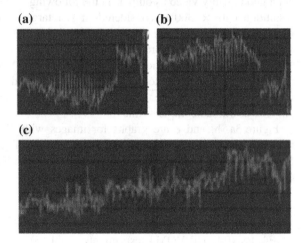

Table 2 Slow motion

Duration	Input video resolution	Comparison parameter	Retargeted video resolution (320 × 240) and its average values	Retargeted video resolution (352 × 288) and its average values	Retargeted video resolution (480 × 360) and its average values
79 s	176 × 144	MSE (R-RGB)	9197.97363	9182.32520	9176.35156
		PSNR (R-RGB)	8.49387	8.50126	8.50695
		VQM (Y-YUV)	15.41782	15.35436	15.39273

Fig. 5 480 × 360 images

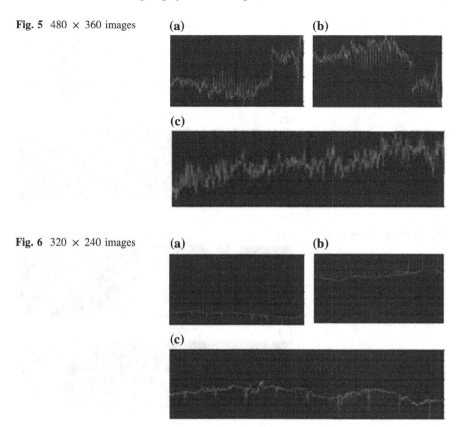

(a)

(b)

(c)

Fig. 6 320 × 240 images

(a)

(b)

(c)

Table 3 Slow motion

Duration	Input video resolution	Comparison parameter	Retargeted video resolution (176 × 144) and its average values	Retargeted video resolution (320 × 240) and its average values	Retargeted video resolution (480 × 360) and its average values
40 s	640 × 360	MSE (R-RGB)	7866.71533	4847.50732	2549.26294
		PSNR (R-RGB)	9.17287	11.27562	14.06662
		VQM (Y-YUV)	15.27699	9.89474	7.75417

Figure 7a, b, and c are graphs for images with resolution 352 × 288 for the different formats such as MSE5, PSNR5, and VQM5 respectively. Figure 8a, b, and c are graphs for images with resolution 480 × 360 for the different formats such as MSE6, PSNR6 and VQM6 respectively.

Fig. 7 352 × 288 images

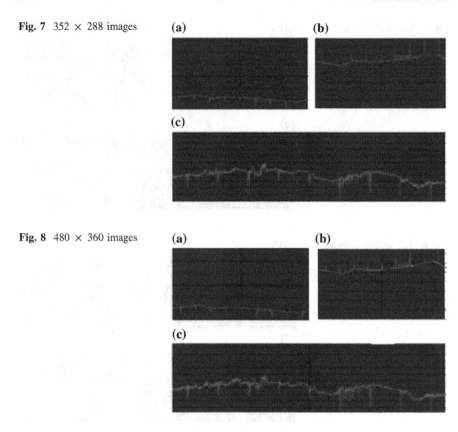

Fig. 8 480 × 360 images

3.1 Observations from the Comparative Study

From the above comparative study we observed that the seam carving algorithm works well with the high motion videos. Value of PSNR for slow motion video is less compared to high motion video. Probable reason for less value of PSNR in case of slow motion videos would be because of slow or rare movement of the objects in the video leading to poor calculation of seams.

In case of MSE, it is observed that MSE (Mean Squared Graph) for slow motion videos MSE value is high as compared to the fast motion videos and one probable reason for this could be the seam calculation for slow motion videos are less, i.e., discontinuous seam carving algorithm works well with high motion videos than slow motion videos.

In case of VQM (Video Quality Measure), it depicts that slow motion videos have higher value of quality measure as compared to high motion videos, one probable reason for which could be in case of high motion videos there will be frequent changes in movement of objects so the number of seams generated are more in case of high motion videos.

Fig. 9 Retargeted video with
various aspect ratios

1920X1080 (16:9)

320X240 (4:3)   1024X576 (16:9)

720X480 (4:3)   720X576 (4:3)

Table 4 Comparative study with the existing systems

Author	Proposed methodology	Dataset	Evaluation method	Result
Benjamin Guthier et al. [6]	Disparity map and temporal consistency-based seam carving	Stereoscopic video (Indoor and outdoor type)	Qualitative (Blind grading system)	Temporal consistency is an important criterion when applying stereo seam carving to video
Yuming Fang et al. [7]	Image resizing based on structural similarity index	Natural scenes	Qualitative (EH, BDW, SIFT and CL model)	Promising quality prediction performance is achieved when testing the proposed IR-SSIM algorithm using two subject-rated image retargeting databases
Proposed method	Graph cut-based seam carving	Natural scenes	Quantitative (PSNR, MSE, and VQM)	The proposed video retargeting method in space and time exhibits greater flexibility, and is scalable for large videos

Figure 9 presents the retargeted video with various aspect ratio adhering to standards ATSC, NTSC- DV, DVD, SD, PAL SD formats. The semantic content and visual quality to user is well within acceptable loss.

Table 4 gives the details of the comparative analysis of the existing system.

4 Conclusion and Future Work

The proposed video retargeting method in space and time exhibits greater flexibility, and is scalable for large videos. We have also demonstrated the benefits of using filter to improve the quality of the retargeted video. The proposed method is applicable for video of longer duration as well as online streaming videos. The challenges we came across with time and memory complexity was that, the time taken to retarget a video given for 1 min of duration was 3 h. And in case of memory complexity, 2 GB memory space is sufficient as we delete the extracted frames once the retargeting is done.

The proposed methodology works considerably fine with still and less motion video as seams detection becomes an easy and efficient task. It also works well with different formats such as Flv, MP4, AVI and it does not work fine with 3gp format and also with very fast motion video; performance reduces because the seam detection becomes difficult. The software can be extended to object detection and removal and motion detection.

From the above comparative study, we conclude that in case of high motion videos there will be frequent changes in movement of objects so the number of seams generated are more in case of high motion videos. The graphs clearly quantify the video quality by three parameters such as PSNR, MSE, and VQM. It is observed that proposed technique works well for fast video as compared to slow video. But still there is scope for improvement. Motion estimation may enhance the performance of proposed technique.

References

1. Matthias Grundmann, Vivek Kwatra, Mei Han, Irfan Essa. Discontinuous Seam-Carving for Video Retargeting, Georgia Institute of Technology, Atlanta, GA, USA and Google Research, Mountain View, CA, USA, 2010.
2. Michael Rubinstein, Ariel Shamir, Shai Avidan Improved Seam Carving for Video Retargeting, ACM Transaction. Graphics, vol 27 No 3, August 2008.
3. Sonawane, Nayana, and B. D. Phulpagar. "Review on Content-Aware Image Re-sizing Using Improved Seam Carving and Frequency Domain Analysis." (2015).
4. Yoon, Jong-Chul, et al. "Optimized image resizing using flow-guided seam carving and an interactive genetic algorithm." Multimedia tools and applications 71.3 (2014): 1013–1031.
5. Décombas, Marc, et al. "Seam carving modeling for semantic video coding in security applications." APSIPA Transactions on Signal and Information Processing 4 (2015): e6.

6. Benjamin Guthier, Johannes Kiess, Stephan Kopf, Wolfgang Effelsberg "SEAM CARVING FOR STEREOSCOPIC VIDEO", Proc. of IEEE Workshop on 3D Image/Video Technologies and Applications (IVMSP), June 2013.
7. Yuming Fang, Kai Zeng, Zhou Wang, Fellow, IEEE, Weisi Lin, Senior Member, IEEE, Zhijun Fang, and Chia-Wen Lin, "Objective Quality Assessment for Image Retargeting Based on Structural Similarity", IEEE journal on emerging and selected topics in circuits and systems, vol. 4, no. 1, march 2014.

Intelligence System Security Based on 3-D Image

K. Anish, N. Arpita, H. Nikhil, K. Sumant, S. Bhagya and S.D. Desai

Abstract In today's world, digital communication plays a vital role. The art of communicating secret information has also evolved. For years, encryption has played a vital role in secure transmission of secret data. But due to lack of covertness, an eavesdropper can identify encrypted data and subject it to crypt-analysis. Here we present a method of hiding information such that its very existence is masked. This study aims to develop an enhanced technique for hiding data in 3-D images ensuring high invisibility. The data will be embedded in 3-D images with .pcd format and the coordinates of the 3-D image are used for data hiding.

Keywords Steganography · Security system · 3-D image · Cartesian coordinate · Stego image

1 Introduction

Steganography comes from the Greek words Stegans (Covered) and Graptos (Writing). It is an art of covert communication by hiding the presence of a message typically in multimedia content. Steganography is applied in internet security, information assurance, copyright protection, authentication, etc. In steganography, discrete wavelet transform (DWT) is used to embed the secret message by modifying the wavelet coefficients of the cover image [1]. Steganography includes the concealment of information within computer les. In digital steganography, electronic communications may include steganographic coding inside a transport layer, such as a document le, image le, program, or protocol [2].

K. Anish · N. Arpita · H. Nikhil · K. Sumant · S. Bhagya · S.D. Desai (✉)
Department of IS & E, B.V.B College of Engineering and Technology,
Hubli 580031, India
e-mail: shree.desai07@gmail.com

S. Bhagya
e-mail: sunagbhagya@gmail.com

© Springer Nature Singapore Pte Ltd. 2017
S.C. Satapathy et al. (eds.), *Proceedings of the 5th International Conference on Frontiers in Intelligent Computing: Theory and Applications*, Advances in Intelligent Systems and Computing 515, DOI 10.1007/978-981-10-3153-3_16

Steganography is divided into two domains such as technical Steganography (image, audio, video, and network) and natural language Steganography. Text steganography is among the most challenging method in steganography domain because it uses limited memory and simple communication as compared to other media [3]. Our objective is to develop time optimized and space optimized, imperceptible and robust 3-D image steganography system [4].

Steganography is at greater advantage than cryptography because the secret message does not grab attention towards itself as an element of scrutiny. Thus, in cryptography the message is protected by encoding it using an encryption key. But steganography involves composing secret messages such that only the sender and receiver know about its existence.

1.1   3-D Image Representation Systems

A coordinate system uses one or more coordinates such as (x, y, z) to identify uniquely the position of geometric element in space as shown in Fig. 1 [5]. The spherical coordinate system is commonly used in physics. It assigns three coordinates to every point in space: radial distance (r), polar angle (θ), and azimuthal angle (phi) [6].

2   Review of Literatures

Various literatures are reported, to address text steganography. Majority of them uses 2-D image. However, better robustness shall be achieved using 3-D image. In this regard, following current research articles are reviewed.

Ekta Walia (2010) et al., presents a paper in which author has analyzed the steganography based on least significant bit and discrete cosine transform. In LSB-based steganography, the secrete message is embedded in the least significant

Fig. 1 3-D representation in Cartesian coordinate system

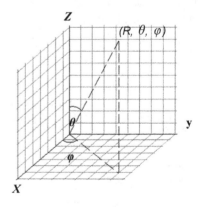

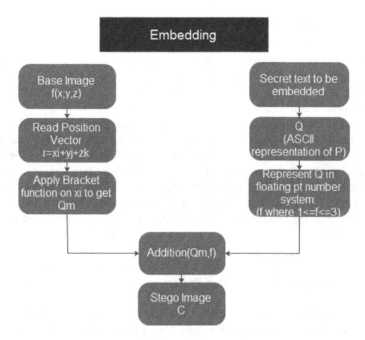

Fig. 2 System model to embed secret information from 3-D image

bits of cover image. In DCT-based steganography, the secrete message is embedded in the least significant bits of DC coefficients are done in an analysis of LSB–DCT-based Steganography [7].

Rama (2011) et al., presents a paper on 3-D models of steganography and conclude with some review analysis of high capacity data hiding and low-distortion 3-D models. It also presents a background discussion on the steganography 3-D deployed in digital imaging. Steganography urges that the cover image must be carefully selected in survey and analysis of 3-D steganography [8].

Thiyagarajan (2012) et al., proposes a paper in which 3-D geometric model is used for high capacity Steganographic scheme. Triangle mesh is manipulated to embed secret message forming a new retriangulated mesh. The author has demonstrated the ability of proposed method to address cropping, rotation and scaling impairments in 3-D image domain [6].

Akhtar (2014) et al., presents a paper in which LSB-based steganography is experimented. Bit inversion technique is used to embed the secret message and thereby improve the quality of stego image [9].

3 Proposed Methodology

Figures 1 and 2 represent the embedding process of the secret text into a 3-D image. Image 2 represents the extraction of the hidden text.

3.1 Embedding Process

A 3-D base image in which the secret text is to be embedded is considered. Embedding in all the three coordinates leads to distortion of image. For efficiency, embedding is done in any of the three coordinates. In the proposed methodology 'X' coordinate is considered.

The values of x coordinates are extracted from the points forming the image. Bracket function (rounds a real number down to the same integer number) is applied to the values. On the other hand, the secret information to be embedded is represented in ASCII format. Since the ASCII values are represented up to three digits, they are expressed in floating point number system with three digits following the decimal point. The corresponding values of the secret text and the result obtained by applying bracket function are added. The values of the result obtained, form the x coordinates of the required stego image [10].

3.1.1 An Example on Embedding Process

The following snippet presents header and first few coordinate details. The one which is underlined is considered for embedding process. The secret message to be embedded is presented in the second snippet. The third snippet presents header and first coordinates of stego image.

Snippet	Content
1	**.PCD file format of original image** .PCD VERSION is 0.7 Fields are x y z Size of each field 4 Type of each dimension is F Count of elements in each dimension 1 Width of dataset is 75422 Height of dataset is 1 viewpoints 0 0 0 1 0 0 0 No of points in an stego image are 75422 Data represented in ascii format 66.284599 50.6171 10.4749 66.556396 20.688299 10.4648...
2	**Secrete information to be embedded** SECRKT In view of the importance which the.......

Steps for embedding process:
ASCII value: S = 083
Apply Bracket Function on 1st 'x' coordinate: 66.284599 => 66
Append ASCII value of S to the above value => 66.083

Snippet	Content
3	**The resulting file after embedding process** PCD VERSION is 0.7 Fields are x y z Size of each field 4 Type of each dimension is F Count of elements in each dimension 1 Width of dataset is 75422 Height of dataset is 1 viewpoints 0 0 0 1 0 0 0 No of points in an stego image are 75422 Data represented in ascii format <u>66.083</u> 50.6171 10.4749 <u>66.069</u> 20.688299 10.4648

3.2 Extraction Process

The values of x coordinates of the stego image are extracted. Bracket function is applied to the values. Each of these values is subtracted from the corresponding values of the x coordinates of the stego image. The resulting values are multiplied by 1000, which give integer numbers. These are the ASCII values of the embedded characters. The ASCII values are then converted into their character representation to obtain the secret message [11] (Fig. 3).

3.2.1 An Example on Extraction Process

The following snippet presents header and first few coordinate details of stego image. The one which is underlined is considered for extraction process.

Snippet	Content
1	**.PCD file format of stego image** .PCD VERSION is 0.7 Fields are x y z Size of each field 4 Type of each dimension is F Count of elements in each dimension 1 Width of dataset is 75422 Height of dataset is 1 viewpoints 0 0 0 1 0 0 0 No of points in an stego image are 75422 Data represented in ascii format <u>66.083</u> 50.6171 10.4749 <u>66.069</u> 20.688299 10.4648......

Fig. 3 System model to
extract secret information
from 3-D image

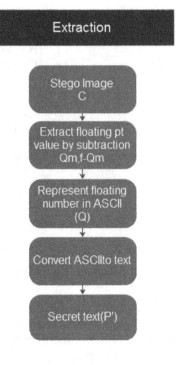

Steps for extraction process:
Apply bracket function on first 'x' coordinate: 66.083 => 66
Subtract the bracket function value from the 'x' coordinate value of
Stego image => 0.083
Multiply it with 1000 => 083
Convert the above value into character => S

4 Result

To test imperceptibility and robustness, following readings are recorded. The
response time for each run is also recorded. Peak signal to noise ratio (PSNR),
structural similarity index metrics (SSIM), and normalized absolute error (NAE) of
stego image in comparison with original image is recorded. For each 3-D image,
readings are recorded for front view, side view, and top view, and averaged scores
are recorded as shown in Figs. 4 and 5, which represents text Steganography for a
3-D image of sports car while Figs. 6 and 7 represent text Steganography over a
flower pot 3-D image. From Table 1, it is observed that the quality of stego image is
excellent with PSNR of 41.95 ± 0.98 dB and SSIM of 0.958 ± 0.005 as well as

Fig. 4 Original car.pcd

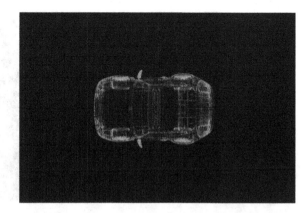

Fig. 5 Stego car.pcd

Fig. 6 Original pot.pcd

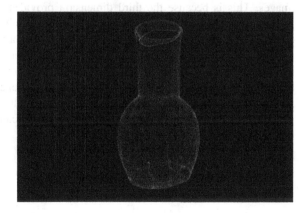

NAE is 0.019 ± 0.001. These values clearly demonstrate the imperceptibility and robustness of proposed methodology. The response time shows computationally efficient attribute of proposed methodology.

Fig. 7 Stego pot.pcd

Table 1 Record of observation

3-D image	File size (characters)	PSNR (db)	SSIM	NAE	Embed time (ms)	Extract time (ms)
Sports car	4473	41.25	0.954	0.0201	2132.7	1660.4
Flower pot	4473	42.65	0.9621	0.018	392.29	286.24

5 Conclusion

3-D images have more provisions to hide the text information as compared to 2-D images. This is because the third dimension provides the better scope. Here the proposed method even though works on spatial domain, but it does not deal with pixel value, rather it deals with its Cartesian coordinate. The coordinates of each pixel are so cleverly used to embed the text data, so as the steganography process becomes more robust. The problem of rotation, cropping or enlarging doesn't arise here. The response time clearly shows computational efficiency of proposed method. The stego image has very close similarity with original image. Overall, the proposed method has emerged as computationally efficient, simplest steganography with robust nature.

References

1. M. Ghebleh and A. Kanso. A robust chaotic algorithm for digital image steganography. Communications in Nonlinear Science and Numerical Simulation 19, 6:1898{1907, 2014.
2. T. Santanam Sumathi, C. P. and G. Umamaheswari. A study of various steganographic techniques used for information hiding. arXiv preprint, 2014.

3. Der-Chyuan Lou and Chen-Hao Hu. Lsb steganographic method based on reversible histogram transformation function for resisting statistical steganalysis. Information Sciences 188, pages 346{358, 2012.
4. H. T. Sencar M. Kharrazi and N. Memon. Cover selection for steganographic embedding. In Proc. IEEE International Conference on Image Processing, pages 110{120, 2006.
5. Haz Malik. Steganalysis of qim steganography using irregularity measure. Proc. of the 10th ACM workshop on Multimedia and security, ACM, pages 149{158, 2008.
6. V. Natarajan G. Aghila V. Prasanna Venkatesan Thiyagarajan, P. and R. Anitha. Pattern based 3d image steganography. 3D Research 4, pages 1{8, 2013.
7. Ekta Walia. Performance improvement of jpeg2000 steganography using qim. Communication and Computer, ISSN. 6:1548{1779, 2009.
8. Sunny Bansal K Rama, Sukhpreet and Rakesh K. Bansal. Steganography and classiffication of image steganography techniques. In Computing for Sustainable Global Development (INDIACom), pages 870{875, 2011.
9. V. Natarajan G. Aghila V. Prasanna Venkatesan Thiyagarajan, P. and R. Anitha. Pattern based 3d image steganography. 3D Research 4, pages 1{8, 2013.
10. Sharifullah Khan Akhtar, Naheed and Prashant Johri. An improved inverted lsb image steganography. In Issues and Challenges in Intelligent Computing Techniques (ICICT)t, pages 74{755, 2014.
11. U. Dinesh Acharya A. Renuka Hemalatha, S. and Priya R. Kamath. A secure and high capacity image steganography technique. Signal Image Processing, page 83, 2013.

PDA-CS: Profile Distance Assessment-Centric Cuckoo Search for Anomaly-Based Intrusion Detection in High-Speed Networks

Kanaka Raju Gariga, A. Rama Mohan Reddy and N. Sambasiva Rao

Abstract The act of network intrusion detection is an obligatory part of network performance under security. Unlike other network security strategies, the act of intrusion detection systems should aware the behavior of the users and signature of the intruded and normal transactions, which is continuous process since the user behavior is not static as well the attack strategies are redefining in magnified speed. Hence, the objective of effective intrusion detection is always a significant factor for research. The bioinspired evolutionary strategies are getting the attention of most of the recent research studies. In order to this, the divergent contexts such as minimal computational complexity, prediction accuracy, ensemble models have been considered as significant objective. The other most significant objective and compatible to current state of art is IDS scalability and robustness in high-speed networks, hence the evolutionary computation approaches are adoptable. In this study, we propose an intrusion detection approach that is based on evolutionary computation technique called Cuckoo search. Further, the proposed detection system is investigated thoroughly in the context of accuracy, robustness, and also from the evolutionary computation point of view.

Keywords Statistical IDS · Anomaly IDS · Bioinspired · Machine learning · Evolutionary computation · Soft computing · Cuckoo search · Hamming distance · High-speed networks · ISCX2012

K.R. Gariga (✉)
Department of CSE, JNTUH, Hyderabad, India
e-mail: kanakarajugariga@gmail.com

A.R.M. Reddy
Department of CSE, SV University College of Engineering, Tirupati, India
e-mail: ramamohansvu@yahoo.com

N.S. Rao
SRIT for Women, Warangal, India
e-mail: snandam@gmail.com

© Springer Nature Singapore Pte Ltd. 2017
S.C. Satapathy et al. (eds.), *Proceedings of the 5th International Conference on Frontiers in Intelligent Computing: Theory and Applications*, Advances in Intelligent Systems and Computing 515, DOI 10.1007/978-981-10-3153-3_17

169

1 Introduction

The network intrusion is a serious cause to down grade the performance of the computer networks. The intrusion detection is an act of identifying the network communication events those compromises the network's privacy, credibility, security, or fair response [1]. This is done by noticing the signs of intrusion through packet signature or anomalies observed. The Anomaly-based Intrusion detection systems (IDSs) attempts to differentiate a bait transaction by an intruder from a normal transaction.

Numerous methods were proposed in past decade that are categorized as statistical approaches [2], mining and machine learning models [3, 4] and bioinspired approaches [5]. All of these approaches fit into the one of the two categories called Anomaly-based detection and signature based detection. The anomaly-based models explores the patterns of normal and intruded transactions to identify the further network transactions are an attempt of intrusion or normal [6]. The signature based detection models define the signatures of the intruded transactions and alarms if a transaction matches that signature.

PDA-CS aims at assessing profile distance of a network transaction with normal and as well as intruded data that given for training. The cuckoo search has been specially customized to assess the profile distance from training data of a given network transaction. The hamming distance technique is used to find the optimal attributes of the given normal and intruded transactions for training. The hamming distance identifies the attributes those differentiating on their values for intruded and normal transactions. The critical factor of the data analysis is to identify the features that significantly influence the data makeover. This is due to the involvement of numerous numbers of insignificant features that causes inefficient, inadequate, and complexed analysis.

This paper suggests the use of customized cuckoo search toward profile distance assessment. The hamming distance is used to reduce the dimensionality of the attributes involved in given training data. The rest of this paper is organized as follows. In Sect. 2, we discuss the associated works of the evolutionary computation based network intrusion detection. Section 3 explores the model PDA-CS proposed here in this paper. Experimental study and performance analysis of the PDA-CS is done in Sect. 4 that followed by Sect. 5, which is concluding the article.

2 Related Work

Soft computing approaches are possible to play a prominent role in the field of intrusion detection. This section explores the contemporary literature about intrusion detection systems that are using soft computing techniques. A flexible neural tree (FNT) [7] that optimizes the neural tree nodes structure using genetic algorithm (GA). The particle swarm optimization [8] was used to tune the node weights and functional parameters. The GA and PSO are recursive till the node structure is optimized with fine-tuned weights. Katar [9] Purposed an ensemble of Naïve Bayes, ANN was proposed [9] that uses PSO and C4.5 to train on multiple sets of data input from the DARPA dataset. The intrusion detection is done with the multiple fusion methods of Bayes average, recognition, substitution, and rejection rates (RSR). The empirical study [10] compared ANN with estimation of distribution algorithm (EDA), with PSO, and a DT performance on DRPA dataset. The distribute IDS [11] is another considerable model that uses an agent system with cooperative intelligence. An ensemble fuzzy classifier is used to notice the intruded network transactions.

A hybrid evolutionary strategy [12] ensembles the genetic algorithm and fuzzy logic. The objective of this model is to identify the anomalies on network traffic. The computational complexity is observed to be nonlinear and anomaly search is not compatible to high-speed networks. The IDS models of fuzzy rule-based classifiers [12], decision trees [13], support vector machines [14] are most cited contributions in recent literature. Empirical results clearly showed that the computational complexity and networks with high-speed transactions are the main constraints of these models. A light weight soft computing IDS [15] that uses a decision tree to discover the optimal features, which were considered further as input to an ensemble classifier formed by linear genetic programming (LGP) and a neural networks.

The observations from the review of these existing models evincing that almost all of these are having different constraints such as expert's involvement, training on static data, and nonlinear computational complexity due to overloaded features. But all of these are limited to low to average speed of network transmission. Hence, it is obvious to conclude that there is significant need to define novel anomaly intrusion detection strategies for high-speed networks. In order to this, we propose an evolutionary computational strategy called profile distance assessment-based Cuckoo search for intrusion detection in high-speed network. The objective of the proposal is to assess the state of network transaction is intrusion or not, which is done by profile distance assessment through Cuckoo search with magnified speed.

3 Profile Distance Assessment-Centric Cuckoo Search (PDA-CS)

The combination of network transactions labeled as normal and attack were considered for training. The profiles of the records are represented by the fields that are used to form the records, such that each field represents a profile. The proposal is initially segmenting the input records into two sets as attack and normal records. In order to identify the optimal features, the distance between each profile of normal and attack records is assessed, which is done using hamming distance (see Sect. 3.1). Further, the selected optimal attributes for attack and normal records are used to form the respective nests to perform Cuckoo search. The nest formation (see Sect. 3.2) is done in the hierarchical order and the nest defined in first level of hierarchy is represented by all optimal profiles (n profiles) selected for respective normal and attack records. The next hierarchy contains the nests represented by $n-1$ optimal profiles, the next level hierarchy contains nests each represented by $n-2$ profiles and last hierarchy contains nests, each represented by one profile. Afterward these nests equipped with eggs represented by the values observed for respective profiles of respective attack and normal records. Further, Cuckoo search (see Sect. 3.3) will be initiated for a given unlabeled record on nests hierarchies of respective attack and normal records.

According to the compatible nests found in each hierarchy for unlabeled record given, the state of the transaction is attack prone or normal will be estimated (see Sect. 3.4).

3.1 Optimal Attribute Selection

- Partition the given network transactions as intruded (I) and normal (N)
- Find the hamming distance between unique values of each attribute of I with the counter part of N
- Select the attributes having the hamming distance more than the given threshold hdt as set of optimal attributes I_a of size n, N_a of size m from I and N, respectively

Assessing Hamming Distance is as follows:

The value of Hamming Distance obtained here is to denote the difference between unique values of same attribute from records labeled true and false. This is one of the significant strategies to assess the difference between to elements in coding theory. This strategy is applied to identify the distance between the unique values observed for an attribute in record set labeled as true and labeled as false.

For a given two vectors $X = \{x_1, x_2, \ldots\ldots\ldots, x_n\}$ and $Y = \{y_1, y_2, \ldots\ldots\ldots, y_m\}$ of size n and m, respectively. Hamming distance can be measured as follows:

Let $Z \leftarrow \phi$ // is a vector of size 0
foreach $\{i \exists i = 1, 2, 3, \ldots..\max(n.m)\}$ Begin
$if\ (\{x_i \exists x_i \in X\} - \{y_i \exists y_i \in Y\}) \equiv 0$ then
$Z \leftarrow \{x_i \exists x_i \in X\} - \{y_i \exists y_i \in Y\}$
Else
$Z \leftarrow 1$
End

$$hd_{X \leftrightarrow Y} = \sum_{j=1}^{|Z|} Z\{i\}$$

//$hd_{X \leftrightarrow Y}$ is the hamming distance between X and Y, $Z\{i\}$ is the ith element of the vector Z, and $|Z|$ is the size of the vector Z.

3.2 Nest Formation

- Prepare $2^n - 1$ unique subsets $\{I_a\}$ such that at least one subset of sizes $\{n, n-1, n-2, \ldots\ldots, 1\}$ and $2^m - 1$ unique subsets $\{N_a\}$ such that at least one subset of sizes $\{m, m-1, m-2, \ldots\ldots, 1\}$ from I_a and N_a respectively
- Prepare a set $\{I_f\}$ such that $\{I_f\}$ contains the set of respective values appeared in 1 or more records of I for attributes of each subset of $\{I_a\}$
- Prepare a set $\{N_f\}$ such that $\{N_f\}$ contains the set of respective values appeared in 1 or more records of N for attributes of each subset of $\{N_a\}$
- Build a hierarchical order TI such that subset $[s \exists s \in \{I_a\}]$ with size of $|I_a|$ as root and subsets with size $|I_a| - 1$ as level 1 nodes and continue further building of that order with all subsets of the $\{I_a\}$ as the subsets of size $|I_a| - (i+1)$ as the next level nodes to the level formed by the subsets of size $|I_a| - i$
- Similarly build hierarchical order TN such that subset $[s \exists s \in \{N_a\}]$ with size of $|N_a|$ as root and subsets with size $|N_a| - 1$ as level 1 nodes and continue further building of that hierarchy with all subsets of the $\{N_a\}$ as the subsets of size $|N_a| - (i+1)$ as the next level nodes to the level formed by the subsets of size $|N_a| - i$
- Further, the nodes at different levels of these hierarchies are considered as nests identified by the combination of level id and node id
- For each node of hierarchy TI that representing a nest, the respective value sets those belongs $\{I_f\}$ are considered as eggs
- $\{I_f\}_{ij}$ is the set of eggs belongs to the nest j in level i of hierarchy TI
- For each node of hierarchy TN, the respective value sets those belongs $\{N_f\}$ are considered to be as eggs in the nest represented by that node
- $\{N_f\}_{ij}$ is the set of eggs belongs to the nest j in level i of hierarchy TN

3.3 Cuckoo Search

For a given network transaction record R, form the 2^n subsets $\{RI_f\}$ from the values of the optimal features reflected by I_a and also form the 2^m subsets $\{RN_f\}$ from the values of the optimal features reflected by N_a.

The $\{RI_f\}$ and $\{RN_f\}$ are considered as set of Cuckoo eggs to be placed in nests represented by TI and TN, respectively.

Cuckoo Search on TI.

Sort the $\{RI_f\}$ in descending order of sizes

$s(I) \leftarrow \phi$ // vector that represents the number of compatible nests in each level of hierarchy TI

For each level $\{l \ni l = 1, 2, 3, \ldots .n\}$ Begin

// l represents the level of the hierarchy from root to leaves, respectively

$s_l = 0$ // possible number of nests in level l accommodates Cuckoo eggs

For each Cuckoo egg $[ce \ni ce \in \{RI_f\} \wedge |ce| \equiv (n-l+1)]$ Begin

//$(n-l+1)$ represents the number of features in Cuckoo egg ce

For each $[j \ni j = 1, 2, 3, \ldots . |TI_l|]$ Begin

// $|TI_l|$ represents the number of nests in level l of hierarchy TI

if$(\{I_a\}_{lj} \subset \{ce_a\} \wedge \{ce_a\} \subset \{I_a\}_{lj})$ Begin

// Attributes representing the nest $\{I_f\}_{lj}$ must be identical to the attributes of the features in ce

$\{t\} \leftarrow \{ce\} \cup \{I_f\}_{lj}$

// forming a temporary set from the union of $\{ce\}$, which is a set with one element and $\{I_f\}_{lj}$ that is a set of eggs in the nest j of level l

If $|\{t\}| \equiv |\{I_f\}_{lj}|$ then $s_l + = 1$

End

End

$s(I) \leftarrow s_l$

End

End

Cuckoo Search on Hierarchy *TN*.

Sort the $\{RN_f\}$ in descending order of sizes

$s(N) \leftarrow \phi$ a vector represents the compatible nests at each level of the hierarchy *TN*

For each level $\{l \ni l = 1, 2, 3, \ldots .m\}$ Begin

l // represents the level of the hierarchy from root to last level, respectively

$s_l = 0$ // possible number of nests in level l accommodates Cuckoo eggs

For each Cuckoo egg $[ce \ni ce \in \{RN_f\} \wedge |ce| \equiv (m-l+1)]$ Begin

// $(m-l+1)$ represents the number of features in Cuckoo egg ce

For each $[j \ni j = 1, 2, 3, \ldots . |TN_l|]$ Begin // $|TN_l|$ represents the number of nests in level l of hierarchy *TN*

if$(\{N_a\}_{lj} \subset \{ce_a\} \wedge \{ce_a\} \subset \{N_a\}_{lj})$ Begin Attributes representing the nest $\{N_f\}_{lj}$ must be identical to the attributes of the features in ce

$\{t\} \leftarrow \{ce\} \cup \{N_f\}_{lj}$ // forming a temporary set from the union of $\{ce\}$, which is a set with one element and $\{N_f\}_{lj}$ that is a set of eggs in the nest j of level l

If $|\{t\}| \equiv |\{N_f\}_{lj}|$ then $s_l + = 1$

End

End

$s(N) \leftarrow s_l$

End

End

3.4 Assessing the State of Transaction

For each level $\{l_{TI} \exists l_{TI} = 1, 2, 3, \ldots .n\}$ of hierarchy TI Begin

$cnr(l_{TI}) = \frac{cnc(l_{TI})}{nc(l_{TI})}$ // $cnc(l_{TI})$ is compatible number of nests in level l_{TI}, $nc(l_{TI})$ is total number of nests in level l_{TI}

End

For each level $\{l_{TN} \exists l_{TN} = 1, 2, 3, \ldots .m\}$ of hierarchy TN Begin

$cnr(l_{TN}) = \frac{cnc(l_{TN})}{nc(l_{TN})}$ // $cnc(l_{TN})$ is compatible number of nests in level l_{TN}, $nc(l_{TN})$ is total number of nests in level l_{TN}

End

For each level $\{l_{TI} \exists l_{TI} = 1, 2, 3, \ldots .n\}$ & $\{l_{TN} \exists l_{TN} = 1, 2, 3, \ldots .m\}$ Begin

$s_{TI} + = \{1 \exists cnr(l_{TI}) \geq cnr(l_{TN})\}$

$s_{TN} + = \{1 \exists cnr(l_{TN}) > cnr(l_{TI})\}$

End

if$(s_{TI} > s_{TN}) || \left(\frac{s_{TI}}{s_{TN}} > \tau\right)$ Begin

// $\frac{s_{TI}}{s_{TN}}$ is the intrusion ratio for each level of TN, τ is the intrusion ratio threshold given

Network transaction R is confirmed to be intrude

Else

Network transaction R is confirmed to be Normal

End

4 Empirical Study and Performance Analysis

4.1 Data Set Description

The renowned dataset that mostly used to assess the anomaly-based intrusion detection systems is ISCX IDS dataset (ISCX [16], which was formed by information security center of excellence, University of New Brunswick (Brunswick, n. d. [17]). This dataset has large number of records and each of that record contains the values for 41 attributes (see Table 1 for list of all attributes). In order to minimize the process complexity of the test bed 16345 labeled (as attack or normal) records selected randomly from the complete dataset. The random selection of the records is done under the criteria of Gaussian distribution [18].

Table 1 Optimal Attributes selected by the upper bound of the hamming distance threshold

>0.594013279 (upper bound of the hamming distance threshold)	
3	0.695252
5	0.676204
6	0.950504
24	0.620774
31	0.822627
32	0.849382
35	0.719548
36	0.675832

4.2 Description of Attribute Selection by Hamming Distance

In order to identify the optimal attributes of the network transactions given, the hamming distance (see Sect. 3.1) is applied on the IDS dataset explored in this (Sect. 4.1), which is done using expression language R. The data preprocessing is done using JAVA. The hamming distance of different attributes under intruded and normal data toward network intrusion detection are explored in Figs. 1, 2 and Tables 1, 2.

The mean of the hamming distances found for all attributes is considered as distance threshold *hdt*. The difference between the distance threshold *hdt* and the mean absolute deviation [19] mad_{hdt} of the hamming distance found for all attributes is considered as the lower bound hdt_l of the distance threshold, and aggregation of the mean absolute deviation and distance threshold is considered as upper bound hdt_u of the threshold.

The mean absolute deviation mad_{hdt} is measured as follows:

Let $hd = \{da_1, da_2, \ldots\ldots, da_{41}\}$ be the set of hamming distances observed for all four attributes

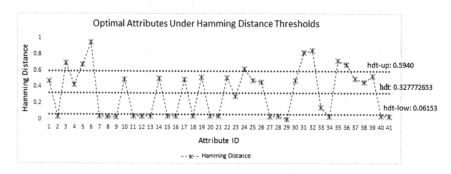

Fig. 1 Optimal attributes of the network transactions labeled as intrude under different hamming distance thresholds

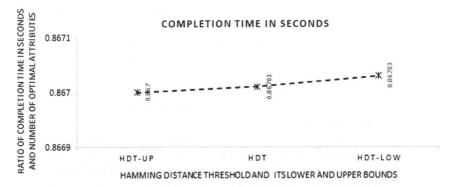

Fig. 2 The ration completion time in seconds observed for PDA-CS under divergent hamming distance thresholds

Table 2 Statistical Metrics and observed values for the results obtained by PDA-CS experiments

Precision	0.954339
Recall	0.970181
Accuracy	0.949388
Sensitivity	0.970181
Specificity	0.908318

$hdt = \frac{\sum_{i=1}^{41} da_i}{41}$ // The hamming distance threshold hdt is measured here, which is the mean of distances observed for all attributes.

$mad_{hdt} = \frac{\sqrt{\sum_{i=1}^{41}(hdt - da_i)}}{41}$ // Here the mean absolute deviation of hd from hdt is measured.

Then the lower bound of the hdt is

$hdt_l = hdt - mad_{hdt}$

The upper bound of the hdt is

$hdt_u = hdt + mad_{hdt}$

4.3 Performance Analysis

Total number of records used for training was 70% (11445) and total records tested were 30% (4900). The optimal attributes selected are under hamming distance threshold 0.3278. The results obtained from the PDA-CS are as follows:

False positives (normal transactions found as attack) are 151, true positive 3156 (records found by PDA-CS as attacks which are actually labeled as attack), the count of true negatives (records labeled as normal are found by PDA-CS as normal)

are 1496 and the count of false negatives (Records labeled as attack are found by PDA-CS as normal) are 97.

The performance assessment metrics [20] called precision, recall, accuracy, and specificity are measured in order to estimate the performance accuracy of the PDA-CS (see Table 2). The other critical factor of the performance is computational complexity of the PDA-CS, which is also measured by considering the different count of optimal attributes under hamming distance threshold and its lower and upper bounds (see Fig. 1).

The result obtained for the statistical metrics (see Table 2) indicating that the sensitivity of the attribute optimization is 97%, which can confirmed by the result obtained for metric recall (ratio of relevant predictions obtained) [20]. The accuracy (see Table 2) of the PDA-CS is identified as 94%, which is the ratio of valid predictions by PDA-CS and actual labels of the records tested. The normal record identification rate of the PDA-CS is 90%. This can be confirmed since the value for metric called specificity is observed is 0.908318154 and this metric indicate the ratio of accuracy toward the true negative detection. Due to the context of network intrusion seriousness, attack detection must be more robust and sensitive than the normal transaction detection. Hence, it is quite obvious to claim that the proposed PDA-CS is robust and scalable since the intrusion detection accuracy observed is 97% that is much greater than the normal transaction detection accuracy, which is 90%.

The experiments also conducted on the same dataset under divergent hamming distance thresholds in order to observe the time complexity. The time complexity found for divergent number of optimal attributes is linear (see Fig. 2).

Hence, it is obvious to conclude that the applying hamming distance toward optimized attribute selection is significant to boost anomaly-based intrusion detection.

5   Conclusion

Profile distance assessment-based Cuckoo search (PDA-CS) uses hamming distance analysis for attribute optimization towards dimensionality reduction. This proposed model is promising in the selection of optimal attributes to simplify the process of network intrusion detection. The devised Cuckoo search amplified the detection accuracy with minimal process complexity. The experiments were done using benchmarking ISCX IDS dataset (ISCX [16]. The exploration of the results concluding that the hamming distance analysis is promising and significant to select optimal attributes of the training records dataset. Further, the training records with optimal attributes are used to define the CUCKOO Search strategy, which is observed to be robust and is with minimal process complexity (which is linear). Hence, the model devised here in this paper is significantly minimized the computational overhead and retains the maximal prediction accuracy. In future hybrid strategy that uses more than one evolutionary technique to define a novel anomaly-based intrusion detection.

References

1. D.S Bauer, M. K. (1988). NIDX- an expert system for real-time network intrusion detection. *Proceedings of the Computer Networking Symposium*, 98–106.
2. Mell, R. B. (2001). Intrusion Detection Systems. *NIST Special Publication on Intrusion Detection System.*
3. A. Sundaram. (1996). An introduction to intrusion detection. *The ACM student magazine.*
4. Denning, D. (1986). An intrusion-detection model. *In IEEE computer society symposium on research in security and privacy,* 118–131.
5. T.Lane. (2000). *Machine Learning techniques for the computer Security.* Purdue University.
6. Stolfo, W. L. (1998). Data mining approaches for intrusion detection. *Proc. of the 7th USENIX security symposium.*
7. W. H. Chen, S. H. (2005). Application of SVM and ANN for intrusion detection. *Comput Oper Res Vol-ume 32, Issue 10,* 2617–2634.
8. Kennedy, J. (2010). Particle swarm optimization. *Encyclopedia of Machine Learning,* 760–766.
9. Katar, C. (2006). Combining multiple techniques for intrusion detection. *Int J Comput Sci Network Security,* 208–218.
10. Chen Y, A. A. (2005). Feature deduction and intrusion detection using flexible neural trees. *Second IEEE International Symposium on Neural Networks,* 2617–2634.
11. A. Abraham, R. J. (2007). D-scids: distributed soft computing intrusion detection system. *J Network Computer,* 81–98.
12. Hassan, M. M. (2013). Current studies on intrusion detection system, genetic algorithm and fuzzy logic. *International Journal of Distributed and Parallel Systems,* 35–48. Retrieved from arXiv.
13. Sindhu, S. S. (2012). Decision tree based light weight intrusion detection using a wrapper approach. *Expert Systems with applications,* 129–141.
14. Li, Y. X. (2012). An efficient intrusion detection system based on support vector machines and gradually feature removal method. *Expert Systems with Applications,* 424–430.
15. Bhatti, D. G. (2012). Conceptual Framework for Soft Computing based Intrusion Detection to Reduce False Positive Rate. *International Journal of Computer Applications,* 1–3.
16. ISCX, U. (2012). *UNB ISCX Intrusion Detection Evaluation DataSet.* Retrieved from Information Security Centre of Excellence: http://www.unb.ca/research/iscx/dataset/iscx-IDS-dataset.html.
17. Brunswick, U. o. (n.d.). *Information Security Centre of Excellence.* Retrieved from University of new Brunswick: http://www.unb.ca/research/iscx/index.html.
18. Goodman, N. R. (1963). Statistical analysis based on a certain multivariate complex Gaussian distribution. *Annals of mathematical statistics,* 152–177.
19. Leys, C. L. (2013). Detecting outliers: do not use standard deviation around the mean, use absolute deviation around the median. *Journal of Experimental Social Psychology,* 764–766.
20. Powers, D. M. (2006). Evaluation: from precision, recall and F-measure to ROC, informedness, markedness and correlation. *23rd International conference on machine learning.* Pitsburg.

Evaluation of Barriers of Health Care Waste Management in India—A Gray Relational Analysis Approach

Suchismita Swain, Kamalakanta Muduli, Jitendra Narayan Biswal, Sushanta Tripathy and Tushar Kanti Panda

Abstract The waste generated by health care units has been contributing a dreadful share in terms of life threatening diseases and environmental pollution. Erroneous management of this waste has not only invited a serious threat to the environment but also to the personnel associated with it; mainly health care experts, patients, workers as well as the general community. A number of studies advocate that there exists certain factors that inhibit effectiveness of health care waste management (HCWM). Prior knowledge of these factors and their relative importance will be helpful for decision makers to better handle these barriers and improve HCWM effectiveness. This research, through the employment of gray relational analysis (GRA) prioritizes 14 barriers identified from literature, according to the degree of their negative impact. The study reveals that "Unauthorized Reuse of Health Care Waste" and Implementation of "Poor Segregation Practices" ranked 1 are perceived as the two most significant barriers while "Lack of Accountability of Authorities of Health Care Facilities towards HCWM" and "Inadequate Awareness and Training Programs" ranked 5 are perceived as the least important barriers of HCWM in India.

Keywords Health care wastes (HCW) · Barriers of waste management · Multi attribute decision making (MADM) · Health care unit (HCU) · Gray relational analysis (GRA)

S. Swain
Dhaneswar Rath Institute of Engineering and Management Studies, Cuttack, India
e-mail: suchismitaswain@rediffmail.com

K. Muduli (✉) · J.N. Biswal · T.K. Panda
CV Raman College of Engineering, Bhubaneswar, India
e-mail: kamalakantam@gmail.com

J.N. Biswal
e-mail: jitendra2000_biswal@yahoo.co.in

T.K. Panda
e-mail: ptusharkpanda@gmail.com

S. Tripathy
School of Mechanical Engineering, KIIT University, Bhubaneswar, India
e-mail: sushant.tripathy@gmail.com

© Springer Nature Singapore Pte Ltd. 2017
S.C. Satapathy et al. (eds.), *Proceedings of the 5th International Conference on Frontiers in Intelligent Computing: Theory and Applications*, Advances in Intelligent Systems and Computing 515, DOI 10.1007/978-981-10-3153-3_18

1 Introduction

Health care units (HCUs) such as hospitals and diagnostic centers are increasing at a faster rate in the country to satisfy the health care requirements of the morbid population. During the immunization or treatment of human beings, diagnosis, surgical procedures, and other activities aimed at abating health problems, HCUs generate significant amount of waste known as health care waste [1]. These wastes including used bandages, syringes, human tissues, and used culture media containing microorganisms [2] are infectious in nature and can transmit the infections to the hospital staff attendants, and the nearby public. Common waste disposal practices adopted by most of the HCUs such as waste dumping in the roadside open bins or low lying areas or directing the untreated waste into the water bodies leads to the growth multiplication of insects, rodents and worms, which in turn lead to the transmission of diseases like typhoid, cholera, HIV, Tuberculosis, and Hepatitis B and C [3]. Further, this kind of indiscriminate disposal of health care waste may lead to contamination of soil, air, and water with harm full metals like mercury, carcinogenic gases like dioxin and toxic chemicals like xylene and formalin [2] and is a matter of concern.

Owing to the rapid growth in population, size and number of healthcare units, and use of disposable medical products, there have been a tremendous rise in the volume of healthcare waste (HCW) generation worldwide over the last few years [4]. Safe disposal of this voluminous amount of HCW has emerged as an issue of major concern not only for the HCUs but also for the environment [5]. Realizing the hazardous nature of healthcare waste, several developed countries have woken up to address this threat, and the HCUs of these countries have adopted many commendable positive steps [6]. However, Indian HCUs do not share the same enthusiasm. This could be due to their lack of knowledge regarding various influential factors of HCWM, particularly the barriers of HCWM implementation. Knowledge of these barriers will help the decision makers to formulate proactive strategies to minimize their impact.

Hence, this research makes an attempt to

- identify various parameters that hinder HCWM adoption in Indian HCUs.
- rank these barriers according to their degree of inhibiting strength using GRA.

2 Literature Review

Owing to the growing consciousness about health care waste and issues associated with its improper management, a number of studies have been conducted in several countries that include Ethiopia [7]; Iran [8]; China [9, 10, 11]; United Kingdom [12, 13]. Many researchers have also addressed health care waste issues in Indian

Table 1 Barriers of HCWM in Indian context

Sl. no.	Barriers	References
1.	Implementation of Poor Segregation Practices	[14, 18, 19]
2.	Adoption of Inappropriate Waste Management Operational Strategy	[20, 21, 19]
3.	Inadequate Assistance from Government Agencies	[20, 21, 17]
4.	Lesser importance to Green Purchasing Practices	[20, 22, 17]
5.	Illegitimate Reuse of Health Care Waste	[20, 23, 21]
6.	Lack of Accountability of Authorities of Health Care Facilities toward HCWM	[24, 21, 19]
7.	Unavailability of Adequate Waste Management Equipment and Facilities	[25, 2, 6]
8.	Financial Constraints	[20, 25, 17]
9.	Inadequate Awareness and Training Programs	[12, 2, 18]
10.	Reluctance to Change and Adoption	[21]
11.	Lack of Awareness among patients, attendants and people at large.	[24, 26]
12.	Lack of coordination between municipality, Pollution Control Board and hospital authorities	[24, 3]
13.	Lesser Prioritization to Waste Management issues in policies of Health Care Units	[24]
14.	Lack of Strict implementation of infection control measures like sterilization and disinfectant techniques	[24, 6]

context [14, 15, 2, 16, 17, 1, 4, 6]. However, studies that exclusively analyze the strength of barriers of HCWM are scarce.

To bridge this gap 14 barriers of HCWM shown in Table 1, are identified from literature

3 Methodology

To determine the most influential barriers causing hindrance in the implementation of SSCM, a standard questionnaire was designed that consists of 14 barriers identified from literature. A survey was conducted by sending the questionnaire by mail and personal contacts to different experts, health workers, and employees of various HCUs. The survey was conducted by using likert scale 1 to 5, (i.e., 1 = "Totally disagree," 2 = "Partialy disagree," 3 = "No opinion," 4 = "Partially agree," 5 = "Completely agree"). The respondent scores were used for prioritization of HCWM barriers using gray relational analysis.

The biggest advantage of the gray system theory is its capability of generating satisfactory outcomes using a relatively low level of information or with great variability in parameters [27]. In gray theory, categorization of the systems depends upon the degree of availability of information; the systems capable of providing adequate amount of required information is termed as white system while a black system refers to the systems with entirely unavailable information; and a system with partially available information is called a gray system [28]. The gray system focuses keenly on what partial or limited information the system can provide, and tries to paint its total picture from this [29]. Problem analysis using GRA involves the following steps.

Step-1: Generation of Gray Relations

Different attributes of a MADM problem might require different units for measurement of their performance. Moreover, certain performance attributes might have a very large range. Further, existence of attributes having different goals and directions also lead to erroneous results in the analysis [30]. GRA addresses these issues by generating gray relational, where all performance values for each alternative is processed into a comparability sequence, in a process analogous to normalization [31].

The ith alternative of a MADM involving k alternatives and p attributes, can be expressed as $B_i = (b_{i1}, b_{i2},\ldots\ldots b_{ij},\ldots b_{ip})$, where b_{ij} represents the performance value of attribute j of ith alternative. Translation of the term B_i can be done into the comparability sequence

$A_i = (a_{i1}, a_{i2},\ldots\ldots a_{ij},\ldots a_{ip})$, by use of one of Eqs. 1, 2 or 3.

$$a_{ij} = \left[b_{ij} - \min\{b_{ij}, i=1,2\ldots k\}\right] / \left[\max\{b_{ij}, i=1,2\ldots k\} - \min\{b_{ij}, i=1,2\ldots k\}\right]$$
For $i = 1,2\ldots k; \ j = 1,2\ldots p$ (1)

$$a_{ij} = \left[\max\{b_{ij}, i=1,2\ldots k\} - b_{ij}\right] / \left[\max\{b_{ij}, i=1,2\ldots k\} - \min\{b_{ij}, i=1,2\ldots k\}\right]$$
For $i = 1,2\ldots m; \ j = 1,2\ldots p$ (2)

$$a_{ij} = 1 - \frac{|b_{ij} - b_j^*|}{Max\left\{Max\{b_{ij}, i=1,2\ldots\ldots k\} - b_{ij}^*, b_{ij}^*, - Min\{y_{ij}, i=1,2\ldots\ldots k\}\right\}}$$
For $i = 1,2\ldots k; \ j = 1,2\ldots p$ (3)

The generated gray relation values are shown in Table 2.

Step-2: Derivation of Reference Sequence

After the normalization process using GRG, all performance values are defined within the range [0, 1]. The alternative with larger value of A_i (equal to unity) is assumed to have highest strength. However, chances of existence of such kind of performance attributes are nil. Hence, a reference sequence A_o whose values are equal to 1 is defined and compared with the generated sequence. The reference sequence is expressed as follows:

$A_o = (a_{01}, a_{02}, a_{03}\ldots\ldots a_{0j}\ldots\ldots a_{0p}) = (1,1\ldots\ldots1\ldots..1)$ [32].

Step-3: Evaluation of gray relational coefficient

Gray relational coefficient is calculated to assess the degree of similarity between x_{0j} and x_{ij}.

Table 2 Results of grey relational generating for HCWM Barriers

Barriers	Values
1.	0
2.	0.25
3.	0.25
4.	0.25
5.	1
6.	0
7.	0.5
8.	0.75
9.	0
10.	0.75
11.	0.75
12.	0.25
13.	0.75
14.	0.75

$$\left(A_{oj}, \ A_{ij}\right) = \left[\Delta_{min} + \xi\Delta_{max}\right] / \left[\Delta_{ij} + \xi\Delta_{max}\right]$$
$$\text{For i} = 1, 2 \ldots k; \quad j = 1, 2 \ldots p \tag{4}$$

$$\Delta_{ij} = \left|A_{oj} - A_{ij}\right|; \ A_{oj} = 1$$
$$\Delta_{min} = \min\{\Delta_{ij}, i = 1, 2 \ldots k; j = 1, 2 \ldots p\}$$
$$\Delta_{max} = \max\{\Delta ij, i = 1, 2 \ldots k; j = 1, 2 \ldots p\}$$

ξ = distinguishing coefficient $\in [0, 1]$
In our case $\xi = 0.5$

Step-4: *Finding Gray Relational Grade*

$$\Gamma(A_0, \ A_i) = W_j \ x \ B(A_{oj}, A_{ij}) \ \text{For i} = 1, 2 \ldots k$$

Step-5: Prioritization of Variables.

The variables are ranked according to their gray relational grades.

4 Results and Discusion

The performance attributes considered here belong to the-larger-the-better category. Hence, Eq. (1) was adopted for the gray relational generating process and the results are shown in Table 2.

Table 3 Results of gray relational coefficient for HCWM Barriers

Barriers	Values
1.	1
2.	0.4
3.	0.4
4.	0.4
5.	1
6.	0.333333
7.	0.5
8.	0.666667
9.	0.333333
10.	0.666667
11.	0.666667
12.	0.4
13.	0.666667
14.	0.666667

The gray relation generating values of the barriers are converted to their gray relation coefficient using step 2 and 3 discussed in methodology section and results are shown in Table 3.

The gray relational grades and ranks of all the barriers were found out following Steps 4 and 5 and are shown in Table 4. The barriers with top rank (here 1) is

Table 4 Gray relational grades and corresponding ranks (Assuming, W as unity)

Sl no.	Barriers	Values	Rank
1.	Implementation of Poor Segregation Practices	1	1
2.	Adoption of Inappropriate Waste Management Operational Strategy	0.4	4
3.	Inadequate Assistance from Government Agencies	0.4	4
4.	Lesser importance to Green Purchasing Practices	0.4	4
5.	Illegitimate Reuse of Health Care Waste	1	1
6.	Lack of Accountability of Authorities of Health Care Facilities toward HCWM	0.333333	5
7.	Unavailability of Adequate Waste Management Equipment and Facilities	0.5	3
8.	Financial Constraints	0.666667	2
9.	Inadequate Awareness and Training Programs	0.333333	5
10.	Reluctance to Change and Adoption	0.666667	2
11.	Lack of Awareness among patients, attendants and people at large.	0.666667	2
12.	Lack of coordination between municipality, Pollution Control Board and hospital authorities	0.4	4
13.	Lesser Prioritization to Waste Management issues in policies of Health Care Units	0.666667	2
14.	Lack of Strict implementation of infection control measures like sterilization and disinfectant techniques	0.666667	2

considered to have highest inhibiting impact, while the barriers having lowest rank (here 5) have lowest inhibiting impact.

5 Conclusion

Effectiveness of HCWM practices of any HCU depends upon its ability to identify the potential barriers and understand the degree of negative influence of these barriers. This study explores 14 potential barriers of HCWM and analyzes their relative strength using GRA. It was observed from the study that "Illegitimate Reuse of Health Care Waste" and Implementation of "Poor Segregation Practices" are the most significant barriers, while "Lack of Accountability of Authorities of Health Care Facilities towards HCWM" and "Inadequate Awareness and Training Programs" are perceived as the least important barriers of HCWM in India. Prioritization of these barriers will help the decision makers to identify the weaker areas (barriers with top ranks) and formulate strategies to bring improvements in those areas.

References

1. Chethana, T., Thapsey, H., Gautham, M. S., Sreekantaiah, P., & Suryanarayana, S. P.. Situation analysis and issues in management of biomedical waste in select small health care facilities in a ward under Bruhat Bengaluru Mahanagara Palike, Bangalore, India. *Journal of community health*. 39(2) (2014) 310–315.
2. Dwivedi, AK, Pandeyand S, Shashi. Fate of hospital waste in India. *Biology and Medicine*. 1 (3) (2009) 25–32.
3. Gupta S, Boojh R, Dikshit AK. Environmental Education for Healthcare Professionals with Reference to Biomedical Waste Management -A Case Study of a Hospital in Lucknow, India. *International Research Journal of Environment Science*). 1(5) (2012) 69–75.
4. Thakur, V., & Ramesh, A. Healthcare waste management research: A structured analysis and review (2005–2014). *Waste Management & Research*, (2015) pp. 1–16, 0734242X15594248.
5. Patan, S., & Mathur, P., Assessment of biomedical waste management in government hospital of Ajmer city–a study. *International Journal of Research in Pharmacy & Science*, 5(1) (2015).
6. Biswal, M., Mewara, A., Appannanavar, S. B., & Taneja, N.. Mandatory public reporting of healthcare-associated infections in developed countries: how can developing countries follow?. *Journal of Hospital Infection*, 90(1) (2015) 12–14.
7. Tesfahun, E., Kumie, A., & Beyene, A., Developing models for the prediction of hospital healthcare waste generation rate. *Waste Management & Research*, 34(1) (2016) 75–80.
8. Jaafari, J., Dehghani, M.H., Hoseini, M. and Safari, G.H. 'Investigation of hospital solid waste management in Iran', *World Review of Science, Technology and Sustainable Development*. (2015) Vol. 12, No. 2, pp. 111–125.
9. Cheng, Y.W., F.C. Sung, Y. Yang, Y.H. Lo, Y.T. Chung, K.-C. Li. Medical waste production at hospitals and associated factors. Waste Management. 29 (2009) 440–444.
10. Yong,Z., X. Gang, W. Guanxing, Z. Tao, J. Dawei. Medical waste management in China: A case study of Nanjing. *Waste Management*. 29(4) 2009 1376–1382.

11. Liu, H. C., You, J. X., Lu, C., & Shan, M. M.. Application of interval 2-tuple linguistic MULTIMOORA method for health-care waste treatment technology evaluation and selection. *Waste Management, 34*(11) (2014) 2355–2364.
12. Tudor,T.L., C.L. Noonan and L.E.T. Jenkin. Healthcare waste management: a case study from the National Health Service in Cornwall, United Kingdom. *Waste Management.* 25 (2005) 606–615.
13. Blenkharn,J.I.: Lowering standards of clinical waste management: do the hazardous waste regulations conflict with the CDC's universal/standard precautions? *Journal of Hospital Infection.* 62 (2006) 467–472.
14. Gupta S, Boojh R. Report: Biomedical waste management practices at Balrampur Hospital, Lucknow, India. *Waste Management Research.* 24 (2006) 584–591.
15. Verma, L. K., Mani, S., Sinha, N., & Rana, S.: Biomedical waste management in nursing homes and smaller hospitals in Delhi. *Waste Management. 28*(12) (2008) 2723–2734.
16. Gupta, S., R. Boojh, A. Mishra, and H. Chandra.: Rules and management of biomedical waste at Vivekananda Polyclinic: A case study. *Waste Management.* 29 (2009) 812–819.
17. Muduli K, Barve A.: Barriers to Green Practices in Health Care Waste Sector: An Indian Perspective. International Journal of Environmental Science and Development. 3(4) (2012) 393–399.
18. Abdulla F, H A Qdais, A Rabi: Site investigation on medical waste management practices in norther Jordan. 28(2) (2008) 450–458.
19. Athavale, A.V., and G. B. Dhumale.: A Study of Hospital Waste Management at a Rural Hospital in Maharastra. *Journal of ISHWM.* 9(1) (2010) 21–31.
20. Patil, AD,. Shekdar AV.: Health-care waste management in India. *Journal of Environmental Management.* 63 (2001) 211–220.
21. Verma LK. Managing Hospital Waste is Difficult: How Difficult? *Journal of ISHWM.* 9(1) (2010) 46–50.
22. Kaiser,B., P.D. Eagan and H. Shaner.: Solutions to Health Care Waste: Life-Cycle Thinking and "Green" Purchasing. *Environmental Health Perspectives.*109 (2001) 205–207.
23. Patil, V. Gayatri, and K. Pokhrel.: Biomedical solid waste management in an Indian hospital: a case study. *Waste Management.* 25 (2005) 592–599.
24. Yadav, M.: Hospital Waste-A major problem. *JK PRACTITIONER. 8*(4) (2001) 276–282.
25. Rao, S.K.M., R.K. Ranyal, S.S. Bhatia and V.R. Sharma.: Biomedical Waste Management: An Infrastructural Survey of Hospitals. *Medical Journal Armed Forces India.* 60(4) (2004) 379–382.
26. Shivalli S. and Sanklapur V.: Healthcare Waste Management: Qualitative and Quantitative Appraisal of Nurses in a Tertiary Care Hospital of India. *The Scientific World Journal* (2014), Article ID 935101: PP 1–7.
27. Rajesh, R. and Ravi V. Supplier selection in resilient supply chains: a grey relational analysis approach, Journal of Cleaner Production. 86 (2015) 343–359.
28. Chang, K. H., Chang, Y. C., & Tsai, I. T.: Enhancing FMEA assessment by integrating grey relational analysis and the decision making trial and evaluation laboratory approach. *Engineering Failure Analysis.* (2013) 211–224.
29. Huang, S. J., Chiu, N. H., & Chen, L. W.: Integration of the grey relational analysis with genetic algorithm for software effort estimation. *European Journal of Operational Research. 188*(3) (2008) 898–909.
30. Huang, J. T., & Liao, Y. S.: Optimization of machining parameters of Wire-EDM bases on grey relation and statistical analysis. International Journal of Production Research. 41 (2003) 1707–1720.
31. Kuo, Y., Yang, T., & Huang, G. W.: The use of grey relational analysis in solving multiple attribute decision-making problems. *Computers & Industrial Engineering. 55*(1) (2008) 80–93.
32. Omoniwa, B.: A Solution to Multi Criteria Robot Selection Problems Using Grey Relational Analysis. *International Journal of Computer and Information Technology* 3(2) (2014): 329–332.

Privacy-Preserving Association Rule Mining Using Binary TLBO for Data Sharing in Retail Business Collaboration

G. Kalyani, M.V.P. Chandra Sekhara Rao and B. Janakiramaiah

Abstract Sharing of data provides mutual benefits for collaborating organizations. Data mining techniques have allowed regimented discovery of knowledge from huge databases. Conversely, in the case of sharing the data with others, knowledge discovery raises the possibility of revealing the sensitive knowledge. The need of privacy prompted the growth of numerous privacy-preserving data mining techniques. In order to deal with privacy concerns, the database is to be transformed into another database in such a way that the sensitive knowledge is concealed. One subarea of privacy-preserving data mining, which got attention in retail businesses, is privacy-preserving association rule mining. A significant feature of privacy-preserving association rule mining is attaining a balance between privacy and precision, which is characteristically conflicting, and refining the one generally reduces the other one. In this paper, the problem has been planned in the perspective of protecting association rules which are sensitive by prudently amending the transactions of the database. To moderate the loss of non-sensitive association rules and to improve the quality of the transformed database, the proposed approach competently estimates the impact of an alteration to the database. The proposed method selects the transactions for alterations using the binary TLBO optimization technique during the concealing process. Experimental outcomes exhibit the efficiency of the proposed algorithm.

Keywords Data sharing · Privacy · Sensitive knowledge · Sanitization · TLBO · Optimization · Association rule mining

G. Kalyani (✉)
Acharya Nagarjuna University, Guntur, India
e-mail: kalyanichandrak@gmail.com; kalyani4work@gmail.com

M.V.P. Chandra Sekhara Rao
RVR & JC College of Engineering, Guntur, India
e-mail: manukondach@gmail.com

B. Janakiramaiah
DVR & Dr HS MIC College of Technology, Vijayawada, India
e-mail: bjanakiramaiah@gmail.com

© Springer Nature Singapore Pte Ltd. 2017
S.C. Satapathy et al. (eds.), *Proceedings of the 5th International Conference on Frontiers in Intelligent Computing: Theory and Applications*, Advances in Intelligent Systems and Computing 515, DOI 10.1007/978-981-10-3153-3_19

189

1 Introduction

Data sharing among the organizations leads to mutual benefits [1]. Data mining techniques are formerly used for discovering the unknown knowledge in continuously increasing assortment of data [2]. In the earlier period, people faced situations where data mining algorithms are intended for revealing the sensitive knowledge which was not intended to be disclosed to others.

An example situation taken from the effort of Verykios et al. [3] encourages the need of association rule hiding algorithms to shield sensitive association rules from confession. The marketing chief of BigMart, which is an out sized supermarket, has made an agreement with the Dedtrees Paper Company. Dedtrees agreed to give its goods with minimum prices, as long as BigMart decided to give permission to operate customer purchase database. BigMart accepted the agreement and Dedtrees began mining the customers' data. They identified an interesting rule that those who were buying skim milk as well bought the Green Paper product. Noticing this, Dedtrees has announced an offer of 30% concession on skim milk with a purchase of Dedtrees product. The offer reduced the sales of Green Paper, and therefore raises the prices to BigMart, because of the reduction in the sales. During the next agreement with Dedtrees, BigMart identified that due to less competition they were disinclined to give the products at minimum price. At the end, BigMart began losing the business to opponents, who were able to get a better deal with Green Paper. The above-mentioned scenario states that BigMart should protect the sensitive knowledge before providing their data to Dedtrees, so that Dedtrees does not control the paper market.

2 Related Work

Data distortion methods work by picking particular items for including into (or excluding from) a number of transactions of the original dataset [4, 10] to enable the protection of association rules which are sensitive.

T.-P. Hong, et al. [5] suggested a method called sensitive items frequency-inverse database frequency (SIF-IDF) to identify the most relevant transactions to the given sensitive itemsets. The TF–IDF value is used for decreasing the frequencies of sensitive itemsets in the hiding process.

Hai Quoc Le, et al. [6] introduced an algorithm using the concept of an intersection lattice of frequent itemsets (FI). By exploring the characteristics of the intersection lattice of FI, itemsets in the generating set of FI (Gen (FI)) are generated during the hiding process.

J. Bonam et al. [7] introduced the sanitization algorithm which preserves the privacy of sensitive rules by applying particle swarm optimization, to maximize the accuracy of the transformed database by minimizing the loss of non-sensitive rules.

3 TLBO Algorithm

Teaching learning-based optimization (TLBO) is a newly projected algorithm which represents the conventional teaching, learning process of a classroom [8, 9]. TLBO models two basic approaches of learning: (i) from the teacher (known as Teaching phase) and (ii) discussion with others (known as Learning phase). The best learner is considered as the teacher in the group. In teaching phase, learners gain knowledge from the teacher and the teacher tries to enhance the knowledge of other individual by increasing the average knowledge of the class towards the position of teacher. In learning phase, algorithm represents the learning of the students with a discussion among themselves. An individual will gain new knowledge if the other individual has more knowledge than him/her. If the updated solution is better, it is accepted into the population. The process will continue until the stopping criterion is satisfied.

4 Problem Formulation

Given with original dataset (DS), comprising transactions which contain a transaction identification number (TID) and purchased list of items, minimum support (MST), and confidence (MCT) threshold given by the data owner, the data owner will identify a subset of association rules mined from DS as sensitive knowledge. The aim of privacy-preserving association rule mining is to create a new, transformed dataset (DS^1) from DS, from which sensitive association rules (SAR) will be protected, by negligibly affecting the non-sensitive association rules (NSAR). The algorithm transforms DS in such a way that, when DS^1 is mined at the equal (or a greater) levels, MST and MCT, only NSARs will be disclosed.

5 Proposed Algorithm

Let SAR is a set of association rules considered as sensitive and $LHS \Rightarrow RHS$ be the format of an association rule in SAR. The proposed method aims at hiding of $LHS \Rightarrow RHS$ by eliminating an item in LHS or RHS from a set of transactions until support $(LHS \Rightarrow RHS) < MST$ or confidence $(LHS \Rightarrow RHS) < MCT$. To balance privacy and accuracy, we propose an association rule hiding algorithm based on an optimization technique which is binary TLBO.

Algorithm 1: Association Rule Hiding using TLBO (ARH-TLBO)

Data: DS, MST, MCT, SAR, Association rules AR.

Result: DS^1.

begin

    repeat

        1. **for** *each rule* $X \Rightarrow Y$ **do**

            a.$T \leftarrow X \cup Y$

            b.$N \leftarrow min\{Sup(X \Rightarrow Y) - MST + 1,$

$$Sup(X \Rightarrow Y) - [Sup(X) * MCT] + 1)\}$$

            c.**for** *every* $I \in T$ **do**

                └ Estimate Impact(I);

        2. Generate *VictimItem* $\leftarrow \{I/I, \forall J \in SAR, Impact(I) < Impact(J)\}$

        3. Generate *VictimRule* $\leftarrow \{R/R, \forall R1 \in SAR, VictimItem \in R$

$$\&Sup(R) < Sup(R1)\}$$

        4. **for** *each rule* $R1 \in SAR$ **do**

            a.**for** *each rule* $R2 \in SAR$ **do**

                └ Compute Similarity(R1, R2);

        5. Group the rules with maximum similarity;

        6. Repeat the steps 4 & 5 until maximum similarity > threshold;

        7. Call BTLBO();

        8. Remove VictimItem from Selected transactions of step 7;

        9. Update(AR) and Update(SAR);

    until *SAR is empty*;

end

Algorithm 2: Function BTLBO

---

begin

    1. Select the initial population;

    2. Compute the Fitness Function(FF) value for each of the population as

$$FF = \frac{Ns}{Ni};$$

    3. Identify the Teacher;

    4. Compute the mean(Avg) of the population;

    5. **Apply Teaching Phase as:**

    a.**for** *each student* $K \in Population$ **do**

        └ $Student_k^d = Student_k^d \cup (Teacher^d \cap T_f * Avg^d)$;

    b.Compute FF value of new population;

    c.Update the population based on FF value of old and new population;

    6. **Apply Learning Phase as:**

    a. **for** *each Student* $k \in Population$ **do**

        Select randomly another student m from the population;

        if *Student* k > *Student* m **then**

            └ $Student_k^d = Student_k^d \cup (Student_k^d - Student_m^d)$;

        else

            └ $Student_k^d = Student_k^d \cup (Student_m^d - Student_k^d)$;

    b.Compute the FF value of new population;

    c.Update the population based on FF value of old and new population;

    7. Repeat the steps from 2 to 6 until maximum number of iterations;

    8. Select top n students and return the corresponding transactions from the database.

end

Table 1 Database considered for demonstration

Transaction ID	Transaction items	Transaction ID	Transaction items
1	1, 0, 0, 0, 0, 1, 1, 0, 0, 1	11	1, 1, 0, 0, 1, 1, 1, 0, 1, 1
2	0, 0, 0, 0, 0, 0, 1, 1, 0, 0	12	1, 1, 0, 1, 1, 1, 1, 0, 1, 1
3	1, 0, 0, 0, 0, 1, 1, 0, 1, 0	13	1, 0, 0, 0, 0, 0, 0, 1, 1, 1
4	0, 1, 0, 0, 1, 1, 1, 0, 0, 1	14	1, 1, 0, 0, 1, 1, 1, 0, 1, 1
5	0, 1, 0, 1, 1, 0, 1, 1, 1, 0	15	1, 0, 0, 1, 1, 1, 1, 1, 0, 0
6	0, 0, 0, 1, 0, 0, 1, 1, 1, 0	16	0, 0, 0, 1, 1, 0, 1, 1, 0, 0
7	0, 0, 0, 1, 1, 0, 1, 1, 0, 1	17	1, 0, 0, 0, 1, 1, 1, 0, 1, 1
8	1, 1, 0, 1, 1, 1, 1, 1, 1, 1	18	1, 0, 0, 0, 0, 1, 1, 1, 1, 0
9	1, 0, 0, 0, 0, 0, 1, 1, 1, 0	19	0, 0, 0, 1, 1, 0, 1, 1, 1, 0
10	0, 1, 0, 0, 0, 0, 1, 0, 0, 0	20	0, 0, 0, 1, 1, 0, 1, 1, 0, 0

The estimation of impact (c of step 1 in Algorithm 1) considers an item I and the set of ARs as input. The procedure first picks the ARs in such a way that the item I must be present in those rules. The modified support of all the selected rules will be calculated by subtracting the n value from its support. At well, support of an item I, for which the impact is required, is also be modified by subtracting n from it and confidence will also be modified based on the modified support. Then the procedure counts the rules for which either the modified support or confidence is less than MST or MCT, respectively. The count value is taken as the impact of the item. After estimating the impact of all items in all the rules, the item with minimum impact will be preferred as victim item and the rule in which victim item is present and has least support is selected as victim rule (steps 1 to 3 of Algorithm 1).

Let us consider a database and a set of SAR to demonstrate the running of the proposed ARH-TLBO (Algorithm 1) method. An example database with 10 items and 20 transactions is shown in Table 1. Here bit 1 indicates the presence and bit 0 indicates the absence of the items in the transaction. Let us consider the set of SAR as $1 \Rightarrow 6, 7$ & $8 \Rightarrow 7$ which were mined from the database with MST = 40% and MCT = 70%. Based on the process shown in Algorithm 1 (Steps 1–3) the impacts of items 1, 6, 7, and 8 were estimated as 13, 6, 9, and 10, respectively. The rule $1 \Rightarrow 6, 7$ was selected as the victim rule because 6 has the least impact which is selected as the victim item. Then, apply BTLBO (Algorithm 2) to select the transactions for alteration from the database. Here, the population size was selected as four (because two transactions need to be altered to hide $1 \Rightarrow 6, 7$ (b of step 1 in Algorithm 1). Hence, randomly four transactions of the database are selected as students, which is initial population. The transactions which have more number of sensitive items and less number of non-sensitive items are considered as the best transactions for alterations. The learner with the highest FF value is considered as a teacher and mean of all the learners is also being evaluated. The initial population, their FF values, teacher, and the mean of the population (steps 1 to 4 of Algorithm 2) are shown in Table 2.

Table 2 Initial population, FF values, teacher, and mean values

S.no	Population	FF value	Teacher	Mean
1	0, 1, 0, 1, 1, 0, 1, 1, 1, 0	2/6 = 0.33		1, 1, 0, 1, 1, 1, 1, 1, 1, 1
2	1, 1, 0, 0, 1, 1, 1, 0, 1, 1	3/7 = 0.43		
3	0, 0, 0, 1, 1, 0, 1, 1, 0, 0	2/4 = 0.5	Selected	
4	1, 0, 0, 0, 0, 0, 1, 1, 1, 0	2/4 = 0.5		

Table 3 Results after the teaching phase (step 5 of Algorithm 2)

S.no	Initial population	FF value	Population after teaching Phase	FF value	Best population
1	0, 1, 0, 1, 1, 0, 1, 1, 1, 0	2/6 = 0.33	0, 1, 0, 1, 1, 0, 1, 1, 1, 0	2/6 = 0.33	0, 1, 0, 1, 1, 0, 1, 1, 1, 0
2	1, 1, 0, 0, 1, 1, 1, 0, 1, 1	3/7 = 0.43	1, 1, 0, 0, 1, 1, 1, 1, 1, 1	4/8 = 0.5	1, 1, 0, 0, 1, 1, 1, 1, 1, 1
3	0, 0, 0, 1, 1, 0, 1, 1, 0, 0	2/4 = 0.5	0, 0, 0, 1, 1, 0, 1, 1, 0, 0	2/4 = 0.5	0, 0, 0, 1, 1, 0, 1, 1, 0, 0
4	1, 0, 0, 0, 0, 0, 1, 1, 1, 0	2/4 = 0.5	1, 0, 0, 0, 0, 0, 1, 1, 1, 0	2/4 = 0.5	1, 0, 0, 0, 0, 0, 1, 1, 1, 0

Table 4 Results after the learning phase (step 6 of Algorithm 2)

S.no	Initial population	FF value	Other learner	Updated population	FF value	Best population
1	0, 1, 0, 1, 1, 0, 1, 1, 1, 0	2/6 = 0.33	2	1, 1, 0, 1, 1, 1, 1, 1, 1, 0	2/8 = 0.5	1, 1, 0, 1, 1, 1, 1, 1, 1, 0
2	1, 1, 0, 0, 1, 1, 1, 1, 1, 1	4/8 = 0.5	3	1, 1, 0, 0, 1, 1, 1, 1, 1, 1	4/8 = 0.5	1, 1, 0, 0, 1, 1, 1, 1, 1, 1
3	0, 0, 0, 1, 1, 0, 1, 1, 0, 0	2/4 = 0.5	4	0, 0, 0, 1, 1, 0, 1, 1, 0, 0	2/4 = 0.5	0, 0, 0, 1, 1, 0, 1, 1, 0, 0
4	1, 0, 0, 0, 0, 0, 1, 1, 1, 0	2/4 = 0.5	1	1, 0, 0, 0, 0, 0, 1, 1, 1, 0	2/4 = 0.5	1, 0, 0, 0, 0, 0, 1, 1, 1, 0

Apply the teaching phase (step 5) of Algorithm 2 on the initial population. In step 5 of Algorithm 2 algorithm, k was considered as dimension number. The example database contains 10 different items; hence, every student of the population will have 10 dimensions. Table 3 shows the results after the teaching phase, i.e., step 5 of Algorithm 2, is implemented. Then apply the learning phase (step 6) of Algorithm 2. In learning phase, every student will interact with other student to improve FF value. Table 4 shows the results after the learning phase is implemented. An iteration of Algorithm 2 is completed with teaching phase and learning phase.

Table 5 Results after 10 iterations of teaching and learning phases

S.no	Population	FF valuve
1	1, 1, 0, 1, 1, 1, 1, 1, 1, 0	4/8 = 0.5
2	1, 1, 0, 0, 1, 1, 1, 1, 1, 1	4/8 = 0.5
3	1, 0, 0, 1, 1, 1, 1, 1, 0, 0	4/6 = 0.67
4	1, 0, 0, 0, 0, 1, 1, 1, 1, 0	4/5 = 0.8

By continuing the process (step 7 of Algorithm 2) up to 10 iterations, the results are shown in Table 5. Select the transactions of the database which has 1, 0, 0, 0, 0, 1, 1, 1, 1, 0 patterns. The transactions with IDs 18 and 8 were selected and returned by BTLBO. Then the victim item 6 will be removed from those transactions to hide the sensitive rule $1 \Rightarrow 6, 7$.

6 Performance Metrics

(a) Hiding Failure (HF): Hiding failure is measured as the percentage of SARs that are discovered from DS^1.

(b) Miss Cost (MC): Miss cost is estimated as the percentage of NSARs that are not discovered from DS^1.

(c) Ghost Rules (GS): Ghost rules is evaluated as the percentage of the revealed rules that are new.

(d) Accuracy of DS^1 (Difference (DS, DS^1)): Accuracy of DS^1 is measured by the percentage of transactions that are altered in DS to generate DS^1.

7 Experiments and Evaluation

The datasets considered in the evaluation have been placed in IEEE ICDM03 with the file name Retail.dat and in FIMI repository with the name Bms1. The characteristics of the datasets are shown in Table 6. In this evaluation, the proposed algorithm is compared with the RSIF-PSOW algorithm presented in [7] to assess the side effects. The RSIF-PSOW algorithm uses particle swarm optimization for selection of sensitive transactions. To observe the performance of the proposed and RSIF-PSOW algorithms, cross-fold validation method is considered with value as 5. The cross-fold validation method randomly distributes the set of association rules mined from the given dataset into a number of groups such that each group contains five rules. The performance of the algorithms is evaluated based on the parameters miss cost, and accuracy of DS^1 (difference between DS and DS^1). The efficacies of these algorithms with respect to the metrics miss cost and accuracy of DS^1. Figure 1 shows

Table 6 Characteristics of the datasets used in the experiments

Dataset	Number of transactions	Number of items	MST(%)	MCT(%)	Number of rules	SAR sets
Retail	88,162	16,469	1	10	236	10
BMS1	59,602	497	0.5	0.01	126	10

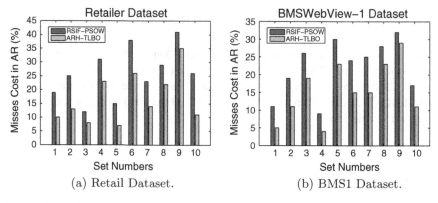

(a) Retail Dataset.

(b) BMS1 Dataset.

Fig. 1 Comparison of miss cost on individual sets of SAR

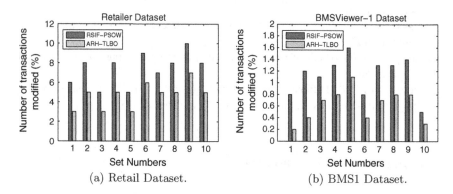

(a) Retail Dataset.

(b) BMS1 Dataset.

Fig. 2 Comparison of accuracy of datasets based on individual sets of SAR

the competence of the proposed algorithm in minimizing the miss cost. In view of that, the proposed algorithm achieved enhanced results in reducing the lost rules (non-sensitive) compared with RSIF-PSOW algorithm. The accuracy of the sanitized dataset increases, as the number of transactions modified in the original dataset decreases. Figure 2 shows the efficiency of the proposed algorithm in minimizing the difference between DS and DS^1. In view of that, the proposed algorithm attained improved results in reducing the difference between the DS and DS^1 when compared with RSIF-PSOW algorithm.

The proposed algorithm attains 0% hiding failure and ghost rules, i.e., the algorithm hides all the sensitive association rules and will not generate new rules from DS^1.

In summary, the evaluation shows that the proposed algorithm based on a computational intelligence technique TLBO yields good results when compared to other algorithms in minimizing the side effects and data distortions.

8 Conclusion

Privacy-preserving data mining has raised its importance in today's information analysis-based research and marketing. The intention of privacy-preserving association rule mining is to conceal some sensitive knowledge with an intention that they cannot be revealed with any association rule mining algorithm. The idea of estimating the impact in the proposed approach was used to reduce the number of alterations in the transactions and to keep the quality of the data intact. The proposed algorithm uses a binary TLBO-based selection of transactions to minimize the loss of non-sensitive association rules. The results indicate that the proposed algorithm based on binary TLBO efficiently selects the transactions for alteration in order to get the transformed dataset with high accuracy.

References

1. A. Amiri. Dare to share: Protecting sensitive knowledge with data sanitization. *Decision Support Systems*, 2007.
2. M.V.P. Chandra Sekhar Rao. Ch Aparna. Dr B Raveendra Babu. Dr A Damodaram. An improved multi-perceptron neural network model to classify software defects. *International Journal on Computer Science and Information Security (IJCSIS)*, 2011.
3. V S Verkios. E Bertino. L P Provenza. Y Saygin. E. Dasseni. Association rule hiding. *IEEE Transactions on on Knowledge and Data Engineering*, 2004.
4. Agrawal R Gehrke. J Evfimievski. A, Srikant.R. Privacy preserving mining of association rules. *In: Proceedings of SIGKDD Conference*, 2002.
5. Yang Kuo-Tung Wang Shyue-Liang Hong Tzung-Pei, Lin Chun-Wei. Using tf-idf to hide sensitive itemsets. *Applied Intelligence*, 2013.
6. Le Hai Quoc. Arch int Sojit. Nguyen Huy Xuan. Association rule hiding in risk management for retail supply chain collaboration. *Computer in Industry*, 2013.
7. J Bonam. A R Reddy. G Kalyani. Privacy preserving in association rule mining by data distortion using pso. *Proc, ICT Critical Infrastruct.*, 2014.
8. R Venkata Rao. Review of applications of tlbo algorithm and a tutorial for beginners to solve the unconstrained and constrained optimization problems. *Decision Science Letters, Growing-Science*, 2016.
9. R V Rao. V J Savsani. Teacher learning-basd optimization: a novel method for constrained mechanical design opttimization problems. *Comput. Aided Des.*, 2011.
10. Xun Yi, Fang-Yu Rao, Elisa Bertino, and Athman Bouguettaya. Privacy-preserving association rule mining in cloud computing. *In Proceedings of the 10th ACM Symposium on Information, Computer and Communications Security.*, 2015.

Performance Analysis of Compressed Sensing in Cognitive Radio Networks

N. Swetha, Panyam Narahari Sastry, Y. Rajasree Rao and G. Murali Divya Teja

Abstract In the recent research, compressive sampling (CS) has received attention in the area of signal processing and wireless communications for the reconstruction of signals. CS aids in reducing the sampling rate of received signals thereby decreasing the processing time of analog-to-digital converters (ADC). The energy minimization is the key feature of CS. In this work, CS has been applied to spectrum sensing in cognitive radio networks (CRN). The primary user (PU) signal is optimally detected using the sparse representation of received signals. The received PU signal is compressed in the time domain to extract the minimum energy coefficients and then applied to sensing. Further, the signal is detected using energy detection technique and recovered using l_1-minimization algorithm. The detection performance for various compression rates is analyzed.

Keywords Energy detection $\cdot$ Compression rate $\cdot$ l_1-minimization

1 Introduction

Dynamic spectrum access alleviates the problem of spectrum scarcity by introducing new wireless networks. CRN provides a promising solution for increasing the spectrum efficiency. Cognitive radio is a smart system that senses the electromag-

N. Swetha (✉) $\cdot$ G. Murali Divya Teja
GRIET, Hyderabad, India
e-mail: swethakarima@gmail.com

G. Murali Divya Teja
e-mail: 1murali5teja@gmail.com

P. Narahari Sastry
CBIT, Hyderabad, India
e-mail: ananditahari@yahoo.com

Y. Rajasree Rao
SPEC, Hyderabad, India
e-mail: yyrao315@yahoo.com

© Springer Nature Singapore Pte Ltd. 2017
S.C. Satapathy et al. (eds.), *Proceedings of the 5th International Conference on Frontiers in Intelligent Computing: Theory and Applications*, Advances in Intelligent Systems and Computing 515, DOI 10.1007/978-981-10-3153-3_20

netic spectrum for unused channels and then adapts them for communication without interfering with the licensed users. The channel is scanned continuously to identify the presence or absence of the signal using spectrum sensing algorithms. Wideband spectrum sensing needs efficient techniques to reduce the processing time. The wideband signal is filtered using several narrow band filters, and then the detection is performed. These signals need multi-channel ADCs sampling at a rate greater than Nyquist rate. Meanwhile, the traditional reconstruction techniques cannot provide necessary statistic using limited measurements. Therefore, a technique is required for sampling the signals at a rate lower than Nyquist frequency. CS accomplishes this task. CS theory refers to perform channel estimation using sparse samples compared to the conventional techniques [1]. A PU localization technique is developed using CS in order to improve the accuracy of sensing [2].

In [3], the signals are compressed using orthonormal basis functions and recovered using a l^p basis pursuit (BP) algorithm. However, the BP and orthogonal matching pursuit (OMP) algorithms use the preliminary information of the sparse nature for solving the problem. A tree-based OMP is proposed in [4] by exploiting the tree-like structure as additional information for reconstructing the signals. The advantage of the tree-like structure is that more elements are considered at a time thereby requiring less computation time than BS and OMP. In contrast with time domain CS, the multi-band detection technique is modeled in frequency domain [5]. Further, CS was applied to recognize the type of digital modulations in communication. The spectrum and moments of higher order are used as a measure to directly identify the different modulated signals without reconstructing the signals. The incoming radar pulses have been recovered using a photon-based CS system in [6].

This paper models a generalized framework for compressed spectrum sensing of a wideband signal and recovery using l_1-minimization algorithm. The wideband signal is first compressed using an orthogonal function in time domain for deriving the energy coefficients. Each narrow band signal is then detected using energy detection technique and recovered linearly. The rest of the paper is organized as follows. Section 2 introduces the compressed spectrum sensing model and Sect. 2 describes l_1-minimization algorithm. The results are discussed in Sect. 3 and finally, Sect. 4 summarizes the conclusions.

2 Compressed Spectrum Sensing Model

Consider a wideband signal x having L non-overlapping channels. The channel occupancy is not balanced at any particular location and time. The number of samples can be decimated using compression scheme.

2.1 Compression

Let x be a discrete time signal of size $N \times 1$. The sparse matrix y of size $K \times 1$ can be obtained as [7]

$$y = Ax \qquad (1)$$

where A is an orthogonal matrix of size $K \times N$, in which N and K denote the total number of samples and number of observations. The condition for compression is $K \ll N$. Hence, y is called as a measurement vector that results in reduced sampling rate. Before applying this signal to sensing, the compressed signal needs to be filtered using a set of narrow band filters.

2.2 Energy Detection Technique

In general, spectrum sensing model can be formulated as a binary hypothesis problem having true and null hypothesis [8].

$$H_0 : s(n) = w(n) \qquad (2)$$

$$H_1 : s(n) = y(n) + w(n) \qquad (3)$$

where $s(n)$ is the received narrow band signal of each channel to the secondary user, $y(n)$ is the compressed signal, and $w(n)$ is the Gaussian noise having zero mean and variance (σ_ω^2). The energy detector test statistic is given as [9]

$$E(s) = \frac{1}{N} \sum_{n=1}^{N} |s(n)|^2 \qquad (4)$$

The average energy $E(s)$ is compared with detection threshold T to evaluate the probability of detection. The performance factors that decide the ability of any sensing technique are probability of detection (p_d) (probability of successful detection under H_1), probability of false alarm (p_{fa}) (probability of incorrectly detecting under H_0), and signal-to-noise (SNR) wall of detection [10] (the minimum SNR value below which the detection is not possible). The detection threshold [8] is evaluated as

$$T = Q^{-1} \left(\frac{P_{fa}(n)}{\sqrt{N}} \right) + 1 \qquad (5)$$

where $Q(x)$ is a complementary error function.

2.3 l_1-minimization algorithm

In recent papers, convex optimization methods are used for reconstructing the signals from insufficient data. Compressed sensing depends on least squares method referred as l_p-norm. In statistics, l_p-norm of x is computed as

$$\|x\| = \sqrt[p]{\sum_i |x_i|^p} \quad p \in \mathbb{R}$$

The mathematical properties of different norms ($p \geqslant 1$) vary dramatically. Compressed sensing scheme finds the sparsest solution for indeterminated systems. The vector with few nonzero entries is called as the sparsest solution. This problem is usually solved using linear programming or optimization methods [11].

The minimum energy x_0 can be evaluated using measurement vector y in Eq. (1) as

$$x_0 = A'y \tag{6}$$

The linear program used for recovery of the original signal is primary dual algorithm [12], stated as

$$min\, x_0 \quad subject\, to \quad Ax = b, \quad f_i(x_0) \leq 0 \tag{7}$$

where (search vector) $x \in \mathbb{R}^N$, $b \in \mathbb{R}^K$, and f_i is a linear function of x_0:

$$f_i(x_0) = <c_i x_0> +d_i \quad c_i \in \mathbb{R}^N \quad d_i \in \mathbb{R} \tag{8}$$

where c_i and d_i are constants. The optimized solution is picked, if Karush–Kuhn–Tucker (KKT) conditions are satisfied [12]. These are first-order conditions that can be used for linear as well as nonlinear programming. The Newton's iterative method arrives at an interior point (x_0, v, λ) provided $f_i(x_0) < 0, \lambda > 0$. The parameters v and λ are duals to x_0. The complementary slackness condition used in our problem is

$$\lambda_i^* f_i(x_0) = -\frac{1}{\tau} \tag{9}$$

The parameter τ is responsible for the iterations in the Newton's method. The increase of τ progresses the interior point toward the solution on the boundary. The proposed problem quantifies the residuals namely primal, dual, and central as the modified KKT conditions.

$$rprimal = Ax_0 - b$$

$$rcentral = \begin{bmatrix} -\lambda_1 f_1 \\ -\lambda_2 f_2 \end{bmatrix} - \frac{1}{\tau} \tag{10}$$

$$rdual = \begin{bmatrix} \lambda_1 - \lambda_2 + A'v \\ 1 - \lambda_1 - \lambda_2 \end{bmatrix}$$

Algorithm 1 Pseudo-code for primary dual algorithm

Set primary dual tolerance value (pdtol) and maximum number of iterations (pdmaxiter) to 0.001 and 50, respectively.
 Step 1: Generate a minimum energy signal from equation (6)
Step 2: Define linear functions as $f1 = x_0 - u$ and $f2 = -x_0 - u$ (Eq. 8)
where $u = 0.95 \underset{\sim}{\jmath} x_0 \underset{\sim}{\jmath} + 0.1 \underset{\sim}{\jmath} x_0 \underset{\sim}{\jmath}_{max}$
Find dual variables $\lambda_i = -\frac{1}{f_i}$ $i \in [1, 2]$
Also find the third dual variable $v = -A(\lambda 1 - \lambda 2)$
Step 3: Derive the surrogate duality gap (sdg) and τ
$sdg = -(f_1' \lambda 1 + f_2' \lambda 2)$
$\tau = \frac{2\mu N}{sdg}$
Step 4: Compute rprimal, rdual, and rcentral from equation (10)
Step 5: Check for the stopping criterion variable *done*
$done = (sdg < pdtol) \underset{\sim}{\jmath} (pditer \geq pdmaxiter)$
Step 6: Newton's method [12]

    while $(\sim done)$ **do**
        Step 6.1:
        Calculate w_1, w_2 and w_3:
        $w_1 = -\frac{1}{\tau}(-\frac{1}{f_1} + \frac{1}{f_2}) - A'v$
        $w_2 = -1 - \frac{1}{\tau}(\frac{1}{f_1} + \frac{1}{f_2})$
        $w_3 = -rprimal$
        Step 6.2:
        Calculate $sigma_1$, $sigma_2$ and $sigma_x$:
        $sigma_1 = -\frac{\lambda_1}{f_1} - \frac{\lambda_2}{f_2}$
        $sigma_2 = \frac{\lambda_1}{f_1} - \frac{\lambda_2}{f_2}$
        $sigma_x = sigma_1 - \frac{sigma_2^2}{sigma_1}$
        Step 6.3:
        Calculate the deviation parameters dx, du, and dv using steps 6.1 and 6.2
        Step 6.4:
        Calculate the partial change in the dual variable λ_i
        Step 6.5:
        Calculate the suitable step size s
        Find the reconstructed signal x_p using back tracing search method
        $x_p = x_0 + sdx$
        $x_0 = x_p$
        repeat until *done* is true
    end while

The feasible step size s specifies the next move direction $(0 < s < 1)$ to derive the minimum energy coefficients of the signal. The complete process of recovering the signal is given in Algorithm 1. This algorithm can be employed for face recognition due to the sparse nature of human features [13].

3 Simulation Results

The wideband PU signal used in simulations is QPSK signal with five channels. The SNR assumption is 0 to -14 dB. The number of samples (N) is 512. The compression rate (K/N) varied from 10 to 100%. The SNR wall is evaluated at $p_d = 0.9$ and $p_{fa} = 0.1$ respectively. The detection performance of each narrow band-compressed signal is analyzed using ROC curves.

Figure 1 illustrates the original, minimum energy, and the recovered signals. The rate of compression assumed here is 50%, where $K = 256$ and $N = 512$. The original signal shown in Fig. 1a is the filtered narrowband signal of one channel. Figure 1b shows the compressed signal in time domain, evaluated using Eq. (6). It consists of minimum energy coefficients of the original signal. The recovery is made using Algorithm 1, shown in Fig. 1c. The figures clearly depict that most of the samples are successfully reconstructed even with 50% compression rate. The sparse nature of the compressed signal enables any spectrum sensing technique to identify the vacant spectrum bands in less time.

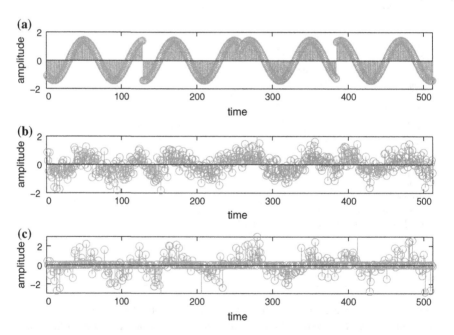

Fig. 1 **a** Received PU signal for single narrowband channel **b** Minimum energy signal **c** Recovery of compressed QPSK signal with $K/N = 50\%$

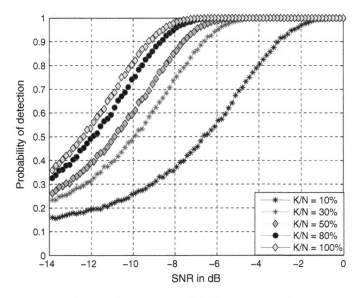

Fig. 2 SNR verses p_d of proposed CSS technique for different compression rates

Table 1 Comparison of SNR wall of the proposed method with variable compression rates

Compression rate in %	SNR wall (dB)
10	−3
30	−6.6
50	−7.6
80	−8.5
100	−9.2

Figure 2 shows the receiver operating characteristics (ROC) for observing the detection performance through p_d for each SNR value. The SNR wall of detection of the proposed technique is shown clearly for various compression rates. The 10% compressed signal is identified at −3 dB, 30% at −6.6 dB, 50% at −7.6 dB, 80% at −8.5 dB, and 100% at −9.2 dB SNR respectively. Table 1 shows these results. The detection strategy increases linearly with compression rate, but the increase in the compression rate increments the number of samples that add up the processing time. On an average, 50% compression can be chosen to detect the PU signals at −7.6 dB which achieves better performance than traditional energy detection technique. The same compression rates are considered for plotting p_d against p_{fa} in Fig. 3.

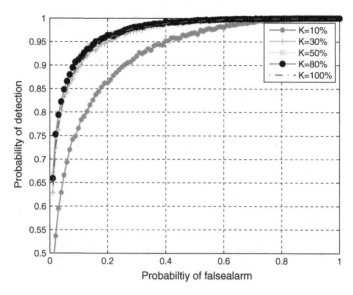

Fig. 3 p_d verses p_{fa} of proposed CSS technique for different compression rates

4 Conclusion

The computational complexity plays a significant role in wideband spectrum sensing. CS provides an optimal solution by compressing the samples at a rate less than 50%. The proposed work performs compression of a wideband signal and senses each narrowband channel using energy detection technique. In this work, a unique measurement matrix is produced to obtain minimum energy coefficients of the received signal. The spectrum sensing is initiated upon these condensed samples to analyze the performance of detection. The minimum energy signal is then reconstructed using l_1-minimization algorithm. The reconstruction can also be made feasible by adopting the optimization methods as a future direction of research. Furthermore, the simulation results are discussed for different compression rates making sparsest detection possible.

References

1. Sharma, S.K., Lagunas, E., Chatzinotas, S., Ottersten, B.: Application of compressive sensing in cognitive radio communications: A survey. (2016)
2. Ye, F., Zhang, X., Li, Y., Huang, H.: Primary user localization algorithm based on compressive sensing in cognitive radio networks. Algorithms **9** (2016) 25
3. Donoho, D.L.: Compressed sensing. IEEE Transactions on Information Theory, **52** (2006) 1289–1306

4. La, C., Do, M.N.: Tree-based orthogonal matching pursuit algorithm for signal reconstruction. In: IEEE International Conference on Image Processing, IEEE (2006) 1277–1280
5. Zhao, Q., Wu, Z., Li, X.: Energy efficiency of compressed spectrum sensing in wideband cognitive radio networks. EURASIP Journal on Wireless Communications and Networking **2016** (2016) 1
6. Guo, Q., Liang, Y., Chen, M., Chen, H., Xie, S.: Compressive spectrum sensing of radar pulses based on photonic techniques. Optics express **23** (2015) 4517–4522
7. Li, S., Wang, X., Zhou, X., Wang, J.: Efficient blind spectrum sensing for cognitive radio networks based on compressed sensing. EURASIP Journal on Wireless Communications and Networking **2012** (2012) 1–10
8. Liang, Y.C., Zeng, Y., Peh, E.C., Hoang, A.T.: Sensing-throughput tradeoff for cognitive radio networks. IEEE Transactions on Wireless Communications **7** (2008) 1326–1337
9. Swetha, N., Sastry, P.N., Rao, Y.R.: Analysis of spectrum sensing based on energy detection method in cognitive radio networks. In: International Conference on IT Convergence and Security (ICITCS), IEEE (2014) 1–4
10. R.Tandra, A.Sahai: Fundamental limits on detection in low snr under noise uncertainty. In: Proceedings of the International Conference on Wireless Networks, Communications and Mobile Computing, IEEE (2005) 464–469
11. Nocedal, J., Wright, S.: Numerical optimization. Springer Science & Business Media (2006)
12. Candes, E., Romberg, J.: l1-magic: Recovery of sparse signals via convex programming. URL: www. acm. caltech. edu/l1magic/downloads/l1magic. pdf **4** (2005) 46
13. Yang, A.Y., Zhou, Z., Balasubramanian, A.G., Sastry, S.S., Ma, Y.: Fast ℓ_1-minimization algorithms for robust face recognition. IEEE Transactions on Image Processing **22** (2013) 3234–3246

Robust Multiple Composite Watermarking Using LSB Technique

S. Rashmi, Priyanka and Sushila Maheshkar

Abstract Digital image watermarking is widely used for enforcing copyright protection and authentication. Color image watermarking has become essential as most of the images used are colored. A novel multiple composite digital image watermarking technique for color images is proposed in this paper. We have exploited the high embedding capacity property of Least Significant Bit (LSB) technique. It is more robust technique of watermarking multiple images in a single color image. Three different binary watermarks are taken as three channels and are combined to form single composite color watermark. The composite color watermark is further embedded in the color image using LSB technique. Simulation results exhibit that our proposed method has higher PSNR values indicating good visual quality of watermarked image. Experimental results show that the proposed scheme is robust under signal processing and geometric attacks.

Keywords PSNR · LSB · Multiple images · Digital image watermarking

1 Introduction

Advancements in computer and communication technologies have made it possible to copy, edit, and share images easily, without degrading its quality [1]. But this also gives rise to the problem of piracy, counterfeiting, and corrupting. Therefore, content authorization and copyright protection of images have become prime concern [2]. Digital image watermarking has evolved as an effective approach to solve this

S. Rashmi · Priyanka (✉) · S. Maheshkar
Department of Computer Science and Engineering, Indian School of Mines,
Dhanbad, Jharkhand, India
e-mail: priyankasingh401@gmail.com

S. Rashmi
e-mail: rashmi.s5991@gmail.com

S. Maheshkar
e-mail: sushila_maheshkar@yahoo.com

© Springer Nature Singapore Pte Ltd. 2017
S.C. Satapathy et al. (eds.), *Proceedings of the 5th International Conference on Frontiers in Intelligent Computing: Theory and Applications*, Advances in Intelligent Systems and Computing 515, DOI 10.1007/978-981-10-3153-3_21

problem [3]. In digital image watermarking, a visible or invisible watermark is embedding into the host/cover image. Watermark is embedded in such a fashion that it can be later extracted for authentication [4]. Watermarking is done either in spatial or frequency/transform domain [5]. Spatial watermarking techniques embed watermark by modifying pixel/intensity values of the cover image directly. Therefore these techniques have low computational complexity and high capacity [6]. Least significant bit (LSB), SSM modulation, patchwork, additive, and correlation-based watermarking techniques are popular methods used in spatial domain [7]. Transform domain watermarking techniques are more robust but have high computational complexity [8]. Some popular transforms such as singular value decomposition (SVD), discrete wavelet transform (DWT), discrete cosines transform (DCT), and discrete fourier transform (DFT) are used to transform image from spatial to frequency domain [9].

Here, we propose a new blind image watermarking scheme for color images. The proposed scheme is based on LSB and yielding good visual quality. Rest of the paper is organized as follows: in Sect. 2, we discuss the basics of LSB. Proposed watermarking scheme is explained in Sect. 3. Section 4 provides simulation results and discussions. Section 5 concludes the paper.

2 Least Significant Bit (LSB) Technique

Least Significant Bit is one of the oldest popular technique. Watermark bits are embedded by substituting the least significant bit of intensity values of the cover image. The watermark can be spread throughout the image or can exist in the selected locations of the cover image. LSB algorithm has higher embedding capacity. Watermark embedding and extraction is less time consuming [10]. LSB techniques are fragile which is advantageous in tamper detection and recovery. Therefore, LSB techniques are still important in watermarking. Tirkel et al. [11] proposed one of the first techniques for digital image watermarking technique. He proposed two techniques based on LSB. The first technique is based on LSB bit plane manipulation. In the second technique watermark is embedded to the image by applying linear addition to provide high security, as decoding is difficult. Celik et al. [12] suggested a lossless watermarking technique based on generalization of LSB.

Kutter et al. [13] proposed a spatial domain method to embed watermark using amplitude modulation. Lu et al. [14] proposed another color image watermarking scheme based on Kutter's scheme and uses neural network in a different way from Yu et al. [15]. A multiple watermarking technique for color images was proposed by Nasir et al. [6] where the cover image is split into four regions, yielding four blocks of size 128 × 128. Encrypted binary watermark is embedded in four regions of the blue component of the cover image. Similarly, Fu et al. [16] proposed to embed the watermark in the blue channel of color image.

3 The Proposed Method

In this method, multiple watermarks are embedded in a single colored image using the spatial domain technique of watermarking. The key advantage of the proposed technique is, that the embedded text image does not introduce substantial distortion in host image. Thus the text watermark cannot be perceived by Human visual system (HVS). Only the LSB bit is extracted from the RGB plane of the watermarked image. Hence the three watermarks are obtained.

The algorithm includes codes to handle attacks such as salt and pepper noise, rotation, and scaling of image. It performs 2D median filtering on the attacked watermarked image to reduce the effect of the attack [17]. And by understanding the local features of the watermarked image, the algorithm finds out the geometric transformation between the images, hence overcoming rotation and scaling attacks on the image. We preserve the watermarks from rotation attack by finding out the original image before the distortion. First we detect features in both the original and distorted watermarked image. These detected features are extracted from both the images. We match the extracted features using descriptors.

The locations of the corresponding points are retrieved. Hence using the point pairs matching, the transformation is found using the MSAC algorithm. It removes the outliers while computing the transformation matrix. Using the geometric transform, we recover the angle and compute its inverse to recover the distortion. After finding the angle, we re-rotate the image to restore it. Hence by finding out the values we can restore back the original watermarked image. After reducing the effects of the above stated attacks from the watermarked image, watermarks are detected from the least significant bit of the watermarked image. Watermarks is preserved from scale attack by finding out the original image before the distortion. Similar to the previous case, we find out the geometric transform, tform, and recover the scale. Using the scale, we rescale the distorted image to recover the original image.

3.1 Watermark Embedding Algorithm

1. Input the color image and the three binary images to be embedded. All the watermark images are of same size as the cover image.
2. Three binary images are combined to form one color image with each image in red, blue, and green plane, respectively.
3. For improved security, the watermark is shifted k times. Where K is a secret key.
4. The composite watermark is embedded in the first bit, which is the LSB of the colored cover image. The red plane of the created watermark is embedded in the LSB of red plane of the cover image. A similar operation is performed on the blue and green planes.

5. If the pixel of the watermark is greater than or equal to 128, the LSB bit of the input image is set to 1, else it is set to 0.
6. The final watermarked image is obtained.

3.2 Watermark Extracting and Attack Handling Algorithm

1. Median filtering is applied on each of the red plane,blue plane, and green plane of the colored watermarked image.
2. The feature descriptors extracted from the original cover image and distorted watermarked image are matched. Locations of such corresponding matched points are retrieved.
3. The transformation corresponding to the matching point pairs are computed.
4. The angle by which the image has been rotated is calculated. Using the angle obtained,the image is rotated back to its original state.
5. The scale by which the image has been scaled is calculated. Using the scale factor obtained,the image is scaled back to its original state.
6. Watermark bits are extracted from the LSBs of the watermarked image.
7. The watermarks obtained is shifted k times in the opposite direction as when it was embedded.
8. The obtained watermarks are displayed.

4 Simulation Results and Discussion

We present simulation results of the proposed technique in this section. Computer simulations were performed on Matlab2013. A set of standard color images, i.e., Pepper,Lena, Mandrill, and Flowers of size $512 \times 512 \times 24$ were taken as test images. We have taken three binary watermark of size 512×512 as shown in Table 1 (Figs. 1 and 2).

We use Peak Signal to Noise Ratio (PSNR) and Mean Square Error (MSE) to evaluate the imperceptibility of the proposed scheme.

$$\text{PSNR} = 10 \log_{10} \frac{max(P(i,j))^2}{\text{MSE}} \qquad (1)$$

where i, j are the coordinates of cover image (P) pixel. MSE is calculated between the cover image (P) and the watermarked image P' is defined as Eq. 2

$$\text{MSE} = \frac{\sum_{j=1}^{N} \left(\sum_{i=1}^{M} (P_{i,j} - P'_{i,j})^2 \right)}{M \times N} \qquad (2)$$

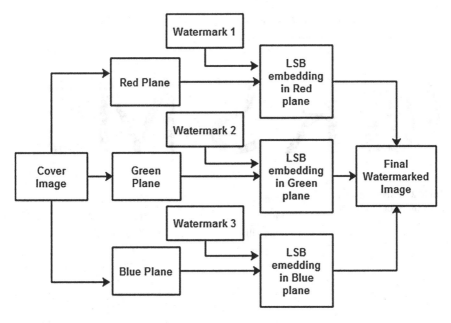

Fig. 1 Proposed LSB embedding scheme

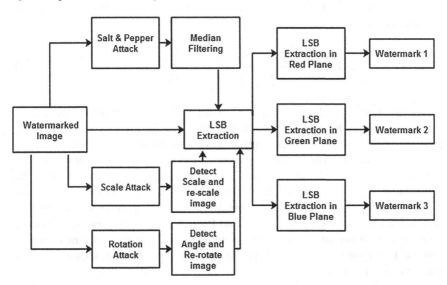

Fig. 2 Proposed LSB extraction scheme

Table 1 (a) Host image, (b)–(d) Watermarks, (e) Composite watermark

Cover image	Watermarked 1	Watermark 2	Watermark 3	Composite watermark
	R	S	M	

Table 2 PSNR (dB) for different color test images

Images	PSNR(dB)
Lena	56.9900
Mandrill	56.9193
Peppers	56.8278
Flowers	57.2740

Normalized Correlation (NC) is used as metric to express the resemblance between original watermark ($W_{original}$) and extracted watermark ($W_{extracted}$). NC can be expressed as Eq. 3.

$$NC = \frac{\sum_{i=1}^{N} \sum_{j=1}^{N} W_{original}(i,j) * W_{extracted}(i,j)}{\sum_{i=1}^{N} \sum_{j=1}^{N} W_{original}(i,j))^2} \tag{3}$$

Table 2 shows PSNR values obtained from test images. Average PSNR is more than 56 dB, which is more than 37 dB (threshold value). Hence, we can say that proposed scheme meets the imperceptibility requirement of watermarking.

NC values of the three different watermarks extracted from different test images are shown in Table 3. Extracted watermark is valid and disgustingly visible as the NC value is more than the threshold (0.75). To assess the robustness of the proposed watermarking scheme, various common attacks like rotation, salt and pepper, scaling, and Median filter were imposed on the watermarked images. Watermarked image under various attacks and corresponding extracted watermarks are shown in Table 4. Experimental results in Table 3 show that the proposed scheme is robust under most of the attacks.

Table 3 NC Values for different attacks

Attack	Lena			Mandrill			Peppers			Flowers		
	W1	W2	W3	W1	W2	W3	W1	W2	W3	W1	W2	W3
Zero attack without scrambling	0.9814	0.9942	0.9912	0.9816	0.9943	0.9913	0.9813	0.9944	0.9916	0.9815	0.9944	0.9916
Zero attack with scrambling	0.8081	0.8322	0.7532	0.8082	0.8319	0.7527	0.8072	0.8318	0.7523	0.8076	0.8320	0.7525
Salt and pepper attack without scrambling	0.9815	0.9945	0.9913	0.9817	0.9942	0.9914	0.9816	0.9944	0.9918	0.9816	0.9942	0.9914
Salt and pepper attack with scrambling	0.8032	0.8322	0.7536	0.8082	0.8322	0.7531	0.8080	0.8321	0.7528	0.8075	0.8321	0.7528
Rotation attack without scrambling	0.9819	0.9949	0.9925	0.9829	0.9958	0.9936	0.9793	0.9922	0.9894	0.9811	0.9941	0.9913
Rotation attack with scrambling	0.8067	0.8314	0.7527	0.8078	0.8319	0.7532	0.8068	0.8306	0.7526	0.8069	0.8314	0.7530
Scale attack without scrambling	0.7436	0.7210	0.7050	0.7850	0.7937	0.7793	0.8252	0.8442	0.8227	0.8390	0.8527	0.8321
Scale attack with scrambling	0.7259	0.7123	0.7020	0.7687	0.7756	0.7295	0.7575	0.7703	0.7442	0.7735	0.7881	0.7545

Table 4 Watermarked image and extracted watermarks under various attacks

Attack	Watermarked image	Extracted watermark 1	Extracted watermark 2	Extracted watermark 3
Zero attack		R	S	M
Salt and pepper attack		R	S	M
Rotation attack		R	S	M
Scale attack		R	S	M

5 Conclusion

Using the above technique, we can embed multiple binary images in a colored image hence exploiting the high capacity property of LSB technique to the maximum. The extraction used in the above technique is robust because it handles major attacks such as rotation, scaling, and salt and pepper noise hence making the technique more robust and useful. The proposed scheme has good imperceptibility also. This method can be utilized in web pages, email, and media.

References

1. Lei-Da Li and Bao-Long Guo. Localized image watermarking in spatial domain resistant to geometric attacks. *AEU-International Journal of Electronics and Communications*, 63(2):123–131, 2009.
2. Ramana Reddy, Munaga VN Prasad, and D Sreenivasa Rao. Robust digital watermarking of color images under noise attacks. *International Journal of Recent Trends in Engineering*, 1(1):334–338, 2009.
3. Qingtang Su, Yugang Niu, Qingjun Wang, and Guorui Sheng. A blind color image watermarking based on dc component in the spatial domain. *Optik-International Journal for Light and Electron Optics*, 124(23):6255–6260, 2013.
4. Sin-Joo Lee and Sung-Hwan Jung. A survey of watermarking techniques applied to multimedia. In *Industrial Electronics, 2001. Proceedings. ISIE 2001. IEEE International Symposium on*, volume 1, pages 272–277. IEEE, 2001.

5. Y. Xing and J. Tan. A color watermarking scheme based on block-svd and arnold transformation. In *Digital Media and its Application in Museum Heritages, Second Workshop on*, pages 3–8, Dec 2007.
6. Ibrahim Nasir, Ying Weng, and Jianmin Jiang. Novel multiple spatial watermarking technique in color images. In *Information Technology: New Generations, 2008. ITNG 2008. Fifth International Conference on*, pages 777–782. IEEE, 2008.
7. Sushila Maheshkar et al. An efficient dct based image watermarking using rgb color space. In *Recent Trends in Information Systems (ReTIS), 2015 IEEE 2nd International Conference on*, pages 219–224. IEEE, 2015.
8. Walter Bender, Daniel Gruhl, Norishige Morimoto, and Anthony Lu. Techniques for data hiding. *IBM systems journal*, 35(3.4):313–336, 1996.
9. Nallagarla Ramamurthy and S Varadrrajan. Effect of various attacks on watermarked images. *International Journal of Computer Science and Information Technologies*, 3:2, 2012.
10. Puneet Kr Sharma. Rajni: Analysis of image watermarking using least significant bit algorithm. *International Journal of Information Sciences and Techniques (IJIST) Vol*, 2, 2012.
11. Anatol Z Tirkel, GA Rankin, RM Van Schyndel, WJ Ho, NRA Mee, and Charles F Osborne. Electronic watermark. *Digital Image Computing, Technology and Applications (DICTA93)*, pages 666–673, 1993.
12. Mehmet Utku Celik, Gaurav Sharma, Ahmet Murat Tekalp, and Eli Saber. Lossless generalized-lsb data embedding. *Image Processing, IEEE Transactions on*, 14(2):253–266, 2005.
13. Martin Kutter, Frank Bossen, et al. Digital watermarking of color images using amplitude modulation. *Journal of Electronic imaging*, 7(2):326–332, 1998.
14. Wei Lu, Hongtao Lu, and Ruiming Shen. Color image watermarking based on neural networks. In *Advances in Neural Networks-ISNN 2004*, pages 651–656. Springer, 2004.
15. Pao-Ta Yu, Hung-Hsu Tsai, and Jyh-Shyan Lin. Digital watermarking based on neural networks for color images. *Signal processing*, 81(3):663–671, 2001.
16. Yonggang Fu, Ruimin Shen, and Hongtao Lu. Watermarking scheme based on support vector machine for colour images. *Electronics letters*, 40(16):986–987, 2004.
17. Vikrant Bhateja, Kartikeya Rastogi, Aviral Verma, and Chirag Malhotra. A non-iterative adaptive median filter for image denoising. In *Signal Processing and Integrated Networks (SPIN), 2014 International Conference on*, pages 113–118. IEEE, 2014.

FOREX Rate Prediction: A Hybrid Approach Using Chaos Theory and Multivariate Adaptive Regression Splines

Dadabada Pradeepkumar and Vadlamani Ravi

Abstract In order to predict foreign exchange (FOREX) rates, this paper proposes a new hybrid forecasting approach viz., Chaos+MARS involving chaos theory and multivariate adaptive regression splines (MARS). Chaos theory aims at constructing state space from the given exchange rate data with the help of embedding parameters, whereas MARS aims at yielding accurate predictions using state space constructed. The proposed model is tested for predicting three major FOREX Rates-JPY/USD, GBP/USD, and EUR/USD. The results obtained unveil that the Chaos +MARS yields the accurate predictions than other chaos-based hybrid forecasting models and recommend it as an alternative approach to FOREX rate prediction.

Keywords Exchange rate prediction · Hybrid algorithm · Chaos theory · MARS · CART · Random forest · Ensembles · TreeNet · LASSO

1 Introduction

FOREX rate prediction plays an important role in revisiting economic policies of a country in order to maintain proper trade relationships with other countries in international monetary market [1]. Usually, FOREX Rate is a nonlinear time series which is noisy, non-stationary, and deterministically chaotic [2]. Its prediction involves the prediction of future given its past and present. As a financial time series, it is also made up of components such as trend, cycle, seasonality, and irregularity. It is a challenge to build good forecasting model that can represent the

D. Pradeepkumar · V. Ravi (✉)
Center of Excellence in Analytics, Institute for Development
and Research in Banking Technology, Hyderabad 500057, India
e-mail: rav_padma@yahoo.com

D. Pradeepkumar
e-mail: dpradeepphd@gmail.com

D. Pradeepkumar
SCIS, University of Hyderabad, Hyderabad 500046, India

© Springer Nature Singapore Pte Ltd. 2017
S.C. Satapathy et al. (eds.), *Proceedings of the 5th International Conference on Frontiers in Intelligent Computing: Theory and Applications*, Advances in Intelligent Systems and Computing 515, DOI 10.1007/978-981-10-3153-3_22

dynamics of FOREX Rate. So FOREX rate prediction attracted many researchers and practitioners over last few decades.

It is popular that combination of time series forecasting models can yield accurate predictions than stand-alone forecasting models. Edmonds et al. [3] states that a financial time series can yield accurate predictions whenever a state space is built from it and the reconstructed time series is trained with a supervised learning mechanism. In this context, chaos theory [4, 5] helps the forecaster in constructing a state space from the scalar financial time series using various embedding parameters: *lag* and *embedding dimension*. Usually, *lag* is the time gap between the autoregressive independent variables and *embedding dimension* is the least count of autoregressive independent variables that are needed to predict chaotic time series. Multivariate adaptive regression splines (MARS) [6] is a simple-to-understand, easy-to-interpret, and quick-to-predict nonparametric regression modeling technique that is best suitable when there is a complex relationship between predictor variables and dependent variable. It deals high-dimensional input data space very well by avoiding curse of dimensionality. In literature, it is found that there are very few chaos-based hybrid models in order to predict FOREX Rate including Pavlidis et al. [7], Huang et al. [8] and Pradeepkumar and Ravi [9]. As there are very few chaos-based hybrids and MARS can represent the dynamics of FOREX Rate well, the current work proposes a new hybrid approach that involves both chaos Theory and MARS.

The study of Pradeepkumar and Ravi [9] forms background for the current work. They had built two two-stage intelligent hybrid models to predict FOREX Rates. These models need one stage of modeling characteristic information and the other stage of modeling residual information in the process of prediction so that accurate predictions are obtained. However, the current study can model the same dynamics of FOREX Rate within only one stage of modeling. In the Chaos+MARS, as a preprocessing step, the state space is constructed from the FOREX Rate data using optimal values of lag and embedding dimension. Later, MARS accepts the remodeled data and yields accurate predictions. The Chaos+MARS is used to predict the three FOREX Rates of Japanese Yen (JPY)/US Dollar (USD), Great Britain Pound (GBP)/USD, and Euro (EUR)/USD.

The rest of the paper is organized as follows. There are very few works related to chaos-based hybrid models and applications of MARS to various financial time series. These are reviewed in Sect. 2. The detailed description of Chaos+MARS is described in Sect. 3. The results obtained are discussed in Sect. 4. Finally, the work is concluded in Sect. 5.

2   Related Work

It is observed that there are very few hybrid FOREX Rate prediction models involving Chaos Theory as preprocessor. Pavlidis et al. [7] proposed a hybrid prediction methodology that works through five different stages in order to predict financial time series and it yielded better predictions for both differential evolution (DE) and particle swarm optimization (PSO) than feedforward neural network (FNN). Then Huang et al. [8] proposed a chaos support vector regression (SVR) model that works through two stages in order to predict FOREX rate. They concluded that chaos-SVR model could extract the FOREX rate dynamics. Finally, Pradeepkumar and Ravi [9] proposed two hybrids that go through two stages involving chaos theory, artificial neural network (ANN) and particle swarm optimization (PSO) in order to predict FOREX Rate. These are: ANN+PSO/polynomial regression (PR) and PSO+ANN/PR. The authors concluded that the proposed two-stage hybrids yielded better predictions than multilayer perceptron (MLP), general regression neural network (GRNN), grouping method of data handling (GMDH), and PSO.

There are also few works that applied MARS to predict various financial time series. De Gooijer et al. [10] forecasted exchange rates using Time series MARS (TSMARS) methodology and concluded that the out-of-sample forecasts generated by TSMARS are better than pure random walk model. Abraham [11] proposed a hybrid CART–MARS technique to predict FOREX Rates and concluded that it outperformed ANN, neuro-fuzzy system, MARS, and classification and regression tree (CART). Lee and Chen [12] proposed a hybrid credit scoring model that goes through two stages using ANN and MARS and concluded that the proposed hybrid outperformed the models of discriminant analysis, logistic regression, ANN, and MARS. Finally, Lu et al. [13] predicted stock index using MARS and concluded that the MARS outperformed BPN, SVR and MLR. However, we found there is no chaos-based MARS to predict FOREX Rates, which consider both optimal lag and embedding dimension in constructing state space.

3 Proposed Hybrid Model

3.1 Overview of Techniques Used

In the prediction methodology of Chaos+MARS, the Saida's method determines whether chaos is present or not in the dataset. Then the Akaike information criterion (AIC) determines the optimal lag. Once the optimal lag is determined, the Cao's method accepts it and obtains the minimum embedding dimension. The both parameters are used to construct state space from scalar FOREX rate time series. The readers are directed to refer to [14–16] for detailed descriptions of these methods.

Various predictive models implemented as part of SPM Predictive Modeler described as follows. Classification and regression tree (CART) [17] obtains predictions by constructing a decision tree by means of recursive partitioning of data. As it constructs decision tree, its results are easily interpretable. It is worth to note that CART works instable when there are small changes in the training data. The concept of bagging can be applied to beat this instability [18]. In bagging, for each sample, a single tree is fit. The overall predictions for test data can be obtained using fitted trees by means of averaging predictions obtained from them. One of these bagged trees is CART Ensembles and Bagger (CART-EB). TreeNet as a part of SPM Predictive Modeler generates many small decision trees built-in a sequential error-correcting process to converge to an accurate model. Random Forest Tree Ensemble (RFTE) as a part of SMP Predictive Modeler is a collection of many independently constructed CART trees. The overall prediction of the RFTE is the sum of the predictions made from decision trees.

Multivariate adaptive regression splines (MARS) [6] is a nonparametric regression procedure that constructs the functional relationship between the dependent and predictor variables from a set of coefficients and basis functions that are entirely driven from the regression data. Generalized least absolute shrinkage and selection operator (LASSO) regression [19] is a regression technique that can handle outliers and can produce extremely sparse solutions. It can also yield predictions of large-scale problems.

3.2 Chaos-Based MARS

In this Chaos+MARS, the chaos theory is used for constructing state space by embedding the time series using embedding parameters of both lag (l) and embedding dimension (m). Once the state space is built, MARS accepts the remodeled data in order to yield accurate predictions.

Let $Y = \{y_1, y_2, \ldots, y_k, y_{k+1}, \ldots, y_N\}$ be a set of N observations (e.g., FOREX Rate) at times $1, 2\ldots, k, k + 1\ldots, N$, respectively. The prediction methodology of Chaos+MARS using the dataset Y, depicted in Fig. 1, proceeds as follows.

1. Check Y whether chaos is present in it or not. Later, remodel Y using both l and m when there chaos is absolutely present.
2. Divide Y into both training set of observations at times $t = lm + 1, lm + 2, \ldots, k$ and test set of observations at times $t = k + 1, k + 2, \ldots, N$, respectively.
3. Submit the remodeled training set of observations to MARS and train it using a nonlinear function $f(y_t)$ which can map the output variable y_t with the input variables $y_{t-l}, y_{t-2l}, y_{t-3l}, \ldots, y_{t-ml}$ and obtain training set predictions $\dot{y}_t$ at corresponding times $t = lm + 1, lm + 2, \ldots, k$.
4. Submit the remodeled test set of observations to the trained MARS so that it yields predictions of test set namely $\dot{y}_t$ at times $t = k + 1, k + 2, \ldots, N$.

4 Experimental Design

The daily datasets of JPY/USD, GBP/USD, and EUR/USD collected from (http://www.federalreserve.gov/releases/h10/hist/) are used here. The data of both JPY/USD and GBP/USD from January 1, 1993 to December 31, 2013 (6036

Fig. 1 Prediction methodology of Chaos +MARS

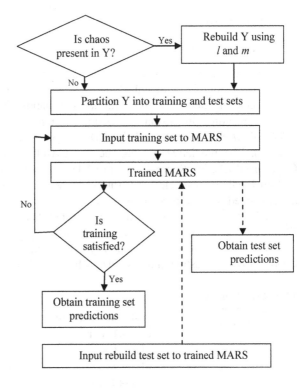

observations each) and EUR/USD from January 3, 2000 to December 31, 2013 (3772 observations), are used as datasets. Each of the datasets is partitioned into training set (80%) and test set (20%). Mean squared error (MSE), a measurement of average of squared errors and mean absolute percentage error (MAPE), a measurement of size of error in terms of percentage, are used to measure performance of the models and are defined in (1) and (2).

$$MSE = \frac{\sum_{t=1}^{N} E_t^2}{N} \tag{1}$$

$$MAPE = \frac{100}{N} \sum_{t=1}^{N} \left| \frac{E_t}{y_t} \right| \tag{2}$$

$$where \quad E_t = y_t - \dot{y}_t$$

In (1) and (2), N is the count of predictions obtained, y_t is original FOREX Rate value at time t, $\dot{y}_t$ is predicted FOREX Rate value obtained at time t.

5   Results and Discussion

The results are obtained by utilizing various tools. The Saida's method is employed to check the presence of chaos, AIC to obtain minimum lag value and the Cao's method to obtain minimum embedding dimension. The readers are directed to refer to Pradeepkumar and Ravi [9] for further information about these. The optimal lag (l) and embedding dimension (m) used in modeling the corresponding data set are JPY/USD ($l = 4$, $m = 20$), GBP/USD ($l = 5$, $m = 16$) and EUR/USD ($l = 1$, $m = 10$). The predictive models such as CART, CART-EB, MARS, TreeNet, Generalized LASSO Regression, and RFTE implemented as part of the tool SPM Salford Predictive Modeler® 8.0 (https://www.salford-systems.com/products/spm) are employed to yield predictions.

The MSE and MAPE values for the test set of datasets JPY/USD, GBP/USD and EUR/USD, respectively, are presented in Tables 1, 2 and 3. These two measures clearly specify that Chaos+MARS yielded better predictions than all other Chaos-based forecasting models, viz., Chaos+CART, Chaos+CART-EB, Chaos

Table 1 Performance measures of proposed model for JPY/USD data

Model	Test set MSE (MAPE)
Pradeepkumar and Ravi [9]	0.35480 (0.4757)
Chaos+CART	10.22168 (2.89)
Chaos+CART-EB	14.16271 (3.318)
Chaos+MARS	**0.34555 (0.468)**
Chaos+TreeNet	9.07775 (2.524)
Chaos+LASSO	3.33874 (1.839)
Chaos+RFTE	5.92272 (7.218)

Table 2 Performance measures of proposed model for GBP/USD data

Model	Test set MSE (MAPE)
Pradeepkumar and Ravi [9]	0.00094 (0.4773)
Chaos+CART	0.00010 (0.483)
Chaos+CART-EB	0.00010 (0.496)
Chaos+MARS	**0.00008 (0.453)**
Chaos+TreeNet	0.00009 (0.459)
Chaos+LASSO	0.00015 (0.628)
Chaos+RFTE	0.00024 (0.737)

Table 3 Performance measures of proposed model for EUR/USD data

Model	Test set MSE (MAPE)
Pradeepkumar and Ravi [9]	0.0000736 (0.4870)
Chaos+CART	0.0008 (0.533)
Chaos+CART-EB	0.00009 (0.525)
Chaos+MARS	**0.00006 (0.447)**
Chaos+TreeNet	0.00007 (0.474)
Chaos+LASSO	0.00027 (1.034)
Chaos+RFTE	0.00008 (0.529)

+TreeNet, Chaos+Lasso, Chaos+RFTE and Pradeepkumar and Ravi [9]. MARS partitions the high-dimensional input space into regions, each with its own regression equation driven from data. As MARS can select best dimensions and build model with them, it helped the proposed model to yield accurate predictions. The comparisons of proposed model with other models are depicted in Figs. 2, 3 and 4. As the other authors did not work on datasets used here, the proposed models cannot be compared with their models in literature.

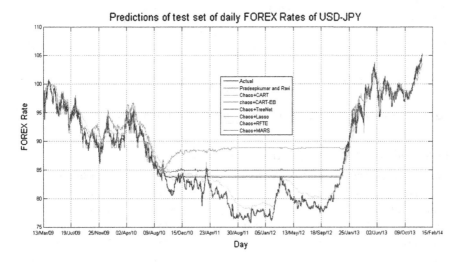

Fig. 2 Comparison of test set predictions of JPY/USD of various models

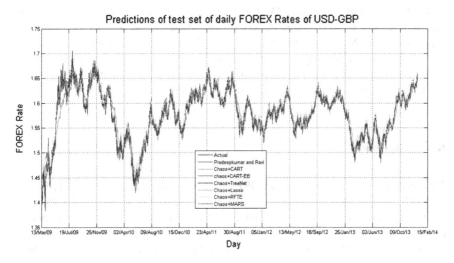

Fig. 3 Comparison of test set predictions of GBP/USD of various models

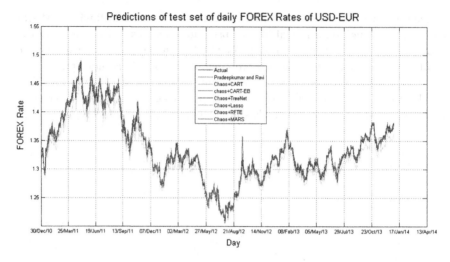

Fig. 4 Comparison of test set predictions of EUR/USD of various models

Table 4 Results of DM Test of predictions of test sets of three datasets

Model *Chaos+MARS* vs	JPY/USD	GBP/USD	EUR/USD
Pradeepkumar and Ravi [9]	0.64844	10.83560	4.118808
Chaos+CART	21.36679	5.99644	6.845096
Chaos+CART-EB	22.64803	7.33954	4.237744
Chaos+TreeNet	20.16983	1.95784	3.818277
Chaos+LASSO	29.84460	12.82946	19.50155
Chaos+RFTE	27.62333	11.89193	6.051142

There is a popular test viz., DM Test [20] that tests the statistical significance of results obtained by forecasting models. It is implemented as part of 'forecast' package in R. According to this test, if $DM\_test\_statistic \leq 0.05$, then Chaos +MARS is not statistically significantly different from the corresponding model. Therefore, from Table 4, it is clear that Chaos+MARS yielded statistically better predictions than all other chaos-based models in three datasets.

6 Conclusion

This work presents a novel hybrid FOREX rate forecasting approach namely Chaos +MARS. The chaos theory helped to construct the state space and MARS built prediction model from selected best independent variables. The proposed model could yield accurate predictions because of systematic modeling of chaos by means of optimal embedding parameters and building a prediction model from best independent variables. The results obtained on test sets of three FOREX rates reveal that Chaos+MARS outperformed the methods of Pradeepkumar and Ravi [9], Chaos+CART, Chaos+CART-EB, Chaos+TreeNet, Chaos+LASSO, and Chaos+RFTE in terms of both MSE and MAPE. The superiority of Chaos+MARS could lead to its application in many other financial and non-financial time series.

References

1. Hoag, A.J., Hoag, J. H.: Introductory Economics, 4th edition, 2006, World Scientific Publishing Co. Ptc. Ltd, Singapore, (2006).
2. Yao, J., Tan, C.L.: A case study on using neural networks to perform technical forecasting of forex. Neurocomputing 12(4), 79–98 (2000).
3. Edmonds, A., Burkhardt, D., Adjei, O.: Simultaneous prediction of multiple financial time series using supervised learning and clustering. In: IEEE World Congress on Computational Intelligence, Vol. 5, pp. 3158 –3163 (1994).
4. Packard, N.H., Crutchfield, J.P., Farmer, J.D., Shaw, R.S.: Geometry from a time series. Phys. Rev Lett, Vol. 45, No. 9, 712–716 (1980).
5. Takens, F.: Detecting strange attractors in turbulence. Lectures notes in mathematics, Vol. 898, 366–381 (1981).
6. Friedman, J.H.: Multivariate Adaptive Regression Splines. The Annals of Statistics, Vol. 19, No. 1, 1–141 (1991).
7. Pavlidis, N.G., Tasoulis, D.K., Vrahatis, M.N.: Financial forecasting through unsupervised clustering and evolutionary trained neural networks. In: The 2003 Congress on Evolutionary Computation, 2003, CEC'03, Vol. 4, pp. 2314–2321 (2003).
8. Huang, S-C., Chang, P-J., Wu, C-F., Lai, H-J.: Chaos-based support vector regressions for exchange rate forecasting. Expert Systems with Applications 37, 8590–8598 (2010).
9. Pradeepkumar, D., Ravi, V.: FOREX Rate Prediction using Chaos, Neural Networks and Particle Swarm Optimization. In: ICSI 2014, Part II, LNCS 8795, pp. 363–375 (2014).
10. De Gooijer, J.G., Ray, B.K., Krager, H.: Forecasting exchange rates using TSMARS. Journal of International Money and Finance. Vol. 17, No. 3, 513–534 (1998).
11. Abraham, A.: Analysis of hybrid soft and hard computing techniques for forex monitoring systems. In: Fuzz-IEEE'02, Vol. 2, pp. 1616–1622 (2002).
12. Lee, T-S, Chen, I-F.: A two-stage hybrid credit scoring model using artificial neural networks and multivariate adaptive regression lines. Expert Systems with Applications, Vol. 28, No. 4, 743–752 (2005).
13. Lu, C.J., Chang, C.H., Chen, C.Y., Chiu, C.C.: Stock index prediction: A comparison of MARS, BPN and SVR in an emerging market. In: 2009 IEEE International conference on Industrial Engineering and Engineering Management, pp. 2343–2347 (2009).
14. Saida, A.B.: Using the Lyapunov exponent as a practical test for noisy chaos (Working paper), http://ssrn.com/abstract=970074.
15. Cao, L.: Practical Method for determining the minimum embedding dimension of a scalar time series, Physica D 110 43–50 (1997).
16. Akaike, H.: A new Look at the Statistical Model Identification, IEEE Transactions on Automatic control, Vol. AC-19, No. 6 (1974).
17. Breiman, L.M., Friedman, J., Olshen, R., Stone, C.: Classification and Regression trees. Wadsworth: Belmont. CA (1984).
18. Breiman, L.: Bagging predictors. Machine Learning 24, 123–140 (1996).
19. Roth, V.: The generalized LASSO. IEEE Transactions on Neural Networks, Vol. 15, No. 1, 16–28 (2004).
20. Diebold, F.X., Mariano, R.S.: Comparing predictive accuracy. Journal of Business and Economic Statistics 13(3) 253–263 (1995).

Gray Scale Image Compression Using PSO with Guided Filter and DWT

Namrata Vij and Jagjit Singh

Abstract The vital goal of the image compression is to abate the insignificant facts of the image. Image compression is acclamatory while uploading and downloading images over the web. Image compression is the concept to compress the multifarious hyper spectral images, landsat images, multispectral images while maintaining the quality of an image and preventing the noise. The prime aim is to get the compressed image with improved radiometric resolution. Existing approaches are also efficient but still suffers from ringing artifacts. So an efficient technique for compressing the grayscale images is introduced. The proposed approach used particle swarm optimization (PSO), discrete wavelet transform (DWT), and guided image filter (GF). The idea behind the proposed technique is to apply PSO on the DWT along with GF to diminish the ringing artifacts, Gaussian noise and improve the radiometric resolution of the images. The overall result shows that proposed technique has improved radiometric information and lesser ringing artifacts than existing methods.

Keywords Image compression · Discrete wavelet transform · Particle swarm optimization · Guided image filter

1 Introduction

Compression ratio depends upon level of decomposition. Scrutinize multifarious wavelet function to compress images using wavelet thresholding. Both local and global thresholding methods give different compression ratio. Using the global threshold method, compression ratio can be high but MSE become high and PSN become less. This technique is difficult in selection of efficient thresholds for the

N. Vij (✉) · J. Singh
Lovely Professional University, Jalandhar, India
e-mail: namrata.20507@lpu.co.in

J. Singh
e-mail: jagjit.singh@lpu.co.in

© Springer Nature Singapore Pte Ltd. 2017
S.C. Satapathy et al. (eds.), *Proceedings of the 5th International Conference on Frontiers in Intelligent Computing: Theory and Applications*, Advances in Intelligent Systems and Computing 515, DOI 10.1007/978-981-10-3153-3_23

image. On the other hand, local threshold method is more flexible in selecting best thresholds and provide high compression ratio with high PSNR values. Wavelets play a vital role in the compression of the images. Biorthogonal wavelet gives high compression ratio as compare to Haar transform [1]. Spatial and spectral redundancy are always create problem. Interrelation between the neighboring pixels is termed as spatial redundancy. Such type of redundancy occurs because of similarity between the images. Interrelation between the multifarious spectral bands is termed as spectral redundancy. Therefore, the prime target of the image compression is to diminish such type of redundancy and improve the radiometric resolution [2]. Particle Swarm optimization is used for optimizing the results. Also compression ratio grows using PSO [3]. A same truncation technique is proposed for rapid compression of remote sensing images using multifarious ground types. Different images have different textures, information, weight, bit rate. First, the compact parameter is assigned for JPEG2000. Then take some relative amount of facts from the image and bit rate is also assigned to an image. Finally, proposed technique increase the compression ratio with some loss of facts [4]. Image Compression diminishes the inessential facts of the image in order to adept to save the facts in the proper form. It also diminishes the transmission time require to send or download images from web. Image compression acquiesce picture archiving and communication to diminish the storage space while maintaining the quality of the image. Teleradiology is advantageous for diminishing the size of images. DCT and DWT compression schemes are compared with this traditional technique for medical image compression. The results shows that improved compression scheme gives high compression ratio as compare to traditional technique [5]. With the growth of remote sensing multispectral camera, its attainment necessities like resolution, view field are also ameliorated. On the other hand, apprehend digital image facts are also raised quickly. However, due to barred satellite bandwidth, it is not easy to accommodate large amount of information of multispectral image. So such type of images must b compacted [6]. Radiometric resolution actuates how exquisitely an image can symbolize differences of intensity and number of bits. The radiometric resolution of a remote sensing system admeasures the number of gray levels between pure black and pure white and depends on the signal to noise ratio. From the multifarious observations, it is concluded that there is no perfect technique for compression of grayscale images and it is very hard to diminish the ringing artifacts of the landsat images. So, a new image compression technique is proposed in this paper based on hybridization of PSO-DWT-GF.

2   Existing Compression Scheme (PSO)

Discrete wavelet domain is used for the squeezing of gray scale images. This approach uses particle swarm optimization to search the favorable thresholds for multifarious subbands in DWT and less deformity and finer compression can be achieved using favorable thresholds. Searching the favorable thresholds diminished

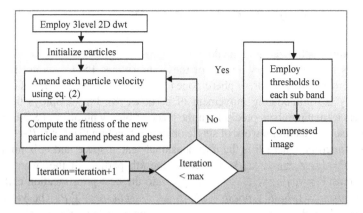

Fig. 1 Flowchart of efficient compression scheme using PSO (Existing method) [3]

the entropy of every subband and also VLC coding approach is used to achieve better compression ratio in results, but this method does not study the consequence of ringing artifacts in the image [3] (Fig. 1).

3 Proposed Compression Scheme (PSO-DWT-GF)

To remove the ringing artifacts and enhance the quality of the image, there are multifarious approaches but proposed technique improves radiometric information and removes ringing artifacts of the grayscale images. This technique includes discrete wavelet transform and particle swarm optimization along with guided image filter for the compression of images. PSO-DWT-GF hybridization is introduced to overcome the disadvantages of existing work, especially for images with ringing artifacts problem. The proposed method will construct the compressed image by using PSO with guided filter instead PSO only. PSO-DWT-GF is more efficient as provides high structure similarity index metric and peak signal to noise. Higher the SSIM means contrast preservation and higher PSNR means less distortion and improvement in the quality of the reconstructed image. To justify the efficiency of the PSO-DWT-GF technique quantitatively, experiments have been carried out on grayscale images [6]. Following subsections includes these steps in detail.

3.1 Particle Swarm Optimization (PSO)

Particle swarm optimization is analogous to genetic algorithm and Ant algorithm, but PSO is more elementary among of these algorithms. PSO does not have any mutation, crossover operator, or pheromone [7, 12]. PSO finds the space of equitable function by accommodating movement of particles [8]. PSO employs the vector parameters as real numbers precisely in the place of binary numbers. PSO is defined more precisely in the framework of single objective function [3]. Let $f: P \to Q$ be the objective function, where P is the d dimensional search space and m is number of particles, $P = \{z_1, z_2, z_3, \ldots z_m\}$. Therefore, the kth particle of the swarm can be symbolize as $Y_j = y_{k,1} y_{k,2} y_{k,3}, \ldots, y_{k,d}) \sum p$ and the best prior location call by z_k, search as: $pbest_k = (pbest_{k,1}, pbest_{k,2}, pbest_{k,3}, \ldots, pbest_{k,d}) \sum P$. The velocity of the kth particle is $R_k = (r_{k,1}, r_{k,2}, \ldots, r)$. Hence, the particle motion is calculated for the $(u + 1)$th Iteration as following [2].

$$z_k(u+1) = z_k(u) + R_k(u+1) \tag{1}$$

$$R_K(u+1) = R_k(u) + P_1 l_{k,1}(u) * (pbest_k(u) - Z_k(u)) + P_2 l_{k,2}(u) * (gbest(u) - z_k(u)) \tag{2}$$

$k = 1, 2, 3 \ldots h$. The kth particle location and acceleration at the uth iteration is implying as $Z_k(u)$ and $Z_k(u)$. At the uth iteration, best location raise by the whole swarm and particle independently so far, presented as *gbest(u) and pbest(u)*. P_1 and P_2 are the two fixed time coefficients, that are called cognitive and social parameters. $l_{k,1}$ and $l_{k,2}$ are two absolute arbitrary shared values extent with [0, 1].

3.1.1 Pseudocode for PSO

1. Objective Function $f(z), z = (z_1, \ldots, z_q)u$
2. Initialize positions z_k and speed R_k of m agents
3. Find O^* from min $\{f(z_1), \ldots, f(z_l)\}$ $bc(u = 0)$
While (condition) $u = u + 1$ (epochs)
4. For epoch over all agents l and dimensions d
5. Estimate new velocity R_k^{u+1} by eq. (1)
6. Compute new positions $z_k^{u+1} = z_k^u + R_k^{u+1}$
7. Evaluate fitness functions at new position z_k^{u+1}
8. Find the current best for each agent z_k^*
End for
9. Calculate the current global best O^*
End while
10. Outputs z_k^* and O^*

3.2 Guided Image Filter

The guided filter accomplishes the filtered output by taking the guidance image.

3.2.1 Pseudocode for Guided Image Filter

Input: Filtering input image t, guidance image M, radius s, regularization eps
Output: Filtered image z
1. Calculate the mean for M, t
   $mu_m = f_{mu}(M)$
   $mu_t = f_{mu}(t)$
   $cor_M = f_{mu}(M.* M)$
   $cor_{Mt} = f_{mu}(M.* t)$
4. Calculate the variance and covariance of (M, t) using the formula:
   $vari_M = cor_M - mu_M.* mu_M$
   $cova_{Mt} = cor_{Mt} - mu_M.* mu_t$
5. Then calculate the value of d, e where d, e are the linear coefficients
   $d = cova_{Mp}./(vari_M + eps)$
   $e = mu_t - d.* mu_M$
6. Then calculate mean of both d and e
   $mu_d = f_{mu}(d)$
   $mu_e = f_{mu}(e)$
7. Finally, get the filtered processed image z.
   $z = mu_d.* M + mu_e$

The guidance image is an unrelated image or it can be input image itself. The guided filter is a continuous model between guidance image M and filtered image z. z is a continuous transformation of M in a window w_c concentrated at the pixel c:

$$z_j = d_c M_k + e_c, \forall w_c$$

where (d_c, e_c) are linear coefficients and constant in the window and is regularization parameter. The s is radius of the window. This model assured that z has an edge only if M has an edge. To estimate the $(d_c e_c)$, some constraints are require from input image t. $z_k = t_k + n_k$ where n_k is noise or nonessential components [9].

3.3 Discrete Wavelet Transform (DWT) and Thresholding

The DWT used individually to the components of the image and maintain the spatial interaction. DWT splits the facts of an image into approximation and details sub signals. Approximation details present the familiar pixel values and use less space to store than original image where as other three details present the vertical, diagonal, and horizontal details [10]. Low-pass filter and high-pass filters are valuable tools for data disintegration. To enhance the performance and to diminish the computational complexities, multifarious filters such as 5/3 filter, 9/7 filter are used. A different threshold is driven for every subband in the DWT. With the increase of thresholds computational complexity, cost and time also increase, so correlative subbands are consolidate to have different threshold value [3].

$$M(i,j) = \left\{ \begin{array}{l} M(i,j) \ if \ M(i,j) > thre \\ thre \ if \ M(i,j) < thre \end{array} \right\}$$

where $M(i, j)$ coefficients of sub band, *thre* represents a threshold value of sub band.

3.4 Artifacts

An unacceptable artifact occurs while compressing facts. These artifacts become the obstacle for the higher compression ratio [7]. Artifacts can be blocking artifacts or ringing artifacts. Deformity of images, video, and audio is a blocking artifact. Counterfeit signal close to pointed transitions is ringing artifacts [11].

3.5 Algorithm: Pseudocode of Proposed Method

Step1: Load the landsat image (ima_1)

Step2: Apply DWT on the selected image using following equations:

$$W_\varphi(m,n) = \frac{1}{\sqrt{P}} \sum_t f(t)\varphi_{m,n}(t)$$

$$W_\varphi(l,n) = \frac{1}{\sqrt{P}} \sum_t f(t)\varphi_{l,n}(t)$$

For $l \geq m$ and $f(t) = \frac{1}{\sqrt{P}} w_\varphi(m,n)\varphi_{m,n} + \frac{1}{\sqrt{P}} \sum_{l=m}^{\infty} \sum_n W_\varphi(l,n)\varphi_{l,n}(t)$

Where $f(t)\varphi_{m,n}(t)$ $\varphi_{l,n}(t)$ are functions of discrete variables t=0,1,2,3...P-1

Step 3: Now apply guided image filter to remove the ringing artifacts and Gaussian noise by using above (2) algorithm

Step4: Initialize the particles $(thre_1, thre_2, thre_3, thre_4, thre_5, \ldots \ldots \ldots thre_{10})$

Step5: Update each particle using Equation (2) and evaluate local best (Pbest) and global best particles (gbest)

Step6: Evaluate the fitness of new particles by using table (1) and update Pbest and gbest

Step7: Move to step (4) if move holds as:

$$Move = \begin{cases} 1 \; if \; iteration < maxi \; iteration \\ 0 \qquad \qquad otherwise \end{cases}$$

Step8: Employ thresholds to each sub-band using the favorable values searched by PSO $(thre_1, thre_2, thre_3, thre_4, thre_5 \; thre_6)$

Step9: Apply Huffman coding approach to get the compressed image with diminished entropy coefficients by using following steps:

a. Put in order all source symbols in descendent
b. order of feasibilities
c. Blend two of the minimum feasibilities
d. Appoint zero to top and one to bottom

$$\begin{cases} go \; to \; step(b) \qquad \quad if \; unblend \; node \; is \; left \\ Generate \; codewords \; form \; top \; to \; bottom \quad otherwise \end{cases}$$

4 Performance Analysis

The performance of the new technique is analyzed in term of structure similarity index metric, mean square error, peak signal to noise ratio, bit error rate (Table 1).

4.1 Experiments Setup

In order to obtain the objectives of the this paper, implement the PSO-DWT-GF technique in MATLAB 2013a tool using image processing toolbox. However,

Table 1 Performance metrics

Metrics	Formula	Metrics	Formula
Mean Square Error (MSE)	$\frac{1}{l*m}\sum_{j=0}^{l-1}\sum_{k=0}^{m-1}[U(d,e)-V(d,e)]^2$ $U(d,e)$ = Original image $V(d,e)$ = Reconstructed image $l*m$ = Size of image	Structured Similarity Index Metric (SSIM)	$\frac{(2\gamma_m\gamma_n+D_1)(2\sigma_{mn}+D_2)}{(\gamma_m^2+\gamma_n^2+D_1)(\sigma_m^2+\sigma_n^2+D_1)}$ γ_m and γ_m = Average of m and n σ_{mn} = Covariance of m and n D_1 and D_2 = constants
Peak Signal to Noise Ratio (PSNR)	$10log_{10}\left(\frac{255^2}{MSE}\right)$ MSE = Mean Square Error	Correlation (COR)	$\frac{\sum_{j=1}^{P}\sum_{k=1}^{Q}(C_{j,k}-\bar{C})(D_{j,k}-\bar{D})}{\sqrt{\sum_{j=1}^{P}\sum_{k=1}^{Q}(C_{j,k}-\bar{C})^2\sum_{j=1}^{P}\sum_{k=1}^{Q}(D_{j,k}-\bar{D})^2}}$ C and d are constants
Bit Error Rate (BER)	$\frac{1}{PNSR}$	Root Mean Square Error (RMSE)	$\sqrt{\frac{1}{S}\sum_{u=1}^{S}\|S_u\|^2}$

Table 2 Existing results (PSO)

Image/Data	Mean Square Error (MSE)	Peak Signal to Noise (PSNR)	Structure Similar Index Metric (SSIM)	Bit Error Rate (BER)	Root Mean Square Error (RMSE)	Correlation (COR)
Image 1	0.0636	60.124	0.74826	0.01663	0.25237	0.98527
Image 2	0.1950	55.263	0.8045	0.01809	0.44166	0.96471
Image 3	0.0305	63.310	0.78609	0.01579	0.17487	0.99411
Image 4	0.0518	61.015	0.90971	0.01638	0.22774	0.99028
Image 5	0.0266	63.912	0.86037	0.01564	0.16316	0.99501
Image 6	0.0793	59.170	0.76177	0.0169	0.28166	0.98513
Image 7	0.1788	55.639	0.85965	0.01797	0.42295	0.96645
Image 8	0.0417	61.955	0.80302	0.01614	0.2044	0.99217
Image 9	0.0118	67.437	0.79453	0.01428	0.10873	0.99778

PSO-DWT-GF is not limited to only grayscale images, this technique can also apply on the other images. Figure 2a is showing original image with ringing artifacts and Fig. 2b is guidance image that can be input image itself. Figure 2c is noisy image and Fig. 2d is filtered image after applying guided filter. Figure 2e is representing results of DWT. Finally, Fig. 2f is our compressed image with enhanced quality.

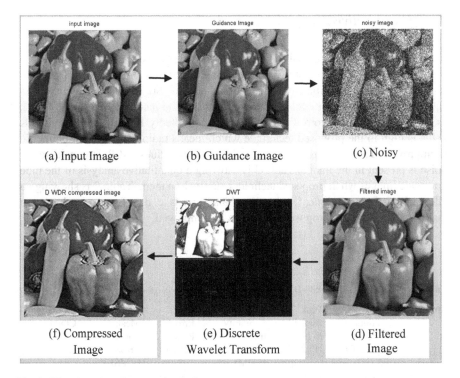

Fig. 2 Visual results of proposed technique

4.2 Qualitative Analysis

Visual analysis means the most predictable evaluation of image quality assessment is biased grading by human. Figure c has shown that quality of the image is degraded due to the effect of ringing artifacts. On the other hand, figure d has shown the filtered image with smooth edges. However when compared with existing method, the PSO-DWT-GF technique has a significant impact on the radiometric information of the image.

4.3 Quantitative Analysis

Evaluated the quantitative analysis in addition to qualitative analysis and compared the PSO-DWT-GF with existing method using multifarious quality metrics. Some important metrics for objective evaluation are structure similarity index, mean square error, nad bit error rate. Our major objective is to increase the value of PSNR as much as possible. From Table 3, it will undoubtedly be observed that PSNR value is high in the proposed technique which means radiometric resolution metric is improved. Table 3 is representing reduced BER values. BER is considered if noise is present in the image. Table 3 is showing a quantitative analysis of the nine grayscale images when proposed technique applied on them. Table 2 is showing an effect of existing technique. On the other hand fig a, b, c, d, e, f have clearly shown that parameters of the proposed technique are improved.

Table 3 Proposed results (PSO-DWT-GF)

Image/Data	Mean Square Error (MSE)	Peak Signal to Noise (PSNR)	Structure Similarity Index Metric (SSIM)	Bit Error Rate (BER)	Root Mean Square Error (RMSE)	Correlation (COR)
Image 1	0.0196	65.188	0.9995	0.01534	0.1403	0.99026
Image 2	0.0465	61.454	0.9985	0.01627	0.2157	0.98312
Image 3	0.0098	68.218	0.9997	0.01465	0.0990	0.99716
Image 4	0.0139	66.673	0.9995	0.01499	0.1183	0.99474
Image 5	0.0041	71.909	0.9999	0.01390	0.0647	0.99578
Image 6	0.0212	64.853	0.9995	0.01541	0.1458	0.98646
Image 7	0.0415	61.9468	0.99824	0.01614	0.2038	0.98622
Image 8	0.0092	68.4877	0.99982	0.01460	0.0960	0.99231
Image 9	0.0025	74.1514	0.99996	0.01348	0.0500	0.99949

5 Graphical Comparison Between Existing and Proposed Work

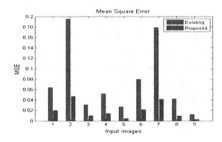

MSE evaluations of different images using existing technique and proposed technique

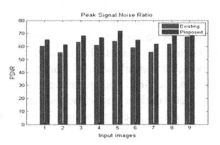

PSNR evaluation of different images using existing technique and proposed technique

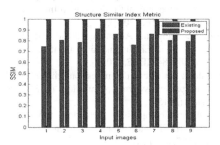

SSIM evaluations of different images using existing technique and proposed technique

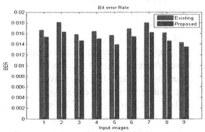

BER evaluations of different images using existing technique and proposed technique

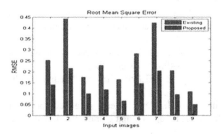

RMSE evaluations of different images using existing technique and proposed technique

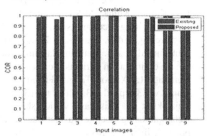

COR evaluations of different images using existing technique and proposed technique

6 Conclusion

Image compression is used to diminish spectral and spatial type of redundancy. The principle goal is to preserve the radiometric information of multifarious images. The existing approaches are also efficient for the grayscale image compression. But suffer from ringing artifacts problem. So to overcome this problem, a PSO-DWT-GF based image compression approach is proposed in this paper. PSO-DWT-GF technique tested on multifarious images. The quantitative analysis of the proposed technique with existing approach has clearly showed that PSO-DWT-GF has better MSE, PSNR, SSIM, correlation, and BER. Radiometric resolution is also improved. This hybridization has not considered the blocking artifacts which are more difficult to reduce. Future directions are to perform image compression using ant colony optimization.

References

1. J.Abirami, K.Narashiman, S.Siva Sankari, S.Ramva.: Performance Analysis of Image Compression using Wavelet Thresholding. In: IEEE International Conference on Information and Communication Technologies, 2013, pp. 194–198:IEEE, (2013).
2. Gui Zhengke, Chen Fu, Yang Jin, Li Xipeng, Li Fangiun.: Automated cloud and cloud shadow removal method for landsat TM images. In: IEEE International Conference on Electronic Measurement and Instrument, vol.3, pp. 80–84, IEEE, (2011).
3. Kaveh Ahmadi, Ahmad Y.Javaid, Ezzatollah Salari.: An efficient compression scheme based on adaptive thresholding in wavelet domain using particle swarm optimization. In: Science Direct Image Communication, pp. 33–39. Elsevier, (2015).
4. S.Zhou, H.Chen, Y.Tao, Y.Zhang, M.Zhang.: A rapid compression technology for remote sensing image based on different ground scenes. In: IEEE International Conference On Geo Science Remote Sensing Symposium, pp. 2555–2557, IEEE, (2013).
5. A.A.Fisunmilola, D.F.Olusayo, A.A.Michael.: Comparative analysis between discrete cosine transform and wavelet transform for medical image compression. In: IEEE International Conference on Computer Vision an Image Analysis Application, pp. 1–6, IEEE, (2015).
6. http://decsai.ugr.es/cvg/CG/base.htm accessed in 2015.
7. P.V. Kranthi Kumar, M.S.R Naidu.: An automated threshold selection using wavelet based PSO for image compression. In: IEEE Journal of Electronics and Communication Engineering, Vol.9, pp. 26–31. IEEE, (2014).
8. Muhammad Imran, Rathiah Hashim, Noor Elaiza Abd Khalid.: An overview of particle swarm optimization variants. Science Direct, pp. 491–496. Elsevier, (2013).
9. Kaiming He, Xiaoou Tang.: Guided image filtering. In: IEEE Pattern and MachineIntelligence, vol.35, pp. 1–13. IEEE, (2013).
10. Rajasekhar V, Vaishnavi V, Kaushik J, Thamarai M.: An efficient image compression technique using discrete wavelet transform. In: IEEE Electronic and communication system, pp. 1–4. IEEE, (2014).
11. M.M.Siddeq, M.A.Rodriguespp .: A novel 2D image compression algorithm based on two levels DWT and DCT transforms with enhanced minimize matrix size algorithm for high resolution. pp. 2–15, Springers, (2015).
12. K.Uma, P.Geetha,Palanisamy, P.Gertha Poornachandran.: Comparison of image compression using GA, ACO, PSO techniques. In: IEEE Recent trends in Information Technology, pp. 815–820. IEEE, (2011).

Graph Partitioning Methods

Prabhu Dessai Tanvi, Rodrigues Okstynn and Fernandes Sonia

Abstract The analysis of large graph plays a prominent role in various fields of research and application area. Initially, we formally define the partitioning scheme based on user needs and requirements. In this paper, we will be dealing with various methods of graph partitioning, its advantages and disadvantages, and from the result we can conclude which is the most effective method of graph partitioning. We can apply the best method in road navigation, stock market, database modeling, and bioinformatics.

Keywords Fiduccia–Mattheyse · Kernighan–Lin · Simulated Annealing

1 Introduction

The graph partition problem is defined on data that is represented in the form of a graph G = (V, E) with V vertices and E edges. A partitioning method should aim at reducing the net-cut cost of the graph. The number of edges that cross the partitioning line is called a net-cut.

Important applications of graph partitioning methods are in scientific computation, and in partitioning various stages of VLSI design circuit. Big hierarchical graph deals with very big volume of data that have to be stored in a way so that their future retrieval and analysis will be efficient [1].

The idea here in this paper we are studying more specific methods for graph partitioning, also comparison of general and specific methods. Vasilis Spyropoulos

P.D. Tanvi (✉) · R. Okstynn · F. Sonia
Padre Conceicao College of Engineering, Verna, Goa, India
e-mail: tanvirit48@gmail.com

R. Okstynn
e-mail: mecta2k7@gmail.com

F. Sonia
e-mail: fdezsonia@gmail.com

© Springer Nature Singapore Pte Ltd. 2017
S.C. Satapathy et al. (eds.), *Proceedings of the 5th International Conference on Frontiers in Intelligent Computing: Theory and Applications*, Advances in Intelligent Systems and Computing 515, DOI 10.1007/978-981-10-3153-3_24

and Yannis Kotidis, have mentioned about general methods like round Robin partitioning and min-split partitioning that can be used in graph partitioning [1].

The methods studied in this paper belong to the class of deterministic algorithm. Both methods begin with random partitioning of a graph, and iterate over to find the most effective partition. Every time we apply the algorithm to the same set of inputs or partitions, it will result in the same result. Hence, both methods are said to belong to the class of deterministic algorithm.

The purpose of this paper is to study in detail the available specific methods for the approach, and find the most prominent one among all. These approaches studied in this paper are also applicable in the VLSI field.

State of the art: True section—In this paper, we are referring to the title "Dynamic Partitioning of Big Hierarchical Graphs" in that they have made a study on general graph partitioning approaches which are not specifically meant for graph partitioning [1]. The properties missing in the reference paper [1] approaches are that they were not guaranteed to find proper partitions.

2   Kernighan–Lin

Kernighan–Lin is an iterative algorithm. This means that the graph or circuit may be already partitioned, but the application of Kernighan–Lin improves the method of partitioning a graph containing nodes and vertices into separate subsets that are connected together in optimal manner.

The Kernighan–Lin runs in O(n) time. This is a greedy algorithm that will make changes if there is a benefit right away without considering other possible ways of obtaining optimal solutions. This method is also deterministic, because the same result will be achieved every time the algorithm is applied with the same set of input values.

The steps to apply the Kernighan–Lin method is as follows,

- Divide the graph as shown in Fig. 1 into two halves with an equal number of vertices in each partition.
- Count the number of edges that cross the line. This number is called the net-cut and the goal is to reduce the number of net-cuts.

Fig. 1 Initial graph

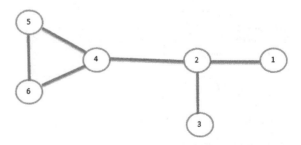

- Find the edge cost of all the vertices in the graph.
- Determine the maximum gain by swapping any two nodes. Maximum gain is given by $G = D_1 + D_2 - C_{21}$, where D denotes the difference between the external and the internal link values. C_{12} denotes if there exist any path between node 1 and node 2.
- Swap the two nodes with the maximum gain.
- Subtract the gain from the original net-cut to get the new net-cut.
- Fix the nodes that where just swapped in place, these nodes cannot be exchanged again.
- Repeat the steps until all nodes have been exchanged.
- Repeat all the steps until the gain is zero or negative.

The advantage of this method is that it is robust. The disadvantages of this method are that the results are random because the algorithm starts with random partitions. This method is computationally intensive, which makes the method very slow. Only two partitions can be created at a time. The partitions have to be equal in size. The method does not work well with weighted edge graph [2].

2.1 Solved Example of Kernighan–Lin

Consider the initial graph as in Fig. 1.

The graph in Fig. 1 is further divided into two partitions. As shown in Fig. 2, each partition has equal number of nodes. The Kernighan–Lin method is well suited for graphs with no weighted edges. In the Kernighan–Lin method graphs can be divided into two subsets having equal number of nodes each.

Now the graph is partitioned into two halves as A = {2, 3, 4} and B = {1, 5, 6}, further computing the D values of each partition (Table 1). The D values are computed as difference between the external (E) value and the internal (I) value. With respect to node 4, node 5, and node 6 denote the external edge outside the box and node 2 denotes the internal edge inside the box (Table 1).

Based on the D values, we calculate the gain of each possible pair and select the one with the highest gain value. Select the nodes with the highest gain value from the initial partition and repeat the process with the new partition.

Fig. 2 Graph after an initial partition

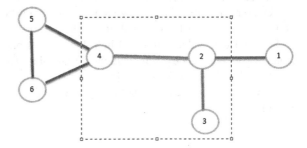

Table 1 Table showing computation of D values

Partition A	E_a	I_a	Compute D_1	Partition B	E_b	I_b	Compute D_2
2	1	2	−1	1	1	0	1
3	0	1	−1	5	1	1	0
4	2	1	1	6	1	1	0

Table 2 Table showing computation of gain for possible pairs of nodes

Possible pair	C(a, b)	Gain
G_{21}	1	−2
G_{25}	0	−1
G_{26}	0	−1
G_{31}	0	0
G_{35}	0	−1
G_{36}	0	−1
G_{41}	0	2
G_{45}	1	−1
G_{46}	1	−1

Table 3 Table showing computation of D values

New D	New gain
D'_2	$D_2 + 2C_{24} - 2C_{21} = -1$
D'_5	$D_5 + 2C_{51} - 2C_{54} = -2$
D'_6	$D_6 + 2C_{61} - 2C_{64} = -2$

Table 4 Gain computation of new partition

Possible pair	C(a, b)	Gain
G_{25}	0	−3
G_{26}	0	−3
G_{35}	0	−3
G_{36}	0	−3

The largest gain value in Table 2 is $G_{41} = +2$. Next, we remove the node 4 and node 1 from the initial partition. The new partition formed are A' = A − {4} = {2, 3} and B' = B − {1} = {5, 6}. Next we compute the new D values for the nodes connected to the {4, 1}, that is 2 in A' and 5, 6 in B' (Table 3).

For every new iteration, a new gain will be computed for every possible pair of nodes, and we select the one with the highest gain.

Since all gain values are equal in Table 4, we arbitrarily choose any one, G_{36}. Again, remove the nodes with the highest gain value. The new partitions that we get are A' − 3 = {2} and B' − 6 = {5}.

Now, compute the gain for the final pair of nodes. We get the gain value as $G_{25} = 1$ (Table 5).

Table 5 Table showing computation of new D values

New D	New gain
D''_2	$D'_2 + 2C_{23} - 2C_{26} = 1$
D''_5	$D'_5 + 2C_{56} - 2C_{53} = 0$

Table 6 Gain computation of new partition

Possible pair	C(a, b)	Gain
G_{25}	0	1

Finally, determine the K value. This is done by summing up the highest gain values selected through iteration, $G_1 = +2$, $G_2 = -3$ and $G_3 = +1$ summing this values we get the K value as $K = 0$ (Table 6).

3 Fiduccia–Mattheyses

Fiduccia–Mattheyses is a linear time heuristic for improving network partitions, and work on O(n) time. This method starts with some random partition. For every iteration, the partition changes. At the beginning of the pass, all the vertices are free to move and each possible move is labeled with an immediate change in total cost, called as gain. The cells are locked once it is moved from one partition to another. Once the cell has been locked it is no longer considered for further iterations.

Selection and execution of the best gain move, followed by the gain update is repeated until all the cells are locked. This method basically consist of three operations, first is the computation of the initial gain value at the beginning of the pass. Second, the retrieval of the best gain move and finally, update of all the affected gain values. This method is an improvement over the Kernighan–Lin method and had many new features such as it aims at reducing the net-cut cost, works well with weighted graphs, and can handle unbalanced partitions.

If all cells on the net are in the same partition, all their gains are decreased. Additionally, to prevent all cells from migrating to a single partition, a balanced condition is enforced.

The steps to apply Fiduccia–Mattheyses method are as follows,

- We start by computing the gains of all the vertices, which is computed as $G_i = FS(i) - TE(i)$, where FS(i) denotes the first partition on the left and TE(i) denotes the partition on the right as shown in Fig. 4.
- Select one base cell or free cell.
- After processing lock the base cell.
- Check for more free cells.
- Choose for best sequence of moves for k-base cells.
- Make k moves permanent and free all the cells.

Fig. 3 Initial graph with
edge sequence

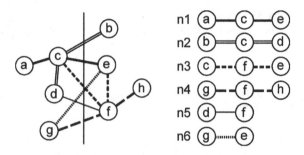

Fig. 4 Final result after all
the last pass

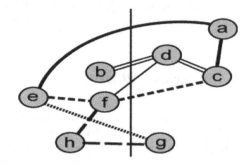

3.1 Solved Example of Fiduccia-Mattheyses

The graph in Fig. 3 shows the initial partition and a different sequence of edges.
Usually, the initial partitions are random. We start by computing the gain value of
each and every node. Based on gain values we select the one with highest gain and
move the node to other partition and lock it. The gain computed for node C in
Fig. 4. C is contained in net N_1, N_2 and N_3. N_3 contain cell C as the only cell from
left partition, so $F(C) = 1$, and none of the nets are entirely located in the left
partition, so $T(C) = 0$. Repeating the same procedure for every iteration, we calculate the gain.

4 Simulated Annealing

Simulated annealing is a general partitioning approach used in many applications
such as network partitioning and widely used iterative technique for solving general
optimization problems. This algorithm belongs to the class of nondeterministic
algorithms that even for same input can exhibit different behavior on different runs,
and guarantees finding optimal solution. It is relatively easy to code for complex
problems. We are limiting our study only to specific graph partitioning approaches.
We have used simulated annealing to show the comparison between general and
specific approaches, based on some criteria.

5 List of Possible Modifications to the Fiduccia-Mattheyses

In this section, we propose three ways of optimizing heuristics. The important changes to the flow of original algorithm are as follows,

- Among moves with same gain, one chooses those which increase the gain of the other move [3]
- By allowing the vertices to move more than once during a pass. That is allowing the nodes even after it is been locked [3],
- Consider vertices with only positive or nonzero cost [3].

6 Comparative Study of Graph Partitioning Algorithm

Graph partitioning is an important problem in an area of VLSI design. The bipartitioning algorithm proposed by Kernighan–Lin randomly starts with two subsets, and a pairwise swapping is iteratively applied. Simulated Annealing is another method used in physical systems for network partitioning. The objective function of simulated annealing is analogs to physical systems [4].

6.1 Performance Analysis

We do a comparative study based on three criteria,

- Estimation of the network area of the graph.
- The cost function.

6.2 Estimation of the Network Area of the Graph

The area is estimated by calculating the Manhattan distance between each possible node in the graph (Table 7).

$$X = (X_1, X_2, \ldots, X_n) \text{ and } Y = (Y_1, Y_2, \ldots, Y_n)$$
$$d = \sum |x_i - y_i|$$

where X and Y denote nodes in two separate regions of the graph. Manhattan distance is a difference between each node of the graph partition, summing them up to get the distance (d). Where x_i denotes the node in X region and, y_i denotes the node in Y region [4].

Table 7 Tabular representation of network area estimation of Kernighan-Lin and Simulated Annealing

Circuit	Number of nodes	Network area	
Actlow	18	66	74
Regfb	21	67	67
Moore	25	102	106
Meealy	37	180	189
Sequene	49	248	283
Dmux1t8	60	373	433
Cntbuf	64	389	437
Decade	71	393	510
Binbcd	*101*	866	979

Table 8 Tabular representation of final cost function in both approaches

F_c (KL)	F_c (SA)
2	3
3.5	4.5
2	2
6	7
5.5	11.5
7.5	13.5
8.5	12.5
9.5	18.5
15.5	29.5

6.3 The Cost Function

T_i and T_f represent the initial cut size and the final cut size, respectively, E_i and E_f represent the initial and final balance number. The cost function F_c is computed using formula,

$$F_c = I_t.T_f + I_e.F_e.$$

The results in Table 8 show that the cost function for Kernighan–Lin is smaller than the Simulated Annealing method [4].

7 Conclusion

Initially, we conducted brief survey on different types of graph in way that would support future analysis. We studied two graph partitioning methods, Kernighan–Lin and Fiduccia–Mattheyses method and found that the second method is prominent one. We also made a brief study on possible modifications to the Fiduccia–Mattheyses method. For comparative study the results show that the Kernighan–Lin method produces the best results.

As a future work, these approaches can be used in many applications like database retrieval, bioinformatics, road navigation, and many more.

References

1. Vasilis Spyropoulos and Yannis Kotidis, "Dynamic Partitioning of Big Hierarchical Graph", IEEE Trans. On First International Workshop on Big Dynamic Distributed Data (DB3), Riva Del Grada, Italy, August 30, 2013.
2. G. A. Ezhilarasi, K. S. Swarup, "Network Decomposition using Kernighan Lin strategy aided harmony search algorithm", Department of Electrical Engineering, Indian Institute of Technology Madras, Chennai, India.
3. Andrew E. Caldwell, Andrew B. Kahng, and Igor. L. Markov, "Design and Implementation of Fiduccia Mattheyses Heuristic for VLSI Netlist Partitioning", UCLA Computer Science Dept., Los Angeles, CA 90095–1596.
4. Zoltan Baruch, Octavian Cret, Kalman Pusztai, "Comparative Study of Circuit Partitioning Algorithm", MicroCAD 2000 International Computer Science Conference, Section F: Electrotechnics-Electronics, February-23–24, 2000, Miskole, Hungary.

Smart and Accountable Water Distribution for Rural Development

Ishaani Priyadarshini and Jay Sarraf

Abstract The Water Distribution Management System is an intuitive approach to eradicate the shortage of water in remote areas by providing requisite amount of water to each and every household by virtue of their uniquely generated water cards. In this paper, the main emphasis is given on the fact that the existing manual water collection systems leading to improper water distribution are replaced to the point where the services provided are efficient and cost effective leading not only to eliminate water scarcity but to encourage water conservation as well. Each household is entitled to a certain amount of water per day and the process of water collection is scheduled automatically. A feedback system has also been incorporated to evaluate the quality of water being supplied.

Keywords Water cards · Water distribution · Feedback system · Water scarcity · Smart card

1 Introduction

It is said that water is the elixir of life. It not only contributes to agriculture and domestic uses but is also crucial for survival. However, it is a not finite and a non-renewable resource, which means that the total amount of water on Earth's surface is constant. Of all water available on the earth's surface, 97% constitutes the oceans, 2% is trapped in the polar caps and the remaining 1% located in lakes or streams makes its way for human use. Thus, the world is experiencing severe water deficiency. Water scarcity is a major issue in several parts of the world [1]. Esti-

I. Priyadarshini (✉) · J. Sarraf
School of Computer Engineering, KIIT University, Bhubaneswar, Odisha, India
e-mail: ishaanidisha@gmail.com

J. Sarraf
e-mail: jaysarraf596@gmail.com

© Springer Nature Singapore Pte Ltd. 2017
S.C. Satapathy et al. (eds.), *Proceedings of the 5th International Conference on Frontiers in Intelligent Computing: Theory and Applications*, Advances in Intelligent Systems and Computing 515, DOI 10.1007/978-981-10-3153-3_25

251

mations reveal that by 2025, around 1.8 billion people will face absolute water scarcity. Moreover, it is expected that two-thirds of the world population will suffer from water crisis [2]. In a densely populated country like India, the situation is even worse. Growing economy, agriculture, and climate changes belittles the water supply [3]. Poor management, unclear laws, and government corruption have caused the water supply crunch. Numerous remote areas witness the struggle of obtaining a bucket of water at the cost of surplus amount of human energy as well as time. The Water Distribution System may be implemented to combat water scarcity issues in rural areas. Since every individual is entitled to drink clean water, the system makes use of system generated water cards for each and every household in a given area. These cards hold the amount of water issued to every individual household, and once read exhibit the amount of water which can be further claimed by the card holders. The process is automated such that the water available is reset every day to make it an efficient daily affair.

2 Case History

Since the past century water consumption has increased at a rampant rate. According to WHO and UNICEF, 663, billion people in the world face acute shortage of drinking water [4]. The water scarce areas are already at risk. In a country like India, out of 632 districts examined only 59 find water safe enough to drink [5]. Northern India is witnessing depletion of ground water rapidly. States like Punjab contribute to more than 50% of grain productions in India and being agriculture plots consume water at an alarming rate leading to its deficiency. The state government has taken initiatives to provide cent per cent individually metered water pipelines to all households in rural Punjab [6]. The largest state of India, Rajasthan, combats water scarcity by means of hydraulic simulations wherein new pipelines are laid parallel to the critical ones [7]. Even though the systems have been implemented accountability of water has not been taken into consideration which leaves several loopholes open for water wastage. Further pipelines have been observed to suffer extensive complications like expensive construction, lack of flexibility, and the inability to increase capacity whenever necessary. They are difficult to repair or detect leakages and often known to transport solids along with water in the form of slurry. The princely state of Odisha, India currently possessing thirty districts and over fifty thousand villages too is familiar with the water scarcity issue. Water scarcity in several villages of Odisha such as Mundapadar, Harinabhata, Ramaguda, Moiliguda and Dharmaguda is mainly because of consumption of contaminated water as a result of non-maintenance of tube wells and water tanks [8]. Ground water containing Fluoride has affected six villages in the district of Khurda [9]. Mining operations, agricultural effluents, and pollution caused by domestic effluents indirectly lead to water containing toxic metals being consumed by many districts of Odisha, hence leading to water-borne diseases such as Malaria,

Diarrhea, Intestinal Parasitism, and skin infections to name a few [10]. Thus, implementation of a proper system is of utmost importance. The solution lies in the fact that a Water Distribution System can be implemented in these places wherein a water booth can be implanted in a locality which is accessible to all residents of a given area. Each house will have a Green Water Card issued which may bear the amount of drinking water allotted to the house depending on the number of members residing in it. The Government of Odisha has latterly been active in setting up water ATMs in the state to provide water by implementing the water card mechanism. The commercialized system provides 20 L of water to a particular house each liter costing 30 paisa. Even though the system is feasible, it does not vouch for the optimum requirement of water per individual and accountability of water is also at stake. Our proposed system combats the above issues by providing water to individuals as per their requirements. As drinking water is provided to each individual on requirement basis, there is no scope for water wastage consequently leading to water conservation as well. Moreover, a feedback system can be enacted using sensors that would certify the quality of water. Canary is one such software [11].

3 Proposed System Design

The system design has been nominated keeping in mind the minute details of water distribution in remote areas. The areas well acquainted with scarcity of water often witness women and children carrying water from pumps over long distances which account for a good amount of time as well as energy consumption. Estimations reveal that farmers walk up to 75 miles a year between their houses and a hand pump roughly situated 30 m apart for water collection [12]. This accounts for spending up to 40 min a day for a can of water which may or may not qualify for drinking. The proposed system design validates conservation of water, time as well as energy. It may be designed to provide just the amount of drinking water required for individuals residing in the area. The system may function by means of a water booth that can be installed in a locality. This water booth would incorporate the amount of water that has been made available to people residing in the zone. A water booth administrator may be in charge of registering water cards to households. The water card may hold the amount of water that has been issued to a given house depending on the number of family members. As the card might be read by a card reader installed in the water booth system, it may show information regarding the water available. The card holder may proceed to withdraw the amount of water required from one of the taps out of a few that have been installed in the system to avoid concurrency. It should be made possible for the card holder to extract water as many times as possible in a day unless the water allotted to him per day has been obliterated. In addition to issuing cards, the water booth administrator may also be provided with the responsibility of sending a request for a water supply. He may be given the privilege of viewing, updating, modifying, and deleting records. The water booth may be entrusted with an additional feature to test the quality of water. Approximately 37.7 million Indians residing in rural areas fall prey to water-borne diseases, wherein 1.5 million children die from diarrhea alone [13]. National Rural Drinking Water Quality Monitoring and Surveillance Programme was introduced in February 2006 [14]. The programme aimed at bringing awareness through broadcasting information related to health hazards due to poor drinking water quality, hygiene, and sanitary survey. The water quality aspect may be taken care by introducing water sensors into the system which would monitor the water quality. A feedback system may be ensued wherein the individuals consuming water could report any abnormalities detected in the water like color, taste, or odor to preserve water quality. The system described above may be replicated to multiple villages facing shortage of water, all constituting a district. The system may be diffused over a number of districts each harboring its many villages. For each district, a district administrator may be employed to monitor the process of water distribution. The water booth administrator may request the district administrator for water supply. The district administrator may be concerned with supplying of water and maintaining master records for safeguarding the accountability of distributed water. The system has been better explained using a diagramming software Edraw UML Diagram version 7.9 to illustrate UML diagrams for the same.

3.1 GREEN Water Card

Green Water cards which have been purveyed to different households are nothing but smart cards which usually contains embedded microprocessors. The very reason microprocessors have been introduced into the smart card is because of security purposes. The card is not only an instrument for authentication, but is also responsible for data storage [15]. Data storage here refers to the amount of water allotted to each individual residence. Smart cards offer durability and reliability. The card may be communicated with a reader through direct physical contact or remote less electromagnetic field whenever the card holder desires to withdraw water. Since the advancement of radio technologies contactless smart cards have become time efficient and are known to reduce maintenance. As the chip is energized, and authenticated data is transferred from the card to the reader and then to an interface which would display the quantity of water available. Based on requirements, the card holder may withdraw the requisite amount of water. As the water is withdrawn, the system is updated with the new quantity of water available to the card holder. For further withdrawal of water, the card holder must again get the card read. Further the Green Water Card may be imparted the potential to convey logs using the 802.11 wireless LAN to a master database which will keep a record of the frequency and quantity of water claimed [16] (Fig. 1).

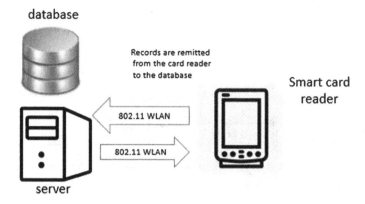

Fig. 1 Proposed architecture for green water card

3.2 Implementation Design

See Figs. 2, 3, 4 and 5.

4 Comparative Analysis

As stated above, the system has not only been designed for the provision of drinking water to every individual, but also to focus on the issue of accountability of water distributed as well as water conservation. Also the quality of water being distributed is preserved. Considering the rural expanse of Odisha, Jharsuguda is one of the many districts facing scarcity of drinking water [17]. The district is known to harbor a total of 83,664 houses in its 356 villages [18]. This accounts for a

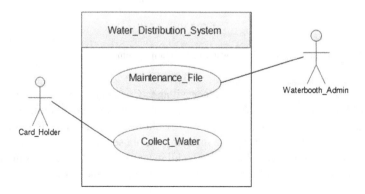

Fig. 2 Use case diagram for water distribution system

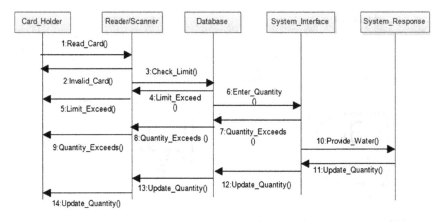

Fig. 3 Sequence diagram for the water distribution system

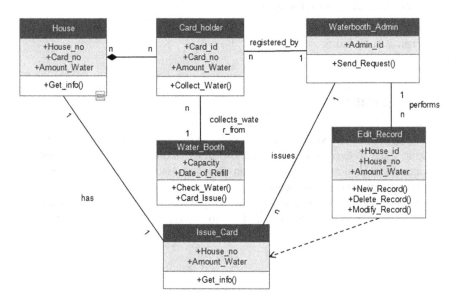

Fig. 4 Collaboration diagram for the water distribution system

relatively 235 numbers of houses per village. Considering a case where each household is provided say for example 20 L of water, per day allotment of water aggregates to 4700 L. Twenty liters of water allotted per day may also exceed the required amount of water by a large margin for a house which has a meagr number of two or three individuals. This would by all means lead to wastage of water due to lack of accountability. However, providing water on the basis of number of

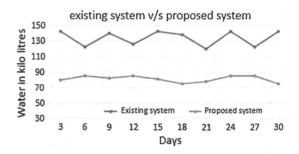

Fig. 5 Graph showing water usage in existing versus proposed system

individuals per house may solve the purpose. Given the number of individuals on an average for each house let us assume is four. According to the European Food Safety Authority, the average intake of water for a human body for sustenance in a given day is approximately 3 L [19]. This amounts for a total of 2820 L of drinking water supplied to the village per day which is reasonably less in comparison to 4700 L of water supplied in the former case. We can easily infer from the above figures that the quantity of water conserved for a given month would clearly agglomerate to 56400 L. Eventually the system will manifest its proficiency by saving massive quantities of water from being exploited and will prove as an aid to areas prone to inadequacy of water in the coming years.

5   Conclusion

To conclude, the proposed Water Distribution Management System is an attainable system design which can be brought into effect in rural areas where drinking water is deficient. The overall design has been carved to reduce the depletion of time and energy invested by people in the process of water collection. The system is an effective means to monitor the consumption of drinking water and hence undertake accountability and conservation of water. It is by all means a sensible system that eliminates the need for laying pipelines which otherwise lack flexibility and possess fixed capacity for water flow. Pipelines are further known to suffer damages and often result in leakages which the water distribution system is comparatively immune to. It abolishes water quality issues by supporting sensors that monitor the quality of water thus making it consumable and hence reducing the risk of people consuming contaminated water and falling prey to health disorders. In summation, as quoted by Thomas Fuller, "We never know the worth of water until the well is dry." With water being limited as well as a necessity, survival of mankind is at stake. Proper distribution of water will ensure conservation of the ecosystem and also ourselves as we are dependent on the finite and nonrenewable resource for our survival.

References

1. Dolatyar, Mostafa; Gray, Tim; Water Politics in the Middle East, St. Martin's Press, Inc. 2000.
2. UTZCertifiedposition paper on water scarcity, Oct 2013, www.utzcertified.org.
3. Nina Brooks, "Imminent Water Crisis in India", August 2007.
4. http://www.wssinfo.org/fileadmin/user_upload/resources/JMP-Update-report-2015_English. pdf.
5. Asian News International (ANI), Shortage of drinking water raises concern of water-borne diseases in Odisha, http://www.business-standard.com/, Rayagada, India, May 6 2014 Regan Murray, Terra Haxton, Sean A. McKenna, David B. Hart, Katherine Klise, Mark Koch, Eric D. Vugrin, Shawn Martin, Mark Wilson, Victoria Cruz, Laura Cutler, "Water Quality Event Detection Systems for Drinking Water Contamination Warning Systems", EPA/600/R-10/036 | May 2010 | www.epa.gov/ord.
6. Government of Punjab, Department of water supply and sanitation, "Punjab State Rural Water Supply and Sanitation Policy 2014" No. 13/168/2013-3B&R2/.
7. Vijay Kumar, MANAGING WATER DISTRIBUTION IN RAJASTHAN, INDIA, DHI CASE STORY, www.dhigroup.com.
8. ANI Rayagada, Shortage of drinking water raises concern of water-borne diseases in Odisha, http://www.business-standard.com, May 6, 2014.
9. Dr. S.K. Sahu, D. Sarangi, & K.C. Pradhan, "Water Pollution in Odisha", Odisha Review January 2016.
10. Progress on sanitation and drinking water, Update and mdg assessment 2015, UNICEF & WHO.
11. Shri J.N. Mathur, Indian Council of Medical Research "Health Status of Primitive Tribes of Odisha" ICMR Offset Press, New Delhi-110 029 R.N. 21813/71.
12. EDMUND G. WAGNER, J. N. LANOIX, "Water supply for rural areas and small opportunities", WORLD HEALTH ORGANIZATION MONOGRAPH SERIES No. 42.
13. Indira Khurana and Romit Sen, WaterAid, Drinking water quality in rural India: Issues and approaches, www.wateraid.org.
14. National Rural Drinking Water Programme (NRDWP), 1.4.2009, http://rural.nic.in/sites/downloads/our-schemes-glance/SalientFeaturesNRDWP.pdf.
15. Md. Kamrul Islam,"Effective use of smart cards" SPM 2012.10, www.diva-portal.org/smash/get/diva2:606154/FULLTEXT01.pdf.
16. Li Bai, Gerald Kane, Patrick Lyons, "Open Architecture for Contactless Smartcard-based Portable Electronic Payment Systems", 4th IEEE Conference on Automation Science and Engineering Key Bridge Marriott, Washington DC, USA August 23–26, 2008.
17. Bikash Kumar Pati, "WATER RESOURCES OF ODISHA ISSUES AND CHALLENGES", Regional Centre for Development Cooperation, December 2010 http://www.rcdcindia.org/PbDocument/8adc57865d55134-7374-401a-97b4-118393445fd2Water%20Resource%20Booklet%20FINALpdf.

Adaptive Huffman Coding-Based Approach to Reduce the Size of Power System Monitoring Parameters

Subhra J. Sarkar, Nabendu Kumar Sarkar and Ipsita Mondal

Abstract For maintaining power system stability, several parameters like voltage, frequency, etc. are monitored sequentially at regular intervals by SCADA, and the informations are transmitted to data centre through suitable communication schemes. If the volume of data can be reduced, then it is possible to reduce the energy and space requirement. This paper emphasizes on the development of an algorithm to compress the monitoring parameters using Adaptive Huffman Coding in MATLAB environment. The compression ratio obtained by this approach is better than what is obtained by other data compression techniques. This results in the reduction of memory requirement by about 60%, thereby enabling it suitable for the data handling of a large volume of monitoring data encountered frequently in a power system.

Keywords Power system · SCADA · Data compression · Adaptive Huffman coding · Data communication · Character string

S.J. Sarkar (✉)
Department of EE, Batanagar Institute of Engineering,
Management and Science, Kolkata, India
e-mail: subhro89@gmail.com

N.K. Sarkar
Department of EE, Haldia Institute of Technology, Haldia, India
e-mail: nsarkares@rediffmail.com

I. Mondal
Department of CSE, Batanagar Institute of Engineering,
Management and Science, Kolkata, India
e-mail: ipsita.mondal@yahoo.com

© Springer Nature Singapore Pte Ltd. 2017
S.C. Satapathy et al. (eds.), *Proceedings of the 5th International Conference on Frontiers in Intelligent Computing: Theory and Applications*, Advances in Intelligent Systems and Computing 515, DOI 10.1007/978-981-10-3153-3_26

1 Introduction

During parallel operation of the alternators, they must be synchronized with each other and the busbar or the National Grid. Grid refers to a total network interconnected by transmission lines where alternators of all power plants are operated in parallel. Parallel operation results service continuity maintenance, for varying load according to the generation and to achieve maximum efficiency. For synchronization of two alternators, voltage, frequency, phase angle and phase sequence of incoming alternator and busbar must be identical [1–6]. Due to some sudden transients or some faults occurring in the system, the system tends to become unstable which results some deviation of voltage and frequency from the predefined value [1, 2, 4]. It is thus important to monitor system voltage and frequency continuously. It can enable the system to take any preventive measures to avoid any loss of synchronism when any one or both the parameters tend to change. This continuous monitoring data must be transmitted to the monitoring station at state, regional and national level by employing suitable highly sophisticated communication system built in within the power system. This communication system is extremely important for modern day power system and is termed as SCADA (Supervisory Control and Data Acquisition) [1, 2].

Data handling of the bulk information of power system parameters collected over a finite time interval is extremely important. The implementation of suitable data compression techniques for storage of the data in the data centre not only

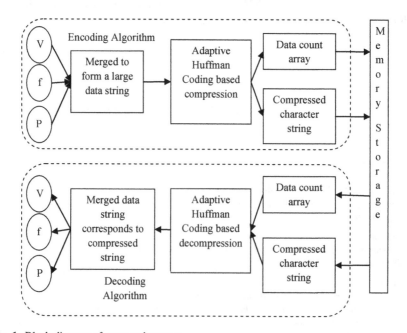

Fig. 1 Block diagram of proposed system

reduces the memory requirement but also provide inherent encryption of the data. In the proposed system, the power system parameters (Voltage, Frequency and Active Power) at any particular time instant are reduced to form a character array by employing adaptive Huffman Coding in MATLAB environment. At the decoding end, this character array, this transmitted character array is decompressed to obtain the parameters being compressed. The block diagram of proposed system is as given in Fig. 1 [1, 2, 7]. This algorithm although tested offline, it is possible to implement the algorithm online where the memory storage is replaced by a suitable wired or wireless communication channel.

2 SCADA in Power System

For maintaining system stability, an exact balance between power generation and load demand is important. SCADA is used for supervision and control of remote field devices using some suitable communication channel. Suitable meters, sensors, process equipment, etc. are connected at remote end for collecting the information and transmit the data to the control centre whenever required. Based on that information, supervisory information is conveyed by the automated or operator driven control centre to the field devices to perform the local operation of opening and closing of circuit breakers or valves, collecting data from sensors, local environment monitoring, etc. It is extremely popular in electricity utilities as it enables the remote operation and control of substations and generating stations [1, 8–10].

Communication between control centre and field devices is extremely important in SCADA, which might be wired, i.e. through Telegraphic or power line, WAN, etc. or might be wireless like Radio, Microwave, Satellite communication (VSAT), etc. [1, 8, 9]. Power line carrier communication employing power line as a communication medium is a good alternative for SCADA due to its zero installation cost. But power line is not meant for communication purpose, so the line characteristics are not suitable for conventional digital communication techniques and requires improved techniques. Field bus can also be used for PLC communication. Wide Area Network (WAN) covers a wide area using leased telecommunication lines [1, 11, 12]. In radio communication, information is impressed over a high frequency carrier signal before transmitting it through air. At the receiving end, antenna tuned at the particular frequency captures the modulated wave to extract the message signal. Microwave communication is used primarily for point to point communication on the earth surface at communication frequency within 3–300 GHz. This property enables two nearby microwave equipment using the same frequency can be operated without any interference. Larger bandwidth of microwave increases, its information capacity increases. But it can be used for line of sight propagation only, and the performance may degrade with atmospheric conditions. In very small aperture terminal (VSAT) communication, for the transmission of data, voice and video signal, microwave is used for the communication between two points through satellite [1, 13, 14].

3   Data Compression Techniques

Data Compression techniques are employed to reduce the number of bits (or bytes) of a given information. It can also be defined as the process of recombination the bits (or bytes) to make the data smaller and compact. This is done by eliminating the identical data bits or continuously recurring data. Similarly at the point of data restoration, decompression is done for decoding the compressed data. Data Compression is very important for the applications involving large volume of information. It is so because there is a reduction in memory requirement for data storage and energy requirement for data transfer with the reduction of bits [1, 15, 16]. A typical data compression system is given in Fig. 2.

Data compression might be lossy and lossless data compression. In lossy data compression, some inexact approximations are done to represent the content to be encoded and have reduced amount of data. It is popular for compressing multimedia files (audio, video, images, etc.), are of extreme importance for applications such as streaming media and internet telephony. Transform coding, Karhunen–Loeve Transform (KLT) coding, wavelet-based coding, etc. are among the few lossy data compression techniques. But in case of lossless data compression, there is no loss of bit and all the information can be reconstructed from the compressed data and can be restored. Similar approach is followed in most of the lossless compression techniques. In the first step, a statistical model for the input data is generated to map input data to bit sequences in such a way that the most probable (frequently encountered) data will produce shorter output than improbable data in the following step. ZIP file format is the most popular application of lossless data compression. Shannon–Fano algorithm, Huffman algorithm, Arithmetic Coding, etc. are among the few lossless data compression techniques [1, 17–20].

David A. Huffman had proposed a method of compression in 1952 where the symbols occurring frequently are assigned shorter code words than those occurring less frequently and the two symbols having least frequently will have the same code word length. For a symbol set $s = \{a, b, c, d\}$ with probability set of symbols, $p = \{0.4, 0.35, 0.15, 0.1\}$, the formation of Huffman tree is given in Fig. 3 from which code word is obtained. For the given s, the code words will be as given in Table 1. Thus for a symbol stream 'bad', the corresponding binary string will be '110100' [19–21] (Fig. 3).

Fig. 2   A simple data compression system

Table 1 Symbol Code

Sl. no.	Symbol	Binary code
1	a	0
2	b	11
3	c	101
4	d	100

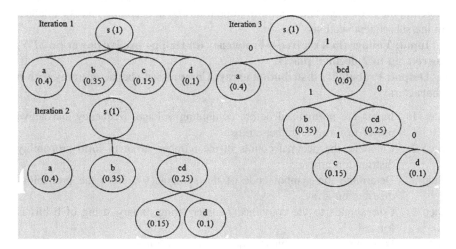

Fig. 3 Formation of Huffman tree

At the receiving end, the encoded data must be decoded to extract the actual string. This can be obtained by following the given algorithm [21].

Step 1 Steps 2–4 are repeated until binary string termination

Step 2 Fetch the first (or next) element, i.e. the twig weight and search for any child nodes of the parent node. If no child nodes found, jump to step 3. Else fetch the next element

Step 3 Determine the symbol corresponding to the binary code from symbol code table and go to step 2

Step 4 End

In case of Adaptive Huffman Coding, the probably of distribution for a particular string is calculated for the string itself. The Huffman tree is then formed by that probability distribution to obtain the binary codes of the characters contained in the string. The advantage of the coding lies in the fact that the binary array being formed will have the least size and thereby best possible compression can be achieved. But the drawback of the method is the requirement of the probability distribution array for each character string thereby limiting the use for large string only [19].

4   Proposed Algorithm

The algorithm is designed to compress the monitoring parameters, voltage, frequency and active power at some time instant by employing Adaptive Huffman Coding in MATLAB environment. The encrypted output will be a character string of reduced size can then be transmitted by using suitable communication channel or stored in the computer. The algorithm to be followed at decoding end is explained in the subsequent sections.

Input: Voltage (in kV); freq—Frequency (in Hz); p—Active power (in MW) correct up to 2 decimal places.

Output: Probability distribution array; Compressed character array with n characters

Step 1   Inputs are normalized before combining voltage, frequency and active power to form a number string

Step 2   Calculate the decimal counts in the number string to form probability distribution array

Step 3   Determine the symbol code of the symbols 0-9 from the probability distribution array

Step 4   Corresponds to the combined number string, binary string of b bits is formed

Step 5   Determine the 3-bit binary equivalent of $(8 * n - b)$

Step 6   Add $(8 * n - b - 3)$ zeros before the binary string to form a new binary string

Step 7   Add the 3-bit binary equivalent before new binary string to form a modified string having $8 * n$ bits

Step 8   Divide the modified string in n equal 8-bit strings and determine the decimal equivalent of all 8-bit strings to form a decimal array

Step 9   Form the character array such that each character will correspond to the ASCII equivalent of the elements of decimal array

Step 10 End

The compressed string is decompressed to obtain the actual transmitted value at the receiving end. The objective can only be achieved when the probability array is available. The algorithm at receiving end is illustrated in the subsequent steps.

Input: Probability distribution array; Compressed character array with n characters

Output: Voltage (in kV); Frequency (in Hz); Active power (in MW) correct up to 2 decimal places

Step 1   Form a decimal array of n elements such that it contains the ASCII value corresponds to each character of the compressed character array

Step 2   Form a binary array having $8n$ bits from char by determining the 8-bit binary for each character of char

Step 3   Remove the first three bits of bin to obtain the number of added zeros (say z)

Step 3   Remove the next z bits of bin to obtain actual Huffman Coded binary array
          with $(8n - 3 - z)$ bits
Step 4   Using the probability distribution array and actual binary array, the
          composite data string is obtained
Step 5   Split the composite array to obtain voltage, frequency and active power
Step 6   End

5   Results and Conclusions

While implementing this proposed algorithm, the obtained results are promising.
With the reduction of the number of characters, the memory requirement for data
storage reduces. Compression ratio (ratio of actual size and compressed size) is an
important parameter of any compression algorithm, which gives the effectiveness of
the algorithm. Higher the compression algorithm be, better will be the compression.
The result obtained by the algorithm for different values of voltage, frequency and
active power is in Table 2. From Table 2, it is clear that the proposed algorithm is
quite effective in terms of the compression ratio with respect to that obtained with
Basic Arithmetic Coding-based compression [2].

It is also visible that there is a variation of the size of compressed array for same
size of actual array. It is so because, the length binary array produced after
employing Adaptive Huffman Coding-based data compression is dependent on the
probability of the data contained in number string. The requirement of probability
distribution for decoding is the main drawback of this algorithm. This algorithm is
tested offline and the time required at encoding and decoding is quite low, typically
in the range of hundreds of milli second. The result obtained by this algorithm is
better than what is obtained with Basic Arithmetic Coding algorithm with equal
probability of occurrence for each data value. The work can also be extended by
employing Adaptive Arithmetic Coding-based compression to determine the best
algorithm for compressing power system monitoring data.

Table 2 Compresses data and compressing output for different input values

System voltage	Frequency	Active power	Actual array size (Bytes)	Compressed array size (Bytes)	Compression ratio	Maximum compression ratio for basic arithmetic coding-based compression
132.01	49.99	500.12	226	98	2.31	2.26
220.02	50.01	499.98	226	82	2.76	
400.02	49.99	498.89	226	82	2.76	
765.05	50.02	510.25	226	82	2.76	

References

1. Sarkar, S. J., Das, B., Dutta, T., Dey, P., Mukherjee, A.: An Alternative Voltage and Frequency Monitoring Scheme for SCADA based Communication in Power System using Data Compression. In: International Conference and Workshop on Computing and Communication (IEMCON), pp 1–7, Vancauver (2015).
2. Sarkar, Subhra J., Sarkar, Nabendu Kr., Dutta, Trishayan, Dey, Panchalika, Mukherjee, Aindrila: Arithmetic Coding Based Approach for Power System Parameter Data Compression. Indonesian Journal of Electrical Engineering and Computer Science, Vol. 2, No. 2, May (2016).
3. Gupta, J. B.: Power System Analysis, 1st Edition, S. K. Kataria & Sons (2011).
4. Grigsby, Leonard L.: Power System Stability & Control, 3rd Edition, CRC Press (2012).
5. Fitzgerald, A.E., Kingsley, Charles, Umans, Stephen. D., Electric Machinery, 6th Edition, http://prof.usb.ve/jaller/Fitzgerald.pdf.
6. Wadhwa, C.L., Electrical Power Systems, 6th ed., New Age International Publishers, New Delhi (2014).
7. Takahashi, Yasuhiro, Matsui, Susumu, Nakata, Yukio, Kondo, Takeshi: Communication Method with Data Compression & Encryption for Mobile Computing Environment. https://www.isoc.org/inet96/proceedings/a6/a6_2.html.
8. ARGHIRA, Nicoleta, HOSSU, Daniela, FĂGĂRĂŞAN, Ioana, ILIESCU, Sergiu Stelian, COSTIANU, Daniel Răzvan: Modern SCADA Philosophy in Power System Operation – A Survey. U.P.B. Sci. Bull., Series C, Vol. 73, Iss. 2 (2011).
9. Patil, C. S., Sonawane, H. M., Patil, K. G.: Overview of SCADA applications in Thermal Power Plant. Proc. of ICMSET-2014 (2014).
10. Singh, Tarlok: Installation Commissioning & Maintenance of Electrical Equipments, 2nd Edition, S. K. Kataria & Sons.
11. Sutterlin, Phil, Downey, Walter, Echelon Corporation: A Power Line Communication Tutorial- Challenges & Technologies. http://www.viste.com/LON/tools/PowerLine/pwrlinetutoral.pdf.
12. Chapter 4: PLC Communication, PLC Hardware User Manual, 4th Ed., Rev. A. https://www.automationdirect.com/static/manuals/c0userm/ch4.pdf.
13. Wikipedia: Radio. https://en.wikipedia.org/wiki/Radio.
14. How broadband satellite Internet works, VSAT Systems. http://www.vsat-systems.com/satellite-internet/how-it-works.html.
15. Wikipedia: Serial Communication. https://en.wikipedia.org/wiki/Serial_communication.
16. Banerjee, Tribeni Prasad, Konar, Amit, Abraham, Ajith: CAM based High-speed Compressed Data Communication System Development using FPGA. http://www.academia.edu/2374082/CAM_based_High-speed_Compressed_Data_Communication_System_Development_using_FPGA.
17. Difference between Lossless & Lossy Data Compression. http://www.rfwireless-world.com/Terminology/lossless-data-compression-vs-lossy-data-compression.html.
18. Theory of Data Compression. http://www.data-compression.com/theory.shtml.
19. Li, Ze-Nian, Drew, Mark S., Liu, Jiangchuan: Fundamentals of Multimedia, 2nd Edition, Springer (2014).
20. Kodituwakku, S.R., Amarasinghe, U. S., Comparisons of Lossless Data Compression Algorithms for Text Data, Indian Journal of Computer Science and Engineering, Vol. 1, No. 4, pp. 406–425.
21. Sarkar, Subhra J., Kundu, Palash K., Mondal, Ipsita: Modified DCSK Communication Scheme for PLCC based DAS with Improved Performance. In: International Conference on Control, Instrumentation, Energy & Communication (CIEC) (2016).

Color Image Visual Cryptography Scheme with Enhanced Security

Prachi Khokhar and Debasish Jena

Abstract Image encryption is one of the most promising fields of research in the conventional scientific society. Visual cryptography is a secured encryption technique which is used to encrypt a secret image based on share generation and superimposition rather than computing. This overcomes the burden of computation but the mammoth risk of attackers is superfluous to the passing of shares in sequence through the communication channel. However, this superimposing problem can be resolved by cracking some additional encryption algorithms alongside the visual cryptography. In this paper, we propose a highly secured visual cryptography scheme which uses encryption and error diffusion halftoning algorithms as intermediate steps in cryptography work. We have also done the comparative analysis in order to select the most optimum technique, among the available algorithms according to the requirement of the system. The proposed work has been tested on various formats of standard color images of varied resolutions and proven more secured than contemporary techniques.

Keywords AES encryption · Correlation · Error diffusion halftoning · Visual cryptography

1 Introduction

In this cybernetic world, acquisition, manipulation, and communication of digital images have been effectively facile due to the popularity of smartphones, android applications, as well as the expeditious growth and extensive availability of internet. Due to increasing digitization, there is a need of preventing pernicious exercises such as unauthorized access and modification of digital data in order to shrink

P. Khokhar (✉) · D. Jena
International Institute of Information Technology, Bhubaneswar, Odisha, India
e-mail: prachikhokhar1919@gmail.com

D. Jena
e-mail: debasish@iiit-bh.ac.in

S.C. Satapathy et al. (eds.), *Proceedings of the 5th International Conference on Frontiers in Intelligent Computing: Theory and Applications*, Advances in Intelligent Systems and Computing 515, DOI 10.1007/978-981-10-3153-3_27

the cybercrime. Hence, it is one of the most promising fields for researchers and scientists to develop techniques and algorithms, which ensures the protection of sensitive data embedded in digital images. Cryptography is an effective approach for secure communications. Primarily, it is an art of writing in the form of secret codes. Thus, the process of cryptography becomes crucial for any secured communication over an unreliable medium such as internet. The method of cryptography preserves data from theft and alteration as well as provides user authentication. In the area of digital data security, there are two domains namely information hiding and secret sharing. Information hiding deals with watermarking and steganography. In the area of secret sharing, Shamir proposed the most widely used cryptographic approach so as to share the input (sophisticated) images, based on visual secret sharing (VSS) scheme and another algorithm was introduced by Naor and Shamir [1, 2] known as visual cryptography (VC). Also, Lin and Lin [3] introduced polynomial-style sharing (PSS) as another alternative to share secret images.

Visual cryptography algorithm encodes the secret data in order to generate different meaningful or meaningless shares and then distributing them into a set of participants. Finally, all or some of the generated shares are stacked up in order to recover the secret image. Shares are especially binary images which are typically offered in transparencies. It is quite simpler and faster than polynomial-style sharing. Additionally, VC is rooted in human visual system (HVS), while PSS is derived from computations performed by digital computers. Hence, because of speedy decoding or recovering properties and perfect cipher, VC method has drawn the attention towards research. Visual cryptographic scheme is extensively employed for secure transmission of sophisticated images and passwords in military, intelligence, E-bill and tax payments, confidential video conferencing, medical imaging system, and internet banking. VC can be used for copyright protection in order to enable the resolution of rightful ownership which is not going to use the original image. Along with all these properties, one of the key features of VC is that during embedding process there is no alteration of the host image. First, VC was applicable to black and white images only as explained in Ateniese et al. [4] but later it has been extended to grayscale (Blundo et al. 2000) [5] and color images (Leung et al. 2009) [6] as well with some additional steps.

Image halftoning is also an interesting field under visual cryptography. Halftoning of the images is a reprographic methodology which is going to simulate continuous tone with the usage of dots into the binary image. Generally, the 8-bit monochrome image represents 256 discrete gray levels which are not much supported by devices like facsimile (FAX), laser and inkjet printing, and electronic scanning and copying, which are capable of displaying only two levels, black or white, for a monochrome image. Hence, halftoning is the technique that simulates multi-tone images into binary (or dual tone) image, i.e., from a gray-level image of 256 levels (8-bit) to single-bit binary image or from a 24-bit RGB image to a 3-bit color halftone image where one bit is associated with each R, G, and B channel. As a result, this becomes the very first pre-processing step of color image visual cryptography scheme.

From security point of view, image scrambling or image encryption algorithms can also be used to rearrange all the pixels in an image to different pixel locations which result in modified non-recognizable image by permutation of the original image. A wide variety of classical encryption algorithms has been proposed such as Arnold transformation, Advanced Encryption System (AES), Hilbert curve, Blowfish, RSA, DES, and Fibonacci transformation. The parameters responsible for the selection of particular algorithm are key space, decrypted image quality, time complexity, immunity against noise, and computational complexity.

Only shares of the secret image are available on communication channel. Hence, an unauthenticated person directly cannot predict the secret information with the availability of a single share, but if the attacker is able to obtain all the shares then there is a risk of retrieval (Hsu and Tu 2008) [7]. Thus to get a shot of this dilemma, we are supposed to increase the security level of shares (Jena and Jena 2008) [8].

This paper presents a highly secured visual cryptography scheme applied on secret color image which uses channel separation and halftoning as pre-processing steps and then shares are generated using (2, 2) visual cryptography scheme. Then apply key-based encryption on generated shares at the transmitter end. At receiver end, initially the transparencies undergo decryption and then stacking up of decrypted shares will prevail the original image. Hence, perceiving any trace about the secret image from an individual share is difficult, due to encryption during transmission phase.

The remaining part of paper has been shaped in such a way that Sect. 2 shows literature review. Section 3 shows proposed work and Sect. 4 covers experimental results and comparisons to exhibit fruitfulness of our proposed work. In Sect. 5 conclusions are presented.

2 Literature Review

For color image visual cryptography, numerous algorithms have been proposed. If someone is able to extract the color or pattern of a section of secret image from a provided share efficiently, the cryptography scheme will become inadequate and insecure. Naor and Shamir [1] proposed a basic 2-out-of-2 visual cryptography scheme, where input image gets divided into two shares and original image is obtained when both the shares are superimposed on each other. Share generation takes place according to the pixel value that is 0 or 1 of input image as presented in Fig. 1. Since pixel value is different for both shares so pixel of the secret image cannot be predicted using a single share. Once both the shares are superimposed together, then only the original image can be exposed.

In a similar manner, we can also generate n shares and out of which we will use only k significant shares in order to generate the secret image, where k <= n. Kandar and Dhara [9] proposed (k, n) secret sharing method for RGB images. Here, for a white pixel, $(n - k) + 1$ number of shares will have the same white pixel at the same pixel location while keeping the pixel as black for remaining shares. These

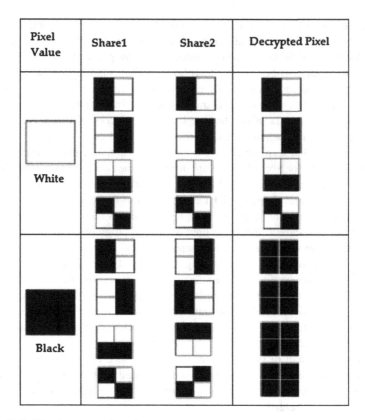

Fig. 1 (2, 2) Visual cryptography scheme

$(n - k) + 1$ shares are located using a random number generator. Hence, it gets easier to predict the pixels in all shares because the secret is not hidden properly. So, one can easily recover the secret image.

Hou [10] proposed a color image visual cryptography using halftoning scheme. Hous technique was not efficient due to two disadvantages. First, one is receiving a color share in CMY color space and another one is the inferior quality of decrypted image. Bani et al. [11] used conjugate error diffusion-based watermarking scheme for data hiding. In this approach, both the shares generated from original image are watermarked into cover image. The stacking of both the shares results in secret and cover image. But the limitations of this technique were requirement of extra storage size for cover image and the probability of attacks on shares due to which quality of extracted image is quite inferior.

Fang and Yu [12] presented a (2, n) visual cryptography method which rely on balancing of the performance between pixel contrast and pixel expansion. Chang et al. [13] also proposed an image encrypting scheme where two secret shares get submerged into two gray-level images. To decrypt the secret data, the shares are superimposed. Han et al. [14] suggested to use XOR operation for superimposing the images. Tan [15] advocated threshold visual secret sharing scheme by mixing XOR and OR operations which used the concept of binary linear error correcting code. Here, methods were limited to the embedding of only single set of secret data.

Chandramathi et al. [16] summarized the overview of all existing visual cryptography schemes and derived the conclusion that one should develop the algorithm which provides better quality and enhanced security with the least possible pixel expansions. Yan et al. [17] recommended a new solution for stacking up of two shares based on the usage of alignment marks in Walsh transform domain. The discovered technique is advantageous all the time since superimposing of two transparencies is the obligatory step for VC decryption. If there is any misalignment then decryption of image is impossible.

Kaur and Khemchandani [18] presented a method for encrypting the shares generated using (2, 2) visual cryptography scheme. For encryption, they have used public key encryption-based RSA algorithm so as to obtain enhanced security of secret document. Due to this encryption process, secret shares are available in encrypted form which can be decrypted only by the authenticated person who is having the private key to decrypt the shares, which avoids the creation of forged shares. This work was limited only to binary and small-sized images. Also, the time complexity of RSA is quite high which is another issue.

Desiha and Kaliappan [19] proposed embedded extended visual cryptography which uses enhanced halftoning technique. The technique uses effective dithering halftone scheme which not only reduces time complexity during halftoning but also enhances the visual quality of input image. But the limitation was first to convert the input image into gray image for applying halftoning. Also, if dimension of dithering matrix falls behind then time complexity gets quite high.

The AES algorithm was standardized by the US National Institute of Standards and Technologies (NIST) as an alternative to the data encryption standard (DES) cryptographic algorithm (National Institute of Standards and Technology 2001) [20]. This algorithm is based on Rijndaels 128-bit symmetric-key algorithm. It is widely used to encrypt/decrypt the data in portable hard drive systems, to implement various processors, satellite imagery, biomedical images, and smart cards. This algorithm provides enhanced security and better performance in both hardware and software; both the sender and receiver utilize the key not only during encryption but during decryption too. AES uses different operations like Sub Bytes, Add Round Key, S-Box Creation, Shift Row, Mix Column, etc. Figure 2 presents the detailed flow of AES encryption algorithm.

3 Proposed Work

As we have seen the issues related to security, time complexity, and quality of image, here we propose an enhanced visual cryptography approach for RGB images which uses (2, 2) basic visual cryptography scheme along with halftoning and encryption for high level security. In this technique, we will generate the shares of

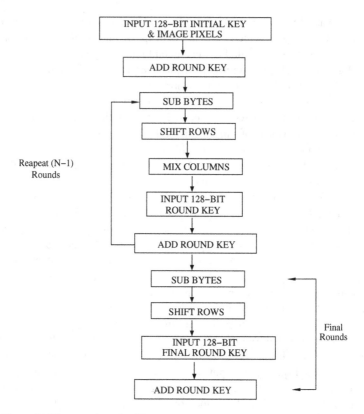

Fig. 2 Flow of AES encryption algorithm

halftone images of each channel and then encryption is applied to preserve from the shares from malware activities which can modify the bit sequences to generate the unauthenticated shares. The proposed work is characterized as follows:

3.1 Transmitter End

At transmitter end following steps are performed, and flow chart for the same is shown in Fig. 4a.

- Step-1: Apply channel separation (RGB) on the color image.
- Step-2: The separated channels are then converted from continuous tone into a series of dots using halftoning algorithm. The denser or darker areas of the image are represented by larger dots and the rarer or lighter areas by smaller dots. We have used two types of error diffusion halftoning algorithms, namely Floyd halftoning and Jarvis halftoning. Figure 3 presents flow for error diffusion halftoning algorithm.

Fig. 3 Block diagram of
error diffusion halftoning
algorithm

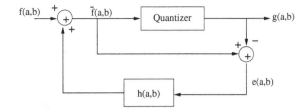

- Step-3: Apply (2, 2) VCS on halftone images of channels to generate two shares for each channel. Hence, we obtain total six shares for the input image.
- Step-4: The generated shares are then encrypted using appropriate image encryption algorithm. Here, we have used 128-bit key-based advanced encryption system (AES) and key generation-based BitXOR encryption algorithms.

3.2   Receiver End

At receiver end following steps are performed, and flow chart for the same is shown in Fig. 4b.

- Step-1: Receiver obtains the encrypted shares on which decryption is performed using the same algorithm and same key which was used on the transmitter side.
- Step-2: The decrypted shares are then stacked up together for individual channels by applying logical XOR operation in order to obtain the original channel.

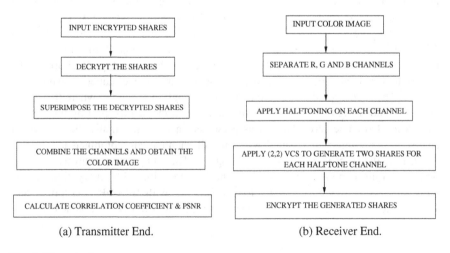

(a) Transmitter End.   (b) Receiver End.

Fig. 4 Flow chart

- Step-3: Concatenation of three channels is performed for obtaining the color image.
- Step-4: We also calculate the correlation coefficient and PSNR as quality measures of obtained image and original image using the following formulas:

$$CC = \frac{\sum_a \sum_b (X_{ab} - \bar{X})(Y_{ab} - \bar{Y})}{\sqrt{(\sum_a \sum_b (X_{ab} - \bar{X})^2)(\sum_a \sum_b (Y_{ab} - \bar{Y})^2)}} \tag{1}$$

where X(a, b) and Y(a, b) are the pixels of original image and output image and $\bar{X}$ and $\bar{Y}$ are the mean pixel value for images X and Y, respectively.

$$PSNR = 10log_{10} \frac{X_{max}^2}{MSE} \tag{2}$$

where X_{max} is the maximum pixel value of original image, P * Q is the resolution of image and

$$MSE = \frac{1}{P*Q} \sum_{a=1}^{P} \sum_{b=1}^{Q} (X_{ab} - Y_{ab})^2 \tag{3}$$

4 Experimental Results and Comparisons

The proposed work has carried out with the usage of MATLAB 8.1 (2013a) software. The programs have been simulated using a computer with 4 GB RAM, 3.07 GHz Intel Core i3 processor having 64-bit Windows 7 OS. For performance analysis, the proposed work has been implemented on a number of images with a variety of formats (.jpg, .bmp, .tiff, etc.) and resolutions (64 × 64, 128 × 128, 256 × 256, etc.). In Fig. 5a, to simulate the results, image House.tiff with 128 × 128 resolution is taken as an input. The halftoning algorithm is then applied to this image and after that shares are generated as shown in Fig. 5b. Further, the shares undergo encryption and the encrypted shares are obtained as shown in Fig. 5c. These encrypted shares are then sent through the channel, where the shares are decrypted using decryption algorithm and the same key as used by the sender. The decrypted shares are obtained as presented in Fig. 5d. Finally, the decrypted shares are superimposed channel wise and all channels are combined in order to reconstruct the color image as shown in Fig. 5e. The comparative analysis has been done for four combinations as Floyd halftoning with AES encryption, Jarvis halftoning with AES encryption, Floyd halftoning with BitXOR encryption, and Jarvis halftoning with BitXOR encryption scheme and has been tabulated for the

(a) Original Input Image.   (b) Generated Shares after halftoning.

(c) Generated Shares after halftoning.   (d) Generated Shares after halftoning.

(e) Generated Shares after halftoning

Fig. 5 Step-by-step process of experimental result

parameters for four different images of 128 × 128 resolution namely Baboon.tiff, House.tiff, Lena.jpg, and Peppers.tiff.

The salient feature of the proposed work is if decryption of the shares is done using the wrong key, the shares will not be decrypted accurately. Hence, the hackers will not be able to decrypt the shares until and unless they have any clue about it. Due to the use of 128-bit key in AES, it is pretty difficult to guess the key as well. This will result in a step forward to achieve the desired security. From comparisons, it is fairly obvious that AES provides more PSNR than BitXOR encryption algorithm as shown in Fig. 6c but with the tradeoff of taking a long time to implement the algorithm, resulting in more time elapsed as shown in Fig. 6b. In halftoning process, Jarvis algorithm proves out to be better in respect of both retrieved image quality (high correlation coefficient and better PSNR) as shown in Fig. 6a and 6c and time elapsed as shown in Fig. 6b. From Tables 1, 2, 3, and 4, it is very much obvious that Jarvis halftoning along with AES encryption technique is

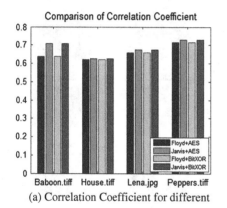

(a) Correlation Coefficient for different Techniques.

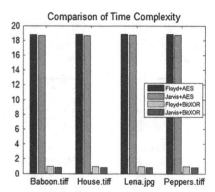

(b) Time Elapsed for different Techniques.

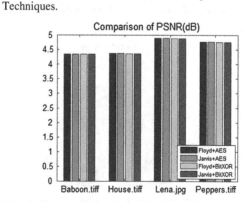

(c) Peak Signal to Noise Ratio Value for different Techniques.

Fig. 6 Comparative analysis

Table 1 Output parameters of BitXOR encryption with Floyd halftoning

Image	Correlation (%)	Time elapsed (ms)	PSNR (dB)
Baboon	63.75	969.922	4.3294
Peppers	62.19	983.754	4.3512
Lena	65.84	975.043	4.8763
House	71.55	981.440	4.7379

Table 2 Output parameters of AES encryption with Floyd halftoning

Image	Correlation (%)	Time elapsed (s)	PSNR (dB)
Baboon	63.75	18.800270	4.3406
Peppers	62.19	18.847479	4.3655
Lena	65.84	18.868430	4.8885
House	71.56	18.842521	4.7515

Table 3 Output parameters of BitXOR encryption with Jarvis halftoning

Image	Correlation (%)	Time elapsed (ms)	PSNR (dB)
Baboon	70.76	841.716	4.3304
Peppers	62.53	861.313	4.3515
Lena	67.44	846.041	4.8768
House	72.77	837.639	4.7384

Table 4 Output parameters of AES encryption with Jarvis halftoning

Image	Correlation (%)	Time elapsed (s)	PSNR (dB)
Baboon	70.76	18.689622	4.3416
Peppers	62.54	18.665042	4.3658
Lena	67.45	18.743785	4.8890
House	72.78	18.738144	4.7521

better in terms of image quality (high correlation coefficient and better PSNR) and Jarvis halftoning along with BitXOR encryption technique is better in terms of time elapsed.

5 Conclusion and Future Work

The proposed work is reliable to reconstruct the color images with exact color transforms when compared with the original image. Also stacking up of encrypted shares cannot retrieve the image; hence, the security is maintained. If encrypted share information is obtained by the attackers, then also original image cannot be retrieved due to unavailability of the key used during encryption. In our work, we have presented the comparative analysis as well which provides the user with a variety of choices in terms of either PSNR, correlation coefficient, or time elapsed. This work can further be extended to do inverse halftoning at the receiver side to improve the image quality, variations in AES algorithms in different modes (ECB, CBC, CFB, OFB, etc.), time efficient 32-bit AES, and compression of images. This technique can further be implemented for video also on the basis of frame processing.

References

1. Naor, M. and Shamir, A., 1995. Visual cryptography. In Advances in Cryptology—EUROCRYPT'94 (pp. 1–12). Springer Berlin/Heidelberg.
2. Naor, M. and Shamir, A., 1996, April. Visual cryptography II: Improving the contrast via the cover base. In Security protocols (pp. 197–202). Springer Berlin Heidelberg.
3. Lin, S.J. and Lin, J.C., 2007. VCPSS: A two-in-one two-decoding-options image sharing method combining visual cryptography (VC) and polynomial-style sharing (PSS) approaches. Pattern Recognition, 40(12), pp. 3652–3666.
4. Ateniese, G., Blundo, C., De Santis, A. and Stinson, D.R., 1996. Visual cryptography for general access structures. Information and Computation, 129(2), pp. 86–106.
5. Blundo, C., De Santis, A. and Naor, M., 2000. Visual cryptography for grey level images. Information Processing Letters, 75(6), pp. 255–259.
6. Leung, B.W., Ng, F.Y. and Wong, D.S., 2009. On the security of a visual cryptography scheme for color images. Pattern Recognition, 42(5), pp. 929–940.
7. Hsu, C.S. and Tu, S.F., 2008. Digital watermarking scheme with visual cryptography. IMECS, Hong Kong (March 2008).
8. Jena, D. and Jena, S.K., 2009, January. A Novel Visual Cryptography Scheme. In Advanced Computer Control, 2009. ICACC'09. International Conference on (pp. 207–211). IEEE.
9. Kandar, S. and Dhara, B.C., 2011. kn Secret Sharing Visual Cryptography Scheme on Color Image using Random Sequence. IJCA (0975–8887) Volume.
10. Hou, Y.C., 2003. Visual cryptography for color images. Pattern Recognition, 36(7), pp. 1619–1629.
11. Bani, Y., Majhi, D.B. and Mangrulkar, R.S., 2008. A Novel Approach for Visual Cryptography Using a Watermarking Technique. In Proceedings of 2nd National Conference, IndiaCom (pp. 08–09).
12. Fang, L. and Yu, B., 2006, August. Research on pixel expansion of (2, n) visual threshold scheme. In Pervasive Computing and Applications, 2006 1st International Symposium on (pp. 856–860). IEEE.
13. Chang, C.C., Hwang, M.S. and Chen, T.S., 2001. A new encryption algorithm for image cryptosystems. Journal of Systems and Software, 58(2), pp. 83–91.
14. Han, Y., Dong, H., He, W. and Liu, J., 2012, November. A verifiable visual cryptography scheme based on XOR algorithm. In Communication Technology (ICCT), 2012 IEEE 14th International Conference on (pp. 673–677). IEEE.
15. Tan, X.Q., 2009, May. Two kinds of ideal contrast visual cryptography schemes. In 2009 International Conference on Signal Processing Systems (pp. 450–453). IEEE.
16. Chandramathi, S., Ramesh Kumar, R., Suresh, R. and Harish, S., 2010. An overview of visual cryptography. International Journal of Computational Intelligence Techniques, ISSN, pp. 0976–0466.

17. Yan, W.Q., Jin, D. and Kankanhalli, M.S., 2004, May. Visual cryptography for print and scan applications. In Circuits and Systems, 2004. ISCAS'04. Proceedings of the 2004 International Symposium on (Vol. 5, pp. V-572). IEEE.

18. Kaur, K. and Khemchandani, V., 2013, February. Securing Visual Cryptographic shares using Public Key Encryption. In Advance Computing Conference (IACC), 2013 IEEE 3rd International (pp. 1108–1113). IEEE.

19. Desiha, M. and Kaliappan, V.K., 2015, January. Enhanced efficient halftoning technique used in embedded extended visual cryptography strategy for effective processing. In Computer Communication and Informatics (ICCCI), 2015 International Conference on (pp. 1–5). IEEE.

20. Pub NF. 197: Advanced Encryption Standard (AES), Federal Information ProcessingStandards Publication 197, US Department of Commerce/NIST, November 26, 2001. Available from the NIST website.

A Comparative Analysis of PWM Methods of Z-Source Inverters Used for Photovoltaic System

Babita Panda, Bhagabat Panda and P.K. Hota

Abstract The photovoltaic cell produces pollution less electricity. It requires almost no maintenance and has long lifespan. Nowadays, the photovoltaic is one of the most promising markets in the world because of these advantages. This paper demonstrates the dynamic model of single-stage three-phase impedance source inverter or Z-source inverter connected to grid. Here ZSI connected to PV is analyzed and designed. As the output of the PV array is very low, in order to commercialize and utilize this, the output voltage must be increased. So to boost up the voltage, Z-source inverter (ZSI) is used instead of VSI or CSI. Different pulse width modulation (PWM) techniques are used to provide pulses for PV connected Z-source converter (ZSI). After this, the final model is simulated using MATLAB/SIMULINK, and THD related to different output waveforms are analyzed for different parameters used.

Keywords PV system · Multidevice boost circuit (MDBC) · Z-source inverter (ZSI) · PWM techniques · MATLAB/Simulink software

1 Introduction

Non-conventional power plants constitute around 27.80% of total installed capacity and the remaining 72.20% is constituted by nonrenewable power plants. India produced around 967 TWh (967,150.32 GWh) of electricity (not including electricity generated from captive power plants and renewable plants) for 2013–2014

B. Panda (✉) · B. Panda
KIIT University, Bhubaneswar, India
e-mail: pandababita18@gmail.com

B. Panda
e-mail: panda_bhagabat@rediffmail.com

P.K. Hota
VSSUT Burla, Sambalpur, India
e-mail: P_hota@rediffmail.com

© Springer Nature Singapore Pte Ltd. 2017
S.C. Satapathy et al. (eds.), *Proceedings of the 5th International Conference on Frontiers in Intelligent Computing: Theory and Applications*, Advances in Intelligent Systems and Computing 515, DOI 10.1007/978-981-10-3153-3_28

period. On March 2013, the per capita consumption of electricity in India was 917.2 kWh. Renewable energy connected to grid capacity in India has reached 29.9 GW, of which wind constitutes 68.9%, while solar PV produces around 4.59% of the renewable energy installed capacity in India.

Dense population and high solar insolation are the two factors that provide a good compounding for generation of solar power in India. As majority regions of the country are not grid connected, the first program of solar power is water pumping. The photovoltaic cell produces pollution less electricity. It requires almost no maintenance and has long lifespan. Nowadays, the photovoltaic is one of the most promising markets in the world because of these advantages. Nevertheless, PV power is quite costly, and the reduction of cost of PV systems is contingent on wide research. From the standpoint of power electronics, this target can be reached by boosting up the output energy of a given PV array. This can be done by Impedance Source Inverter or Z-source inverter. Section 2 of this paper describes solar PV module, Sect. 3 describes about ZSI and its modes of operation. Section 4 describes about the different PWM techniques. Section 5 describes about Simulation model for different techniques and their results. Section 6 states about Conclusion of my work.

2 Solar PV System

Solar cells are the basic constituents of photovoltaic panels. Maximum solar cells are manufactured using silicon also other materials are employed. Solar cells have property of photoelectric effect: where some semiconductors have capability of changing electromagnetic radiation precisely to electrical current. The charged particles produced using incident radiation is distinguished smoothly to develop an electrical current by using suitable layout of the solar cell. The electricity generated by solar cell depends on the intensity of sunlight. When the incidence of sunlight is perpendicular to the front side of PV cell, the power generated by the solar cell is optimum. The large number of solar cells is combined in series and parallel to form PV array. A single solar cell can be modeled can using a diode, two resistor, and current source named as a single diode model of solar cell. Figure 1 represents the equivalent circuit diagram of photovoltaic system.

Fig. 1 Single diode model of a solar cell

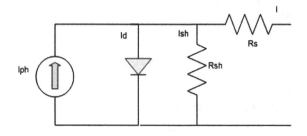

$$I_{PV} = N_P * I_{PH} - N_P * I_O * [\exp\{q * \frac{V + I_{PV} * Rs}{N_S * A * k * T}\} - 1]) \tag{1}$$

$$I_O = I_{RS} * (\frac{T}{Tr})^3 * [\exp\{q * E_G * \frac{\frac{1}{T_R} - \frac{1}{T}}{A * k}\}] \tag{2}$$

$$I_{PH} = I_{SCR} + K_I * (T - T_R) * \frac{S}{1000} \tag{3}$$

$$I_{RS} = I_{SCR}/[\exp(q * V_{OC}/N_S * A * k * T) - 1] \tag{4}$$

where I_{Ph} is the light generated Current, I_d is the diode current, I_{sh} is the current through shunt resistance, R_s and R_{sh} represents series and shunt resistance, respectively. I_{pv} is the current at the output terminal, I_o is the reverse saturation current of diode, N_s is number of cells connected in series, N_P represents number of cells connected in series, q is the electron charge, k Boltzmann's constant and a the ideality factor modified, S represents the Irradiance. In MATLAB/SIMULINK the PV array is modeled by using these equations [1, 2].

The KC200GT solar is being chosen for modeling and simulation using MATLAB (Table 1) [3].

3 ZSI and Its Control Techniques

Impedance source inverter or Z-source inverter or Z-source converter is taken into consideration to overcome the limitations and barriers of traditional voltage source Inverters and Current Source Inverters [4]. The impedance network of ZSI couples the converter main circuit to power source, inverter or load of the circuit. It can perform both buck and boost operation. So it can be implemented for different forms of power conversions, i.e., from AC-to-AC, AC-to-DC, DC-to-AC, and DC-to-DC. The impedance network consists of two inductors L_1 and L_2 and two capacitors C_1 and C_2 and they are connected in X-shape [5]. This feature in case of inverter gives boosting of voltages. As the traditional VSIs and CSIs, it also have

Table 1 Specifications of PV Panel

Temperature	25 °C
V_{OC}	32.9 V
I_{SC}	8.21 A
V_{MPP}	26.3 V
I_{MPP}	7.61 A
Maximum power, MPP	200.143 W
Temperature coefficient of I_{SC}	0.0032 A/K
Temperature coefficient of V_{OC}	−0.1230 V/K
N_s	54

six active states and two zero states, but it has seven more states known as shoot-through states which causes when both the upper and lower leg of a phase or more than one phase get shorted. ZSI operates in two states: shoot-through state and non-shoot-through state [6, 7]. This causes the buck or boost operation of the converter. In this paper, the ZSI is integrated with the PV array to get boosting of voltage from low input voltage (Fig. 2).

The boosting of voltages is carried out with the formula;

$$\hat{v}_s = M * B * V_{IN}$$
$$B = \frac{1}{1 - 2D_s} \tag{5}$$

where

B = boosting factor
D_s = shoot-through duty ratio
M = modulation index

The ZSI has three control methods [8] known as

(a) simple boost control
(b) constant boost control
(c) maximum boost control

In simple boost control, some part of the total zero state is converted into shoot-through state for voltage boost up. In case of maximum boost control the total zero state of the states are converted into shoot-through state which gives a result of maximum boosting of input voltage to the inverter [9].

The formulas followed for maximum boost controls are given below:
For space vector analysis,

$$D_{sh} = 1 - \frac{2\sqrt{3}M}{\Pi} \tag{6}$$

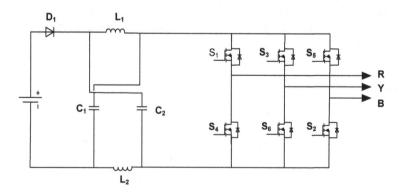

Fig. 2 Impedance source inverter

For carrier-based analysis,

$$D_{sh} = 1 - \frac{3\sqrt{3}M}{2\Pi}$$
$$B = \frac{\Pi}{(3\sqrt{3}M) - \Pi}$$

(7)

In this paper, all the operations are done in maximum boost control as it gives better boost up than the simple boost control.

4 PWM Techniques

Following are some PWM techniques, which are carried out in maximum boost control for ZSI. Here both carrier-based PWM as well as space vector-based PWM techniques are discussed and also both conventional and some Advanced PWM techniques are implemented in both the cases [4, 10].

4.1 Carrier Based PWM Techniques

In carrier-based PWM techniques, for all the techniques, the carrier signal is a triangular signal and it is of a very high frequency compared to the modulating signal. The modulating signal is always a sine wave that has to be compared with the carrier signal to generate desired pulse, which is having the frequency same as fundamental frequency. The modulating signal gets changed by different carrier based techniques added with it at different modulation index. In this way, we get different desired PWM gate pulses from this technique. Below, some numbers of carrier-based techniques are discussed. They are as follows:

4.1.1 Sine-Triangle PWM (ST PWM)

In this type of PWM technique, the modulating signal is a sine wave and it gets compared with the carrier signal to generate pulse.

4.1.2 Third Harmonic Injected PWM (THI PWM)

Here the modulating signal or sinusoid signal is injected with another sinusoid signal which is called third harmonic injection and this signal is having frequency exactly three times of it and having amplitude generally of one-sixth of it. Then the new modulating signal is compared with the carrier signal to get pulse [10].

4.1.3 Triplen-Injected PWM

Here the sinusoid modulating signal is injected with repeating signal which has different magnitude and time period according to the selection. This technique is somehow similar with the third Harmonic injected PWM technique.

4.1.4 Bus-Clamping 30° PWM and Bus-Clamping 60° PWM (BC 30° and 60° PWM)

Here each phase of the modulating sinusoid signal is clamped to one-one leg of DC bus terminal for a period of 30° and 60°, respectively, to create the desired modulated pulse. Then these new generated signals are compared with carrier signals and gate pulses are generated.

4.2 Space Vector-Based PWM Techniques

4.2.1 Conventional Space Vector PWM (CSVPWM)

In CSVPWM, the six active vectors are placed in such a way that they form a hexagon and they are apart from each other on 60° interval from the center. Two zero vectors are placed at the center. The area bounded by any two vectors is known as sector. They form six sectors. The reference vector is created from any two adjacent vectors applied for a particular time period. The switching sequence stars here from (0127-7210) for the first sector and likewise next sectors follow. In this way this technique works.

4.2.2 Bus-Clamping 30° and 60° PWM (BC 30° PWM and BC 60° PWM)

In Space Vector approach, the bus-clamping technique [10] is carried out as every phase of the inverter is clamped to one of the DC bus terminal for a duration of 30° and 60° for BC 30° and BC 60° PWM, respectively. In both the cases, each sector is divided into two subsectors. Thus, we get 12 numbers of subsectors. The switching sequence of these techniques are (012-210) and (721-127) for 30° and 60° clamping, respectively. Thus these techniques work accordingly.

5   Analysis of Simulation Result

Simulation studies are performed on PV ZSI implemented with all proposed PWM techniques in MATLAB/Simulink. The results obtained from the simulations are taken with some specific values of the parameters taken. Here the input voltage from the PV cell to the inverter is 100 V. For ZSI, the DC side inductors L1 and L2 are 37.5 μH, the capacitors values are as C1 = C2 = 470 μF and the AC-side three-phase load resistors are 200 Ω and inductors are 300 μH. In this topology the results from the experimental studies are taken as 0.8 modulation index for both carrier-based and space vector-based PWM techniques. Figure 3 shows the block diagram of the proposed system.

5.1   Simulink Block Diagram

The figure shows simulink block diagram of the proposed system. PV system is connected to Voltage source inverter through MDBC impedance source network. The RL load is connected at the output. A comparative analysis is made by using different PWM techniques to the voltage source inverter. The system is modeled in MATLAB/simulink environment (Fig. 4).

5.2   Wave Forms

Figures 5 and 6 show the line current and voltage waveforms of the proposed system when conventional space vector PWM technique is applied.

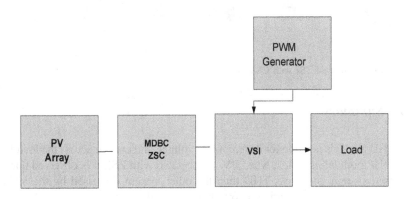

Fig. 3 Block diagram of the proposed system

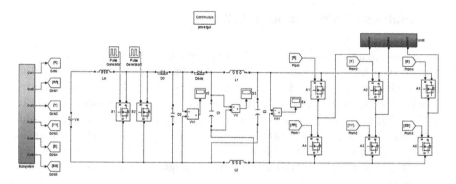

Fig. 4 Simulink block diagram

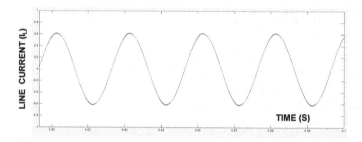

Fig. 5 Line current of ZSI with CSVPWM

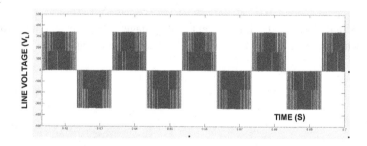

Fig. 6 Line voltage of ZSI with CSVPWM

5.3 Summary

The THD analysis is made for different line voltage and current across the load for a particular modulating index, when PV is integrated with ZSI. It is observed that the proposed system gives less THD than the other system discussed by the authors [10]. Table 2 gives summarize report about the THD analysis.

Table 2 PV Array integrated space vector-based simulation results of conventional ZSI

	C-ZSI (Maximum boost control)			
$Ma = 0.8$	I_L (A)	THD (%)	V_L (V)	THD (%)
ST	0.627	0.57	240.2	71
THI	0.635	0.53	241.3	69.5
TRIPLEN	0.637	0.525	241.6	69.4
BC-30°	0.693	0.505	273	63.48
BC-60°	0.671	0.51	270.1	65.41
CSVPWM	0.625	0.61	240.7	75.2
BC-30°	0.66	0.55	255.4	68.8
BC-60°	0.642	0.58	0.58	71.6

6   Conclusion

In this paper, the ZSI is integrated with the PV array to get boosting of voltage from low input voltage. In all the cases the bus-clamping techniques are giving better results. In carrier-based techniques, the triplen-injected harmonic PWM and BC-30° and BC-60° PWM techniques are giving less THD compared to others. Among BC 30° and BC 60°, BC 30° gives better result. In case of space vector-based PWM techniques, the bus clamping 30° is giving the best output and less THD among the three proposed PWM techniques.

References

1. M. G. Villalva, J. R. Gazoli and E. R. Filho, "Comprehensive Approach to Modeling and Simulation of Photovoltaic Arrays," in *IEEE Transactions on Power Electronics*, vol. 24, no. 5, pp. 1198–1208, May 2009.
2. G. Farivar and B. Asaei, "A New Approach for Solar Module Temperature Estimation Using the Simple Diode Model," in *IEEE Transactions on Energy Conversion*, vol. 26, no. 4, pp. 1118–1126, Dec. 2011.
3. A. Yazdaniet al., "Modeling guidelines and a benchmark for power system simulation studies of three-phase single-stage photovoltaic systems," IEEE Trans. Power Delivery, vol. 26, no. 2, pp. 1247–1264, April 2011.
4. Kun YU, Fang Lin LUO, Miao ZHU"Space Vector Pulse-Width Modulation Based Maximum Boost Control of Z-Source Inverters" *978-1-4673-0158-9/12/$31.00 ©2012 IEEE.*
5. Fang Zheng Peng, "Z-source inverter," in *IEEE Transactions on Industry Applications*, vol. 39, no. 2, pp. 504–510, Mar/Apr 2003.
6. B. Y. Husodo, M. Anwari, S. M. Ayob and Taufik, "Analysis and simulations of Z-source inverter control methods," *IPEC, 2010 Conference Proceedings*, Singapore, 2010, pp. 699–704.

7. Poh Chiang Loh, Feng Gao, Pee-Chin Tan. Three-Level AC-DC-AC ZSource Converter Using Reduced Passive Component Count [J]. IEEE Transactions on Power Electronics, 2009, 24(7), pp: 1671–1681.

8. F. Z. Peng, M. Shen and Z. Qian, "Maximum boost control of the Z-source inverter," *Power Electronics Specialists Conference, 2004. PESC 04. 2004 IEEE 35th Annual*, 2004, pp. 255–260 Vol.1.

9. Poh Chiang Loh, D. M. Vilathgamuwa, Y. S. Lai, Geok Tin Chua and Y. Li, "Pulse-width modulation of Z-source inverters," in *IEEE Transactions on Power Electronics*, vol. 20, no. 6, pp. 1346–1355, Nov. 2005.

10. Shaojun Xie, Yu Tang and Chaohua Zhang, "Research on third harmonic injection control strategy of improved Z-source inverter," *2009 IEEE Energy Conversion Congress and Exposition*, San Jose, CA, 2009, pp. 3853–3858.

Fault Mitigation in Five-Level Inverter-Fed Induction Motor Drive Using Redundant Cell

B. Madhu Kiran and B.V. Sanker Ram

Abstract Induction motor was very often used machine in industries. Recent developments in electronics led to use of induction motor-driven electrical vehicles. Faults in inverter-fed induction motor can lead to unusual operation. The knowledge of faults in inverter circuit is as important. Fault identification is as much important as prior knowledge to mitigate the faults that might occur in inverter-driven induction motor drives. In this paper, five-level multi-level inverter was taken up for testing of mitigation using redundant cell. In this simulation model only two fault cases are considered for voltage source inverter (VSI)-fed squirrel cage induction motor drive. Those two cases are switch open fault and switch short fault. In this work redundant cell comes into operation when switch is open fault, short fault in drive system. The Matlab/Simulink-based model clearly explains the effect of adding redundant cell during fault.

Keywords H-bridge redundant cell · Voltage source inverter · Mitigation of fault · Switch opens fault and short fault

1 Introduction

Designing of protection system plays vital role in planning and operation of drive systems. Adjustable induction motor drives widely used in manufacturing and EVHV applications due to their simplicity and rugged construction. In sensitive applications like EVHV fault tolerant capability plays vital role; so designing of fault tolerant induction motor for such applications needs better fault detection and mitigation [1–3]. Most commonly three-level voltage source inverter (VSI)-fed induction motor drive is used in industrial applications. But for high power

B. Madhu Kiran (✉)
PSRCMR College of Engineering and Technology, Vijayawada AP, India
e-mail: madhukiran1.eee@gmail.com

B.V. Sanker Ram
JNTUH College of Engineering, Hyderabad TS, India

© Springer Nature Singapore Pte Ltd. 2017
S.C. Satapathy et al. (eds.), *Proceedings of the 5th International Conference on Frontiers in Intelligent Computing: Theory and Applications*, Advances in Intelligent Systems and Computing 515, DOI 10.1007/978-981-10-3153-3_29

multilevel inverter (MLI)-fed induction motor drive is very popular because of its low THD and stress on winding insulation. But due to increase in number of switches fault tolerance becomes complex.

There are commonly six types of faults in any voltage source converter (VSI), but out of those switch open and switch short faults are commonly occurred. When particular switch is open or shorted, in normal three-level VSI, the pulsation in torque is more but in case of five-level multilevel inverter-based VSI this fault will create less torque ripple. For complete mitigation of fault here we are using redundant cell in each phase leg. Whenever there is a fault in particular phase leg, the faulted H-bridge cell is short circuited and redundant cell comes into operation. This paper discusses short and open fault of redundant cell simulated for five-level inverter.

2 Multilevel Inverter for Electrical Vehicle

Multilevel inverter (MLI) is classified into three types named as follows:

1. Diode clamped multilevel inverter
2. Flying capacitor multilevel inverter
3. Cascaded H-bridge multilevel inverter (CHB).

Out of those, cascaded H-bridge multilevel inverter (CHB) is very popular because of its simplicity. Based on the number of levels required, H-bridge cells are connected in series for a five-level inverter where each phase required two H-bridges in series [1–3]. As the number of levels increases, dv/dt decreases due to this stress on winding insulation decreases. Each H-bridge cell voltage is measured and calculates the average value of each H-bridge cell voltage. If there is any fault the average value is not equal to zero and we can identify it as faulty cell. The block diagram 3 of phase CHB inverter-fed induction motor drive is shown in Fig. 1.

3 Fault Analysis

Commonly occurred various faults in VSI are continuous gate ON, continuous gate off, switch open, switch short, diode open, and diode short. Out of this, more than 60% switch open and short fault cases are commonly occurred. So in this study two major faults are studied: switch open and switch short faults.

Whenever the fault is identified in particular phase, the fault line is short circuited and redundant cell comes into picture. In normal operation redundant cell output is short circuited. The block diagram of fault mitigation using redundant cell for five-level multilevel inverter-fed induction motor drive is shown in Fig. 2.

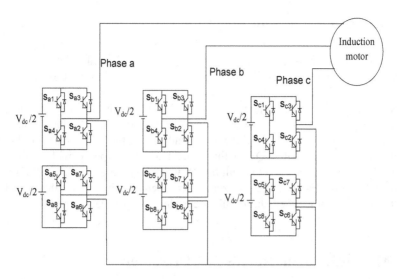

Fig. 1 Cascaded H-bridge five-level inverter-based induction motor

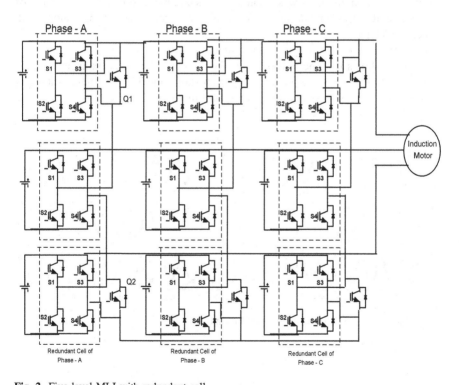

Fig. 2 Five-level MLI with redundant cell

4 Redundant Cell with Logic

To mitigate the fault for each cell in the H-bridge, bi-directional switches have been connected, and pulses developed from the fault detection will operate these switches to short the leg of the fault occurred in switch.

Redundant cell is an extra bridge circuit added to the inverter circuit. This redundant cell acts and deactivates during fault and when no fault condition. Figure 2 shows the five-level MLI with redundant cell for fault mitigation. During fault condition, the switch Q1 in the upper part of the phase will be ON and vice-versa. The lower switch Q2 will be ON when no fault and will be OFF during faulty conditions. Fault is equal to zero when open circuit fault condition and the fault is equal to one, when short fault occurs in phase of the inverter.

5 Matlab/Simulink Results and Discussions

For the mitigation, the fault was created at 0.4 s and made to persist up to 0.6 s. At 0.6 s the fault mitigation was done. The results were shown accordingly along with respective THDs.

Case 1: five level with open fault mitigation

Figure 3 shows the line voltages and Fig. 4 shows the phase voltages of open fault mitigation of five-level inverter. Figure 5 shows the THD in phase voltage of five-level inverter with switch open after fault mitigation.

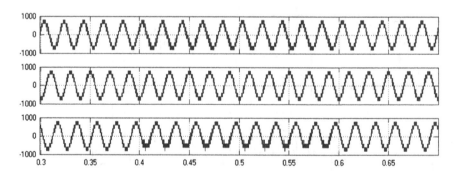

Fig. 3 Line voltage with switch open fault mitigation

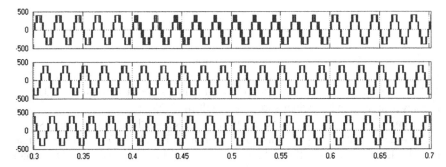

Fig. 4 Phase voltage with switch open fault mitigation

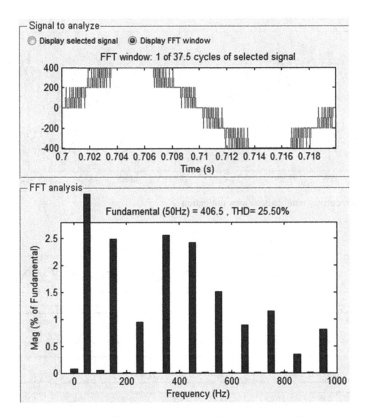

Fig. 5 THD in phase voltage with switch open mitigation (after fault mitigation)

Figure 6 shows the line currents and Fig. 7 shows the stator currents, speed, and torque characteristics of induction motor. THD in stator current is shown in Fig. 8 after fault mitigation.

Case 2: five level with short fault mitigation

Figure 9 shows the line voltages and Fig. 10 shows the phase voltages of short fault mitigation of five-level inverter. Figure 11 shows the THD in phase voltage of five-level inverter with switch short after fault mitigation.

Figure 12 shows the line currents and Fig. 13 shows the stator currents, speed, and torque characteristics of induction motor and Fig. 14 THD after fault mitigation. Table 1 represents THD in phase voltage and Table 2 represents THD in stator current.

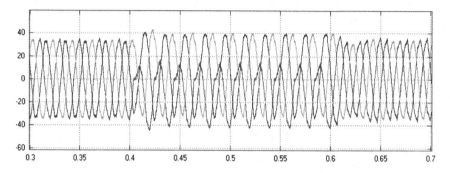

Fig. 6 Line current with switch open mitigation

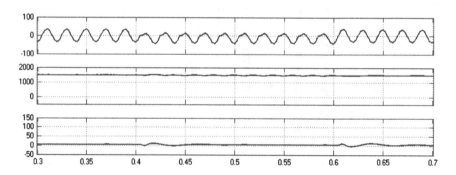

Fig. 7 Stator current, speed, and torque of induction motor with mitigation

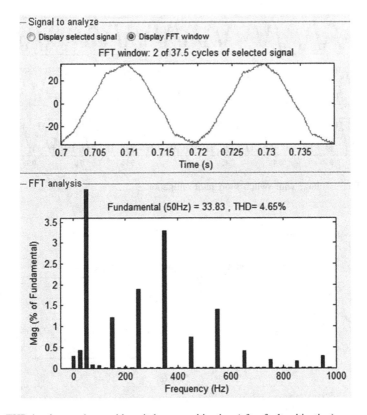

Fig. 8 THD in phase voltage with switch open mitigation (after fault mitigation)

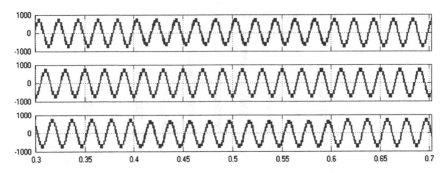

Fig. 9 Line voltage with switch short fault mitigation

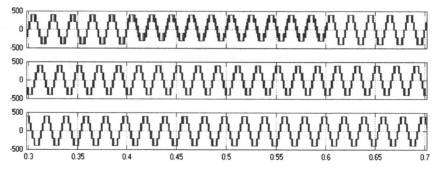

Fig. 10 Phase voltage with switch short fault mitigation

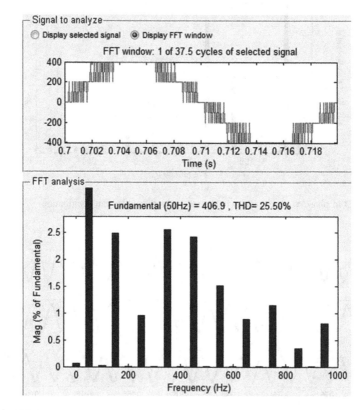

Fig. 11 THD in phase voltage with switch short mitigation (after fault mitigation)

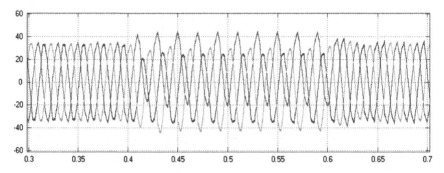

Fig. 12 Line current with switch short mitigation

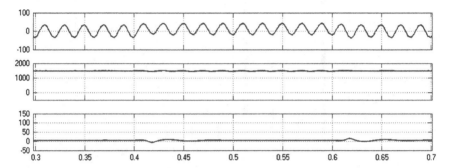

Fig. 13 Stator current, speed, and torque of induction motor with mitigation

6 Conclusion

In this paper redundant cell was used to mitigate the faults that occur in inverter which feed induction motors for electrical vehicles. Fault identification is essential to mitigate the faults. Redundant cell comes handy and operate when fault is present in the inverter circuit. This paper discusses the fault mitigation with redundant cell for five levels. THD in phase voltages and THD in stator current of induction motor are reduced as the level of output is increased. This was verified from the tabular column. The use of multilevel inverter and redundant cell can effectively reduce the fault and mitigates with reduction in THD in stator currents and phase voltages.

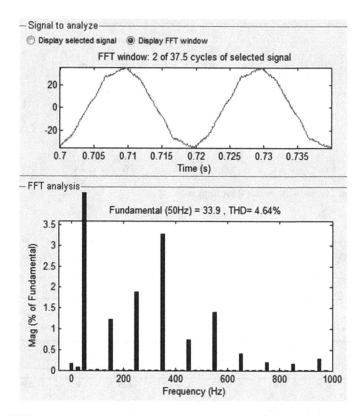

Fig. 14 THD in stator current with switch short mitigation (after fault mitigation)

Table 1 THD in phase voltage

Level	Switch condition	Before fault (%)	During fault (%)	After fault mitigation (%)
5-level	Open	26.04	36.49	25.5
	Short	26.04	28.77	25.5

Table 2 THD in stator current

Level	Switch condition	Before fault (%)	During fault (%)	After fault mitigation (%)
5-level	Open	5.43	19.69	4.65
	Short	5.43	5.49	4.65

References

1. C. Kral and K. Kafka, "Power electronics monitoring for a controlled voltage source inverter drive with induction machines," in Proc. IEEE 31st Annu. Power Electron. Spec. Conf., 2000, vol. 1, pp. 213–217.
2. F. W. Fuchs, "Some diagnosis methods for voltage source inverters in variable speed drives with induction machines—A survey," in Proc. IEEE Ind. Electron. Conf., 2003, pp. 1378–1385.
3. A. M. S. Mendes, A. J. M. Cardoso, and E. S. Saraiva, "Voltage source inverter fault diagnosis in variable speed.
4. R. L. A. Ribeiro, C. B. Jacobina, E. R. C. Silva, and A. M. N. Lima, "Fault detection of open-switch damage in voltage-fed PWM motor drive systems," IEEE Trans. Power Electron., vol. 18, no. 2, pp. 587–593, Mar. 2003.
5. D. Diallo, M. E. H. Benbouzid, D. Hamad, and X. Pierre, "Fault detection and diagnosis in an induction machine drive: a pattern recognition approach based on concordia stator mean current vector," IEEE Trans. Energy Conv., vol. 20, no. 3, pp. 512–519, Sept. 2005.
6. K. Rothenhagen and F. W. Fuchs, "Performance of diagnosis methods for IGBT open circuit faults in voltage source active rectifiers," IEEE PESC proc., 2004, pp. 4348–4354.
7. AC drives, by the average current Park's vector approach," IEEE IEMDC Proc., 1999, pp. 704–706.

A Web-of-Things-Based System to Remotely Configure Automated Systems Using a Conditional Programming Approach

Debajyoti Mukhopadhyay, Sourabh Saha, Rajdeep Rao
and Anish Paranjpe

Abstract A system has been designed and implemented to remotely configure generic automated systems using only conditional logic statements and a web-based application for intuitive user interaction. Current automation systems use company-specific applications and require manual configuring in order to make any changes as per the user's needs. The purpose of this system is to give users the ability to self-configure and control all their systems from anywhere, using a simple internet-based application, thus requiring no external intervention. Our system is built using open-source software and hardware, thus rendering it cost-effective to implement in real time. Conditional programming, or If-then-Else logic, has been used as it is an easily understandable logic construct to maintain and configure the automated systems.

Keywords Web of things · Automation · Mobile · Conditional programming

D. Mukhopadhyay (✉) · S. Saha · R. Rao · A. Paranjpe
Department of Information Technology, Maharashtra Institute of Technology,
Pune 411038, India
e-mail: debajyoti.mukhopadhyay@gmail.com

S. Saha
e-mail: sourabhsaha16@gmail.com

R. Rao
e-mail: rajdeeprrao94@gmail.com

A. Paranjpe
e-mail: anish.paranjpe@gmail.com

© Springer Nature Singapore Pte Ltd. 2017
S.C. Satapathy et al. (eds.), *Proceedings of the 5th International Conference on Frontiers in Intelligent Computing: Theory and Applications*, Advances in Intelligent Systems and Computing 515, DOI 10.1007/978-981-10-3153-3_30

303

1 Introduction

In this paper, we have put forth a web-of-things-based system consisting of multiple automated systems in multiple domains (home, industry, agriculture) and a web application to control and configure the said systems. These automated systems, through the Internet, would be able to communicate with the users by providing real-time updates and the also the ability to control their behaviour using the application, from any geographical location (with internet access) [1]. The protocol used to configure these systems is based on an *If-then-Else* programming framework (IfE), for which we have filed a patent application at the Indian Patent Office (file no. 4623/mum/2015). This framework is used to generate a semantic tree which is stored and evaluated at our servers, whose output results in dynamically changing the targeted automation system behaviour. In order to obtain the proof of concept, we implemented the system using Raspberry pi, several sensors (temperature, proximity, flame and light) and actuators (dc motor, LEDs). The intention of this work is to make remote configuration of several inter-connected/independent automation systems cost-effective and effortless for the end users.

2 Present Approach

An automated system is one that senses and controls an environment by making decisions based on available data [2]. The present approach is such that most automation systems developed are proprietary and inflexible. Due to this, they are extremely domain centric. These systems make the users heavily dependent on the organization implementing it, to make the tiniest of changes. This freedom of customization and the existence of a cross platform automation system are missing. Today, a wide range of programming languages exist for achieving different types of programming tasks. Different platforms are required for the execution of these programming languages. The web has the capability of facilitating nearly all kinds of programming tasks of what we can imagine. The ubiquity of a platform like the web is not being exploited to its fullest. In order to exploit this platform, there needs to be a solid system that abstracts all the minute problems present in the implementation of a language on this platform.

3 Proposed Approach

This approach is aimed at providing a seamless integration of various domains like home, industry, etc., and monitoring and modifying the configuration using across platform mobile application, thereby making it domain independent and easily accessible. Users are allowed to create and modify rules as per their convenience thereby providing customizability. This system would only need the organization to install the required hardware devices at first. Any change following this requires no intervention from the organization. Also, the use of open-source technologies eliminates any proprietary content.

3.1 Target Entities

3.1.1 Developers

Anyone, having basic knowledge of how to use one of the supported boards and sensors can create wonders. Simplistic, intuitive interface gives everyone a chance to revolutionize their environment.

3.1.2 Industries

Industry-scale automation can also be made cost-effective, and user-friendly. Real-time monitoring of lights, sound levels, air-pollution and temperature can lead to eco-friendly and energy saving workplaces [3]. Open-source technology gives in-house workers ability to create independent solutions. Companies can register multiple branches and can view data on single platform.

3.1.3 Common Man

End-to-end solutions can transform any home into a "smart" living space. Wide range of combinations possible with multiple sensors along with multi-platform compatibility gives user ability to use the system from Tablets, Mobile phones and Desktop environments [4].

3.2 System Architecture

The system architectures shown in Fig. 1 provides a comprehensive insight into the implementation and the different domains the system can handle. The components are as follows:

1. **Source**: These nodes generate the data in the form of real-time continuous status updates. For example, a flame sensor would stream its current state, whether ON or OFF to the server

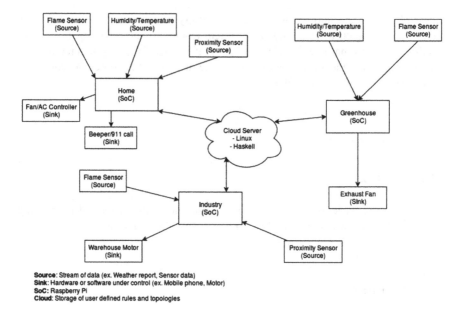

Source: Stream of data (ex. Weather report, Sensor data)
Sink: Hardware or software under control (ex. Mobile phone, Motor)
SoC: Raspberry Pi
Cloud: Storage of user defined rules and topologies

Fig. 1 System architecture

2. **Sink**: These nodes are the ones affected by the modifications made by the user. The behaviour of the sink changes with the variation in configuration, for example, consider the sink to be an exhaust fan, whose speed would dynamically change based on the value input by the user, or depending upon the current temperature.
3. **Cloud server**: This is the brain of the system, which evaluates the semantic tree generated, as per user modifications, using our If-then-Else (IfE) framework.

Sample rules are shown in Fig. 2.

3.3   Software Used

3.3.1   Socket.IO

Socket IO allows real-time duplex communication between the client (Mobile or SoC and server (Cloud). It has the ability to switch between different types of transports back-end depending on the device capabilities.

3.3.2   A Custom-Tailored Storage System (IfEDB)

IfEDB is a Fault-Tolerant, In-Memory Abstract Syntax Tree Storage and Evaluator System implemented in Haskell. Following are a few comparison factors with traditional Database Management Systems (Table 1).

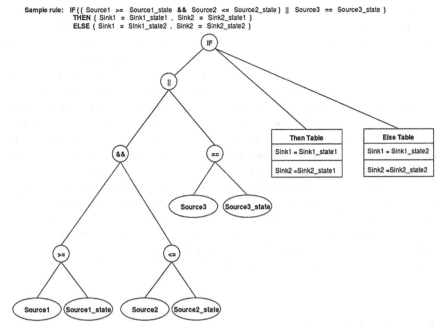

Sample rule: IF ((Source1 >= Source1_state && Source2 <= Source2_state) || Source3 == Source3_state)
THEN (Sink1 = Sink1_state1 , Sink2 = Sink2_state1)
ELSE (Sink1 = Sink1_state2 , Sink2 = Sink2_state2)

Fig. 2 Sample rule

Table 1 Comparison between RDBMS and IfEDB

	RDBMS	IfEDB
Interface to system	SQL queries	JSON messages (sent using SocketIO)
Building blocks	Tables comprised of rows and columns	Abstract syntax trees name-spaced by the user that created it
Primitives	String, Number, Date, Time, List, Map, etc.	Sensor and actuator id and values respectively
Storage	Tables stored on disk. Rows maybe cached for faster access	Abstract syntax trees are present in memory
Operations	Create Table, Delete Table, Insert Row(s), Update Row(s), Delete Row(s)	Create Rule,* Edit Rule, Evaluate Rule, Delete Rule

*If-Then-Else statement

4   IfE

This architecture revolves around an "If Then Else" (IfE) programming language. Unlike other programming languages, where the user writes a program in a text file and instructs the programming language binary to execute it, IfE takes a different approach. The user is presented with a Web-based user interface, which can be accessed from any web browser present on any type of electronic device. There is no binary program that the user needs to execute. The "If Then Else" program that the user constructs is then stored on the cloud, where it eventually gets executed when need be. As mentioned before, IfE consists of inputs and outputs. The input is a set of source(s) and the output is a set of sink(s). A program in IfE is known as a topology.

A topology consists of "*If Then Else*" statements, where the sources appear in the "*if*" condition and the sinks appear in the "*then*" and "*else*" clauses. A user can create multiple topologies. A certain source or sink can appear in more than one topology. IfE makes use of standard logical (not, and, or) and comparison (==, <, <=, >, >=) operators along with variables whose values the user can manipulate to achieve Turing Completeness. The variables in IfE are the source or sink identifiers. The values in IfE depend on the source or sink. An Integer can take values from −Infinity to +Infinity. The same way the values that the user can assign to a sink or the values that the user can compare with the source, are those values that the source or sink can take. We will refer to each such value as a state to imply that the source or sink is in that particular state.

4.1 IfE Specification

```
Topology := if  LogicalExpr then ThenClause else ElseClause

LogicalExpr   :=  ComparisonExpr  and  ComparisonExpr
              |   ComparisonExpr  or  ComparisonExpr
              |   not  ComparisonExpr
              |   ComparisonExpr


ComparisonExpr  :=  <Src_ID>   ==    <Src_STATE>
                |   <Src_ID>  <      <Src_STATE>
                |   <Src_ID>  <=    <Src_STATE>
                |   <Src_ID>  >      <Src_STATE>
                |   <Src_ID>  >=    <Src_STATE>

ThenClause, ElseClause  :=  Statement

Statement  :=  <Sink_ID>  =  <Sink_STATE>
           |   <Sink_ID>  =  <Sink_STATE>  Statement
```

4.2   Internal Representation

Each topology is internally (on the cloud) stored as an abstract syntax tree [5] as illustrated in Fig. 1. There are three types of nodes encountered on the abstract syntax tree:

1. **Language Nodes**: These are IfE specific nodes, like *"if"*, *"logical"*, *"comparison"*, *"statements"* and *"assignments"*.
2. **State Nodes**: These are nodes that contain the states associated with a source or sink.
3. **Source Nodes**: These are nodes representing a certain Source.
4. **Sink Nodes**: These are nodes representing a certain Sink.

   The semantics of the *if* expression are:

1. If the first child (condition) of the if node evaluates to true, then evaluate the second child (then clause).
2. If the first child (condition) of the if node evaluates to false, then evaluate the third child (else clause).

5   Implementation

The implementation has been split into the following parts:

5.1   User Interface

We have hereby developed an internet-based cross platform application for Android, iOS and Windows Phone. The application allows the user to monitor and change the behaviour of his/her automation system(s). The user interface allows the following functionalities:-

1. Create/Edit/Delete Rule: Rules are the IF-Else statements which govern the systems.
2. Live sensor/actuator status: Real-time streaming of the current states and values of the sensors and actuators which constitute the automation system.

5.2   Sensors and Actuators

A few of the sensors used in the real-world scenarios are—Temperature, Flame, Light and Proximity.

The actuators used correspondingly with the above are—Buzzer (Alarm), Motor (Exhaust fan or AC) and LED (Lighting)

5.3 Cloud (Server)

The server that is made available to the users is responsible to store and evaluate the rules created by them based on every sensor update it receives. This allows for continuous monitoring and deploying actuators accordingly.

5.4 Controller (System on Chip)

The controller serves as an abstraction between the sensors, actuators and the cloud. We created a scenario where Home was governed by the Raspberry pi and the green house was governed by the Intel Edison. The rules stored in the cloud were evaluated for every sensor update through these controllers and appropriate actuators were manipulated.

6 Performance Evaluation

In order to evaluate performance during the practical implementation of this approach, we deployed six rules registered under three different users. Latency, in this approach is defined as the time elapsed after the sensor senses data and before its corresponding actuator is deployed. Each of their sensors would emit data at an interval of 1 s. Hence, after every second the SoC would emit 50 bytes per sensor. With this configuration, the net latency observed was 0.2 s. This was followed by performing stress testing, wherein, physical sensors were simulated by software threads. A net total of 1000 sensors along with 2000 actuators functioning as a part of 500 rules were deployed. The latency now observed was 0.5 s.

7 Conclusion

The approaches to develop automation systems are evolving constantly but they are proprietary, expensive and difficult to manage. The framework proposed in this paper effectively provides an open-source and cross platform solution for the aforementioned problem. In this approach, we have proposed a prototype that uses the web as a platform to integrate all of the users' device thereby making it a cross platform automation framework which is intuitive and cost-effective. We have proposed a centralized server and a web interface which is accessible to everyone with Internet connectivity to create their rules as they desire thereby making it a very flexible and independent automation framework. This approach can easily be adopted by any organizations seeking to automate their systems with universal access.

References

1. Federica Paganelli, Stephano Turchi, Dino Giuli, "A Web of Things Framework for RESTful Applications and Its Experimentation in a Smart City, DOI 10.1109/JSYST.2014. 2354835, IEEE Systems Journal, 2014.
2. Olfa Mosbahi, Mohamed Khalgui, Hans-Michael Hanish, Zhiwu Li, "A Component-Based Approach for the Development of Automated Systems", DOI 10.1109/TSMCA.2010.2093885, IEEE Transactions on Systems, Man and Cybernetics – Part A, 2011.
3. Roberto Zafalon, "Smart System Design: Industrial Challenges and Perspectives", DOI 10. 1109/MDM.2013.106, IEEE 14th International Conference on Mobile Data Management, 2013.
4. Dhiraj Sunehra, Ayesha Bano, "An Intelligent surveillance with cloud storage for home security", DOI 10.1109/INDICON.2014.7030567, Annual IEEE India Conference, 2014.
5. Andrei Arusoaie, Daniel IonutVicol, "Automating Abstract Syntax Tree Construction for Context Free Grammars", DOI 10.1109/SYNASC.2012.8, 14th International Symposium on Symbolic and Numeric Algorithms for Scientific Computing, 2012.
6. Simon Duquennoy, Gilles Grimaud, Jean-JaquesVandewalle, "The Web of Things: Interconnecting Devices with High Usability and Performance", DOI 10.1109/ICESS.20009.13, International Conference on Embedded Software and Systems, 2009.

On the Security of Chaos-Based Watermarking Scheme for Secure Communication

Musheer Ahmad and Hamed D. AlSharari

Abstract A new digital image watermarking scheme based on chaotic map was proposed to hide the sensitive information known as watermark. The authors claimed that the scheme is efficient, secure, and highly robust against various attacks. In this paper, the inherent security loopholes of the watermarking embedding and extraction processes are unveiled. The cryptanalysis of watermarking scheme is presented to demonstrate that the scheme is not robust and secure against the proposed attack. Specifically, with chosen host image and chosen watermarks, the successful recovery of securely embedded watermark from received watermarked image is possible without any knowledge of secret key. The simulation analysis of proposed cryptanalysis is provided to exemplify the proposed attack and lack of security of anticipated watermarking scheme.

Keywords Image watermarking · Chaotic map · Security · Cryptanalysis · Chosen-plaintext attack

1 Introduction

Digital watermarking is a method of embedding an imperceptible digital watermark into a host digital media, like still images, audio signals, and video streams such that it does not perceptually degrade the quality of host media and the watermark could be recovered easily by its reverse process. The goal of digital watermarking is to provide practical platforms for designing security systems to accomplish copy-

M. Ahmad (✉)
Faculty of Engineering and Technology, Department of Computer Engineering,
Jamia Millia Islamia, New Delhi 110025, India
e-mail: musheer.cse@gmail.com

H.D. AlSharari
Department of Electrical Engineering, College of Engineering, AlJouf University,
Sakakah, AlJouf, Kingdom of Saudi Arabia
e-mail: hamed_100@hotmail.com

© Springer Nature Singapore Pte Ltd. 2017   313
S.C. Satapathy et al. (eds.), *Proceedings of the 5th International Conference on Frontiers in Intelligent Computing: Theory and Applications*, Advances in Intelligent Systems and Computing 515, DOI 10.1007/978-981-10-3153-3_31

right protection, broadcast monitoring, tamper detection, digital fingerprinting, intellectual property protection, etc. [1, 2]. Watermarking systems for digital media involve two stages, namely (i) watermark embedding—to indicate copyright and (ii) watermark detection—to identify the owner [3]. Depending on the requirements and purpose, a reliable and dependable watermarking is characterized by four properties [4]. These properties include: Robustness—the watermarking scheme must be capable to resist different kinds of attacks, imperceptibility—watermark is embedded in a way such that the degradation in the quality of digital media is imperceptible, capacity—It refers to the number of bits a watermark encodes within a unit of time for an image, it would refer to the number of bits embedded within the image, and security—it refers to the fact that an unauthorized person should neither detect nor read the watermark. The watermark is sensitive piece of information, which is to be hidden/embedded within the cover data secretly. The watermark can be of various types: text, audio, images, logos, authentication, verification codes, etc.

The digital watermarking schemes are broadly divided into fragile watermarking and robust watermarking. Fragile watermarking is aimed to detect the integrity, authenticity, and modifications of digital media [5], whereas the robust watermarking prevents someone to remove the watermark to determine the owner of digital media even if it is altered [6]. Most of the multimedia applications require imperceptible, secure, and robust watermarking schemes to protect the ownership of digital media [7]. In the last decade, a lot of digital image watermarking schemes have been proposed for providing content authentication, rightful ownership, copyright protection, etc. [8–18]. Contemporarily, the attempts have been also made by the researchers to analyze and assess the security of individual watermarking schemes. It has been performed with intent to arrive and design more robust, secure and efficient schemes. Consequently, the customized attacks are framed by the cryptanalysts to unveil the inherent flaws and weaknesses. It has been found that some watermarking and encryption schemes are not secure and incompetent to withstand possible attacks, as exposed by the cryptanalysts in [19–30].

Jamal et al. [16] recently proposed a digital image watermarking scheme, specifically for grayscale images with the use of a grayscale watermark image. They employed the chaotic logistic map to locate random embedding positions inside the host image. The authors have claimed that based upon certain simulations carried by them, the algorithm is highly robust against any possible attack. However, a careful security analysis of the scheme reveals its inherent weaknesses, which facilitates unauthorized recovery of watermark. The main flaw of the scheme is that the generation of random embedding positions is independent of the host image. It only depends on the secret key, namely the initial conditions of chaotic logistic map. As a result, the scheme will always yield the same embedding positions with same secret key, regardless of the host image used. In this paper, we have exploited this flaw and proposed an attack to recover the watermark without the knowledge of security key.

The organization of the rest of this paper is as follows: Sect. 2 reviews and analyzes the Jamal et al. image watermarking scheme. In Sect. 3, the watermarking scheme under study is scrutinized and cryptanalyzed with proposed attack which is followed by its simulation illustrations. Section 4 is concerned with conclusions and future scope of the work is put in Sect. 5.

2 Jamal et al. Watermarking Scheme

This section deals with the review and description of the grayscale image watermarking scheme proposed in [16]. Jamal et al. make use of the features of chaotic logistic map in the scheme. The chaotic logistic map is a simple nonlinear system, which has very complicated dynamics. It is defined as

$$x(n+1) = \mu.x(n)[1 - x(n)] \tag{1}$$

where $x(n) \in (0, 1)$ for all $n \geq 0$ and the map exhibits chaotic behavior for all values of system parameter $\mu \in [3.45, 4]$. The initial values of $x(0)$ and μ constitutes the secret key of the scheme provided in Ref. [16]. The sequence, obtained by applying the iterations to map (1) has high sensitivity to initial conditions and parameter, pseudorandomness; nonperiodicity and unpredictability [31]. The features of chaos mentioned earlier have motivated cryptographers to use them in the design of crypto and watermarking systems. The core of Jamal et al. watermark embedding procedure has two simple steps. First, the chaotic logistic map is employed to identify the random positions to embed watermark information in the host image. Then, the four least significant bits of the grayscale host image pixel at identified positions are replaced with the four most significant bits of the grayscale watermark pixel. The four least significant bits (LSBs) of the watermark pixel are ignored, which subsequently results in the loss of watermark information. The watermark extraction procedure obtains the watermark by extracting pixels four most significant bits only by applying the inverse of embedding procedure and the lost four LSBs are assumed to be 0.

To simplify the description of watermarking scheme under study, the algorithm with minor variations is provided in the subsequent subsection under the condition that the main structure of the scheme is not altered. The following notations and symbols are used in the rest of the paper.

W	Grayscale watermark image of size $E_1 \times E_2$
H	Host grayscale image of size $F_1 \times F_2$
$L\_W(i)$	Four LSBs of watermark pixel i
$M\_W(i)$	Four MSBs of watermark pixel i
$L\_H(i)$	Four LSBs of host image pixel i
$M\_H(i)$	Four MSBs of host image pixel i
wH	Watermarked image

L_wH(i)   Four LSBs of watermarked image pixel *i*
M_wH(i)   Four MSBs of watermarked image pixel *i*
recW   Recovered watermark image
L_recW(i)   Four LSBs of recovered watermark image pixel *i*
M_recW(i)   Four MSBs of recovered watermark image pixel *i*

The Jamal et al. watermarking scheme embeds a watermark as per the following procedure:

E.1. Take grayscale watermark image *W* and grayscale host image *H*
E.2. Convert decimal value of each pixel of *W* and *H* into 8-bits
E.3. Split the nibbles of corresponding byte of each individual pixel of *W* and *H* to get *M_W*, *L_W*, *M_H* and *L_H*
E.4. Reshape *M_W*, *L_W*, *M_H* and *L_H* into 1D row vectors
E.5. Iterate logistic map (1) to generate a random embedding position *k*
E.6. Replace the four LSBs of host image pixel with four MSBs of watermark pixel as:

$$L\_H(k) = M\_W(n)$$

where *n* is iteration count varying from 1 to $E_1 \times E_2$ and $k \in [1, F_1 \times F_2]$.

E.7. Apply Steps **E.5–E.6** to embed all pixels of watermark *W*
E.8. Apply the inverse of Steps **E.4–E.2** to obtain watermarked image *wH*

The watermark extraction procedure is as follows:

X.1. Take watermarked image *wH*
X.2. Create grayscale image *recW* of given size $E_1 \times E_2$
X.3. Convert decimal value of each pixel of *wH and recW* into 8-bits
X.4. Split the nibbles of corresponding byte of each individual pixel of *wH and recW* to get *M_wH*, *L_wH*, *M_recW and L_recW*
X.5. Reshape *M_wH*, *L_wH*, *M_recW and L_recW* into 1D row vectors
X.6. Place four zero bits as *L_recW(i)* for all *i* = 1 to $E_1 \times E_2$
X.7. Iterate chaotic logistic map (1) to find position *k*
X.8. Pick four LSBs *L_wH* and place them as *M_recW* i.e.,

$$M\_recW(n) = L\_wH(k)$$

where *n* is iteration count varying from 1 to $E_1 \times E_2$ and $k \in [1, F_1 \times F_2]$.

X.9. Apply Steps **X.7–X.8** to extract all values of *M_recW*
X.10. Apply inverse of Steps **X.5–X.3** to get watermark image *recW*

For a detailed explanation of this watermarking scheme, readers are advised to refer to Ref. [16].

3 Cryptanalysis

3.1 Weaknesses

The main weakness of Jamal et al.'s scheme lies in the method applied to generate random embedding positions through chaotic logistic map. The generation of embedding positions is purely independent to the host image and has nothing to do with its information. However, the method solely depends upon the secret key needed to iterate the chaotic logistic map and the initial conditions of logistic are significant in determining the distribution of embedding positions. Still for different host images, the method generates same embedding positions with same secret key (initial conditions). The attacker can recover the watermark if the embedding positions are known to him. The embedding positions could be obtained by choosing a special host image and set of cleverly designed watermark images. The detailed description of the attacker's methodology for finding embedding positions is discussed in upcoming section.

Second, in Jamal et al.'s scheme, neither the calculation of embedding position out of chaotic logistic map variable x is mentioned, nor it was said that generated embedding positions are "unique," which is extremely necessary for extraction of watermark image with minimum loss. If the embedding positions are not unique then it may be possible that two or more watermark image pixels (four MSBs only) would get embedded at same position in host image, this will results in loss of watermark information during the extraction of watermark. This is because the direct extraction of embedding positions out of logistic map produces redundant values with finite probability [31]. The probability increases with the size of watermark, which in turn grows amount of loss of watermark information while extraction. Hence, some sort of prechecking is needed to ensure the generation of unique embedding positions $k \in [1, F_1 \times F_2]$. However, we assumed that the authors in Ref. [16] generate unique embedding positions.

Third, an inherent inefficiency of Jamal et al.'s scheme is that the embedding procedure ignores the information content of four LSBs of each pixel of watermark image. This causes a loss of $\approx 15/256 = 5.88\%$ information in the extracted watermark.

Lastly, because of 5.88% loss of watermark information during embedding, the original watermark image, as claimed by the authors in Ref. [16], cannot be extracted.

3.2 Proposed Attack

In cryptanalysis, a fundamental assumption enunciated by Kerckhoff is that the cryptanalyst knows exactly the details, design and working of the cryptographic algorithm [32]. Based on the above loopholes analysis and scrutiny of the security scheme under study, the proposed is attacked is designed to justify the lack of inherent security of scheme. The proposed attack involved the selection of chosen host image and chosen watermark image to recover the watermark from a watermarked image. The attack is completely depicted in Fig. 1. In this attack we use Jamal et al. watermarking scheme as a black box illustrated in Fig. 1 as "Watermark embedding machine." We use a special grayscale host image of size $F_1 \times F_2$ and a different grayscale watermark image W of size $E_1 \times E_2$ over a number of iterations. The host image consists of $F_1 \times F_2$ pixels and each pixel has value zero, in order to identify the positions where embeddings are made in the watermarked image (which will become nonzero after watermarking) (Fig. 2).

To generate our chosen watermark image W of size $E_1 \times E_2$, we make use of the dynamics of Jamal et al. algorithm. As the algorithm discards the four LSBs, we keep these LSBs as zero. Now to locate the embedding positions from watermarked image we need to generate different pixel values which must be (i) nonzero and (ii) distinct from each other. As we can only use the four MSBs, we can generate only 15 (24 − 1) such values in each iteration. These 15 values are 16(00010000), 32(00100000), 48(00110000), 64(01000000), 80(01010000), 96(01100000), 112 (01110000), 128(10000000), 144(10010000), 160(10100000), 176(10110000), 192 (11000000), 208(11010000), 224(11100000), and 240(11110000). All these values have their LSBs as 0 and unique upper nibble with values lying between the range 1 and 15. For first iteration, we place these values in the starting 15 pixels of our chosen watermark image, next 15 pixels for second iteration and so on for *floor* [$(E_1 \times E_2)/15 + 1$] iterations. The remaining pixels are zero for each of these iterations. After watermarking using Jamal watermarking scheme, we get watermarked image that have all zero pixel values except the 15 pixels, which will have their values lying in between range 1 and 15. We can extract the embedding positions of first 15 pixels by locating these values in the watermarked images in

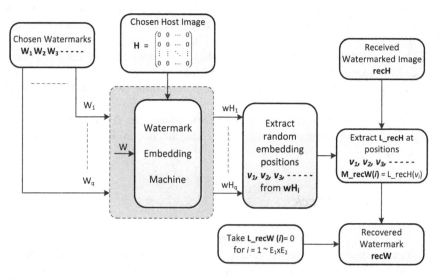

Fig. 1 The proposed attack to recover watermark image

(a)

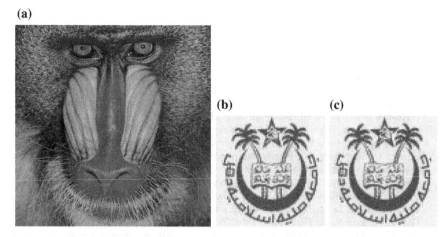

(b)   (c)

Fig. 2 Simulation of cryptanalysis: **a** received watermarked *Baboon* image which is watermarked by Jamal et al. scheme with grayscale logo of *Jamia Millia Islamia University* **b** recovered watermark logo of *Jamia Millia Islamia University* obtained by proposed attack without the knowledge of keys, **c** extracted watermark logo of *Jamia Millia Islamia University* obtained by Jamal et al. extraction procedure with the knowledge of keys

this first iteration. Proceeding this way, the secret positions of next 15 pixels can be obtained in the second iterations and so on. Eventually, we get all $E_1 \times E_2$ embedding positions on $floor[(E_1 \times E_2)/15 + 1]$ iterations.

4 Conclusion

In this paper, the security of the chaos-based watermarking scheme is scrutinized. The investigations have found that the scheme has certain loopholes in the embedding procedure. It has been also shown that the attacker can successfully recover the crucial concealed watermark by selecting special host image and set custom designed watermarks. The proposed attack is successfully simulated and tested. The presented cryptanalysis of Jamal et al. watermarking scheme, proving it is not robust against the anticipated attack and inappropriate for use in secure communication.

5 Future Work

The security analysis, loopholes finding, and break of the watermarking scheme can be persuaded to design a more robust and improved watermarking scheme that not only inherits the features of Jamal et al.'s scheme, but also exhibits the robustness against the proposed and other types of attacks.

References

1. Potdar, V.M., Han, S., Chang, E.: A survey of digital image watermarking techniques. In: 3rd IEEE International Conference on Industrial Informatics, 709–716 (2005).
2. Lei, C.L., Yu, P.L., Tsai, P.L., Chan, M.H.: An efficient and anonymous buyer–seller watermarking protocol. IEEE Transaction on Image Processing 13(12), 1618–1626 (2004).
3. Swanson, M.D., Zhu, B., Tewfik, A.H., Boney, L.: Robust audio watermarking using perceptual masking. Signal Processing 66(3), 337–355 (1998).
4. Cox, I.J., Miller, M., Bloom, J., Fridrich, J., Kalker, T.: Digital watermarking and steganography. Second Edition, The Morgan Kaufmann Publishers (2008).
5. Zhang, X., Shuozhong W.: Fragile watermarking scheme using a hierarchical mechanism. Signal processing 89(4), 675–679 (2009).
6. Wang, X.Y., Yang, H.Y., Cui, C.Y.: An SVM-based robust digital image watermarking against desynchronization attacks. Signal Processing 88(9), 2193–2205 (2008).
7. Cox, I.J., Killian, J., Leighton, T.: A secure, robust watermark for multimedia. In: Information Hiding. Springer Berlin Heidelberg, 185–206 (1996).
8. Liu, J.L., Lou, D.C., Chang, M.C., Tso, H.K.: A robust watermarking scheme using self-reference image. Computer Standards & Interfaces, 28(3), 356–367 (2006).
9. Shieh, J.M., Lou, D.C., Chang, M.C.: A semi-blind digital watermarking scheme based on singular value decomposition. Computer Standards & Interfaces, 28(4), 428–440 (2006).
10. Tao, P., Eskicioglu, A.M.: A robust multiple watermarking scheme in the discrete wavelet transform domain. In: Proceedings of SPIE 5601 Internet Multimedia Management Systems V, 133–144) (2004).
11. Wang, M.S., Chen, W.C.: A majority-voting based watermarking scheme for color image tamper detection and recovery. Computer Standards & Interfaces 29(5), 561–570 (2007).
12. Mohammad, A.A., Alhaj, A., Shaltaf, S.: An improved SVD-based watermarking scheme for protecting rightful ownership. Signal Processing 88(9), 2158–2180 (2008).
13. Bhatnagar, G., Raman, B.: Encryption based robust watermarking in fractional wavelet domain. In: Recent Advances in Multimedia Signal Processing and Communications, Springer Berlin Heidelberg, 375–416 (2009).
14. Rawat, S., Raman, B.: A chaotic system based fragile watermarking scheme for image tamper detection. AEU-International Journal of Electronics and Communications 65(10), 840–847 (2011).
15. Chang, C.C., Chen, K.N., Lee, C.F., Liu, L.J.: A secure fragile watermarking scheme based on chaos-and-hamming code. Journal of Systems and Software 84(9), 1462–1470 (2011).
16. Jamal, S.S., Shah, T., Hussain, I.: An efficient scheme for digital watermarking using chaotic map. Nonlinear Dynamics 73(3), 1469–1474 (2013).
17. Aslantas, V., Dogru, M.: A new SVD based fragile image watermarking by using genetic algorithm. In: Sixth International Conference on Graphic and Image Processing, 94431H-94431H (2014).
18. Ansari, I.A., Pant, M., Ahn, C.W.: ABC optimized secured image watermarking scheme to find out the rightful ownership. Optik 127(14), 5711–5721 (2016).
19. Das, T.K., Maitra, S., Mitra, J.: Cryptanalysis of optimal differential energy watermarking (DEW) and a modified robust scheme. IEEE Transactions on Signal Processing 53(2), 768–775 (2005).
20. Chou, J.S., Chen, Y., Chan, C.J.: Cryptanalysis of Hwang-Chang's a time-stamp protocol for digital watermarking. IACR Cryptology ePrint Archive (2007).
21. Ting, G.C.W., Goi, B.M., Heng, S.H.: Attacks on a robust watermarking scheme based on self-reference image. Computer Standards & Interfaces 30(1), 32–35 (2008).
22. Ting, G.C.W., Goi, B.M., Heng, S.H.: Attack on a semi-blind watermarking scheme based on singular value decomposition. Computer Standards & Interfaces 31(2), 523–525 (2009).
23. He, H., Zhang, J.: Cryptanalysis on majority-voting based self-recovery watermarking scheme. Telecommunication Systems 49(2), 231–238 (2012).

24. Teng, L., Wang, X., Wang, X.: Cryptanalysis and improvement of a chaotic system based fragile watermarking scheme. AEU-International Journal of Electronics and Communications 67(6), 540–547 (2013).
25. Caragata, D., Mucarquer, J.A., Koscina, M., El Assad, S.: Cryptanalysis of an improved fragile watermarking scheme. AEU-International Journal of Electronics and Communications, 70(6), 777–785 (2016).
26. Ahmad, M.: Cryptanalysis of chaos based secure satellite imagery cryptosystem. In: Contemporary Computing. Springer Berlin Heidelberg, 81–91, (2011).
27. Sharma, P.K., Ahmad, M., Khan, P.M.: Cryptanalysis of image encryption algorithm based on pixel shuffling and chaotic S-box transformation. In: Security in Computing and Communication. Springer-Verlag Berlin Heidelberg, CCIS 467, 173–181, (2014).
28. Ahmad, M., Ahmad, F.: Cryptanalysis of Image Encryption Based on Permutation-Substitution Using Chaotic Map and Latin Square Image Cipher. In: 3rd International Conference on Frontiers of Intelligent Computing: Theory and Applications. Springer International Publishing Switzerland, AISC 327, 481–488 (2015).
29. Ahmad, M., Khan, I.R., Alam, S.: Cryptanalysis of Image Encryption Algorithm Based on Fractional-Order Lorenz-Like Chaotic System. In: Emerging ICT for Bridging the Future. Springer International Publishing Switzerland, AISC 338, 381–388 (2015).
30. Sharma, P.K., Kumar, A., Ahmad, M.: Cryptanalysis of Image Encryption Algorithms Based on Pixels Shuffling and Bits Shuffling. In: International Congress on Information and Communication Technology. Springer Singapore, AISC 439, 281–289 (2016).
31. May R.M.: Simple mathematical model with very complicated dynamics. Nature 261, 459–467, (1967).
32. Schneier, B.: Applied Cryptography: Protocols Algorithms and Source Code in C. New York, Wiley (1996).

Neighborhood Topology to Discover Influential Nodes in a Complex Network

Chandni Saxena, M.N. Doja and Tanvir Ahmad

Abstract This paper addresses the issue of distinguishing influential nodes in the complex network. The k-shell index features embeddedness of a node in the network based upon its number of links with other nodes. This index filters out the most influential nodes with higher values for this index, however, fails to discriminate their scores with good resolution, hence results in assigning same scores to the nodes belonging to same k-shell set. Extending this index with neighborhood coreness of a node and also featuring topological connections between its neighbors, our proposed method can express the nodes influence score precisely and can offer distributed and monotonic rank orders than other node ordering methods.

Keywords Influential nodes · k-shell index · Neighborhood coreness · Topological connections

1 Introduction

The most authoritative nodes in a network can accomplish the speed and span of information diffusion when matched with other nodes [1–4]. Locating these influential nodes is of theoretical and practical significance in controlling the spread of information [5], ranking reputation of scientists [6], finding social leaders [7], developing efficient strategies to control epidemic spreading [8], promoting new products [9], and so on. Years of innovation have refined approaches to identify highly influential nodes to the consequences of spreading process on a given network. Many centrality indicators have been proposed to measure the estimated

C. Saxena (✉) · M.N. Doja · T. Ahmad (✉)
Department of Computer Engineering, Jamia Millia Islamia, New Delhi, India
e-mail: cmooncs@gmail.com

T. Ahmad
e-mail: tahmad2@jmi.ac.in

M.N. Doja
e-mail: ndoja@yahoo.com

© Springer Nature Singapore Pte Ltd. 2017
S.C. Satapathy et al. (eds.), *Proceedings of the 5th International Conference on Frontiers in Intelligent Computing: Theory and Applications*, Advances in Intelligent Systems and Computing 515, DOI 10.1007/978-981-10-3153-3_32

importance of nodes within the network, such as degree centrality (dc), closeness centrality (cc), betweenness centrality (bc) and eigenvalue centrality (ec), etc. The important issue is how to determine and distinguish the spreading capability of a node. Nodes having high centrality scores are notified to be more competent in the spreading process. Among these measures, dc is elementary and efficient, but based only on local structure information hence fails to identify influential nodes. Based on global link information bc is difficult to apply on large-sized networks. If there are disconnected components in networks, then cc has limitations to measure true centralities. If there are two or more nonidentical elements, the eigenvectors of the nodes adjacency matrix will mark only one of the elements, as a result ec interprets wrong results. Other centrality criteria are also on hand, such as neighborhood centrality [10], page rank [11], HITS [12], leader rank [7], semi-local centrality [13], and so on.

Recently, Kitsak et al. [5] found that the most prestigious nodes are those situated within the core of the network, which is obtained when a network is mouldered with the k-shell decomposition method. However, this method assigns same k-core score to many nodes lying in the same core set, although their spreading potential may not be the same. By taking into account the disposed degree in this disintegration process, Zeng et al. [14] proposed a mixed degree decomposition method to find distinct values of nodes spreading influence within same k-core node set. Considering the next nearest neighborhood, Chen et al. [13] devised a semi-local centrality index. Liu et al. [15] exhibited an improved ranking of influential nodes by considering the shortest path distance between a given target node and the nodes from the same k-core set. Summing up all neighbor's k-shell values, Bae et al. [16] proposed an improvement to k-shell centrality index. Looking at the topological connections among neighbors of a node, Gao et al. [17] proposed local structure centrality which considers degree of neighbor and neighbors of neighbor, and the topological connections among neighbors.

However, while using k-shell indices to score the nodes influence, only immediate and two-hop neighbors of a node are considered, but concentration of the topological connections among neighbors are entirely ignored. It has been manifested in majority actual world networks, and in specifically social networks, that the nodes are inclined to create firmly integrated groups qualified by a comparatively high density of ties [18]; this likeliness incline to be larger than the expected average probability of a link arbitrarily accomplished between any two nodes. Since the clustering coefficient measures the concentration of triangles [18] as it gives insight into how well the neighborhood of a node in connected [19], hence denser connected neighbors get more chance to influence each other. Therefore, investigation of nodes will be more prominent when integrating the clustering coefficient along with the k-shell value. To explore this notion, we propose a centrality measure using merits of k-core values and clustering coefficient as topological feature of node and its neighborhood. To appraise strength of the proposed method, we apply simple SIR (susceptible–infected–recovered) [20] epidemiological model for investigating the spreading process. Results of experimentation show that proposed method effectively rank the most influential nodes than other measures considered by evaluating the rank correlations with Kendall's tau.

The rest of this paper is organized as follows: We briefly review previous studied centrality measures in Sect. 2. We introduce new centrality measure in Sect. 3. In Sect. 4, we have reported the experimental results. Section 5 offers conclusions of this research.

2 Review of Centrality Measures

We consider an unweighted, undirected, and simple network $G = (V, E)$ where $n = |V|$ vertices and $m = |E|$ links.

2.1 Degree Centrality [21]

The degree centrality (dc), denoted by $C_D(i)$, of a node 'i' can be calculated as:

$$C_D(i) = d_i / (n - 1) \tag{1}$$

Where d_i is number of edges linked with the node 'i'. It shows the efficiency of a node 'i' to interact with other nodes in the network, straightaway. The greater the value of $C_D(i)$, the more authoritative node 'i' is.

2.2 Closeness Centrality [22]

Assume l_{ij} is the shortest path length from node 'i' to node 'j'. The closeness centrality (cc) of node 'i', denoted by $C_C(i)$ can be stated as the inverse of the total shortest paths length from node 'i' to all other nodes in the network. It is given by the following equation:

$$C_C(i) = (n - 1) / \sum_{j=1}^{n} l_{ij} \tag{2}$$

The higher the value $C_C(i)$ is, the nearer the node 'i' lies to the network core, this depicts the important position of node in the network.

2.3 The k-Shell Decomposition [5]

Nodes are ascribed to the kth shell in accordance with their unexhausted degree, which is achieved by sequential elimination of nodes having degree smaller than 'k' and the removed nodes get their k's values at the same time. S = (G, E|G) is induced subgraph of graph G by k-shell decomposition and it is denoted as the k-core of G if and only if the degree of all the nodes in S are greater than 'k'. Each node in the network has a k's value.

2.4 Improved k-Shell Decomposition

Following previous studies [16], neighborhood (one-hop) coreness $C_{nc}(i)$ and extended (two- hop) neighborhood coreness $C_{nc+}(i)$ of node 'i' are defined as:

$$C_{nc}(i) = \sum_{j \in N(i)} ks(j) \tag{3}$$

Where N(i) is the set of adjacent neighbors of node 'i' and ks(j) is the k-shell value of neighbor.

$$C_{nc+}(i) = \sum_{j \in N(i)} C_{nc}(j) \tag{4}$$

Where $C_{nc}(j)$ is the neighborhood coreness of 'j', $j \in N(i)$, is a set of adjacent neighbors of node 'i'.

3 Proposed Two-Hop Connected Coreness Centrality

Considering topological connections among neighborhood of a node and its extended coreness, a new centrality measure called *connected coreness* is defined as:

$$CC_{nc+}(i) = \lambda(C_{nc+}(i)) + (1 - \lambda)(C_{1+}(i)) \tag{5}$$

Where $C_{nc+}(i)$ is defined in Eq. (4) and $C_{1+}(i)$ is specified as sum total of local clustering coefficient of both immediate and next two-hop immediate neighbors of node 'i'. A tunable parameter λ ranges between 0 and 1, balances the two factors in Eq. (5). Based on the trial runs on different values of λ, we have selected $\lambda = 0.7$ at its best.

In Fig. 1, the nodes in the core (k = 3) are having two-hop neighborhood coreness score as $C_{nc+}(a) = 22$, $C_{nc+}(b) = 27$, $C_{nc+}(c) = 24$, $C_{nc+}(d) = 22$, also

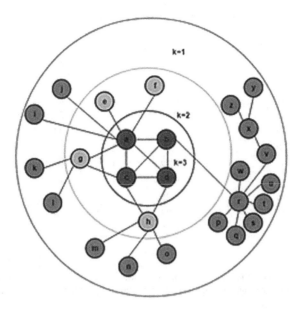

Fig. 1 An example network ref. [5], representing core with k = 3 and periphery of network with k = 1, 2. Node a, d and g having same score by extended (2-hop) neighborhood coreness and different degree centralities, connected (2-hop) coreness can monotonously rank these nodes

$C_{nc+}(g) = 22$ in other core (k = 2). One can observe that ranking order does not differentiate the spreading powers of nodes 'a' and 'd'. These largest core nodes differ in topological connections among their neighbors. The *connected coreness* (two-hop) can efficiently differentiate the ranks of these nodes with numeric scores of $CC_{nc+}(a) = 16.68$, $CC_{nc+}(b) = 20.25$, $CC_{nc+}(c) = 17.94$, $CC_{nc+}(d) = 16.50$, $CC_{nc+}(g) = 16.45$.

Summary of various centrality scores for top 10 nodes from the network given in Table 1 is presented with their average spreading ability s^β values experimentally drawn from simulation of SIR model. For epidemic threshold $\beta = 0.25$ (also epidemic threshold β_{rand}^c for the example network), s^β is estimated by 10^5 runs of SIR model taking each node as seed and its average spreading ability calculated from all runs. It can be observed that spreading ability score and connected coreness score are mutually convincing, also connected coreness is able to rank the nodes with more resolution assigning nonoverlapping orders to the nodes having different topological features. Hence we can argue that the connected coreness is likely to be effective for detecting influential nodes, because it balances the coreness centrality with topological connectedness of locally connected nodes in neighborhood of a target node to find its centrality score.

Table 1 The top-10 nodes from Fig. 1 with high spreading ability score s^β and their respective centrality values by degree centrality (dc), closeness centrality (cc), k-shell decomposition (ks), (two-hop) extended coreness (C_{nc+}) and connected coreness (CC_{nc+})

Node	s^β	dc	cc	ks	C_{nc+}	CC_{nc+}
b	5.78	4	0.45	3	27	20.25
c	4.97	6	0.51	3	24	17.94
a	4.62	8	0.51	3	22	16.68
d	4.39	4	0.43	3	22	16.55
g	4.20	4	0.33	2	22	16.45
r	4.07	8	0.32	1	20	14.60
i	2.70	1	0.36	1	19	14.52
j	2.70	1	0.25	1	19	14.52
f	2.56	2	0.33	2	18	13.50
e	2.57	2	0.47	2	18	13.50

Table 2 The basic topological features of the networks structural properties number of nodes (n), number of edges (m), average degree (‹k›), average shortest path length ‹d›, degree heterogeneity (H) and epidemic threshold (β^c_{rand})

Network	n	m	‹k›	‹d›	H	β^c_{rand}
C elegan	453	4596	20.287	2.664	7.625	0.0064
Karate club	34	77	4.629	2.424	1.669	0.1323
Dolphin	62	158	5.097	3.359	1.334	0.147
Jazz	198	2741	27.687	2.235	1.395	0.026

4 Experimental Results and Analysis

In this paper, we assess the performance of connected coreness in four networks including the C. elegans, Jazz, Zachary karate club, and Dolphin. The C. elegans is metabolic network of roundworm, having proteins as nodes and edges as interactions between them. Jazz is a collaboration network between musicians. Karate club is dataset of members of club and ties between them. Dolphin dataset is social network of bottlenose dolphin association. Basic statistics of these networks can be found in Table 2.

4.1 Evaluation Analysis

We evaluate the resolution of proposed method on the basis of monotonicity of ranking vector obtained, correctness of proposed method on the basis on Kendall tau defined as a correlation coefficient, performance of proposed method based on epidemic SIR model to investigate spreading efficiency of nodes.

Table 3 The monotonicity value M for different centrality methods applied on experimental datasets. The value M(.) corresponds to the centrality mentioned

Network	M(dc)	M(cc)	M(ks)	M (C_{nc+})	M(CC_{nc+})
C elegans	0.015	0.31	0.20	0.71	0.90
Karate club	0.10	0.44	0.34	0.40	0.93
Dolphin	0.002	0.52	0.23	0.75	0.99
Jazz	0.001	0.46	0.83	0.94	0.99

Monotonicity M of a ranking vector is defined as follows:

$$M(R) = \left[1 - \frac{\sum_{r \in R} n_r(n_r - 1)}{n(n-1)} \right]^2 \tag{6}$$

Where n is the size of ranking vector, n_r is the number of nodes having same rank r.

Monotonicity of a ranking vector quantifies the resolution of ranking method, Table 3 summarizes the monotonicity score of different ranking measures upon given experimental networks. The one-hop extended coreness and *connected coreness* are the competitive measures. The *connected coreness* is the best among all measures considered here. Further, to elucidate the ranking distribution of nodes, we plot a complementary cumulative distribution function (CCDF), as shown in Fig. 2. These plots verify the goodness of our method on mentioned four real data sets. Our proposed method can efficiently differentiate the spreading efficiency of the influential nodes and distribute the ranking monotonically.

The ranking vectors obtained by applying different methods should be maximal consistent with the ranking vector generated by real spreading process. To simulate real spreading process, we use the SIR model. In the SIR epidemic model, all nodes of a network can be in one of the three states, defined as susceptible, infected, and recovered. At the beginning all the nodes in network are in the susceptible state except one node, which is in infected state acting as seed node for infection spread. At each time stamp depending upon network configuration and states of neighbors a node becomes infected with probability β and then it enters to recovered state. This spreading process ends when no more infected node left in the network. Finally spreading ability of a node, acted here as seed node is specified as the number of recovered nodes after infected states at the end of spreading process which is originated from seed node. We have obtained spreading ability (influence) of a node denoted as s^β, which is average ability of node on entire range of β over sufficient number of runs on spreading process with same seed node.

Furthermore, to check the performance of our method and other ranking methods, Kendall's τ is introduced. Kendall's τ measures the correlation between two ranking lists and its score lies between −1 and 1. The score close to 1 indicates

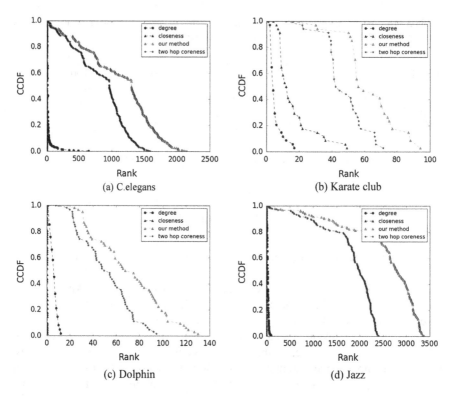

Fig. 2 Rank distribution in the experimental networks: (a) c elegans, (b) karate club, (c) dolphin, and (d) jazz datasets

Table 4 The Kendall τ coefficient of correlation of different centrality measures with spreading influence (s^β). Epidemic threshold is β_{rand}^c and infection probability β of SIR epidemic simulation is listed

Network	β_{rand}^c	B	$\tau\,(s^\beta,dc)$	$\tau\,(s^\beta,cc)$	$\tau\,(s^\beta,ks)$	$\tau\,(s^\beta,C_{nc+})$	$\tau\,(s^\beta,CC_{nc+})$
C elegans	0.0064	0.05	0.6237	0.6723	0.7243	0.8127	0.8223
Karate club	0.9323	0.25	0.6905	0.7123	0.7276	0.8256	0.8330
Dolphin	0.147	0.20	0.7689	0.7578	0.7734	0.9067	0.9220
Jazz	0.026	0.03	0.7902	0.7609	0.8035	0.9122	0.9507

strong correlation. Table 4 summarizes this correlation score for SIR spreading influence rank vectors with all ranking vectors generated by mentioned centrality measures. Considering ranking vectors X and Y, Kendall's tau of X and Y, $\tau\,(X,Y)$ is defined as follows:

$$\tau(X, Y) = \frac{n_c - n_d}{\sqrt{(n_c + n_d + n_{t1})*(n_c + n_d + n_{t2})}} \tag{7}$$

Where n_c are the numbers of concordant pairs and n_d are the numbers of discordant pairs, and n_{t1} is the number of ties only in X and n_{t2} is the number of ties only in Y. If a tie occurs for the same pair in both X and Y, it is not added to either n_{t1} or n_{t2}.

From Table 4, it can be seen that Kendall's τ generated by *connected coreness* is much larger than other centralities. This indicates the ranking accuracy of our method.

5 Conclusions

By considering the neighbors resource of a node, taking into account their k-shell score featuring its embeddedness and clustering coefficient brimming their topological property; we propose *connected coreness* as a new centrality method. Combining these two features, spreading influence of a node can be effectively measured in a given network. We evaluate the proposed method for its monotonicity assigning various ranks to the nodes with good resolution. It is experimentally verified that it holds a positive and higher quantified coefficient of correlation index according to Kendall's tau when compared with SIR simulations. Incorporating two hop neighborhood features into nodes score, our method outperforms mentioned methods.

Focusing on more topological features can further extend this work covering local information in the neighborhood of nodes apart from its embeddedness score in the network. The generalization of coreness score in weighted and directed graph may be interesting and an important open problem.

References

1. Tan, Y.J., Wu, J. and Deng, H.Z.: Evaluation method for node importance based on node contraction in complex networks. Systems Engineering-Theory & Practice, 11(11), pp. 79–83 (2006).
2. Wang, L. and Zhang, J.J.: Centralization of complex networks. Complex systems and complexity science, 3(1), pp. 13–20 (2006).
3. He, N., Li, D.Y., Gan, W. and Zhu, X.: Mining vital nodes in complex networks. Computer Science, 34(12), pp. 1–5 (2007).
4. Sun, R. and Luo, W.B.: Review on evaluation of node importance in public opinion. Jisuanji Yingyong Yanjiu, 29(10), pp. 3606–3608 (2012).
5. Kitsak, M., Gallos, L.K., Havlin, S., Liljeros, F., Muchnik, L., Stanley, H.E. and Makse, H.A.: Identification of influential spreaders in complex networks. Nature physics, 6(11), pp. 888–893 (2010).
6. Jiang, X., Sun, X. and Zhuge, H.: Graph-based algorithms for ranking researchers: not all swans are white!. Scientometrics, 96(3), pp. 743–759 (2013).
7. Lü, L., Zhang, Y.C., Yeung, C.H. and Zhou, T.: Leaders in social networks, the delicious case. PloS one, 6(6), p.e21202 (2011).

8. Albert, R., Jeong, H. and Barabási, A.L.: Error and attack tolerance of complex networks. Nature, 406(6794), pp. 378–382 (2000).
9. Feng, X., Sharma, A., Srivastava, J., Wu, S. and Tang, Z.: Social network regularized Sparse Linear Model for Top-N recommendation. Engineering Applications of Artificial Intelligence, 51, pp. 5–15 (2016).
10. Ma, L.L., Ma, C. and Zhang, H.F.: Identifying influential spreaders in complex networks based on gravity formula. arXiv preprint arXiv:1505.02476 (2015).
11. Page, L., Brin, S., Motwani, R. and Winograd, T.: The PageRank Citation Ranking: Bringing Order to the Web (1999). In Stanford InfoLab.
12. Kleinberg, J.M.: Authoritative sources in a hyperlinked environment. Journal of the ACM (JACM), 46(5), pp. 604–632 (1999).
13. Chen, D., Lü, L., Shang, M.S., Zhang, Y.C. and Zhou, T.: Identifying influential nodes in complex networks. Physica a: Statistical mechanics and its applications, 391(4), pp. 1777–1787 (2012).
14. Zeng, A. and Zhang, C.J.: Ranking spreaders by decomposing complex networks. Physics Letters A, 377(14), pp. 1031–1035 (2013).
15. Liu, J.G., Ren, Z.M. and Guo, Q.: Ranking the spreading influence in complex networks. Physica A: Statistical Mechanics and its Applications, 392(18), pp. 4154–4159 (2013).
16. Bae, J. and Kim, S.: Identifying and ranking influential spreaders in complex networks by neighborhood coreness. Physica A: Statistical Mechanics and its Applications, 395, pp. 549–559 (2014).
17. Gao, S., Ma, J., Chen, Z., Wang, G. and Xing, C.: Ranking the spreading ability of nodes in complex networks based on local structure. Physica A: Statistical Mechanics and its Applications, 403, pp. 130–147 (2014).
18. Lawyer, G.: Understanding the influence of all nodes in a network. Scientific Reports, (2015) 5.
19. Narayanam, R. and Narahari, Y.: A shapley value-based approach to discover influential nodes in social networks. *Automation Science and Engineering, IEEE Transactions on, 8*(1), pp. 130–147 (2011).
20. Keeling, M.J. and Eames, K.T.: Networks and epidemic models. Journal of the Royal Society Interface, 2(4), pp. 295–307 (2005).
21. Albert, R. and Barabási, A.L.: Statistical mechanics of complex networks. Reviews of modern physics, 74(1), (2002), p. 47.
22. Sabidussi, G.: The centrality index of a graph. Psychometrika, 31(4), pp. 581–603 (1966).

Venn Diagram-Based Feature Ranking Technique for Key Term Extraction

Neelotpal Chakraborty, Sambit Mukherjee, Ashes Ranjan Naskar, Samir Malakar, Ram Sarkar and Mita Nasipuri

Abstract Classification of text documents from a pool of huge collection of the same is performed usually on the basis of certain key terms present in the said documents that distinguish a particular document set from the universal set. Generally, these key terms are identified using some feature sets, which can be statistical, rule-based, linguistic, or hybrid in nature. This paper develops a simple technique based on Venn diagram to prioritize the different standard features available in the literature, which in turn reduces the dimension of the feature sets used for document classification.

Keywords Document clustering · Key term extraction · Venn diagram · Feature dimension reduction

N. Chakraborty (✉) · A.R. Naskar · R. Sarkar · M. Nasipuri
Jadavpur University, Kolkata, India
e-mail: neelotpal_chakraborty@yahoo.com

A.R. Naskar
e-mail: naskar.ashes@yahoo.com

R. Sarkar
e-mail: raamsarkar@gmail.com

M. Nasipuri
e-mail: mitanasipuri@gmail.com

S. Mukherjee
Future Institute of Engineering and Management, Kolkata, India
e-mail: sambit.m94@gmail.com

S. Malakar
MCKV Institute of Engineering, Howrah, India
e-mail: malakarsamir@gmail.com

© Springer Nature Singapore Pte Ltd. 2017
S.C. Satapathy et al. (eds.), *Proceedings of the 5th International Conference on Frontiers in Intelligent Computing: Theory and Applications*, Advances in Intelligent Systems and Computing 515, DOI 10.1007/978-981-10-3153-3_33

1 Introduction

Data explosion [1] within the past few decades has led to the necessity of storing and managing data in clustered form with utmost efficiency and dexterity. Also the information retrieval [2] needs to be done most accurately. Time consumed by the system to perform these operations requires significant consideration as well. Data management operations can be constructed on the basis of some concrete statistical/probabilistic [3] or linguistic [4] or using hybrid model. In this paper, our focus would revolve around some model to ensemble the different candidate term sets generated by each individual feature and prioritizing the overlaps [5] of these term sets on the basis of features' importance. This operation is required to cluster a particular set of data from the universal pool (contains all types/classes of data) based on their content, subject, topic, etc.

Clustering [6] of text documents usually involves extraction of KTs from each individual text document and monitoring the density of these KTs in a certain number of documents in the universal document pool. KTs are relatively small finite phrases or words that aids in identifying document topic/subject/type. If a certain number of documents share similar KT chunk, then those documents are classified as belonging to a particular class, different from other documents in the universal set. Candidates for KTs are in most cases nouns/noun phrases (NP) [7], preferably the named entities (NE) [7]. A number of research articles claim with convincing statistical evidence that those candidates having their presence in the heading/title, introductory paragraph, abstract, and sometimes in the conclusive paragraph are having high potential to be qualified as KTs. In this work, the features are combined by intersection and these intersections are evaluated to find optimal feature combination, which gives KTs comparable to human selected KTs (here, human selected KTs are used to validate the proposed work).

2 Related Works

Most of the works emphasize on the need for automatic KT extraction for document classification or information retrieval based operations. We find many features commonly used in these works can be considered as relatively standard while qualifying a candidate term as KT for a particular document.

The work reported in [7] gives a comparative analysis of C4.5 decision tree induction algorithm and a custom designed algorithm *GenEx*. In the work [8], *KEA* algorithm uses lexical methods to generate candidates whose feature values are evaluated using *Naïve Bayes* algorithm. This algorithm is compared with *decision tree* and *artificial neural network (ANN)* algorithms in the work [9]. Approaches based on *conditional random field (CRF)* are experimented in [10] and its improved version, *semi-Markov CRF*, is applied in the work [11] that claims to give better results. The works in [3, 4, 12–15] give a detailed survey on the various supervised and unsupervised techniques which explore the linguistic and statistical features and combining them to get better quality terms. Most of the works apply machine learning techniques to automatically extract KTs and yield better results. However, the issue of importance of various features is still a critical issue while qualifying a term as KT or not.

3 Corpus Description

The corpus is prepared here to conduct experiment on KT extraction from English text documents. The text document set comprises of 100 news articles from three columns (Politics, Entertainment, and Sports) of popular English newspapers in India: The Telegraph, The Times of India, and The Statesman. A hierarchy of these columns is prepared as shown in Fig. 1.

These columns are further divided into more specific classes (*Politics*: Indian, International; *Entertainment*: Music, Film, Fashion; *Sports*: Cricket, Football, Badminton, and Tennis) for the entire document pool. In this work, the categorization is purely authors' choice. This can be either broader or it can have finer details depending on the requirement of the application. The newspaper-wise distribution of documents is shown in Fig. 2.

Three different newspapers are selected since each newspaper has a writing style distinct from others, thus giving us more insight into factors affecting KT selection procedure.

4 Proposed System

All the NPs from a text document are considered as candidates for qualifying as *KTs*; since non-NPs (conjunctions, prepositions, adjectives, verbs, etc) are highly common and frequent in all documents but each class of documents usually contain a specific set of NPs. The NPs are extracted from the documents using segregation/splitting method as reported in [1]. Here, all the sentences are extracted and split using punctuation symbols and stop words sequentially in order of their

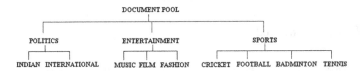

Fig. 1 Hierarchical categorization of documents used in the present work

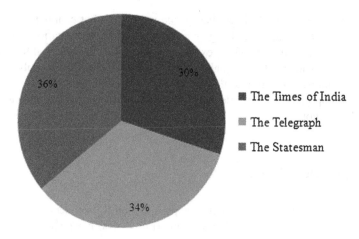

Fig. 2 Newspaper-wise document distribution used in present work

functionality. However, this NP list is reduced to a list of specific terms where a certain number of candidate KTs has their reference to the same term, i.e., they may share among themselves the same words/subphrase. These candidate KTs is then clustered into a single candidate KT. Conditions applied to achieve this reduced candidate KT list are as follows:

(1) Identify abbreviations and their full forms and cluster them as one term (either in full or abbreviated form).
(2) Cluster all those candidates sharing common words/subphrase.
(3) Remove candidates having single occurrence throughout the document.

4.1 Feature Description

Now the candidates are qualified as KTs on the basis of certain features. These features may be statistical/probabilistic, machine learning scheme driven, structural or position/content/topic based. Name and description of some popular/standard features applied in our work are given in Table 1. We have taken most of the features from literature except two of them viz. *Term Word Frequency Summation* (modified) and *Term with Numerals or Symbols* (developed).

4.2 Feature Intersection

Each of these features (say F_1, F_2, ..., F_n) are individually applied on the candidate KT list as F_1(candidate KT list), F_2(candidate KT list), ..., F_n(candidate KT list)

Table 1 Features applied

Sl #	Features	Description
1.	Term Frequency (TF) [7]	Number of occurrences of the term in the whole document. A standard TF (usually mean TF) is calculated from the TF values of all the terms based on which each term is evaluated to be qualified as KT or not. A log based function is used to determine the weight (= $\log_2$TF) of each term
2.	Term Presence in Heading/Title (TPIH/TPIT) [15]	Decides whether a term occurs in heading/title of the document or not. Boolean weight (1/0) is assigned to each term. A document's title gets significant weightage since it summarizes the same in a best possible way
3.	Term's First Occurrence (TFO) [7]	All sentences in a document are indexed. A term first time occurred in which sentence, is evaluated. Inner bell-shaped fuzzy function is considered for weight measure. A term's weight decreases as its sentence index increases toward the mid index and, increases as its sentence index increases beyond the mid index
4.	Term Paragraph Position (TPP) [16]	Usually in any text document, the introductory and the concluding paragraphs describe the document's subject. So a term found in any of these paragraphs carries more weight than those occurring in other paragraphs. A Boolean weight is assigned to each term; where a term present in first or last paragraph is assigned 1 and others are assigned 0
5.	Term Word Frequency Summation (TWFS) Score [5]	Mentioned as Substring Frequencies Sum (SFS) in [5], the frequency of each word in a term is determined w.r. t. the document and divided by document frequency. Modification includes log value of result of summation of word frequencies in a term that becomes the weight (= $\log_2$TWFS) of the term
6.	Term with Capital Letters (TCL) [16]	Term having all the letters or first letter as capital is given top preference as KT
7.	Term with Numerals or Symbols (TNS)	Numerals may be year, code, serial/version number etc. Alphanumeric/alpha-symbolic/alpha-numero-symbolic terms may be the name of any event, product, etc. thus increasing importance for KT extraction
8.	Named Entity (NE) [6]	Not all NPs are NEs. NE may be the name of any organization, object, place, etc. NEs are more preferred among NPs to be qualified as KTs
9.	Word Count in Term (WCT) [7]	A KT is considered to have a relatively fixed number of words. The standard number is determined with respect to word count of all candidate terms in a document
10.	Term Length (TL) [9]	A KT is considered to have a relatively fixed number of characters. The standard length is determined w.r.t length of all candidate terms in a document

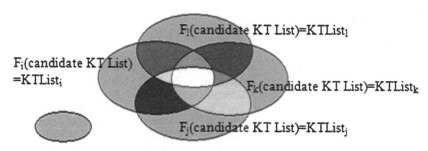

Fig. 3 Venn diagram representation of KT sharing among features

that returns $KTList_1$, $KTList_2$, ..., $KTList_n$ as their respective KT list. To get the final KT list, set intersection ($\cap$) operation is then applied on these lists as follows:

$$KTList_{Final} = KTList_1 \cap KTList_2 \cap \cdots \cap KTList_n$$

A Venn diagram (also known as a set diagram or logic diagram) is a diagram that shows all possible logical relations between finite collections of different sets. These diagrams represent elements as points in the plane, and sets as regions inside curves. Say we have sets A and B, then the region in both A and B, where the two sets overlap, is called the *intersection* of A and B, denoted by A $\cap$ B.

Note that there may be cases, although rare, where any individual KT list(s), say $KTList_m$, may not share any of its KTs with other KT lists as shown in Fig. 3. This situation is dealt with by some priority-based conditions as follows:

(1) A KT present in all KT lists is given the top priority.
(2) Priority of each KT decreases as its commonality among the KT lists reduces.
(3) A KT present in only one KT list is given the lowest priority.

The KTs are listed on the basis of priority as produced in Fig. 3. In this work, since we get a minimum of six terms with maximum overlapping from each document, so these terms are qualified as KTs for each respective document. Other terms with lower overlapping are discarded. The priority given to an overlapping plays a crucial role in determining good quality KTs. Here, although we have given top priority to KTs reside in the white region (see Fig. 3), still the intersection TF $\cap$ TPIH/T $\cap$ TFO $\cap$ TPP $\cap$ TWFS $\cap$ NE gives a much better quality KTs when compared to human generated KTs.

4.3 Document Classification

A correct KT set is able to classify the document correctly. However, not all KTs in a *correct KT set*, produced by any algorithm, may be "correct." Again a subset of the KT set from a particular document is supposed to determine the "uniqueness" of this document from all other individual documents in the universal document set/pool.

As we move toward more specific class, the more specific a KT has to be for each document. Some classes maybe highly interrelated to one another, e.g., the classes "Music," "Fashion," and "Film" are highly interrelated. Accurate classification of documents under "Entertainment" class thus becomes more critical.

5 Result and Discussion

An average of 150 NPs is identified from each document using the technique described in [1]. This NP set is reduced to 90–100 distinct terms. For each document, out of ten features implemented, the maximum number of feature intersections is found to be eight and the minimum is found to be six. In case of short documents (<= 300 words), a maximum of fivr terms are identified to have the maximum overlapping, whereas for long documents a minimum of 6 terms are identified to have the maximum overlapping. Sample evaluation of documents is given in Table 2. A detailed evaluation is found at http//www.termsetintersection. oonmad.com. From an independent human evaluation of KTs from the given term set, we find that the number of KTs is more or less the same as that of those generated from the proposed system. On an average three out five KTs match the human evaluated KTs, which is considered to be satisfactory.

From our experiments, we observe that the features WCT and TL do not provide much information about a term. Features similar to these are restrictive in nature. However, terms with very low word count and TL are usually preferred as KTs as per human evaluation. Also we see that features like TNS generates very rare terms. It shows a tendency to generate null set. Hence, prioritize the feature plays a crucial role in determining the quality of a KT. Furthermore, the number of features should be limited as large number of features may lead to greater ambiguity in determining good quality KTs and reduce system efficiency. Also person names are usually given very low priorities to qualify as KTs.

Table 2 Sample document evaluation

Doc #	TERM	TF	TPIH/T	TFO	TPP	TWFS	TCL	TNS	NE	WCT	TL
1	Luis Suarez	5	1	1	1	7	1	0	1	2	10
	Barca	8	1	1	1	8	1	0	1	1	5
2	Saleh Abdeslam	4	1	1	1	5	1	0	1	2	13
	Brussels	3	1	1	1	3	1	0	1	1	8
3	Lal Rang	3	1	1	1	6	1	0	1	2	7
	Movie	2	1	1	0	2	1	0	1	1	5
	film	3	0	9	1	3	0	0	1	1	4

6 Conclusion

This paper proposes a modest Venn diagram-based methodology to prioritize the different standard features, available in the literature, along with our own features used for extracting the KTs from English text documents. Till date, researchers have developed various features but all of those features may not be equally important. When we apply any combination of the features without any knowledge, it is observed that sometimes the final outcome of the combination downgrades because some of the features among the set may not give complimentary information to each other. Also, the number of features selected should be limited, as applying a large number of features may considerably reduce the system efficiency. Though the proposed system is simple but it produces satisfactory results considering the ambiguity level of the documents used in the present work.

There remains enough possibility to improve the quality of the feature ranking strategy. One of the future aims is to explore some evolutionary algorithms like genetic algorithm, harmony search to intelligently select the optimum number of features. Another plan is to incorporate varied categories of documents to ascertain the adaptability of the current KT extraction methodology.

Acknowledgements The authors are thankful to the Center for Microprocessor Applications for Training Education and Research (CMATER) of C.S.E. Dept., JU, for providing infrastructural facilities during progress of the work. The current work, reported here, has been partially funded by Technical Education Quality Improvement Programme Phase–II (TEQIP-II), Jadavpur University, Kolkata, India.

References

1. Chakraborty, Neelotpal, Samir Malakar, Ram Sarkar, Mita Nasipuri. "*A Rule based Approach for Noun Phrase Extraction from English Text Document.*" *2016 Seventh International Conference on CNC.* CNC, 2016.
2. Han, Jiawei, Micheline Kamber, and Jian Pei. *Data mining: concepts and techniques.* Elsevier, 2011.
3. Hasan, Kazi Saidul, and Vincent Ng. "Automatic Keyphrase Extraction: A Survey of the State of the Art." *ACL (1).* 2014.
4. Mangina, Eleni, and John Kilbride. "Evaluation of keyphrase extraction algorithm and tiling process for a document/resource recommender within e-learning environments." *Computers & Education* 50.3 (2008): 807–820.
5. Haddoud, Mounia, and Saïd Abdeddaïm. "Accurate keyphrase extraction by discriminating overlapping phrases." *Journal of Information Science* (2014): 0165551514530210.
6. Jurafsky, Dan, and James H. Martin. *Speech and language processing.* Pearson, 2014.
7. Turney, Peter D. "Learning algorithms for keyphrase extraction." *Information Retrieval* 2.4 (2000): 303–336.
8. Witten, Ian H., et al. "KEA: Practical automatic keyphrase extraction." *Proceedings of the fourth ACM conference on Digital libraries.* ACM, 1999.

9. Sarkar, Kamal, Mita Nasipuri, and Suranjan Ghose. "Machine learning based keyphrase extraction: comparing decision trees, naïve Bayes, and artificial neural networks." *Journal of Information Processing Systems* 8.4 (2012): 693–712.

10. Yu, Feng, Hong-Wei Xuan, and De-quan Zheng. "Key-Phrase Extraction Based on a Combination of CRF Model with Document Structure." *Computational Intelligence and Security (CIS), 2012 Eighth International Conference on.* IEEE, 2012.

11. Sarawagi, Sunita, and William W. Cohen. "Semi-markov conditional random fields for information extraction." *Advances in neural information processing systems.* 2004.

12. Beliga, Slobodan, Ana Meštrović, and Sanda Martinčić-Ipšić. "An Overview of Graph-Based Keyword Extraction Methods and Approaches." *Journal of Information and Organizational Sciences* 39.1 (2015): 1–20.

13. Dharmadhikari, Shweta C., Maya Ingle, and Parag Kulkarni. "Empirical Studies on Machine Learning Based Text Classification Algorithms." *Advanced Computing* 2.6 (2011): 161.

14. Jiang, Xin, Yunhua Hu, and Hang Li. "A ranking approach to keyphrase extraction." *Proceedings of the 32nd international ACM SIGIR conference on Research and development in information retrieval.* ACM, 2009.

15. Siddiqi, Sifatullah, and Aditi Sharan. "Keyword and Keyphrase Extraction Techniques: A Literature Review." *International Journal of Computer Applications* 109.2 (2015).

16. Kaur, Jasmeen, and Vishal Gupta. "Effective approaches for extraction of keywords." *Journal of Computer Science* 7.6 (2010): 144–148.

Bangla Handwritten City Name Recognition Using Gradient-Based Feature

Shilpi Barua, Samir Malakar, Showmik Bhowmik, Ram Sarkar and Mita Nasipuri

Abstract In recent times, holistic word recognition has achieved enormous attention from the researchers due to its segmentation-free approach. In the present work, a holistic word recognition method is presented for the recognition of handwritten city names in Bangla script. At first, each word image is hypothetically segmented into equal number of grids. Then gradient-based features, inspired by Histogram of Oriented Gradients (HOG) feature descriptor, are extracted from each of the grids. For the selection of suitable classifier, five well-known classifiers are compared in terms of their recognition accuracies and finally the classifier Sequential Minimal Optimization (SMO) is chosen. The system has achieved 90.65% accuracy on 10,000 samples comprising of 20 most popular city names of West Bengal, a state of India.

Keywords Gradient-based feature · Handwritten word recognition · Holistic approach · Bangla script

S. Barua (✉) · S. Bhowmik · R. Sarkar · M. Nasipuri
Jadavpur University, Kolkata, India
e-mail: shilpibarua2@gmail.com

S. Bhowmik
e-mail: showmik.cse@gmail.com

R. Sarkar
e-mail: raamsarkar@gmail.com

M. Nasipuri
e-mail: mitanasipuri@gmail.com

S. Malakar
MCKV Institute of Engineering, Howrah, India
e-mail: malakarsamir@gmail.com

© Springer Nature Singapore Pte Ltd. 2017
S.C. Satapathy et al. (eds.), *Proceedings of the 5th International Conference on Frontiers in Intelligent Computing: Theory and Applications*, Advances in Intelligent Systems and Computing 515, DOI 10.1007/978-981-10-3153-3_34

343

1 Introduction

In modern times, the world is getting smaller and smaller day by day due to the rapid advancements in science and technology. There is a tremendous urge of having everything within palmtop, which has initiated the era of digital world. One of the requirements of digital world is digitization of documents, which could be office documents, manuscripts, bank cheques, historical documents, etc. Literature on document recognition confirms that the recognition of printed document has achieved huge success [1], whereas handwritten document recognition is yet to achieve the same progress [2].

In recent years, the research in Handwritten Word Recognition (HWR), a branch of Optical Character Recognition (OCR) system, has advanced to a new level and such applications are available on different gadgets and handheld devices. However, these works belong to online category of HWR as the input is taken from the movement of pen on the touch panel from which the stroke patterns can be easily extracted and recognized. But, for the offline HWR the same cannot be said. Offline HWR comes under the purview of both image processing and pattern recognition and it needs effort and different mechanism all together than its printed counterpart due to various writing styles of individuals, as illustrated in Table 1.

Till date many researches have been conducted on HWR. Two approaches have been significantly identified to be well versed in recognizing handwritten words, playing a vital role in OCR. These approaches are analytical [2] and holistic [3]. Analytical approach has been in use for a longer time than holistic approach. The analytical approach comprises of identification of segmentation points in a handwritten word and to recognize the segmented characters. Though significant success is achieved in recognizing isolated characters, yet the analytical HWR fails to achieve acceptable success. It is so because there is a pre-requisition of having a proper segmentation algorithm [4], but in real world it is still a distant dream for handwritten words. Designing a segmentation algorithm is difficult due to the indistinct spacing and hefty variation in strokes and less scope of being disjoint.

The authors in [5] show that human mind takes less time in recognizing a word rather than recognizing each letter within. They also prove that the regular readers recognize the words as a whole which may support recognition of a handwritten word as a single pattern. Holistic approach removes the complexity of segmentation of word and concentrates on recognition of the entire word. The less complex

Table 1 Sample images of same city names, written in Bangla, have varied styles

City name	Sample 1	Sample 2	Sample 3
Kolkata			
Alipur			
Bardhaman			

outlook towards the said problem has made holistic approach popular. The work [6] shows that holistic approach for HWR performs better for small lexicon size over analytical approach.

One of the important requirements of advanced society is automation technology for all repeatable and hectic works to reduce the processing time. One such requirement is postal automation. The pre-requisite for postal automation is city name recognition written in different regional scripts. Mostly, mail sorting is done using the zip/pin code information inscribed in the postal documents. But, sometimes it is seen that complete pincode is not written (one/two digit(s) is/are missing) or these are too cursive to be recognized by any system. In that type of cases, city name recognition system could be a great help in mail sorting. Bangla is an important regional script with large number of speakers around the world [7]. Therefore, it is really needful to have a practical HWR system for the large section of populace using Bangla script in their daily life.

1.1 Related Work

The recognition of handwritten word has attracted many researchers due to the enormous deviations in word formation which makes the said problem challenging. A number of researches have already been conducted on the holistic approach towards HWR [3, 5, 7–12]. The works [3, 7, 8] are on handwritten Bangla word recognition. The work [2] uses convex hull and concentric rectangle-based features to classify 54 distinct word classes and achieves 84.74% of recognition rate in best case. In [7, 8], the authors have introduced elliptical features and HOG descriptor-based features respectively to recognize 20 city names of West Bengal, India. These works are experimented over 1020 number of word samples and have achieved 87.35% and 85.88% recognition rate respectively in best case using 3-fold cross validation mechanism.

The works [5, 6, 9] have dealt with English word recognition in a holistic way. In [5], authors have employed scalar features such as height, width, aspect ratio, area, number of descendants and ascendants in word and profile-based features like upper and lower word profile. The words are collected from handwritten historical manuscript. The work [6] has described a word image generation model using likelihood model to recognize words. In [9], longest run features are computed from hypothetically segmented sub-images. It has used the most frequent word from CMATERdb 1.2.1 [13].

Among different Indic scripts, there is an attempt in recognizing Devanagari word using Hidden Markov Model (HMM) and directional chain-code feature [10]. It uses the histogram of chain-code directions in the image-strips, scanned from left to right by a sliding window, as the feature vector. The authors in [11] have used curvelet transform to recognize handwritten words written in Devanagari script. In recognizing Arabic words, a holistic approach using discrete HMM is presented in [12].

1.2 Motivation

The world of art has opened up a new dimension for regional languages, i.e., with days passing by, all the regional languages are getting accepted worldwide. That has made Bangla language well known due to its world-renowned literature. Also the restriction of choosing analytical approach for HWR is also true for Bangla like other scripts.

In general, local feature describes an object (word image) more precisely than global one. In the work [8, 14], HOG descriptor has been used as local feature extractor. Not only this, an accuracy of 87.35% for handwritten Bangla city name recognition using HOG descriptor is found in [7]. HOG descriptor extracts local gradient information of an object at its preliminary stage. So there remains unfolded possibility that local gradient feature generation might generate good feature for Bangla word recognition in holistic way. Not only this, as mentioned earlier, it comes with less overhead. Therefore, local gradient-based feature extraction mechanism is exercised here for handwritten Bangla city name recognition in holistic way.

2 Present Work

The entire work can be divided into the following modules *namely*, data preparation, feature extraction, and classification, which are described below.

2.1 Data Preparation

First, handwritten word images of 20 city names, based on population of cities [15] are collected on pre-formatted datasheets from 500 individuals varying in age, sex, and educational qualification. The datasheets are then scanned using flat-bedded scanner with 300 dpi resolution. The cropped word images are binarized using Otsu's thresholding [16] mechanism and then morphological close operator [17] with rectangular structuring element of size 3×2 is applied to fill holes and to join short discontinuities which may occur during binarization. Then minimum bounding box enclosing the word images are estimated to avoid needless processing time overload (see Fig. 1).

(a)   (b)

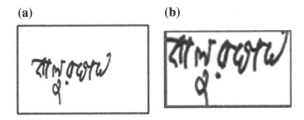

Fig. 1 **a** Cropped image **b** minimal bounding box of the image

2.2 Feature Extraction

For extracting the features, gradient orientation information from each of the word images is extracted which is purely a concept motivated by HOG descriptor. HOG is initially invented for object detection. It counts occurrences of gradient orientation in localized portions of an image. The descriptor is similar to Scale-Invariant Feature Transform (SIFT) descriptors. But it differs in that it is computed on a dense grid of uniformly spaced cells. The working principle of HOG is described in [14].

To calculate the features, each word image is hypothetically segmented into $n \times n$ sub-images. 3 samples of same word image, segmented in different grid size, are depicted in Fig. 2. Selection of optimal grid size for handwritten word images is a challenging issue. Hence, in the present work, experiment is conducted with varied number of grid sizes to decide the best fitting grid size. These images (see Fig. 2) also provide the evidence for requirement of local information extraction from word images. The images show that a word belonging to the other class bears dissimilar data distribution in the corresponding grid. This simple idea, which is the prime difference from [8], makes the model so powerful. In the work [8] each word image is segmented into 10 vertical segment of equal width.

Each sub-image is smoothed using Gaussian filter [14]. Gradient orientation is then extracted from each of the n^2 sub-images which are accumulated in

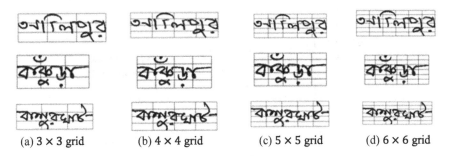

(a) 3 × 3 grid   (b) 4 × 4 grid   (c) 5 × 5 grid   (d) 6 × 6 grid

Fig. 2 **a** Illustration of different grid sizes; each row shows different instances of the same city name which are segmented into varied number of grids

8-directional bins [8]. The mathematical preliminaries regarding the feature extraction model is described in [8]. Therefore, each image is represented with feature vector of size $8n^2$.

2.3 Classifier Selection

There are a number of good classifiers present in literature. The choice of best classifier is itself a research problem. In the present work, the selection procedure is carried out experimentally on the basis of the classifiers' performance.

3 Experimental Results

For experimentation, 20 city names of West Bengal, India are considered. Five hundred sample images for each of the city names are collected which implies a database of 10,000 handwritten words. The total database is partitioned in 4:1 ratio as training and test data sets (i.e., 400 training samples and 100 test samples for each of the 20 city names) to perform 5-fold cross validation.

For the classification of city names, written in Bangla script, using the said feature vector, five well-known classifiers *namely*, SMO, Bayes Network, Random Forest, Bagging, and Multilayer Perceptron (MLP) are considered. To perform the actual experiment, an eminent machine learning tool, called WEKA (Waikato Environment for Knowledge Analysis) [18], is used. The recognition accuracies of all the five classifiers for 5-fold cross validation while recognizing the handwritten city names are shown in Fig. 3. Here, the word images are hypothetically divided into 6×6 grid size.

From the Fig. 3, it is observed that SMO outperforms other classifiers in the experiment. The possible reasons for this are (1) SMO efficiently takes care of

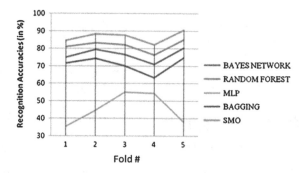

Fig. 3 The comparative result of 5 different classifiers using 6×6 grid

quadratic programming problem that arises during the training of Support Vector Machine (SVM) and (2) SMO also breaks the problem into a series of smallest possible subproblems, and solves analytically.

Thus, in the present work, SMO is used as final classifier to recognize words and more experiment is carried out thereafter for estimating the optimal grid size to achieve the best recognition accuracy. So, different feature vectors obtained by varying the grid numbers are then fed to SMO and 5-fold cross validation is performed for the said problem. The plot of average recognition accuracy versus number of grids is shown in Fig. 4. This diagram shows that the best result is achieved for grid size of 6×6 and increasing the grid size beyond that results into more misclassification. The possible reason behind this is that increasing grid dimension results into smaller sub-images which consequently provide insufficient discriminative information out of this small sub-region.

The recognition accuracies for 5-fold cross validations are reported in Table 2. It is noted that the best performance achieved using 6×6 grid size is 90.65%. A diligent observation of the misclassified patterns renders that the 5 city names are affected with more than 10% of the entire misclassification.

The reasons behind misleading identification are strong pattern similarity among two/more city names (See Fig. 5), alternative spelling used for same city name (see Fig. 6), largely skewed data (see Fig. 7) and error occurred during auto-matic cropping of word images from the pre-formatted datasheets (see Fig. 8).

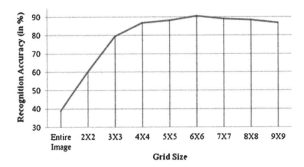

Fig. 4 Average recognition rate with varying grid size

Table 2 The recognition accuracies given by SMO in 5-fold cross validation

Run#	# of images per class in train database	# of images per class in test database	Recognition accuracy (in %)
1	400	100	84.65
2			88.40
3			87.70
4			82.05
5			90.65

Fig. 5 Rectangular boxes show structural similarity of words belong to different classes

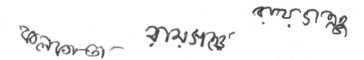

Fig. 6 Same city name with different spellings

Fig. 7 Examples of extremely skewed word images

Fig. 8 Due to automatic cropping, some portions of ascendants are missing

To ascertain the soundness of the present work, the same is compared with two previously developed holistic HWR techniques [7, 8] (see Table 3). It is worth mentioning that these works were conducted with same set of city names with less number of sample images per class. In the said techniques, the best recognition results were achieved in 3-fold cross validation using MLP-based classifier. Therefore, they are re-evaluated on the present database using SMO as classifier in 5-fold cross validation mechanism. From the result, it can be concluded that the present method comprehensibly outperforms the previous methods on the current database.

Table 3 Comparison of present work with previously developed mechanism

Technique	Best recognition rate (in %)	Average			
		Recall	Precision	F-measure	ROC area
Bhowmik et al. [7]	82.45	0.82	0.82	0.82	0.97
Bhowmik et al. [8]	81.05	0.81	0.83	0.81	0.98
Present Work	**90.65**	**0.90**	**0.90**	**0.91**	**0.99**

4 Conclusion

The present work deals with HWR of 20 popular city names of West Bengal, India written in Bangla script. For this purpose gradient-based feature, motivated by HOG descriptor, has been extracted after hypothetically dividing the images into a number of grids. Finally, SMO is chosen as classifier and recognition accuracy achieved in best case is 90.65% using 5-fold cross validation. Though the work outperforms state-of-art techniques, yet it suffers from a number of shortfalls. The length of feature vector to represent a word image is relatively large which takes considerable amount of time to train the classifier. Hence, one of the future aims is to use some feature dimension reduction techniques. Also, a non-symmetric zoning scheme around the structurally similar regions could be applied on the images which suffer heavy misclassification among each other. Another plan is to incorporate more number of city names to make a realistic postal automation system.

References

1. Chaudhuri, B.B., Pal, U.: A complete printed Bangla OCR system. Pattern Recognition. 31 (5), 53–549(1998).
2. Basu, S., Das, N., Sarkar, R., Kundu, M., Nasipuri, M., Basu, D.K.: A hierarchical approach to recognition of handwritten Bangla characters. Pattern Recognition, 42(7), 1467–1484 (2009).
3. Bhowmik, S., Polley, S., Roushan, M.G., Malakar, S., Sarkar, R., Nasipuri, M.: A holistic word recognition technique for handwritten Bangla words. International Journal Applied Pattern Recognition, 2(2), 142–159(2015).
4. Malakar, S., Ghosh, P., Sarkar, R., Das, N., Basu, S., Nasipuri, M.: An improved offline handwritten character segmentation algorithm for Bangla script. In: 5th Indian International Conference on Artificial Intelligence (IICAI-11), 71–90(2011).
5. Lavrenko, V., Rath, T.M., Manmatha, R.: Holistic word recognition for handwritten historical documents. In:Proceedings of First International Workshop on Document Image Analysis for Libraries,278–287, IEEE(2004).
6. Ishidera, E., Lucas, S.M., Downton, A.C.: Top-Down Likelihood Word Image Generation Model for Holistic Word Recognition. In: Proceedings of the 5th International Workshop on Document Analysis Systems V, 82–94, Springer-Verlag(2002).
7. Bhowmik, S., Malakar, S., Sarkar, R., Nasipuri, M.: Handwritten Bangla Word Recognition Using Elliptical Features. In: International Conference on Computational Intelligence and Communication Networks (CICN), 257–261, IEEE(2014).

8. Bhowmik, S., Roushan, M. G., Sarkar, R., Nasipuri, M., Polley, S., Malakar, S.: Handwritten Bangla Word Recognition Using HOG Descriptor. In:Fourth International Conference of Emerging Applications of Information Technology (EAIT), 193–197, IEEE (2014).

9. Acharyya, A., Rakshit, S., Sarkar, R., Basu, S., Nasipuri, M.. Handwritten Word Recognition Using MLP based Classifier: A Holistic Approach. International Journal of Computer Science Issues, 10(2), 422–427(2013).

10. Shaw, B., Parui, S.K., Shridhar, M.: Offline Handwritten Devanagari Word Recognition: A holistic approach based on directional chain code feature and HMM. In:International Conference onInformation Technology, 203–208, IEEE(2008).

11. Singh, B., Mittal, A., Ansari, M.A., Ghosh, D.: Handwritten Devanagari Word Recognition: A Curvelet Transform Based Approach. International Journal on Computer Science and Engineering, 3(4), 1658–1665(2011).

12. Dehghan, M., Faez, K., Ahmadi, M., Shridhar, M.: Handwritten Farsi (Arabic) word recognition: a holistic approach using discrete HMM. Pattern Recognition, 34(5), 1057–1065 (2001).

13. Sarkar, R., Das, N., Basu, S., Kundu, M., Nasipuri, M., Basu, D.K.: CMATERdb1: a database of unconstrained handwritten Bangla and Bangla–English mixed script document image. International Journal on Document Analysis and Recognition, 15(1), 71–83(2012).

14. Dalal, N.,Triggs, B.: Histograms of oriented gradients for human detection. In: Computer Society Conference on Computer Vision and Pattern Recognition (CVPR),1, pp. 886–893, IEEE (2005).

15. List of cities in West Bengal by population: https://en.wikipedia.org/wiki/List_of_cities_in_ West_Bengal_by_population

16. Otsu, N.: A threshold selection method from gray-level histograms. IEEE Transactions on Systems, Man and Cybernetics, 9,62–66 (1979).

17. Gonzalez, R.C.: Digital Image Processing. Pearson Education, India(2009).

18. M. Hall, M., Frank, E., Holmes, G., Pfahringer, B., Reutemann, P., Witten, Ian H.: The WEKA Data Mining Software: An Update; SIGKDD Explorations, 11(1)(2009).

Shortest Path Algorithms for Social Network Strengths

Amreen Ahmad, Tanvir Ahmad and Harsh Vijay

Abstract In social media directed links can represent anything from close friendship to common interests. Such directed links determine the flow of information and hence indicate an individual influence on others. The influence of a person X over person Y is defined as the ratio of Y's investment that Y makes on X. Most contemporary networks return source–target paths in an online social network as a result of search ranked by degrees of separation. This approach fails to reflect tie of social strength (i.e., intimacy of two people in terms of interaction), and does not reflect asymmetric nature of social relations (i.e., if a person X invests time or effort in person Y, then the reverse is not necessarily true). In this paper, it is proved that in social graph result can prove to be more effective by incorporating the concept of directed and weighted influence edges taking into account both asymmetry and tie strength. The study is based on two real-world networks: Twitter capturing its retweet data and DBLP capturing its author–coauthor relationship. The experiments have been conducted based on two algorithms—Dijkstra shortest path algorithm and influence-based strongest path algorithm. Then a comparative study was done capturing different cases in which strongest path algorithm was better than shortest path algorithm in different cases.

Keywords Influence-based strongest path algorithm · Shortest path algorithm · Asymmetric influence · Strength of ties

A. Ahmad (✉) · T. Ahmad · H. Vijay
Department of Computer Engineering, Jamia Millia Islamia, New Delhi, India
e-mail: amreen.ahmad10@gmail.com

T. Ahmad
e-mail: tahmad2@jmi.ac.in

H. Vijay
e-mail: iharsh234@gmail.com

S.C. Satapathy et al. (eds.), *Proceedings of the 5th International Conference on Frontiers in Intelligent Computing: Theory and Applications*, Advances in Intelligent Systems and Computing 515, DOI 10.1007/978-981-10-3153-3_35

1 Introduction

The universality of social search in social network has gained much popularity over time, where the aim is to search for a "sequence" of people or search a particular person who will prove beneficial for person X in finding a job or introduce him to some particular person. Most contemporary social networks such as Twitter and LinkedIn model social relations as binary relationship (i.e., two people are either related through some common relation or not). Hence, in these types of social networks the available shortest path is returned between the source and the target as a result of social search. This paper is based on the hypothesis that by taking into account the influence of a person over another, social search results can be made more effective, which has varying strength and is inherently asymmetric.

1.1 Asymmetric Influence and Strength of Ties in Social Network

A large number of relationships are maintained by people in real life with varying tie strength: family, work colleagues, friends, close friends, etc. In real-life social networks, weak ties are extremely important (e.g., in finding jobs) [1, 2]. Therefore, information is lost in a network such as LinkedIn where a person is linked to another person whom he knows well and turns down the invitation from unknown people. Despite of this sometimes LinkedIn users prefer to connect with slightly known people. For capturing ties of varying strength, online social networks have proved to be beneficial. Besides tie strength, relationship asymmetry has to be considered in conducting a social search. For example, in a coauthorship network it is not necessary that if advisor has more influence on his researcher then vice versa will be true.

1.2 Contributions

Following work has been done in this paper:

Experiment is conducted on two large datasets of Twitter (retweet data) and DBLP (author–coauthor network) based on two algorithms for calculating shortest path between two nodes and comparative study is done. The two algorithms are Dijkstra shortest path algorithm and influence-based strongest path algorithm based on influence measure [2].

The remaining paper proceeds as follows: concerned literature survey field has been discussed in detail in Sect. 2. In Sect. 3 an influence metric has been defined. The two different approaches for calculating shortest path between two nodes are discussed in Sect. 4. Next the experiments and results are discussed in Sect. 5, providing a detailed discussion in Sect. 5.3 and conclusion in Sect. 6.

2 Related Work

A lot of work conducted in social network has focused on the local search aspect. Users were asked to forward a message to target person in a social search experiment performed by Doddas et al. [3] through their acquaintances. Here users and target person need not know each other. It was found that professional ties can vary from medium to weak in strength and successful social search depended a lot on it. Similar experiments were performed on email data and online social network by Adamic et al. [4]. The above research work used social hierarchies as a framework for performing social search and was based on earlier. Consideration of tie strength with different information such as homophily, geographical proximity is the common concept that has been used in the above research.

Aardvark (i.e., http://www.vark.com), a social network site, connects people who have specific questions in their mind to those people who are most qualified or have a higher chance of answering them. To find correlation between users symmetric measure of affinity is used internally by this site [5]. In social graphs a lot of work has been conducted on the concept of incorporating edge weights. One of the common approaches is to use threshold for converting interaction data into a binary edge weight (e.g., an edge is defined between X and Y if at least five messages have been exchanged between them) [6, 7]. In signed networks, it is considered that edges can be either negative or positive or absent [8]. A lot of work has been done in social network domain for determining relationship strength [4].

There have been other works done outside the domain of search based on diffusion models that employ measure of influence.

3 Influence

In a social network the task of routing requests from source to target is the main aim of global social search. Hence, the most influential path is considered the best path from source to destination. As discussed above, if there exists a path where each node has considerable influence over other succeeding node, then global search problem can be successful.

3.1 Social Interaction

Several types of social interactions are modeled by social networks and such interactions can be undirected (such as coauthorship network in DBLP) or directed (such as Twitter). Based on social interactions of the people involved influence can be modeled where interactions can be in terms of effort and time, and hence interaction count can be used for measurement of tie strength.

3.2 Asymmetry of Influence

Influence has asymmetric nature that is if Y is highly influenced by X, then it is not guaranteed that the reverse will be true. In case of undirected interactions there is a possibility of existence of asymmetry.

3.3 Influential Ties

A quantitative definition of influence is described here [2]. The influence from X to Y is defined as the proportion of X's investment on Y. Let Invests(X, Y) be defined as the investment that Y makes on X, then

$$\text{Influence } (X, Y) = \frac{\text{Invest}(Y, X)}{\sum_{Z} \text{Invest}(Y, Z)} \tag{1}$$

where $\sum_{Z}$ Invests (Y, Z) is defined as the sum of all investments that Y makes on all other nodes (z = sum of all nodes present in the network) in the network. In a social graph the influence of an edge always lies between 0 and 1. In Fig. 1(a) an undirected weighted graph is shown where the edges represent the investments and the same graph is shown in Fig. 1(b) where influence is represented by weighted edges.

Figure 1 (a) depicts the non-directional interaction of authors–coauthors in a DBLP network. The interpretation of the above graph is as follows:

Suppose there are four nodes in the network A, B, X, Y. Node X is supervisor, and nodes A, B, and Y are his students. Number of coauthorships between pair of nodes is represented by edge weights. The interpretation of the above figure is that

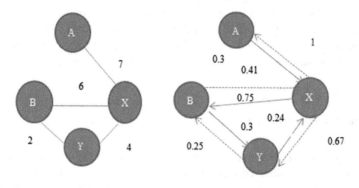

(a) Before Calculating Influence   (b) After calculating Influence

Fig. 1 a Before calculating influence 1 b. After calculating influence

A and X together have published papers in seven different conferences. It is clear from the above figure that students A, B, and Y have interacted more with their supervisor than with their peer groups. This is due to the fact that every student includes the name of his supervisor in his paper but it is not necessary that he will include the name of his peer group.

In Fig. 1b, the dotted line depicts the influence of supervisor on his students and the solid line depicts the influence of student on his supervisor. It is clear from Fig. 1b that students are more influenced by their supervisor. Moreover, student Y is more influenced by student B, which is due to the involvement of student B in a large proportion of student Y's publication.

4 Algorithm

Shortest path between two nodes is computed using Dijkstra algorithm. Influence-based strongest path is computed based on the algorithm discussed in [2].

Algorithm 1:

```
G(V,E);
All_paths (Source, Dest, Paths = [])
      Paths = [Source]
      If Source == Dest:
            Return [Paths]
      Paths = []
      For vertex in G[start]
         Vertex not in path
            Extend_paths = All_paths(vertex,Dest,Paths)
            For path in Extend_paths
               Paths.append (path)
      Return Paths
   Strongest_paths(Paths)
      D = 0.95
Influence;
For path in Paths
Influence_p = 1
For edges in path
Influence_edge = influence(edges)
Influence_p *= influence_edge
Total_influence = log(1/D)+log(1/influence_p)
            Influence[Total_influence] = path
      Return path with minimum Total_influence
```

Algorithm 2:

```
Dijkstra(G,start,Dest)
    Queue = [(0,start,[])]
    Seen = set()
    While True:
(cost,vertex,path) = heappop(Queue)
            If vertex not Seen:
    Path.push(vertex)
  Seen.add(vertex)
 If vertex == end:
                        Return path
  For (next , c) in G[vertex]
 heappush(Queue,(Cost+c,next,path))
```

5 Experiments

Performance of global search on two large networks: Twitter and DBLP are evaluated when influenced edge weight is incorporated. The goal is to compare the performance of influence-based strongest path algorithm [2] and traditional Dijkstra shortest path algorithm. In both real-world network DBLP and Twitter, influence can be defined as the interaction between individuals. Both datasets are large and depict realistic and global view of social data in the network.

5.1 DBLP Computer Science Bibliography

The DBLP dataset used in our experiment is a subset of DBLP dataset (i.e., http://dblp.uni-trier.de/xml/) and includes approximately 38,942 unique author names and 16 conferences names (selected Data Mining conferences) held between January 2001 and February 2010. In a social graph $G = (V, E)$, V is the set of all authors and $E = \{(v_i, v_j): i \neq j, v_i, v_j \in V$ and v_i, v_j are coauthors of paper presented in a conference}.

Papers (v_i, v_j) represent the number of papers coauthored by v_j and v_i and presented in a conference, then

$$\text{Influence}(v_i, v_j) = \frac{\text{Papers}(v_i, v_j)}{\text{Papers}(v_j, v_k)} \tag{2}$$

where v_k is the number of authors with whom v_j has coauthored a paper and presented it in conference.

To analyze the effectiveness of influence edge weights in global search, randomly 500 source–destination node pairs are selected and computed:

(1) S (P_{short})—the strength of the shortest path based on the number of hops using Dijkstra algorithm. There may exist more than one shortest path with same length; so randomly one of them is chosen.

(2) S (P_{strong})—the strength of the influence-based strongest path algorithm discussed above.

5.2 Twitter Retweets

The Twitter dataset used in our experiment consists of 15 days of tweets crawled from Twitter. There are approximately 23,491 unique users and 32,132 retweet-directed edges. The influence weights are assigned over edges using the following concept: if Retweets(Y, X) is the number of times Y retweeted X, then

$$\text{Influence}\,(X, Y) = \frac{\text{Retweets}(Y, X)}{\sum\limits_{Z} \text{Retweets}(Y, Z)} \tag{3}$$

where Z is the total number of other nodes in the network. Using the influence model, a line graph of the node influence distribution is shown in Fig. 2.

5.3 Discussion

It can be observed from Fig. 2a, b that influence of most of the nodes in given datasets: DBLP and Twitter lies below one, this is due to the fact that influence has

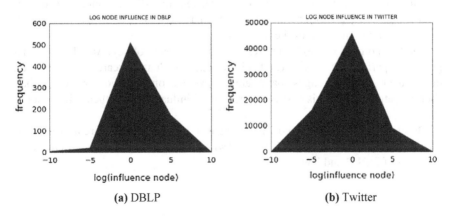

Fig. 2 a DBLP **2 b.** Twitter

Table 1 Results of experiments conducted on Twitter and DBLP datasets. $|P_{strong}|$ and $|P_{short}|$ denote lengths of strongest and shortest path. There are 500 total nodes in the network

Dataset	$|P_{strong}| > |P_{short}|$	$|P_{strong}| = |P_{short}|$
Twitter	323(64.6 %)	177(35.4 %)
DBLP	225(45.0 %)	275(55.0 %)

Table 2 Results of experiments conducted on Twitter and DBLP datasets. μ denotes average influence of a node in the network. Avg.$|P_{strong}|$ denotes average strongest path length and Avg.$|P_{short}|$ denotes average shortest path length in the network

Dataset	Avg. $|P_{strong}|$	Avg. $|P_{short}|$	μ
Twitter	6.8	4.9	0.67
DBLP	4.7	3.3	0.60

asymmetric nature in real life. Figure 2a, b shows real-life scenario in which a handful of nodes are highly influential, besides there are other nodes present in the network who play a key role in increasing the influence of other nodes rather than having an influential status. As can be seen from Fig. 2a, b, the highly influential nodes in Twitter are present at the tail of influence distribution graph, whereas in DBLP the same scenario is not there. This is due to the fact that influence in DBLP depends on contribution or collaboration of two people and hence it is evenly distributed in DBLP.

In the Twitter and DBLP graph, shortest path is smaller than their corresponding strongest path in 64.6 and 45% of the cases, whereas they are equal in 35.4 and 55.0% of the cases, respectively, as can be observed from Table 1. The average $|P_{strong}|$ length in Twitter and DBLP datasets is 6.8 and 4.7, whereas average $|P_{short}|$ length in Twitter and DBLP datasets is 4.9 and 3.3 as observed from Table 2. It is also seen that average influence of a node in Twitter is 0.67 and it is almost nearly same in DBLP, which is 0.6. In the Twitter dataset, the shortest paths are smaller than the strongest path by one or two hops whereas in DBLP the shortest paths are smaller than strongest path by one hop. There may exist more than one shortest path in a graph; randomly one of them is chosen.

The key point to note from the result is that inclusion of edge weights in social graph for performing social search gives better result than search performed on unweighted graph. Strongest paths are on average one or two hops more than shortest paths, so if at this cost if we are getting influential path then it is not a big cost.

Different results are obtained when the above two algorithms were implemented on two different datasets: DBLP and Twitter. This is due to the difference in structure of DBLP and Twitter.

6 Conclusion

In this paper it was proved that tie strength concept in weighted graph gave better result than unweighted graph. An experiment was conducted on two large datasets: DBLP and Twitter using two algorithms, influence-based strongest path algorithm and Dijkstra shortest path algorithm. The result from the experiment proved that shortest path was on an average one or two hops less than strongest path. There were many cases in which the strongest path and shortest path were same, but the selection of an influential path based on strongest path would yield better result than a random path based on shortest path. It is therefore concluded that influence-based strongest path based on social interactions would boost the performance of social search algorithms in online social networks as compared to traditional Dijkstra algorithm.

References

1. M. Granovetter, The Strength of Weak Ties, The American Journal Sociology, 78(6):1360–1380, 1973.
2. S. Hangal, D. MacLean, M. S. Lam, and J. Heer, All Friends are Not Equal: Using Weights in Social Graphs to Improve Search, 4th SNA-KDD, Workshop'10, Washington D.C, USA, 2010.
3. P. Dodds, R. Muhammad, and D. Watts, An study of search in global social networks, Science, 301(5634):827, 2003.
4. L. Adami and E. Adas, How to search a social network, Social Networks, 27(3):187–203, 2005.
5. D. Horowitz and S. Kanvar, The Anatomy of a Large-Scale Social Search Engine 2010.
6. Leskovec and E. Horwitz, Planetary-scale views on a large instant messaging network, In WWW'08: Proceedings of the 17th international conference on World Wide Web, pages 915–924, NewYork, USA, 2008, ACM.
7. J.R. Tyler, D.M. Wilkinson, and B. A. Huberman, Email as spectroscopy: Automated Discovery of Community Structure within Organizations, pages 81–96, Kluver, B.V., Deventer, The Netherlands, 2003.
8. J. Leskovec, D. Huttenlocher and J. Kleinberg, Signed networks in social media, Proc. 28th CHI, 2010.

Improvised Symbol Table Structure

Narander Kumar and Shivani Dubey

Abstract Symbol table is the environment where the variables and functions/methods exist according to their scope and the most recent updated values are kept for the successful running of the code. It helps in code functioning. It is created during compilation and maintained, used during running of the code. Adding a utility called common file can help in conversion of one code to another code. As common file can be explained as the file containing the common functionalities of different languages, say, every language has a print function but with different syntax; these different syntax of print are added in common file which help in the conversion. In this paper, we present the compilation process mechanism with the help of common file in the symbol table. It also explains how a code is converted into another code.

Keywords Symbol table · Data structure · Hash table · Binary tree and compilation

1 Introduction

Symbol table is the collected data of the compiler which helps in the functioning of the compiler. Token generation and formation of parse tree helps in the Analysis of code, i.e., whether the variable or function declared or not and errors are reported for the same. This data is helpful for further working. It deals with scope too. The functions such as insert (), object lookup (), begin scope (), and end scope () performs the functioning of the symbol table. Insert () adds the token into the symbol table, initialize them with the values specified and report error if any symbol

N. Kumar (✉) · S. Dubey
Babasaheb Bhimrao Ambedkar University (A Central University),
Lucknow, India
e-mail: nk_iet@yahoo.co.in

S. Dubey
e-mail: shishubham85@gmail.com

© Springer Nature Singapore Pte Ltd. 2017
S.C. Satapathy et al. (eds.), *Proceedings of the 5th International Conference on Frontiers in Intelligent Computing: Theory and Applications*, Advances in Intelligent Systems and Computing 515, DOI 10.1007/978-981-10-3153-3_36

is not initialized. Object lookup () searches the symbol while running the code and update the values side by side. Begin scope () and end scope () deals with the scope of the variable and function. It also helps in data overloading and overridding. By providing scope management, it deals with global and local scope of variables and provide effective functioning. Linear ordered and unordered data structures are used as symbol table structure for small codes. Hash table and binary trees are used for large codes. Multiple symbol tables are chained in order to have proper functioning and provide memory management.

If a common file containing the functionalities of each language grouped together helps in determining the similarities of different language and help in conversion of code. Structure of common file can be binary tree which contains the functionality according to the levels of the tree.

2   Review of Literature

In [1], compiler working in Java is described. The symbol table is used to map identifiers to its type, location, and scope for proper functioning. It involves disassembly, lifting, data flow, control flow, and type analysis. Symbol table uses, i.e., remembrance of declarations, type checking, and mapping with the operations are specified in [2]. Imperative programming languages, their features, problems/ambiguity, translators, and automated compiler generator are discussed in [3]. In [4] describes the usefulness of symbol table in error detection and correction, also the ways for organizing and accessing symbol tables. Importance of studying compiler and its uses and functioning is described in [5]. A mechanism to remove the deficiencies such as poor handling of indirect jumps, poor recovery of parameters and return values, calls have been discussed in [6]. Symbol table goals and approaches are discussed and implemented which is flexible and language independent in [7]. Various structures of symbol table are discussed in [8] which are created during lexical, syntax, and semantic analysis phase and used by further phases for proper running. In [9], shows symbol table uses in a compiler, object-files and executables with their operations and structures. Functionality and working of different structures of symbol table are explained in [10]. Symbol table interactions with all phases of compiler are explained in detail in [11]. In [12] semantic checks, its scope, type, and symbol table are discussed. Symbol table chaining, unique and multiple symbol table, grammars, YACC are explained in [13]. In [14], Symbol table implementation techniques and operations with definition and examples are explained.

From the extensive review of work, findings are that symbol table helps in the compilation and decompilation process. If we add a commonality feature, it could help in linking all the languages and it is beneficial in its conversion.

3 Research Design and Proposed Model

We can define the structure of symbol table as the multiple chained hash table and the common object file can be a binary tree.

To design and explain its detailed structure let take an example:-

```
int a=5;b=10;
void method1()
    {
            int var1;int var2;
                { int var3;int var4;}
            int var5;
                { int var6;int var7;}
    }
void method2()
    {
            int char1;int char2;
                { int char3;int char4;
                        { int char5;}
                }
    }
```

According to the example the generalized symbol table that is used by every data structure is depicted in Table 1 given below. The symbols, their token, data type, and whether they are initialized or not are in generalized symbol table. It helps in telling the errors regarding the initialization.

The common object file added can be a binary tree with the structure defined in Fig. 1. It takes the directives and code as input and then it is divided into the levels. Each level describes about the common functionality of all languages.

Table 1 Generalized symbol table

Symbol	Token	Data type	Init?
a	Id	Int	Yes
b	Id	–	No
method1	Id	Procedure name	Yes
var1	Id	Int	Yes
var2	Id	Int	Yes
var3	Id	Int	Yes
var4	Id	Int	Yes
var5	Id	Int	Yes
var6	Id	Int	Yes
var7	Id	Int	Yes
method2	Id	Procedure name	Yes
char1	Id	Int	Yes
char2	Id	Int	Yes
char3	Id	Int	Yes
char4	Id	Int	Yes
char5	Id	Int	Yes

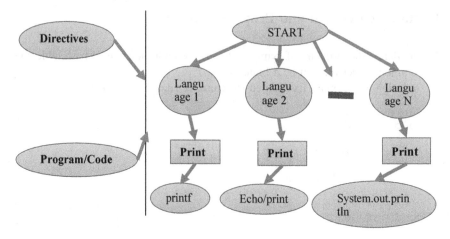

Fig. 1 Common object file structure defined as a binary tree

To understand the scope and chaining of multiple symbol tables is explained in the Fig. 2. Global functions and methods can be used by the entire code but the local functions and variables work only in its block of code. And the symbol table is generated using the hash function and common object file is linked with symbol table.

4 Working Example

Let take an example to find factorial of a given number using recursion. This code can be written in C, compiled, and then converted into PHP or Java.

Let the C code be:-

```
#include<stdio.h>
int factorial(intnum);
void main()
{    intnum, result;
     printf("Enter number to find its factorial");
     scanf("%d",&num);
     if(num<0){ printf("Factorial of negative number not
possible"); }
     else { result=factorial(num);
          printf("Factorial    of    number    %d    is
     :%d",num,result);}}
int factorial(intnum)
{    if(num==0 || num==1)
          { return 1; }
     else
{ return (num*factorial(num-1));} }
```

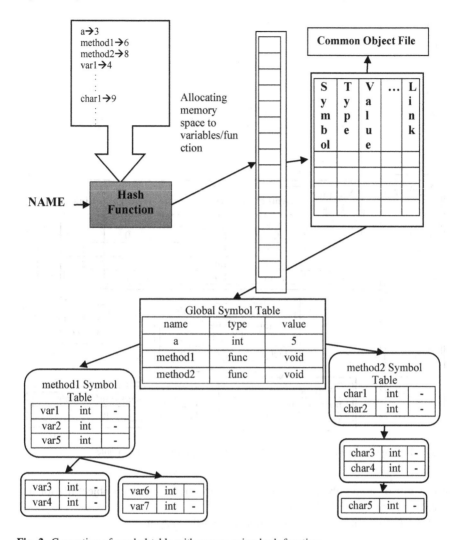

Fig. 2 Generation of symbol table with scopes using hash function

The generalized symbol table for the given example is shown in Table 2. The symbol table created using hash function for the above example is shown in Fig. 3 and its common object file is described in Fig. 4 where each level consists of similar type of information like Print level tells about the printing function of the language.

The common object file is linked to symbol table. It takes the original code as input and divides it in various levels according to its functionality like Taking Input, printing, etc., of that language. Other languages levels are filled according to the

Table 2 Generalized symbol table for the given example

Symbol	Token	Data type	Init?
factorial	Id	Procedure name	Yes
num	Id	Int	Yes
main	Id	Main procedure name	Yes
num	Id	Int	Yes
result	Id	Int	Yes

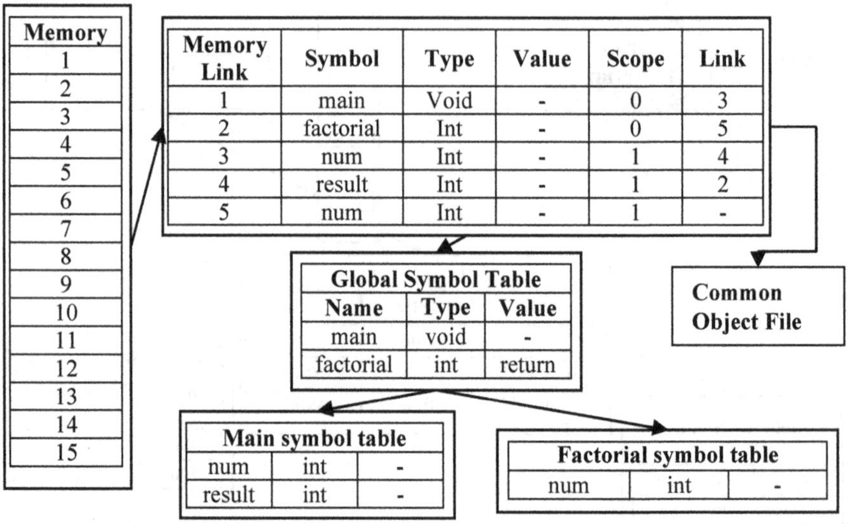

Fig. 3 Hash representation of the symbol table with scopes for given example

rule set of common object file and we get summarized common object file for the given code as shown in Fig. 4.

Now if the following C code is run, we get the output:-

Enter number to find its factorial 5
    Factorial of number 5 is: 120

Now if the following code has to be converted into Php/Java language then it follows following steps and the code generated is shown in Fig. 5:

1. Add necessary elements.
2. According to C code make changes with the help of common object file.

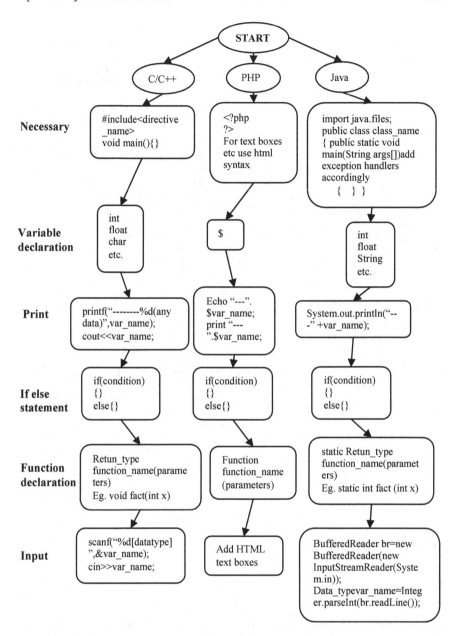

Fig. 4 Common object file generated for the given example

```
  <html>
  Enter number to
find its factorial
:< Input type=
"text" name=
"text1"/>
  <input type=
"submit" value=
"submit"/>
  <?php
  $x=$_GET["text1"];
  $num=(int)$x;
  if($num<0)
  {echo "Factorial
of negative number
not possible";}
  else{
  $result=factorial
($num);
  Echo "Factorial of
number".$num. " is
:".$result;}
  function
factorial($num) {
  if ($num==0 ||
$num==1) {
  return 1;
    } else {
  return ($num *
factorial($num-1));
} }
  ?> </html>
```

PHP Code

```
  importjava.io.BufferedReader
;
  importjava.io.IOException;
  importjava.io.InputStreamRea
der;
  public class factorial{
  public static void main
(String args[]) throws
NumberFormatException,IOExcept
ion{System.out.println("Enter
number to find its
factorial");
  BufferedReaderbr=new
BufferedReader(new
InputStreamReader(System.in));
  intnum=Integer.parseInt(br.r
eadLine());
  int result=factorial(num);
  if(num<0)
  {System.out.println("Factori
al of negative number not
possible"); }
  else
  {System.out.println("Factori
al of number "+num+ "is
:"+result);}}
  static int factorial(int
num)
  {if(num==0 || num==1)
  {   return 1;   }
  else
  {    return
(num*factorial(num-1));}}}
```

Java Code

Fig. 5 PHP and Java code

5 Conclusion

Symbol table was designed for the proper functioning of compiler by adding help and keeping record of the scope of variables and functions/methods. It keeps the updated value of each variable according to their scopes, add variables with the beginning of scope and then delete once scope is over. By this, it maintains memory and notify if any variable is used beyond scope. It provides the correct output. It not only helps in compilation process but also in running of code. Hence, it can be used in decompilation process too.

References

1. Appel, A.W., Palsberg, J.: Modern Compiler Implementation in Java. By: Cambridge University Press, Second Edition, ISBN: 052182060x, 501 pages (2002).
2. Singh, R., Sharma, V., Varshney, M.: Design and Implementation of Compiler. ISBN-978-81-224-2398-3(2008).
3. Terry, P.D.: Compilers and Compiler Generators an introduction with C++. Rhodes University (1996).
4. Afolorunso, A.A., Dr. Awodele, O., Prof. Obidairo, K.: Principles and techniques of Compilers. In: CIT 753 Internet concepts and Web Design, ISBN: 978-058-470-6, National Open University of Nigeria, Course Guide (2013).
5. Fischer, Charles, LeBlanc, R.: Crafting a Compiler. By: Pearson Education, ISBN: 978-81-317-0813-2 (2008).
6. Emmerik, M.J.V.: Static Single Assignment for Decompilation. PhD thesis, The University of Queensland (2007).
7. Brown, P.J.: Writing Interactive Compilers and Interpreters: A Modern Software Engineering Approach Using Java. By: Wiley Publishing, Inc., ISBN: 978-0-470-17707-5 (2009).
8. Chattopadhyay, S.: Compiler Design. By: PHI Learning Pvt. Ltd., ISBN-978-81-203-2725-2 (2005).
9. O'Donnell, M., Ortega, A.: The symbol table and block structure. Compilers, Autonomous University of Madrid, Madrid, Spain.
10. Fritzson, P., Kessler, C.: Compilers and interpreters, Compiler Construction, Symbol Tables. IDA, Linköpingsuniversitet (2011).
11. Siegfried, R.M.: Compiler Construction: The Symbol Table. Lecture 3, Adelphi University, New York (2003).
12. Teitelbaum, T.: Introduction to Compilers: Symbol Table. Spring, 978-1-935528, (2008).
13. Hsu, T.S.: Symbol Table. Academia Sinica, Twaiwan. http://www.iis.sinica.edu.tw/~tshsu/compiler2006/slides/slide5.pdf [18.5.16].
14. Mudawwar, M.: Symbol tables, hashing and hash tables-compiler design. American University, Cairo, Egypt. http://www.cse.aucegypt.edu/~rafea/csce447/slides/table.pdf [18.5.16].

Digital Watermarking Using Enhanced LSB Technique

Narander Kumar and Jaishree

Abstract Nowadays, there is a huge requirement of multimedia data security with an advent use of internet that is being more important to save the data confidentiality and save from various attacks. This paper represents the improvement of LSB (Least Significant Bit) watermarking mechanism by using two host images instead of one image to embed the watermark using some logical operations on the bits of the images. It provides better security and on the other and it preserves the originality of the host image. To compare the image quality we used some image quality parameters like PSNR, MSE, NAE, AD, MD, NCC, and SC.

Keywords Watermarking · LSB · JPEG · Security · Matlab

1 Introduction

Although the huge use of internet every data is available to every person easily, sometimes it is very beneficial for everyone but sometimes not. It has been difficult to protect the data which is on internet from tempering and from being misused by unauthorized person. To secure ones digital property from being copied or misused, various techniques are used. The on trend technique is digital watermarking; there are so many techniques being used for watermarking as DWT, DFT, DCT, etc. The watermarking can be done on various document like text, audio, and video. In this paper, we are working on one of the technique of watermarking that is LSB. The technique is implemented on an image. The traditional technique of watermarking was very simple and can be easily traced down, and the hidden data can easily be

N. Kumar (✉) · Jaishree
Babasaheb Bhimrao Ambedkar University (a Central University),
Lucknow 226025, India
e-mail: nk_iet@yahoo.co.in

Jaishree
e-mail: jaishreebansal99@gmail.com

© Springer Nature Singapore Pte Ltd. 2017
S.C. Satapathy et al. (eds.), *Proceedings of the 5th International Conference on Frontiers in Intelligent Computing: Theory and Applications*, Advances in Intelligent Systems and Computing 515, DOI 10.1007/978-981-10-3153-3_37

extracted so in order to improve the technique we add some encryption method to the technique.

Different parameters are calculated in order to judge the resultant watermark image. We are using MATLAB tool for this implementation, image processing toolbox is being used in order to get better results. Simple watermarking is done by the secret watermark which is to be inserted on a cover image.

2 Related Work

Many theories had been given by many researches which include various method of improving the watermarking techniques to provide more security. In [1], the DWT domain and chaotic system based medical image is being proposed to hide the patient details into the medical images as a watermark. In order to improve the security details of the patient, chaotic system is being used the division of the low frequency sub band is done into 3*3 nonoverlapping blocks. Spread spectrum technology is used in [2]; the strength is automatically adjusted according to the original image. This proposed method improves the anti-attack capabilities also the hidden natures of the image enhance the security of the image. Recovery of the watermark and to perform tamper detection using modified LSB is done in [3]. The results show that 100% accuracy can be achieved by detecting tamper and can recover up to 100%. Spatial LSB modified technique is used in [4]; it is also robust against moderate JPEG compression [5]. States that implement the tamper detection method by three LSBs technique the embedding of 12 bit watermark of each block of host image into the last 3 significant bit of each block. Preprocessing done by image segmentation comparison between RS method and SPA method is shown in [6]; the proposed method gives the better detection result. The method of stereo image compression is done by using block matching in [7], which gives the high compression rate. Also in order to increase the robustness of the watermarking technique, the two level of hamming code is used. In order to give the high copy right protection fuzzy watermarking system is used in [8]. Host image is embedded with a secret key and second phase is the extraction phase in which output of first decomposed by the Haar wavelet transform in [9]. The LSB schemes used but with the additional elements by adding secrete key and a hash function to the watermark. The technique mentioned in [10] is of embedding watermark in two unused bit in alpha channel in order to maintain the luminance and the chrominance parameter. Adaptive blind digital watermarking is discussed in [11], and the cover image is partitioned equally. Genetic Algorithm used for the selection of image block, which preserves the image quality. Jigsaw Puzzle solver (JPS) used for image reconstruction. The method uses in [12] is chaotic system with a three step procedure, location matrix is generated first then the embedding location is decided with the help of mapping image, technique also used for temper detection. By using two identical copies of watermark, then copies are decomposes in bit planes. The watermark embedded in a center-point-symmetric manner, in order to recover the watermark this method is introduced in [13].

From the extensive review of work, findings are these methods somehow affect the quality of the image; too much of manipulation with the data can be led to the hidden data loss. So to preserve the originality of the data, a simpler methodology which used less computation and easy to execute and provide better security to the data is proposed.

3 Proposed Mechanism

The LSB watermarking comes under the spatial domain of watermarking techniques; in this technique, the chosen pixel bits are replaced by the pixel bits of the embedding data which we can also say the hidden data or image.

If we take a still image, every pixel have three components that is red, green, and blue, so we allocate 3 bytes for each pixel then each color component have 8 bits. These are the following rules which are used to encode the hidden data in the host image.

The first step will be to find the least significant bit of both cover images one and two and then perform the following operation on them (Fig. 1 and Table 1).

Hiding process:

1. Find the pixel value of cover image 1 and 2.
2. Find the randomized LSB of each cover image.

```
For i=1: row
    j=round (1+rand*2);
    k=round (1+rand*2);
    If    ((original_image_r    (row,    j)    ==0)    &&
    (original_image1_r(row, k)==0))
    original_image_r (row, 1) =1;
    End
    If        ((original_image_b(row,        j)==0)        &&
    (original_image1_b(row, k)==0))
    original_image_b(row,1) =1;
    End
    If              ((original_image_g(row,              j)==0)
    &&(original_image1_g(row, k)==0))
    original_image_g (row,1) =1;

    End
```

3. Find the binary value of the image to be hidden.
4. Perform the operation with the help of rules between the LSB of each cover image.
5. Find the resultant value of LSB.
6. Perform same operations between the resultant LSB and the LSB of the watermark image.
7. After getting the value, replace that bit with the LSB of the watermark image.

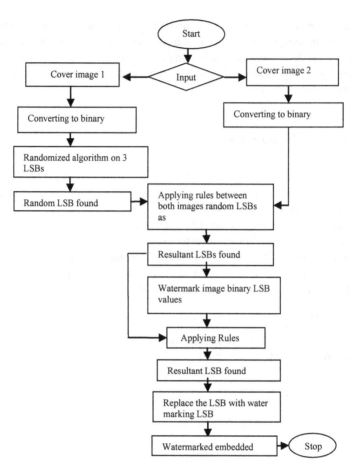

Fig. 1 Proposed process model

Table 1 This table shows the rules which are used to perform the operations between the LSBs of the two cover images

Least significant bit of C1	Least significant bit of C2	Resultant least significant bit to be used to replace with blue LSB bit
0	0	1
0	1	0
1	0	0
1	1	0

Extracting process:

1. Find the pixel value of watermarked image.
2. Find the value of secret bit LSB and LSB of each pixels of watermarked image.
3. Perform the operation with the help of rules between the value secret LSB, and
4. LSBs of each pixels of cover image.

```
For i=1: row
    j=round (1+rand*2);
    k=round (1+rand*2);
If    ((original_image1_r    (row,    j)    ==0)    &&
(original_image_r (row,k)==0))
        original_image1_r (row, 1) =1;
End
If    ((original_image1_b    (row,    j)    ==0)    &&
(original_image_b (row, k)==0))
        original_image1_b (row, 1) =1;
End
If        ((original_image1_g        (row,        j)==0)
&&(original_image_g(row,k)==0))
        original_image1_g(row,1) =1;
End
```

5. Find the resultant bits of hidden message.

The results are discussed in the below section, parameters are calculated of different jpeg images of MATLAB. The image dataset is taken from MATLAB image processing toolbox of dimensions 794×632. The results of the proposed method are also compared with the other watermarking techniques.

4 Results and Discussion

After performing the enhanced LSB method, we found the resultant watermark image. The following images which are being used as the cover image 1 and cover image 2 (Figs. 2, 3, 4 and 5).

The actual hidden data are in gray scale which needed to be converted into binary format in order to embedded it into cover image and also to perform binary operation with the bits, so the binary image first compared then the size of the cover image and hidden image and then tile up the hidden binary image according to the size of the cover image. This is done in case the hidden image is small. But to minimize the chances of data loss in presence of any attack like cropping, tiling up the hidden data is beneficial. After performing the enhanced LSB method on all the three components of image, we found the resultant images are (Figs. 6, 7, 8, and 9).

Fig. 2 Cover image one

Fig. 3 Cover image two

Fig. 4 The secret image

Fig. 5 Binary form of secret Image

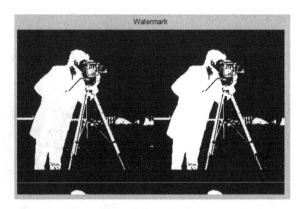

Fig. 6 *Blue* Channel watermarked image

Fig. 7 *Green* Channel watermarked image

After obtaining all the resultant images, now some of the parameters are calculated on both the original image and the final watermarked image and its component. The image and different color component quality can be judged by some parameters like PSNR, MSE, NCC, AD, SC, MD, and NAE. The value of PSNR and MSE are inversely proportional to each other. PSNR high means good quality of image and low PSNR depicts the bad quality image (Table 2).

Fig. 8 *Red* Channel
watermarked image

Fig. 9 The final
watermarked image

Table 2 The following table
shows the values of different
parameters of watermarked
image and original image

PSNR original	37.55
MSE original	12.36
PSNR watermarked	38.740 (PSNR Red)
	38.3622 (PSNR Green)
	39.75 (PSNR Blue)
MSE watermarked	8.75 h(MSE Red)
	9.5 (MSE Green)
	8.20 (MSE Blue)
Normalized Cross Correlation (NCC)	0.9999
Structural Content (SC)	0.9482
Normalized Absolute Error (NAE)	0.0448
Maximum Difference (MD)	16
Average Difference (AD)	-0.0802

Table 3 This technique is performed on various jpeg images and different parameters are observed as follows

Jpeg images	PSNR	NCC	AD	SC	MD	NAE
football	36.46	0.99	0.00	0.99	17	0.04
Ngc	41.27	0.99	−1.26	0.99	17	0.13
Pepper	36.51	0.99	0.01	0.99	16	0.03
streets	36.57	0.99	0.09	1.00	17	0.02

Table 4 Comparison between proposed method and other methods on the basis of PSNR value

Methods	PSNR values of jpeg images
Haar discrete wavelet [15]	33.58
Inter wavelet transform [16]	31.8
DCT, DW [17]	32.07
DCT, DWT, SVD [18]	35
Quantization [19]	34.84
Video watermarking [20]	36.75

$$PSNR = 10.\log_{10}\left(\frac{MAX_I^2}{MSE}\right)$$
$$= 20.\log_{10}\left(\frac{MAX_I}{\sqrt{MSE}}\right) \quad (1)$$
$$= 20.\log_{10}(MAX_I) - 10.\log_{10}(MSE)$$

$$MSE = \frac{1}{mn}\sum_{i=0}^{m-1}\sum_{j=0}^{n-1}[I(i,j) - K(i,j)]^2 \quad (2)$$

Now, by studying all the parameters that are Normalized Cross Correlation, Peak Signal-to-Noise Ratio, Structural Content, Average Difference, Maximum Difference, and Normalized Absolute Error, we can say the effect on the original image is less as compared to some of the other techniques [14]. We can easily see the minimal difference of values in original images and watermarked images (Tables 3, 4 and Figs. 10, 11, 12, 13).

Fig. 10 Technique applied
on football.jpg

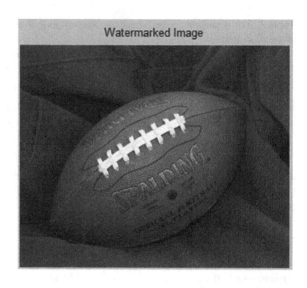

Fig. 11 Technique applied
on Ngc.jpg

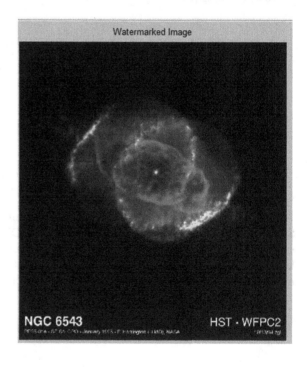

Fig. 12 Technique applied on Streets.jpg

Fig. 13 Technique applied on Pepper.jpg

5 Conclusion and Future Scope

In this paper, the new method is implemented using two cover images instead of one host image. The result shows that by providing better security it also takes less computation as being of simpler methodology. The technique puts less effect on the original image and preserving the originality, also concluded that the blue component is showing comparatively better values then other components, i.e., red or green, the less effected component we can say. The highest value of PSNR is obtained and lowest value of MSE is found in blue component. We can also work on the image restoration after the attacks are performed in the network in future in this technique.

References

1. M. Moniruzzaman, M. A. K. Hawlader, M. F. Hossain.: Wavelet based watermarking approach of hiding patient information in medical image for medical image authentication: 17th International Conference on Computer and Information Technology (ICCIT), 2014 DOI:10.1109/ICCITechn.2014.7073133.
2. Qing Liu, Jun Ying.: Grayscale image digital watermarking technology based on wavelet analysis. In: IEEE Symposium on Electrical & Electronics Engineering (EEESYM), DOI:10. 1109/EEESym.2012.6258734 (2012).
3. Agung, B. W. R., Adiwijaya, F.P. Permana,.: Medical image watermarking with tamper detection and recovery using reversible watermarking with LSB modification and run length encoding (RLE) compression. In: IEEE International Conference on Communication, Networks and Satellite (ComNetSat), pp: 167–171, DOI:10.1109/ComNetSat.2012.6380799 (2012).
4. Tao Chen, Hongtao Lu,.: Robust spatial LSB watermarking of color images against JPEG compression. In: IEEE Fifth International Conference on Advanced Computational Intelligence (ICACI), pp. 872–875, DOI:10.1109/ICACI.2012.6463294 (2012).
5. S. Dadkhah, A.A. Manaf, S. Sadeghi.: Efficient Two Level Image Tamper Detection Using Three LSBWatermarking. In: Fourth International Conference on Computational Intelligence and Communication Networks (CICN), pp: 719–723, DOI:10.1109/CICN.2012.108 (2012).
6. Hu Lingna, Jiang Lingge.: Blind Detection of LSB Watermarking at Low Embedding Rate in GrayscaleImages. In: Second International Conference on Communications and Networking in China, CHINACOM pp: 413–416, DOI:10.1109/CHINACOM.2007.4469415 (2007).
7. El Kerek B, El Baba H, El Hassan M.: A New Technique to Multiplex Stereo Images- LSB Watermarking and Hamming Code. In: 1st International Conference on Artificial Intelligence, Modelling and Simulation (AIMS), pp: 267–271, DOI:10.1109/AIMS.2013.49 (2013).
8. Umaamaheshvari, A., Thanushkodi, K.: Robust image watermarking based on block based error correction code. In: International Conference on Current Trends in Engineering and Technology (ICCTET) 2013.
9. Bandyopadhyay, P., Das S., Paul S., Chaudhuri A., Banerjee M.: A Dynamic Watermarking Scheme for Color Image Authentication. In: International Conference on Advances in Recent Technologies in Communication and Computing, ARTCom pp: 314–318, DOI:10. 1109/ARTCom.2009.94 (2009).
10. Bandyopadhyay P., Das S., Chaudhuri A., Banerjee M.: A new invisible color image watermarking framework through alpha channel. In: International Conference on Advances in Engineering, Science and Management (ICAESM), pp: 302–308 (2012).
11. Anwar M.J., Ishtiaq M., Iqbal M.A., Jaffar, M.A.: Block-based digital image watermarking using Genetic Algorithm. In: 6th International Conference on Emerging Technologies (ICET), pp: 204–209, DOI:10.1109/ICET.2010.5638488 (2010).
12. Liu Pei-pei, Zhu Zhong-liang, Hong-Xia Wang, Yan Tian-yun.: A Novel Image Fragile Watermarking Algorithm Based on Chaotic MapImage and Signal Processing. In: CISP Congress on Image and Signal Processing, CISP'08, Volume: 5 pp: 631–634, DOI:10.1109/CISP.2008.636 (2008).

13. Subudhi Asit Kumar, Kabi Krishna, Subudhi Abhilipsa, Patra Himanshu, Panda Jyoti Bikash.: An enhanced technique for color image water marking using HAAR transform. In: International Conference on Electrical, Electronics, Signals, Communication and Optimization (EESCO), pp: 1–5, DOI:10.1109/EESCO.2015.7253844 (2015).

14. A Novel Full-Reference Image Quality Index for Color Images," Proc. of the (Springer) International Conference on Information Systems Design and Intelligent Applications (INDIA 2012), pp. 245–253 (2012).

15. Lai B, Chang L (2006) Adaptive data hiding for images based on haar discrete wavelet transform. In: Lecture Notes in Computer Science, Springer-verlag Berlin Heidelberg, vol 4319, pp 1085–1093.

16. El Safy RO, Zayed HH, El Dessouki A.: An adaptive steganography technique based on integer wavelet transform. In: International Conference on networking and media convergence. Pp-111–117 (2009).

17. Mei Jiansheng1, Li Sukang1 and Tan Xiaomei2.: A Digital Watermarking Algorithm Based On DCT and DWT. In: International Symposium on Web Information Systems and Applications, pp. 104–107, ISBN 978-952-5726-00-8 (2009).

18. K.Chaitanya1,Dr E. Srinivasa Reddy2,Dr K. Gangadhara Rao3.: Digital Color Image Watermarking In RGB Planes Using DWT-DCT-SVD Coefficients. In: International Journal of Computer Science and Information Technologies, ISSN: 0975-9646 (2014).

19. Chin-Chen Chang a,*, Tung-Shou Chen b,1, Lou-Zo Chung a.: A stenographic method based upon JPEG and quantization table modification. In: Elsevier Information Sciences 141 123–138 (2002).

20. Guochuan Shi1, Guanyu Chen1,Binhao Shi2, Jiangwei Li1 and Kai Shu1.: Research and Implementation of an Integrity Video Watermarking Authentication Algorithm. In: International Journal of Security and Its Applications Vol.8, No.4 pp. 239–246, (2014).

Fuzzy-Based Adaptive IMC-PI Controller for Real-Time Application on a Level Control Loop

Ujjwal Manikya Nath, Chanchal Dey and Rajani K. Mudi

Abstract Internal Model Control (IMC) technique is one of the well accepted model-based controller designing methodologies which is widely accepted in process industries due to their simplicity and ease of tuning. For controlling non-linear processes IMC controllers are designed based on the linear approximation of nonlinear models. As a result IMC controllers sometimes fail to provide satisfactory performance under model uncertainty and large load variations with its fixed settings. Here we propose an adaptive IMC-PI controller for a level control process where the IMC tuning parameter, i.e., the close-loop time constant (λ) is varied based on a set of predefined fuzzy rules depending on the process operating conditions in terms of process error (e) and change of error (Δe). Two sets of rule bases are used consisting of 25 and 9 rules for online fuzzy tuning of the IMC-PI controller. Widely different choice of the rule bases defined on two distinct fuzzy partitions justify the effectiveness as well as general applicability of the proposed scheme.

Keywords Model identification · IMC-PI controller · Fuzzy tuner · Level control process

U.M. Nath (✉) · R.K. Mudi
Department of Instrumentation and Electronics Engineering,
Jadavpur University, Kolkata 700098, India
e-mail: um.nath29@gmail.com

R.K. Mudi
e-mail: rkmudi@yahoo.com

C. Dey
Department of Applied Physics, Instrumentation and Control Engineering,
University of Calcutta, 92 A.P.C. Road, Kolkata 700009, India
e-mail: chanchaldey@yahoo.co.in

© Springer Nature Singapore Pte Ltd. 2017
S.C. Satapathy et al. (eds.), *Proceedings of the 5th International Conference on Frontiers in Intelligent Computing: Theory and Applications*, Advances in Intelligent Systems and Computing 515, DOI 10.1007/978-981-10-3153-3_38

1   Introduction

Model-based controllers are widely accepted in industrial applications [1] due to their simple design and straight forward tuning methodology. Internal Model Control (IMC) [2] technique is one of the well-known model-based controller designing methodologies. The most important feature of IMC controller is that it has a single tuning parameter. Early works on IMC-based PID controller (IMC-PID) is reported by Rivera et al. [3] for first order process and thereafter a good number of research findings [4–7] are published. By choosing an appropriate value of the sole tuning parameter (i.e., close-loop time constant λ) of IMC controller desired process response can be obtained. To achieve improved load rejection behavior Skogestad [5] proposed a modified IMC method for processes with large time constant. Further works toward enhanced load regulation is reported by Shamsuzzoha and Lee [7]. From the basic understanding of IMC design technique it is found that a larger value of λ gives good set point tracking with acceptable overshoot while an improved disturbance rejection can be achieved for a smaller value of λ. Hence, it is evident that IMC controllers with a fixed value of λ (large or small) fail to provide improved responses during transient and steady state phases simultaneously. Moreover, an IMC controller with static value of λ has linear nature and hence fails to perform satisfactorily for processes with nonlinear behavior. To overcome this limitation Datta and Ochoa [8] first proposed an online adaptation scheme for IMC controller. Further works on adaptive IMC design are reported for specific class of processes in [9, 10].

Here our main objective is to design an adaptive IMC-PI controller for maintaining water level of an overhead tank at the desired position during set point change and load disturbances where water is supplied from a reservoir by a pump with nonlinear flow characteristics. In the proposed adaptation scheme the sole tuning parameter λ get modified continuously depending on the current process states (i.e., error (e) and change of error (Δe)) based on predefined fuzzy rule sets [11]. Two rule bases are designed with 25 and 9 rules to provide the necessary adaptation. Performance of the proposed scheme is tested on a laboratory scale tank level process manufactured by Feedback Instrument Ltd. [12]. From the real time performance analysis it is found that noticeable improvement can be obtained during both the set point tracking and load rejection phases. In addition, it is found that the proposed adaptive IMC-PI controller with larger rule base (25 rules) offers superior process response compared to smaller rule base with 9 rules only.

2   Model Identification

Block schematic of a typical level control process is shown in Fig. 1. Dynamic relationship of the tank level [13] is given by:

Fig. 1 Schematic diagram of
a typical level control loop

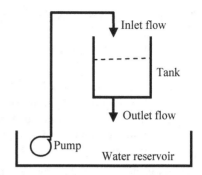

$$\frac{dh_1(t)}{dt} = -\frac{a_1}{A}\sqrt{2gh_1(t)} + \eta_1 u_1(t) \tag{1}$$

After linearization we get the transfer function (TF) as follows:

$$\frac{\Delta H_1(s)}{\Delta U_1(s)} = \frac{\eta_1}{s + \left(\frac{a_1}{A}\right)^2\left(\frac{g}{\eta_1 u_{10}}\right)} \tag{2}$$

In Eq. (2) $\Delta H_1(s)$ and $\Delta U_1(s)$ are the incremental liquid level of the tank and the corresponding change in control action. A is the area of the tank, a_1 is area of the tank outflow path, η_1 is the pumping rate, and g is the gravitational acceleration. After substituting the given values [12] in Eq. (2), we get the mathematical TF of the tank. This empirical TF is further validated against the model obtained through Process Reaction Curve (PRC) method [14]. We found that both the models are quite close as given in Table 1.

Table 1 Experimental response and corresponding empirical model of the tank

PRC response of the tank
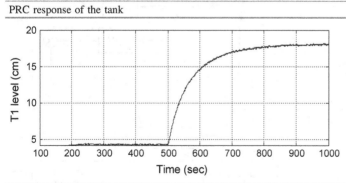

Experimental model	Empirical model
$\dfrac{12.7}{(58s + 1)}$	$\dfrac{12.12}{(50.50s + 1)}$

3 Proposed Adaptive Controller Design

3.1 IMC-PI Controller

Basic structure of IMC control [15] technique is shown in Fig. 2. Simplifying the loop (enclosed by dotted line in Fig. 2) we get the PI structure of the IMC controller. Detailed steps are described in [13].

After simplification the tuning parameters of IMC-PI controller [6] are given by: proportional gain $k_c = \frac{\tau_p}{k_p \lambda}$, and integral time $\tau_I = \tau_p$, where k_p is the open-loop process gain and τ_p is the process time constant.

3.2 Proposed Fuzzy Rule-Based Adaptive IMC-PI Controller

Block diagram of the proposed adaptive IMC-PI controller is shown in Fig. 3. According to the adaptation strategy the only tuning parameter, i.e., the close-loop time constant (λ) get modified continuously based on two fuzzy rule sets depending on the process operating conditions in terms of current process error (e) and change of error (Δe). In the proposed fuzzy-based tuner, rules are formed using e and Δe as the antecedent part of the rules while λ represents the consequent part of the rule [16]. Center of gravity method [17] is used for defuzzification purpose. Depending on the instantaneous value of λ, IMC-PI controller calculates the new PI tuning parameters.

Fig. 2 Basic block diagram of IMC structure

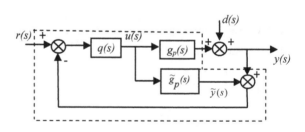

Fig. 3 Block diagram of fuzzy adaptive IMC-PI controller

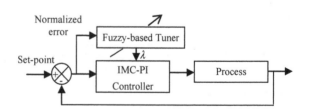

Fig. 4 Triangular MFs for **a** inputs e and Δe with 5 fuzzy sets and, **b** output λ with 3 fuzzy sets

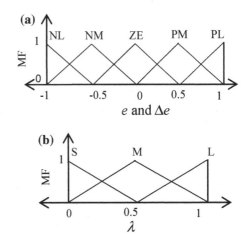

3.3 Membership Function

Membership Functions (MFs) for fuzzy tuner inputs e and Δe are defined by 5 fuzzy sets (NL, NM, ZE, PM, and PL). All the MFs are symmetrical triangles having 50% overlap and defined over the common normalized domain $[-1, 1]$ [18] as shown in Fig. 4a. Figure 4b shows MFs for fuzzy tuner output, i.e., λ. Here, three triangular MFs (S, M, and L) are used with equal base and 50% overlap defined over the domain $[0, 1]$. Rule bases of the reported fuzzy tuner consisting of 25 and 9 rules are given in the following section.

3.4 Rule Base

Rules of the proposed fuzzy tuner has the form "IF e is E AND Δe is ΔE THEN λ is $\Delta \lambda$." Here we used Mamdani inference technique [15] for calculating the rule output. Proposed rule bases and the corresponding nature of surfaces with 25 rules and 9 rules are shown in Fig. 5a and b, respectively.

4 Real Time Experimentation

Our experimentation setup is a laboratory scale quad-coupled tank process manufactured by Feedback Instrument Ltd. [12] as shown in Fig. 6. For our experimental purpose we have chosen only the first tank (Tank 1 in the Fig. 6) to maintain the liquid level at the desired position irrespective of set point change and load variation. Advantech PCI 1711 data acquisition interface card is used to interface the

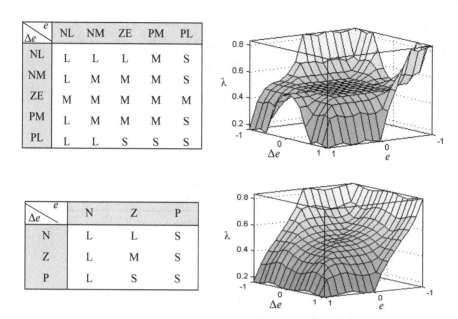

e $\backslash$ Δe	NL	NM	ZE	PM	PL
NL	L	L	L	M	S
NM	L	M	M	M	S
ZE	M	M	M	M	M
PM	L	M	M	M	S
PL	L	L	S	S	S

e $\backslash$ Δe	N	Z	P
N	L	L	S
Z	L	M	S
P	L	S	S

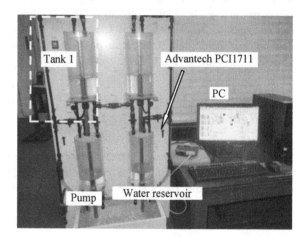

Fig. 5 Rule bases and control surfaces for **a** 25 rules and **b** 9 rules

Fig. 6 Level control of Tank 1

experimental rig and the PC. We design our controller in PC with the help of MATLAB/SIMULINK. To maintain the liquid level at desired value the control action (0–5 V) goes to the submerged pump which regulates the flow rate.

At the commencement of our experiment we have filled the tank manually nearer to the desired value. Responses under set point change are shown in Fig. 7. In Fig. 8 three set point step changes are given (two positive steps and one negative step), and once the liquid level reaches the desired value load disturbance is also applied.

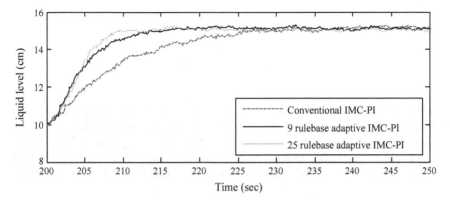

Fig. 7 Responses of conventional IMC-PI and fuzzy-based adaptive IMC-PI controllers under set point change

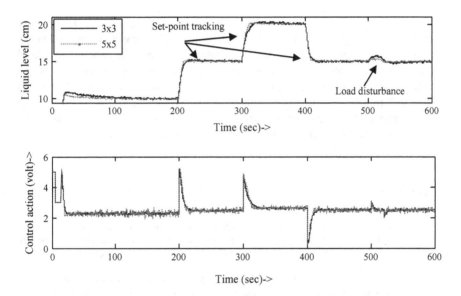

Fig. 8 Performance analysis of the proposed fuzzy-based adaptive IMC-PI controllers with 25 and 9 rules under set point change and load variation

5 Result

Responses of fuzzy rule-based adaptive IMC-PI controllers with 25 and 9 rules are shown in Fig. 8 and the related performance indices are depicted in Table 2. The responses for 25 and 9 rules are plotted in dotted red line and solid black line, respectively. Integral Absolute Error (IAE) and Integral Time Absolute Error (ITAE) [14] are calculated for quantitative performance analysis. From Table 2 it is

Table 2 Performance analysis

Controller	IAE	ITAE
Conventional IMC-PI	40.70	14140
9 rule-based adaptive IMC-PI	23.56	7168
25 rule-based adaptive IMC-PI	21.50	6340

found that the values of IAE and ITAE are lesser in case of our proposed controller compared to conventional IMC-PI. In addition, adaptation mechanism with 25 rules show improved performance compared to 9 rules. But, it is to note that our proposed scheme is capable to provide performance improvement even with 9 rules (more than 60% lesser in size than 25 rules) which justifies its general applicability.

6 Conclusion

IMC controllers are useful for industrial applications for their simplicity and easy tuning methodology. Conventional IMC-PI controller is linear in nature and hence fails to perform satisfactorily for nonlinear processes. Here, the experimental level control loop is not linear in nature due to its nonlinear pumping characteristics. An adaptive fuzzy-based online tuner is suggested in the proposed work with the help of two different fuzzy rule bases defined on two distinct fuzzy partitions. In both cases a considerable amount of performance improvement can be observed from the experimental results under set point change and load disturbances.

References

1. Garcia, G.E., Morari, M.: Internal model control- 1. A unifying review and some new results, Industrial Engineering Chemical Process Design Development, vol. 21, pp. 308–323, (1982).
2. Chien, I.L.: IMC–PID Controller Design-An Extension, IFAC Proceeding Series, 6, pp. 147–152, (1988).
3. Rivera, D.E., Skogested, S., Morari, M.: Internal Model Control for PID Controller Design, Industrial Engineering Chemical Process Design and Development, vol. 25, pp. 252–265, (1986).
4. Morari, M., Zafiriou, E.: Robust Process Control, New Jersy: Prentice—Hall, (1989).
5. Skogestad, S.: Simple analytic rules for model reduction and PID controller tuning, Journal of Process Control, vol. 13, pp. 291–309, (2003).
6. Nath, U.M., Datta, S., Dey, C.: Centralized auto-tuned IMC-PI controllers for a real time coupled tank process, International Journal of Science, Technology and Management, 4(1), pp. 1094–1102, (2015).
7. Shamsuzzoha, M., Lee, M.: IMC-PID controller design for imrproved disturbance rejection of time delayed processes, Ind. Chem. Eng. Res., 46(7), pp. 2077–2091, (2007).
8. Datta, A., Ochoa, J.: Adaptive Internal Model Control: Design Stability and Analysis, Automtica, vol. 32, pp. 261–266, (1996).
9. Rupp, D., Guzzella, L.: Adaptive internal model control with application to fueling control, Control Engineering Practice, 18(8), pp. 873–881, (2010).

10. Silva, G.J., Datta, A.: Adaptive internal model control: The discrete-time case, International Journal of Adaptive Control and Signal Processing. 15(1), pp. 15–36, (2001).
11. Lee, C.C.: Fuzzy Logic in Control Systems: Fuzzy Logic Controller, Part II, IEEE Transactions On Systems, Man. And Cybernetics, 20(2), pp. 419–435, (1990).
12. Feedback Instruments Ltd.,East Sussex, UK.
13. Nath, U.M., Datta, S., Dey, C.: Centralized Auto-tuned IMC-PI Controllers for Industrial Coupled Tank Process with Stability Analysis, 2nd International Conference on Recent Trends in Information Systems, IEEE, pp. 296–301, (2015).
14. Seborg, D.E., Edgar, T. F., Melichamp, D. A.: Process Dynamic and Control, Wiley, 2nd edn., (2004).
15. Datta, S., Nath, U.M., Dey, C.: Design and Implementation of Decentralized IMC-PI Controllers for Real Time Coupled Tank Process, Michael Faraday IET International Summit: MFIIS, pp.93–98 (2015).
16. Mudi, R.K., Pal, N.R.: A Robust Self-tuning Scheme for PI Type Fuzzy Controller, IEEE Transactions on Fuzzy Systems, 7(1), pp. 2-16, (1999).
17. Dey, C., Mudi, R.K., Mitra, P.: A self-tuning fuzzy PID controller with real-time implementation on a position control system, Emerging Applications of Information Technology (EAIT), IEEE, pp.32–35 (2012).
18. Dey, C., Mudi, R.K.: Design of a PI-Type Fuzzy Controller with on-line Membership Function Tuning, Proc. 12th Int. Conf. on Neural Information Processing – ICONIP, (2005).

Face Recognition Using PCA and Minimum Distance Classifier

Shalmoly Mondal and Soumen Bag

Abstract Face is the most easily identifiable characteristic of a person. Variations in facial expressions can be easily recognized by humans, while it is quite difficult for machines to recognize faces portraying varying facial expressions, pose, and illumination conditions efficiently. Face recognition works as a combination of feature extraction and classification. The selection of a combination of feature extraction technique and classifier to obtain maximum accuracy rate is a challenging task. This paper presents a unique combination of feature extraction technique and classifier that yields a satisfactory and more or less same accuracy rate when tested on more than one standard database. In this combination, features are extracted using principle coponent analysis (PCA). These extracted features are then fed to a minimum distance classification system. The proposed combination is tested on ORL and YALE datasets with an accuracy rate of 95.63% and 93.33%, respectively, considering variations in facial expressions, poses as well as illumination conditions.

Keywords Eigenface · Face datasets · Minimum distance classifier · Face recognition · Principle component analysis

1 Introduction

Face recognition system is an application of biometric technology, which is equipped for distinguishing or confirming a man from a picture or a video outline from a video source. Face recognition is one of the most successful applications of image analysis and has received significant attention, during the last several years. A face recognition system is comprised of two parts: feature extraction and recognition. There are several factors that affect the face recognition process like illumination conditions,

S. Mondal (✉) · S. Bag
Department of Computer Science and Engineering, ISM, Dhanbad, Dhanbad, India
e-mail: shalmolymondal@gmail.com

S. Bag
e-mail: bagsoumen@gmail.com

© Springer Nature Singapore Pte Ltd. 2017
S.C. Satapathy et al. (eds.), *Proceedings of the 5th International Conference on Frontiers in Intelligent Computing: Theory and Applications*, Advances in Intelligent Systems and Computing 515, DOI 10.1007/978-981-10-3153-3_39

pose, and various facial expressions. Various methods for face recognition are categorized as holistic methods, feature-based methods, and hybrid methods [6]. In the holistic method, the face recognition system takes the whole face as an input. One of the most widely used representations of the face region is eigenfaces, which are based on principal component analysis (PCA) [2]. In the feature-based methods, facial features such as eyes, ears, and nose are extracted and their locations are given as input to a classifier. The hybrid method makes use of both local features as well as the whole face as input to recognize a face.

Turk and Pentland [6] have developed an eigenface based face recognition method, which uses PCA to decompose face images into a small set of characteristic feature images called eigenfaces. These eigenfaces are used as feature vectors for the purpose of face recognition by comparing the features of the test face with those of known individuals.

Zhao et al. [8] have given a detailed survey of face recognition from still images as well as from a video frame. Later et al. [5] have proposed a way of recognizing faces using the concept of two thresholds (acceptance and rejection) in order to increase the recognition rate and decrease the rejection rate of the existing eigenface method.

Kukreja and Gupta [3] have proposed a PCA with KNN classifier based method for face recognition. Experiment is done on ORL database [9] and YALE database [10]. PCA and KNN when tested on the ORL database has given an accuracy rate of 92% whereas the same combination when tested on the YALE database has given a lower accuracy rate of 81.33%.

Latha et al. [4] have proposed a way to deal with ways to detect frontal view of faces. The dimensionality of face image is reduced by PCA and the recognition procedure is carried out by the back propagation neural network (BPNN). This neural network-based face recognition approach has better performance of more than 90% acceptance ratio.

Face recognition is basically done in two steps: feature extraction followed by classification [1]. The selection of a combination of feature extraction technique and a classifier to obtain maximum accuracy rate is a challenging task. Many such combinations have been obtained and tested on various face datasets. Some of these combinations have given very high accuracy on one database and low accuracy when the experiment with the same combination is performed on some other database. In our proposed method, we have tried to find out a combination of a feature extraction technique and a classifier that will yield a satisfactory and more or less same accuracy rate when tested on more than one databases.

This paper is organized as follows. Section 2 describes the proposed methodology of face recognition. The experimental results on two independent databases are shown in Sect. 3. We conclude this paper with some remarks on the proposed method in Sect. 4.

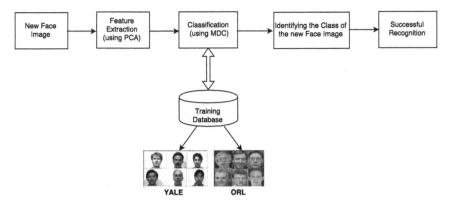

Fig. 1 System architecture of the proposed method

2 Proposed Method

Our objective is to recognize a person with various facial expressions, pose, and illumination conditions using a statistical pattern recognition approach. Our proposed system aims to implement face recognition with the combination of PCA as a feature extraction technique and MDC as the recognition method. We then compare the performance on two independent face datasets named YALE [10] and ORL databases [9]. Figure 1 shows the system architecture of the proposed methodology for face recognition.

2.1 Computing Eigenfaces Using PCA

Eigenface approach utilizes the idea of PCA and decomposes face images into a small set of feature images called eigenfaces. These eigenfaces are used for mapping images from image space to face space. A new face when obtained is also projected to a low dimensional face space. Recognition is performed by computing the minimum distance between the projection of the new image in the face space and the projections of known faces in the training database. The steps of computing eigenfaces using PCA are as follows:

1. We create a training set of N face images and represent each image I_i as a vector Γ_i. A 2-D facial image can be represented as 1-D vector by concatenating each row (or column) into a long vector. The training set now becomes as $\Gamma_1, \Gamma_2, \ldots, \Gamma_M$, where M is the number of images.
2. After forming column vectors, we calculate the mean face ψ (Fig. 2a) from the training set using Eq. 1.

$$\psi = \frac{1}{M} \sum_{i}^{M} \Gamma_i \qquad (1)$$

3. We normalize the training set (Fig. 2b) by subtracting the mean face from all vectors corresponding to the original faces using Eq. 2.

$$\Phi_i = \Gamma_i - \psi \qquad (2)$$

where Φ_i is the ith normalized training image.

4. A covariance matrix is then computed in order to extract a limited no of eigenvectors corresponding to largest eigenvalues. Covariance Matrix is given as

$$Cov = \frac{1}{M} \sum_{n=1}^{M} \phi_n \phi_n^T = AA^T \qquad (3)$$

where $A = [\phi_1, \phi_2, \ldots, \phi_M]$

AA^T is a very large matrix of dimension $N^2 \times N^2$. So the computation of eigenvectors of this matrix is very difficult. To avoid this computational complexity, we instead form a covariance matrix $A^T A$ of reduced dimensionality which is a $M \times M$ matrix as shown in Eq. 4.

$$Cov = A^T A \qquad (4)$$

5. Covariance matrix which was of size $N^2 \times N^2$ is now of size $M \times M$, where $M \ll N^2$. We now compute 'M' eigenvalues and 'M' eigenvectors of this covariance matrix to form the eigenspace using Eq. 5. Eigen decomposition of covariance matrix C is performed to determine the eigenvectors (eigenfaces) u_i and the corresponding eigenvalues λ_i. The eigenfaces (Fig. 2c) must be normalized so that they are unit vectors, i.e., they are of length one.

$$u_i = Av_i. \qquad (5)$$

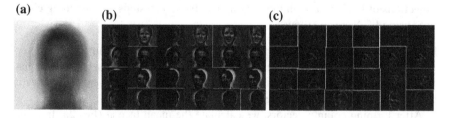

Fig. 2 **a** Mean face; **b** Normalized images of the training set; **c** Eigenfaces of the training set

6. We keep only 'k' eigenvectors (corresponding to the k largest eigenvalues). Number of eigenfaces is always less than or equal to the number of original images, i.e., $k \ll M$. The first few eigenfaces shows the most dominant features of training set of images. Among M eigenvectors (eigenfaces) the one with the highest eigenvalues are chosen and rest are discarded. After 'k' eigenfaces are determined, the training phase of the algorithm is completed. Therefore, the training set of 'M' images is now represented by only 'k' eigenfaces.

7. In order to carry out face recognition, we now find out the training images projection by mapping the training images from image space to face space. Mapping is done by the eigenfaces obtained in step 6 using Eq. 6.

$$\Omega_i = u^T \Phi_i. \tag{6}$$

where $i = 1$ to M, $u = [u_1, u_2, u_3, \ldots, u_k], k \ll M, \Omega_i = $ face space training images. These 'k' eigenfaces can safely represent the whole training set because they depict the major features that make up the dataset.

2.2 Face Classification Using Minimum Distance Classifier

The following steps demonstrates the process of face recognition.

1. Normalization of test image is done by Eq. 7.

$$\phi_{test} = \Gamma_{test} - \psi \tag{7}$$

where Γ_{test} represents the vector form of our test image, ϕ_{test} is the normalized test image, and ψ is the mean of the training set.

2. Mapping test image into face space by Eq. 8.

$$\Omega_{test} = u^T \Phi_{test} \tag{8}$$

where $u = [u_1, u_2, u_3, \ldots, u_k,]$ and $\Omega_{test} = $ mapped test image in the face space.

3. Finding the difference between test image projection and the projections of the training set.

$$Distance = \Omega_{test} - \Omega_i \tag{9}$$

where, Ω_{test} is the test image projection and Ω_i is the projection of images of the training set.

4. Computing the minimum value among all the distance computed in Step 3 using the Eq. 10.

$$D_{min} = Min(Distance) \tag{10}$$

where $D_{min} = $ minimum value among all distances

5. Recognizing known or unknown faces.
   If D_{min} is less than or equal to the threshold (Θ), then it is a known image; hence, display the class which has minimum difference value. Otherwise it is an unknown image.

$$IfD_{\min} <= \Theta, \qquad (11)$$

then Ω_{test} belongs to the database of images.

3 Experimental Analysis

3.1 Face Databases

We have carried out our experiment on two independent face databases, the YALE Database [10] and the ORL Database [9]. The YALE Database contains 165 images of 15 individuals (each person has 11 different images) under various facial expressions, lighting conditions, and images with and without glasses. All images are in grayscale with a resolution of 320×243 pixels. The ORL Database contains 10 different images each of 40 distinct subjects. The facial expressions (open/closed eyes, smiling/non smiling) as well as the facial details (glasses/no glasses) are varied. The images are taken with a tolerance of some tilting and also a rotation of about $20°$. All images are in grayscale and normalized to a resolution of 92×112 pixels. All the programs are written using Matlab R2015a.

3.2 Recognition Results

The experiment is first performed on the YALE database by considering 150 images. The database has been divided into training and test sets. Out of 10 images of a person, we have used six images for training and the rest four images for testing. Therefore, the training set consists of 90 images and the test set consists of 60 images. Figure 3a, b show the set of images of a person from the YALE database that has been used for training and testing, respectively. The features of the database are extracted using PCA. The features are classified using MDC and the classification accuracy is 93.33% for test images of the YALE Database.

(a)   **(b)**

Fig. 3 **a** Sample images from YALE training database; **b** sample images from YALE test database

(a)   **(b)**

Fig. 4 **a** Sample images from ORL training database; **b** Sample images from ORL test database

The experiment is again performed on the ORL database. The experiments are performed with six training images and four test images per person. Figure 4a, b show the set of images of a person that has been used for training and testing, respectively. There are no overlaps between the training and test sets. Since the recognition performance is affected by the selection of the training images, the reported results are obtained by randomly selecting six images per subject, out of the total images. The overall classification accuracy is 95.63% for test images of the ORL database. The proposed method is computationally efficient in terms of time complexity.

3.3   Comparison with Other Methods

We have done a comparative study on the basis of the rate of recognition accuracy in between our proposed method and other existing methods. We have observed that the performance of these combinations highly depends on the experimental databases. For example, the combination of PCA and KNN [3] gives an accuracy rate of 92% for ORL database, whereas the same combination performs less significantly (81.33%) for the YALE database. Our main focus was to select a unique combination of feature extraction technique and classifier, which can perform similarly well for more than one databases. In our proposed method, we have chosen the combination of PCA and MDC for face recognition. We have shown that this proposed combination gives an accuracy rate of about 93.33% and 95.63% for YALE and ORL databases, respectively. So, we can conclude that our proposed method can handle different datasets in a similar way. Table 1 shows the performance measurement among other methods. It shows that our method has achieved the highest recognition rate compared to the other methods. Comparative results obtained by testing PCA and MDC on both YALE and ORL databases are shown in Fig. 5.

Table 1   A comparative study of experimental results of the two databases

Database	Method	Feature extraction	Classification	Recognition rate (%)
ORL [9]	Kukreja and Gupta [3]	PCA	KNN	92
	Yang and Zhang [7]	ICA	MDC	85
	Proposed method	**PCA**	**MDC**	**95.63**
YALE [10]	Kukreja and Gupta [3]	PCA	KNN	81.33
	Yang and Zhang [7]	ICA	MDC	71.52
	Proposed Method	**PCA**	**MDC**	**93.33**

Fig. 5 Comparison of the accuracy rate of our proposed system on two databases using the combination of PCA and MDC

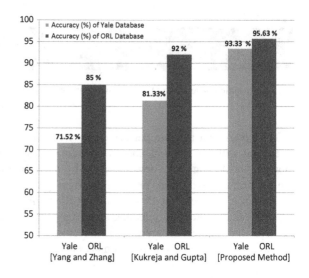

4 Conclusion

This paper presents a unique combination of feature extraction technique and classification method that perform well on more than one standard face datasets. We have performed feature extraction and classification using PCA and MDC, respectively. We have also successfully handled the cases where the input face is an unknown face, i.e., the face of a person is not present in the database. The experiment is successfully performed on two databases named YALE and ORL databases. The experimental results show that our selected combination gives more or less same recognition accuracy for both the datasets.

References

1. Bag, S., Sanyal, G.: An efficient face recognition approach using PCA and minimum distance classifier. In: International Conference on Image Information Processing, pp. 1–6 (2011)
2. Bouzalmat, A., Kharroubi, J., Zarghili, A.: Comparative Study of PCA, ICA, LDA using SVM Classifier. Journal of Emerging Technologies in Web Intelligence. 2, 64–68 (2014)
3. Kukreja, S., Gupta, R.: Comparative study of different face recognition techniques. In: International conference on Computational Intelligence and Communication Systems, pp. 271–273 (2011)
4. Latha, P., Ganesan, L., Annadurai, S.: Face Recognition using Neural Networks. An International Journal on Signal Processing. 3, 155–157 (2000)
5. Ma, Y., Li, S.: The modified eigenface method using two thresholds. International Journal of Signal Processing. 2, 236–239 (2006)
6. Turk, M., Pentland, A.: Eigenfaces for recognition. Journal of Cognitive Neuroscience. 3, 71–86 (1991)

7. Yang, J., Zhang, D.: Two-Dimensional PCA: A New Approach to Appearance-Based Face Representation and Recognition. IEEE Transactions on Pattern Analysis and Machine Intelligence. 26, 131–137 (2004)
8. Zhao, W., Chellappa, R., Phillips, P.J, Rosenfeld, A.: Face recognition: A literature survey. ACM Computing Surveys. 35, 401–458 (2003)
9. http://www.cl.cam.ac.uk/Research/DTG/attarchive:pub/data/att_faces.tar.Z.
10. http://vision.ucsd.edu/datasets/yale_face_dataset_original/yalefaces.zip.

A Soft Computing Approach for Modeling of Nonlinear Dynamical Systems

Rajesh Kumar, Smriti Srivastava and J.R.P. Gupta

Abstract A procedure based on the use of radial basis function network (RBFN) is presented for black box modeling of nonlinear dynamical systems. The generalization ability of RBFN is invoked to approximate the mathematical model of the given unknown nonlinear plant. This approximate model will then be used to predict the output of the plant at any given time instant. The parameters associated with RBFN are updated using the recursive equations obtained through the gradient-descent principle. The other benefit of using gradient descent principle is that it exhibits the clustering effect while adjusting the radial centers of RBFN. Real-time data of two benchmark problems: Box-Jenkins gas furnace data and Chemical process (polymer production), were used to show the application of RBFN for modeling purpose. Simulation results show that RBFN is well suited as a modeling tool for capturing the unknown nonlinear dynamics of the plant.

Keywords Radial basis function networks · Box-Jenkins gas furnace data · Chemical process · Modeling · Gradient descent

1 Introduction

Depending upon what we know about the given system, the modeling (or identification) process has been color coded as: white box modeling and grey box modeling. The grey box modeling is used when we have some information regarding the mathematical structure of the plant. The only thing missing is the value of some of its parameters. These parameters values will be then approximated by the RBFN

R. Kumar (✉) · S. Srivastava · J.R.P. Gupta
Netaji Subhas Institute of Technology, Sector 3, Dwarka, New Delhi 110078, India
e-mail: rajeshmahindru23@gmail.com

S. Srivastava
e-mail: smriti.nsit@gmail.com

J.R.P. Gupta
e-mail: jairamprasadgupta@gmail.com

© Springer Nature Singapore Pte Ltd. 2017
S.C. Satapathy et al. (eds.), *Proceedings of the 5th International Conference on Frontiers in Intelligent Computing: Theory and Applications*, Advances in Intelligent Systems and Computing 515, DOI 10.1007/978-981-10-3153-3_40

identification model (identifier) by using the plant's input–output training data. The other much more difficult modeling is the black box modeling [7]. In this case, no physical insight of the given plant is available. The only thing available is the training data obtained by conducting experiment on the plant [8, 9]. In this paper, black box modeling based on RBFN is described. For online and adaptive applications of RBFN models, however, we require some kind of recursive identification algorithm which is developed using the powerful concept of gradient descent principle. Artificial neural networks (ANNs) are widely used for modeling purpose [4] but they suffer from the problems like slow learning and poor process interpretability and sometimes stucking in the local minima. RBFN is used as an alternative to the ANN. They possess simpler structure with only three layers as compared to the ANN. The other advantages of RBFN over ANN are the smaller extrapolation errors, higher reliability, and faster convergence. The performance of RBFN depends upon the values of its parameters. These parameters are adjusted using the recursive identification algorithm which derived using the gradient-descent principle [6, 10].

2 Dynamics of Radial Basis Function Network

The RBFN architecture is shown in Fig. 1. Vector X denotes the m-inputs training samples $X = (x_1, x_2 \dots x_m)$. The hidden layer nodes are known as radial centers. They are associated with Gaussian radial basis function as their activation function [3], though other choices for radial basis functions are also available. Every radial center, C_i, is a representative of one of the input training sample and generates an appreciably high output if the current input training sample is present in its vicinity. This is the property of the Gaussian radial basis function. The weights connecting inputs to hidden layer nodes are of unity values [5]. For SISO or MISO systems, RBFN output layer contains only single neuron with induced field (input) equal to the sum of weighted outputs coming from the hidden layer nodes. The count of radial centers present in the hidden layer of RBFN is equal to q, where $q \ll m$ [1] and these radial centers have different functionalities than the neurons present in the MLFFNN. RBFN adjustable output weight vector is denoted by $W = [w_1, w_2, w_3 \dots w_q)$. Its output is calculated as:

$$y_r(k) = \sum_{i=1}^{q} \phi_i(k) W_i(k) \tag{1}$$

where $y_r(k)$ is the RBFN output at kth instant. The ith radial center output at any kth instant is determined using the below expression:

$$\phi_i(k) = \phi \, \|X(k) - C_i(k)\| \tag{2}$$

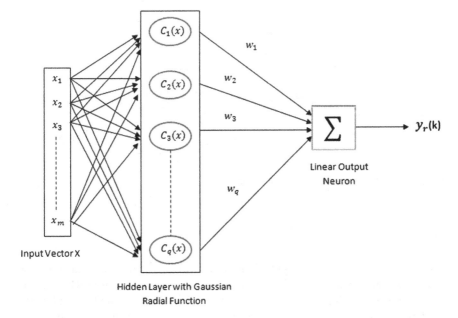

Fig. 1 Structure of RBFN

where $\|X(k) - C_i(k)\|$ represents the Euclidean distance present between $X(k)$ and $C_i(k)$ at the kth time instant. The $\phi(.)$ represents a Gaussian radial basis function having the following definition [1]:

$$\phi(t) = \exp\left(\frac{-t^2}{2\sigma^2}\right) \tag{3}$$

where $t = \|X(k) - C_i(k)\|$ and $\sigma = (\sigma_1, \sigma_2 \ldots \sigma_q)$ denotes the widths of each radial center.

3 Update Rules for RBFN Identification Model

The general identification structure based on RBFN is shown in Fig. 2. The RBFN adjustable parameters includes: radial center, radial center's width, and the weights in the output weight vector. They are updated in a sequential manner using training samples so as to reduce the cost function (which is taken to be an instantaneous mean square error (MSE)). To make adjustments in these parameters, update equations for each of them are obtained using the gradient descent principle. The computation is as follows: Let $E(k)$ denotes the cost function value at any kth time instant and is defined as follows:

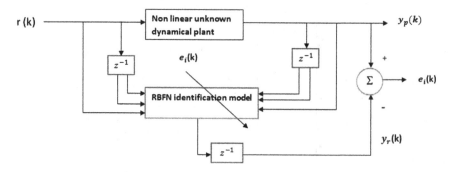

Fig. 2 General modeling structure based on RBFN

$$E(k) = \frac{1}{2}(y_p(k) - y_r(k))^2 \tag{4}$$

where $y_p(k)$ represents the output of the plant whose dynamics was unknown and would be approximated by RBFN identifier and $y_r(k)$ is the output of RBFN at kth instant. Now on taking the derivative of $E(k)$ w.r.t $C_{ij}(k)$ (which also exhibits the clustering effect) will give the rate at which E(k) undergone change with respect to changes in each element of each radial center, where $i = 1$ to m, $j = 1$ to q. Using the chain rule:

$$\frac{\partial E(k)}{\partial C_{ij}(k)} = \left(\frac{\partial E(k)}{\partial y_r(k)} \times \frac{\partial y_r(k)}{\partial \phi_j(k)} \times \frac{\partial \phi_j(k)}{\partial C_{ij}(k)} \right) \tag{5}$$

or

$$\frac{\partial E(k)}{\partial C_{ij}(k)} = \left(\frac{\partial E(k)}{\partial y_r(k)} \times \frac{\partial y_r(k)}{\partial \phi_j(k)} \times \frac{\partial \phi_j(k)}{\partial t_j(k)} \times \frac{\partial t_j(k)}{\partial C_{ij}(k)} \right) \tag{6}$$

After simplifying the partial derivative terms in Eq. 6, the update equation obtained for radial centers with the momentum term (momentum term was added in order to increase the speed of learning without the danger of instability) is:

$$C_{ij}(k+1) = C_{ij}(k) + \Delta C_{ij} + \alpha \Delta C_{ij}(k-1) \tag{7}$$

here $\Delta C_{ij} = \eta e_i(k) W_j(k) \frac{\phi_j(k)}{\sigma_j^2(k)} \left(X_j(k) - C_{ij}(k) \right)$ and η is the learning rate which have a value lying in $(0, 1)$ and $\alpha \Delta C_{ij}(k-1)$ denotes the momentum term where, α, called as momentum constant, is a positive number which lies between 0 and 1 range. Similarly, the update equation for output weights is:

$$\frac{\partial E(k)}{\partial W_j(k)} = \left(\frac{\partial E(k)}{\partial y_r(k)} \times \frac{\partial y_r(k)}{\partial W_j(k)} \right) \tag{8}$$

Thus, every element of $W = [w_1, w_2, w_3 \ldots w_q]$ is updated as:

$$W_j(k+1) = W_j(k) + \Delta W_j + \alpha \Delta W_j(k-1) \tag{9}$$

where $\Delta w_j = \eta e_i(k)\phi_j(k)$. Similarly, for widths of radial centers the update equation can be found as:

$$\frac{\partial E(k)}{\partial \sigma_j(k)} = \left(\frac{\partial E(k)}{\partial y_r(k)} \times \frac{\partial y_r(k)}{\partial \phi_j(k)} \times \frac{\partial \phi_j(k)}{\partial \sigma_j(k)} \right) \tag{10}$$

So, each element in $\sigma = (\sigma_1, \sigma_2 \ldots \sigma_q)$ is updated as:

$$\sigma_j(k+1) = \sigma_j(k) + \Delta \sigma + \alpha \Delta \sigma(k-1) \tag{11}$$

where $\Delta \sigma = \eta e_i(k) W_j(k) \frac{\phi_j(k) r_j^2(k)}{\sigma_j^3(k)}$

4 Simulation Results

The identification algorithm based on RBFN was tested on two real-time systems. The two systems which were considered for simulation purpose were:

1. Box Jenkins gas furnace data
2. Chemical Process data.

4.1 Modeling of Box-Jenkins Gas Furnace Data

The training data is obtained from a single-input and single-output system and is taken from [2]. The input is the gas rate and output is the percentage of CO_2 in it. A total of 296 input–output data were generated from the real gas furnace system and were used for sequential training of RBFN. Let $y_p(k)$ and $r(k)$ denotes output and input of Box-Jenkins gas furnace system. The dependency of output is an unknown nonlinear function, f, of input and needs to be identified using the RBFN identification model. The number of radial centers was taken to be 20, learning rate and momentum term constant were set to 0.08 and 0.04 respectively. Figure 3 shows the response of RBFN identification model (identifier) after 110 iterations. It can be seen from the figure that RBFN response (dotted purple curve) is very much closer to the desired response (solid green line curve) which indicate that RBFN identifier was able to capture the hidden dynamics governing the input–output relationship of the gas furnace system. Figure 4 shows the plot of MSE during the online training. It can be easily seen from the plot that as the training progressed, the error between

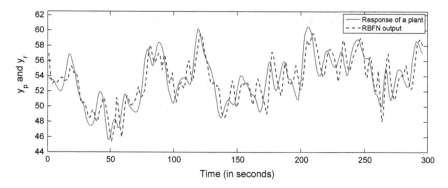

Fig. 3 RBFN identification model response

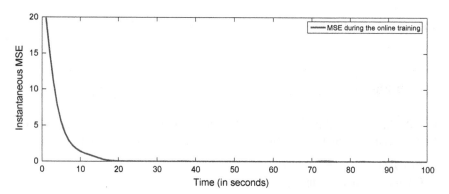

Fig. 4 MSE of RBFN identification model during the online training

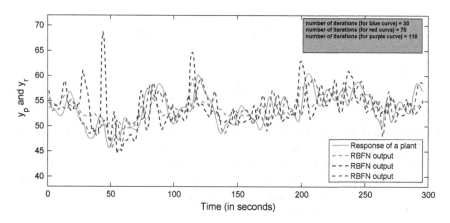

Fig. 5 RBFN identification model response with different number of iterations

plant output and RBFN output was gradually reduced to zero which means that the parameters of RBFN were reached to their desired values. In other words, the dynamics of plant under consideration is now embedded in the form of parameters values of RBFN identification model. To show the effect of number of iterations on the improvement of the RBFN identifier response, the response of RBFN with different number of iterations chosen was shown in Fig. 5. It can be seen from the figure that with the increment in the number of iterations, the response of RBFN improved since more iterations means more training and in each iteration, every training sample was presented again to the RBFN identifier, thus further improving its parameters values and hence, its response.

4.2 Modeling of Chemical Plant Data

This example uses the input–output training data obtained from the plant [11] whose output is a polymer which was produced from the polymerization of some monomer. The human operator decides the set point, which in this case is a monomer flow rate. The actual value of monomer flow rate was inputted by the PID controller ti the plant (as a control signal). There are 5 input variables available to the human operator and he accordingly decides the monomer flow rate value (set point for the plant). Thus, monomer flow rate decided by the human operator is the output of the training data and is denoted by $y_p(k)$. So, this is a MISO case (multiple-input single output). The 5 inputs are described below:

1. $x_1(k)$ = monomer concentration
2. $x_2(k)$ = change of monomer concentration
3. $x_3(k)$ = monomer flow rate
4. $x_4(k)$ and $x_5(k)$ = local temperatures inside the plant

A total of 70 input–output training data pairs, $[x_1, x_2, x_3, x_4, x_5, y_p]$, were generated for RBFN identification model online training. The hidden nonlinear input–output relationship, G, was to be identified (approximated) by RBFN identifier.

Figure 6 shows the response of RBFN identification model after learning was continued for 75 iterations. The response of RBFN shows that its parameters reached close to their desired values as output of RBFN approached very near to the desired output. The MSE plot was shown in Fig. 7. In this case also, the learning ability of RBFN can be seen as the error gradually dropped down to zero (with the progression of online training).

Finally, the response of RBFN identification model with different number of iterations chosen was shown in the Fig. 8. Here also it can be easily seen that the response of identifier improved as the number of iterations were increased. The reason is that RBFN parameters have undergone adjustments number of times and with every new iteration their values got further improved. In more number of iterations, each sample interacts with the RBFN identification model number of times, thus, improving its corresponding response.

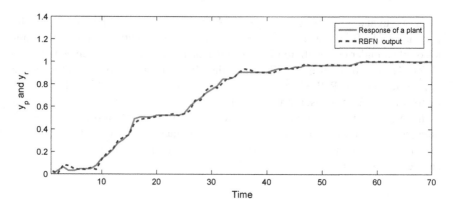

Fig. 6 RBFN identifier response

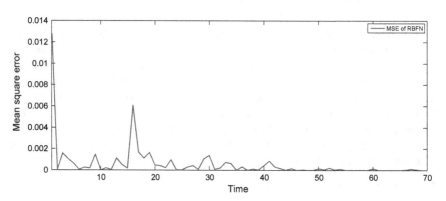

Fig. 7 RBFN MSE plot

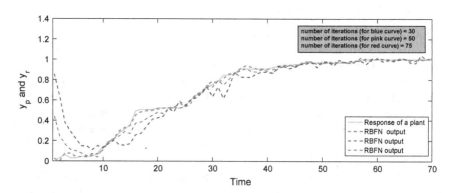

Fig. 8 RBFN identifier responses with different number of iterations chosen

5 Conclusion

In this paper, RBFN parameters were adjusted using the powerful gradient-descent principle. The update equations derived were recursive in nature and hence, were able to do the online training of the parameters of RBFN. The other advantage of gradient descent is that it shows the clustering effect which is very important from the training point of view as it ensures that the radial centers would be placed in those regions of input training data space where the density of training data is more. Two benchmark nonlinear systems were considered which were modeled successfully by the RBFN-based identification model. This shows the approximation ability of RBFN and its potential in handling the complex and nonlinear systems. The learning ability of RBFN is much more better than that of the ANN and its simpler structure adds one more advantage of it over the ANN. The speed of convergence of RBFN parameters was also fast as its response approaches the plant's response very quickly. Simulation results obtained show the effectiveness of RBFN as a modeling tool.

References

1. Behera, L., Kar, I.: Intelligent Systems and control principles and applications. Oxford University Press, Inc. (2010)
2. Box, G.E., Jenkins, G.M., Reinsel, G.C., Ljung, G.M.: Time series analysis: forecasting and control. John Wiley & Sons (2015)
3. Du, K.L., Swamy, M.: Radial basis function networks. In: Neural Networks and Statistical Learning, pp. 299–335. Springer (2014)
4. Esfe, M.H., Saedodin, S., Bahiraei, M., Toghraie, D., Mahian, O., Wongwises, S.: Thermal conductivity modeling of mgo/eg nanofluids using experimental data and artificial neural network. Journal of Thermal Analysis and Calorimetry 118(1), 287–294 (2014)
5. Kitayama, S., Huang, S., Yamazaki, K.: Optimization of variable blank holder force trajectory for springback reduction via sequential approximate optimization with radial basis function network. Structural and Multidisciplinary Optimization 47(2), 289–300 (2013)
6. Singh, M., Srivastava, S., Gupta, J., Handmandlu, M.: Identification and control of a nonlinear system using neural networks by extracting the system dynamics. IETE journal of research 53(1), 43–50 (2007)
7. Sjöberg, J., Zhang, Q., Ljung, L., Benveniste, A., Delyon, B., Glorennec, P.Y., Hjalmarsson, H., Juditsky, A.: Nonlinear black-box modeling in system identification: a unified overview. Automatica 31(12), 1691–1724 (1995)
8. Srivastava, S., Singh, M., Hanmandlu, M.: Control and identification of non-linear systems affected by noise using wavelet network. In: Computational intelligence and applications. pp. 51–56. Dynamic Publishers, Inc. (2002)
9. Srivastava, S., Singh, M., Hanmandlu, M., Jha, A.N.: New fuzzy wavelet neural networks for system identification and control. Applied Soft Computing 6(1), 1–17 (2005)
10. Srivastava, S., Singh, M., Madasu, V.K., Hanmandlu, M.: Choquet fuzzy integral based modeling of nonlinear system. Applied Soft Computing 8(2), 839–848 (2008)
11. Sugeno, M., Yasukawa, T.: A fuzzy-logic-based approach to qualitative modeling. IEEE Transactions on fuzzy systems 1(1), 7–31 (1993)

Optimization of Workload Scheduling in Computational Grid

Sukalyan Goswami and Ajanta Das

Abstract Computational grid houses powerful resources to execute computation-intensive jobs, which are submitted by the clients. Resources voluntarily become available in the grid, as a result of which, this collaborative computing becomes more cost effective than traditional HPC. In the grid, since, the participating resources are of varying capabilities, load balancing becomes an essential requirement. This workload distribution mechanism among available resources aims to minimize makespan, optimize resource usage, and prevent overloading of any resource. Eventually, the resources need to be prioritized based on their capability and demand in the current scenario. Thus, prioritization of resources balances workload in grid. In the proposed workload scheduling algorithm, nearest deadline first-scheduled (NDFS), resource ranking, and subsequent job scheduling maintains balanced load across the grid. The ranking of resources in computational grid is achieved using analytic hierarchy process (AHP) model. The primary objective of this paper is to optimize the workload of grid environment while executing multiple jobs ensuring maximum resource utilization within minimum execution time. Service quality agreement (SQA) is met through proper scheduling of jobs among ranked resources. The grid test bed environment is set up with the help of Globus toolkit 5.2. This paper presents the simultaneous execution results of the benchmark codes of fast Fourier transform (FFT) and matrix multiplication in order to balance the workload in grid test bed.

Keywords Computational grid · Load balancing · Resource ranking · Analytic hierarchy process (AHP) · Workload scheduling algorithm

S. Goswami
Institute of Engineering & Management, Salt Lake, Kolkata, India
e-mail: sukalyan.goswami@gmail.com

A. Das (✉)
Birla Institute of Technology, Mesra, Kolkata Campus, Ranchi, India
e-mail: ajantadas@bitmesra.ac.in

© Springer Nature Singapore Pte Ltd. 2017   417
S.C. Satapathy et al. (eds.), *Proceedings of the 5th International Conference on Frontiers in Intelligent Computing: Theory and Applications*, Advances in Intelligent Systems and Computing 515, DOI 10.1007/978-981-10-3153-3_41

1   Introduction

Computational grid coordinates sharing of resources in large scale and caters to several autonomous groups in problem solving [1]. Grid computing has gained popularity mainly because of three reasons: (1) *a given amount of computer resources can be used more cost-effectively*, (2) *as a way to problem solving which requires enormous computational capability*, and (3) *it ensures that cooperative usage of resources is always helpful toward achieving a common objective.* Computational grid will ensure effective resource scheduling, reliability, parallel CPU capacity, and access to additional resources as well.

Ensuring balanced load across the grid is a challenging task. Because, to exploit the capability of the computational grid, the resource broker is responsible for resource scheduling, resource optimization and client's request scheduling, in a balanced way. Grid, being a dynamic heterogeneous environment, each resource can not be evenly utilized always. Based on the capacity of resources and current scenarios, the workloads are distributed or allocated among the resources up to its optimum level. Hence, allocation of jobs and load balancing of resources become an optimisation problem.

Optimization of workload scheduling means, jobs need to be scheduled among the active resources in grid environment, such that the loads of the resources should be balanced. Jobs can be allocated on a "first come first serve" basis, but in that case, some resources could become overloaded or remain under utilized in the grid environment. Hence, this research uses technique of resource ranking and based on this ranking, the jobs are scheduled. The ranks of the resources are finalized by Saaty's AHP model [2] with all necessary metrics. Once the ranking is finalized, the scheduling is done primarily based on submitted jobs' individual deadlines. Since the job deadline is the most important parameter of prioritization, hence the algorithm is named as n*earest deadline first-scheduled* (NDFS) [3] algorithm. The simulations of NDFS and other existing algorithms are run in GridSim [4]. Simulation results show that the performance of NDFS is better than majority of the other algorithms [5]. The scheduling also takes into account the current load scenarios of the resources and mapping of tasks-to-resources happens depending upon the capability and current load of respective resources. In this proposed workload scheduling approach, client is involved and solely aware of allocation of resources for completion of its job. Hence, the client and the broker need to sign service quality agreement (SQA) prior to the submission of job in grid. Therefore, objective of this paper is to optimize the workload of grid environment while executing multiple jobs. At the same time, it also maintains SQA and checks whether Quality of Service is achieved.

Organization of this paper is as follows: related works in this field are represented in Sect. 2. Section 3 presents proposed workload scheduling algorithm in detail. Section 4 presents the globus implementation details and results. Section 5 concludes the paper.

2 Related Works

Efficient job scheduling is of prime importance for successful operation of a computational grid. Among the available approaches [6, 7], the ones relevant with this research work are presented here.

Michael Stal [8] proposed a client/server system. Implementation of this framework has several dependencies resulting in few limitations. Load balancing among resources is not ensured in this approach. After the jobs are submitted in the grid, the resources often end up in being differently loaded, hampering the overall load balancing of the computational grid.

Two different broker architectures, namely, handle-driven broker and forwarding broker, have been found by Omotunde Adebayo et al. [9]. The handle-driven broker performs like a name server. A service handle comprising of resource's working request format along with name and network address (sent by the broker to the client). On the other hand, for any possible transaction, clients and resources have forwarding broker as the mediator. After execution of job is complete, result is forwarded from resource via this broker to the client.

Scheduling decision is a multicriteria decision-making (MCDM) process. In [10], AHP's effectiveness as a MCDM tool has been established. Attributes' relative significance play crucial role in finalizing the assigned weightages logically in AHP calculations. After studying numerous schemes for workload balancing in computational grid environment, a novel framework for workload balance and an algorithm, NDFS, were proposed in [11, 12], where jobs are prioritized on the basis of their deadlines. In comparison, FPLTF [3] algorithm schedules largest task to the fastest processor. The job having minimum completion time (MCT) is allocated first by min–min algorithm [3], whereas max-min algorithm [3] allocates job which has maximum completion time. Sufferage algorithm [3] allocates jobs on the basis of calculation of each job's sufferage value. But these widely used scheduling algorithms do not rank the resources and individual job deadline is not the prioritization criteria for scheduling in those cases. This proposed algorithm gives equal priorities to each job submitted in the grid and accordingly it treats the deadline as well. Novelty of this approach is that the fastest processor based resources are also not reserved for any high prioritized job. Moreover, the jobs are always allocated to the resources based on their current ranking in the grid.

3 Proposed Workload Scheduling Algorithm

The jobs submitted can have differing deadline and processing power requirements. The responsibility of smooth operation of the grid is bestowed upon the broker, which works as the middleware.

In the proposed workload scheduling algorithm, NDFS, the clients are given highest priority, as parameters related to the submitted jobs play an important role for proper workload balancing. The broker then searches for the resources which are having matching or higher capacity of these parameters. Major phases of the proposed algorithm are presented below:

Phase I: Submission of Job Parameters

Initially, the clients submit the job requirements, specifically, the number of cores, current CPU utilization, the clock frequency of the CPU, network utilization and the available RAM (the primary memory criteria has been introduced in this research work to improve the $Job_{weightage}$ calculation). Then the broker calculates $Job_{weightage}$ using AHP. AHP decision matrix for resource ranking is presented in Table 1. Equation (1) presents $Job_{weightage}$ in details. Saaty's AHP model has been used for finalization of Eq. (1). Then broker creates and maintains the job queue.

$$f\left(Job_{weightage}\right) = 0.464x_1 + 0.195x_2 + 0.195x_3 + 0.073x_4 + 0.073x_5 \qquad (1)$$

where,

x_1 = Number of cores of the processor
x_2 = Free CPU usage = $(1 - CPU\ utilization/100)$
x_3 = RAM availability
x_4 = Clock frequency
x_5 = Free network usage = $(1 - Network\ utilization/100)$

Phase II: Ranking of Resources

Ranking of resources is a dynamic process, where the values of each parameter can get changed at any point of time based on the current load scenarios of the resources. Hence, the resources submit their system parameters to the broker as and when jobs are submitted. The broker calculates $Resource_{weightage}$ for every resource in the same manner according to Eq. (2):

$$f\left(Resource_{weightage}\right) = 0.464x_1 + 0.195x_2 + 0.195x_3 + 0.073x_4 + 0.073x_5 \qquad (2)$$

(where x_n are as specified in phase 1)

AHP model is also used to finalize Eq. (2). Table 1 is prepared using AHP and presents decision matrix for resource ranking. It also shows the eigenvector values. Eigenvectors are actually the calculated weight for each parameter. Higher the $Resource_{weightage}$ value boils down to higher ranking of that particular resource. $Resource_{weightage}$ is calculated at the broker end by the broker after receiving the system parameter values from the resource. The resource having highest $Resource_{weightage}$ value is assigned as $Res_{rank} = 1$, next high $Resource_{weightage}$ valued resource is assigned as $Res_{rank} = 2$ and so on. It then places the resources on a priority queue according to their ranks. Higher a loaded resource, lesser will be the value, lesser will be the resource rank.

Phase III: Signing of SQA

In the next phase, the broker checks for the possibility of meeting the job requirements by matching $Job_{weightage}$ and $Resource_{weightage}$ values. If matching found, the client and broker sign a bipartite agreement, SQA, regarding that particular job.

Phase IV: Job Allocation

Once the SQA is signed, the broker allocates the submitted job to that resource which has Resource$_{weightage}$ value just immediate closer to the submitted Job$_{weightage}$ value.

The resources are heterogeneous and scenarios can be changing dynamically in grid environment. So in the execution of any phase of the proposed algorithm, there is high chance of submission of other new jobs in grid. Hence, it is a continuous process of scheduling of jobs based on the ranking of available resources. Therefore, maintaining job queue against time and QoS is always challenging.

Phase V: Compliance of SQA

After completion of job, the broker checks whether SQA has been satisfied or not and sends the execution results to the corresponding client.

These above-mentioned phases explain the process starting from submission of single job to allocation to resource and execution of the same job in computational grid environment. However, in grid, heterogeneous types of jobs are getting submitted simultaneously and almost continuously. Therefore, the job queue is always updated and remaining jobs in the job queue will be handled accordingly.

4 Execution of Multiple Jobs

A grid test bed is set up, for the purpose of this research work, consisting of three clients, three resources, and a grid broker. Globus Toolkit 5.2 [13] is used to set up the real grid environment. Java has been used as the programming language for implementation of this research work because of its extensive and robust support in the distributed environment. The system information gatherer and reporter (SIGAR) [14] API is used to retrieve system parameters.

Moreover, to ensure non-trivial performance of the grid, heterogeneous benchmark codes of fast Fourier transform (FFT) [15] and matrix multiplication [16] are executed. Both of these are highly computation-intensive jobs. Hence, for these experiments, specifically two jobs from two clients and two resources, which execute those jobs, are considered to explain the scenarios in the next two sections.

Table 1 AHP decision matrix for resource ranking

	x_1	x_2	x_3	x_4	x_5	Px$_i$ (n = 5)	Eigenvector (W$_i$)
x_1	1	3	3	5	5	2.954	0.464
x_2	1/3	1	1	3	3	1.246	0.195
x_3	1/3	1	1	3	3	1.246	0.195
x_4	1/5	1/3	1/3	1	1	0.467	0.073
x_5	1/5	1/3	1/3	1	1	0.467	0.073

4.1   Execution of FFT Benchmark Code

Broker received "FFT.java" as job file with its specific parameters from the Client1 (192.168.30.3). Job requirement specification for each parameter is represented in Table 2. Broker then calculates $Job_{weightage}$ (2.32) and sends into job pool. Figure 1 shows the $Resource_{weightage}$ of individual resources in computational grid environment. Upon request, Broker received specifications from three resources, Resource1 (192.168.30.4), Resource2 (192.168.30.5), and Resource3 (192.168.30.6). Then it calculates rank metrics and assigns subsequent ranks to all these resources.

At this time, bipartite agreement, SQA is signed between Broker and Client1.

According to the proposed workload scheduling algorithm, "FFT.java" job is scheduled to Resource1 (192.168.30.4) comparing the $Resource_{weightage}$ and $Job_{weightage}$. At the end of its execution, broker receives execution outputs from Resource1 and it sends the details to Client1. Finally, the SQA is met for "FFT.java" job.

4.2   Execution of Matrix Multiplication Benchmark Code

While the client and broker have signed SQA for the job "FFT.java," then another job, "MatMul.java" is submitted by Client2 (192.168.30.8) in computational grid

Table 2 Job submission specification

Requirement specification	FFT.java	MatMul.java
No. of cores of the processor (x_1)	4	4
Free CPU usage (x_2)	0.7	0.02
RAM availability (x_3)	0.63	0.03
Clock frequency (x_4)	2	1.6
Free network usage (x_5)	0.8	0.1
$Job_{weightage}$	2.32	1.98

Fig. 1 $Resource_{weightage}$ values for simultaneous job submission in grid

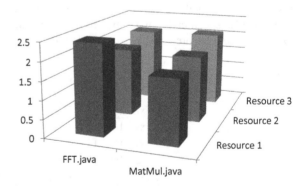

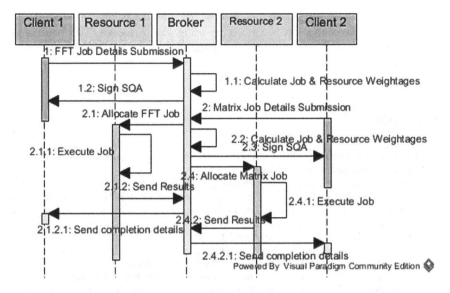

Fig. 2 Sequence diagram of multiple job execution

environment. So, broker receives all the job requirement parameters relevant to this newly submitted job (shown in Table 2) and calculates Job$_{weightage}$. As a regular phenomenon, it also calculates Resource$_{weightage}$ at that instant and presented in Fig. 1. Since, matching is found between Job$_{weightage}$ and Resource$_{weightage}$, bipartite agreement, SQA is signed and the matrix multiplication job is scheduled to Resource3 (192.168.30.6). Finally Broker receives execution outputs from Resource3 and it sends the details to Client2. It is found that agreed QoS is achieved for this job also.

Discussion
Figure 2 represents the simultaneous execution sequences of the above-mentioned two jobs. Each significant step of the algorithm is marked and explained in the diagram. Upon completion of its execution, broker forwards these completion details to respective clients after verifying respective SQA.

   This experimental result demonstrates that at the time of scheduling of the specific jobs load of each resource is balanced in computational grid environment.

5 Conclusion

Major challenge faced in computational grid environment is workload balancing among the participating resources. This research aims to find a solution to this problem. This paper deals with allocation of resources for simultaneous job submission in computational grid environment. Jobs are submitted at random and those are allocated to ranked resources for execution. Ranking procedure of resources has also been improvised by considering all possible resource parameters. In order to test the algorithm, a grid test bed is set up using globus toolkit. This paper also presents the execution results of the computation-intensive benchmark codes of FFT and matrix multiplication.

References

1. Foster, I., Kesselman, C., Tuccke, S.: The Anatomy of the Grid. International Journal of Supercomputer Applications, (2001).
2. Saaty, T.L.: The Analytic Hierarchy Process (AHP). (1980).
3. Goswami, S., De Sarkar, A.: A Comparative Study of Load Balancing Algorithms in Computational Grid Environment. In: Fifth International Conference on Computational Intelligence, Modelling and Simulation, pp 99–104, (2013).
4. Buyya, R., Murshed, M.: GridSim: a toolkit for the modeling and simulation of distributed management and scheduling for Grid computing. The Journal of Concurrency and Computation: Practice and Experience, vol. 14 pp. 13–15, (2002).
5. Goswami, S., Das, A.: Deadline Stringency Based Job Scheduling in Computational Grid Environment. In: Proceedings of the 9th INDIACom; INDIACom-2015, pp 531–536, (2015).
6. Erdil, D., Lewis, M.: Dynamic grid load sharing with adaptive dissemination protocols. The Journal of Supercomputing, pp 1–28, (2010).
7. De Sarkar, A., Roy, S., Ghosh, D., Mukhopadhyay, R., Mukherjee, N.: An Adaptive Execution Scheme for Achieving Guaranteed Performance in Computational Grids. Journal of Grid Computing, pp 109–131, (2009).
8. Stal, M.: The Broker Architectural Framework. (2003).
9. Adebayo, O., Neilson, J., Petriu, D.: A performance study of client broker server systems. In: Proceedings of CASCON'97, pp 116– 130, (1997).
10. Saaty, T.L.: Decision making with the analytic hierarchy process. International Journal of Services Sciences, Vol. 1, No. 1, pp. 83–98, (2008).
11. Goswami, S., Das, A.: Handling Resource Failure towards Load Balancing in Computational Grid Environment. In: Fourth International Conference on Emerging Applications of Information Technology (EAIT 2014), pp 133–138, (2014).
12. Goswami, S., De Sarkar, A.: Service Oriented Load Balancing Framework in Computational Grid Environment. International Journal of Computers and Technology, Volume 9, Number 3, pp 1091 – 1098, (2013).
13. Globus Toolkit. http://toolkit.globus.org.
14. SIGAR. https://github.com/hyperic/sigar.
15. FFT benchmark code. http://introcs.cs.princeton.edu.
16. Matrix multiplication benchmark code. http://introcs.cs.princeton.edu.

Cloud Security and Jurisdiction: Need of the Hour

Tamanna Jena and J.R. Mohanty

Abstract Features of cloud services that users' data is remotely in obscure machine is neither owned nor controlled by user. For adoption of services, from users' point of view, cloud security and arbitration are significant. Security gap among different tiers causes privacy issues, customer concern about losing sensitive data available in a cloud computing infrastructure. Disagreement among different components of cloud can be mitigated by Online Dispute Resolution (ODR) to great extent. In this paper, we are focusing mainly on three factors, first trying to identify the customer threats, concerns and cross-border conflicts while using cloud computing. Second on ODR (Online Dispute Resolution) and its mechanisms. Finally on required regulatory framework between consumer, industry, and geographic boundaries. An accepted regulatory framework across participants/consumers/providers is currently at a premature stage but as imperative adoption to cloud computing increases across industry there will be paradigm shift in effort sooner than later to build the same.

Keywords Consumer protection · Security threat · Cloud computing · Cross-border conflict · Online dispute resolution · Jurisdiction

1 Introduction

The Internet is one of the greatest innovations of the Information Age, and with the explosion of wireless communication mankind is now almost entirely connected... be it car, smart phone, television, electrical/electronic appliances to everything. Invent of cloud computing and increase in adoptions will lead to many more

T. Jena (✉)
School of Computer Engineering, KIIT University, Bhubaneswar, India
e-mail: tamanna_jena@yahoo.com

J.R. Mohanty
School of Computer Applications, KIIT University, Bhubaneswar, India
e-mail: jmohantyfca@kiit.ac

© Springer Nature Singapore Pte Ltd. 2017
S.C. Satapathy et al. (eds.), *Proceedings of the 5th International Conference on Frontiers in Intelligent Computing: Theory and Applications*, Advances in Intelligent Systems and Computing 515, DOI 10.1007/978-981-10-3153-3_42

425

technological innovations to help on decision-making for the industry and mankind. Across geographies/industries cloud computing adoptions is seen in silos so far and it is predicted will see significant jump in 2016 onwards where it will be the year when cloud adoption and implementation will be in mainstream for any organizations discussions or implementations.

The digital business environment leads overwhelming opportunity in cloud platform. Tangible features like high computability, scalability, obligation free, usage-based pricing are quiet tempting for users of all capacities, there exist critical issues like privacy and security of sensitive data. This data is an essential component of decision-making, ongoing operations, market competitiveness and regulatory compliance. Hence as seen security is still the major concerns across industry and CIO's/CEO's out of key issues for cloud adoption, e.g., security concerns, availability, data protection, cost, standards, regulations, performance, intellectual assets. Current Gartner report predicts beyond 2016 majority of the security breach will not from the cloud service provider's rather internal reason e.g., employee, stolen passwords, etc.

Historically, on premise infrastructure has been seen as a more secure setting for sensitive company information. It is typically easier to manage because no responsibility is shared with third parties, which largely facilitates certain processes. Big data databases are usually hybrid environments which involve traditional databases, in house applications and cloud deployments (if any). Since there are no common points raising potential entry points of threat to access data through multiple entry points. Cloud is basically an abstract system from user's point of view. When an user belongs to one territory, cloud provider may belong to another and data may be processed in a third territory, in such cases many more territorial elements are included for completion of one applications. Law and enforcements of different territories are not aligned; the "gap" gives a big room of error, especially when dispute arises between core tiers. DPD (Data Protection Directives) is European Union (EU) directive which regulates personal data processing, mainly about law of human right and privacy [1]. Similar FISMA (Federal Information Security Management) is followed by Americans.

2 Related Research

Cloud computing having three models: IaaS, PaaS, and SaaS. Each model has its own security issues. SLA (Service Level Agreement) is the documented contract between cloud provider and user, it should give detail information about customer's requirement, providers charge rate, accountability of provider, framework of understanding, mention all stake holders with their responsibility, legalities in the event of dispute, security management, customer grievance handling, and monetary claim. It is found that present SLA is limited to services provided and waivers associated if the service not met the agreement. Waivers are insufficient, accountability of providers is missing and security policies are overlooked [2]. Across world, people are using webmail and online data backup sites. For the sake of security, data confidentiality, integrity, and availability (CIA) are difficult to obtain, if not impossible. Researchers have raised concerns like "Who has jurisdiction over data flow across borders?" [3]. Even mentioned cloud computing as an easy target for cybercrime, age-related sensitive actions can be compromised. Giants are capable of deflecting cyberattacks to a large extent. But all cloud is not equally capable. Single malicious activity can compromise hundreds of sites, by identifying weakest link of cloud. Importance to leverage research on multi-domain policy integration and secure-service composition is detailed in [4], in order to build comprehensive policy-based management framework in cloud environment. Accountability is cloud organizations are scrutinized in detail in [5]. The responsibility carried by cloud organization, in which it will be answerable for one's conduct and will be held responsible for reward and penalty for its actions. The identified core attributes are: transparency, responsibility, appropriateness, and effectiveness. [6] Proposed accountability framework for the policy languages (machine understandable) and mapping to enforcement policy. Researchers identified five privacy attributes: confidentiality, integrity, availability, accountability, and security. Each attribute is detailed but overlooked the enforcement needed to back up the concept [7]. Decentralized information accountability is suggested, using JAR (Java Archive, it is an independent file format, authors can digitally sign entries to authenticate and automate logs) [8]. Security issues of cloud computing is researched extensively by [9, 10–13]. [14] Focuses on the battle between tangible aspects of cloud and risks involved with sensitive data as well as accountability, for customers/organizations judging to act cloud.

3 Cloud Customers Concern

Customers of all capacities have security concerns. We are outlining few common threats of cloud user as follows:

(1) Data abuse: Cloud users are highly vulnerable to fraud and data loss, registration system is comparatively weak and fraud detection capabilities are limited. Loss of core intellectual property could cause competitive and financial implications.

(2) Security Leakage: Security arrangements are difficult to understand, associated with usage, orchestration, managements and API integrity.

(3) Shared Technology Issues: Recent targets on shared technology in cloud computing environment suggest strong compartmentalization is needed, in CPU caches, GPU, Disk partition etc.

(4) Noxious Insiders: It is critical for user to understand, cloud provider responsibility and accountability to detect, deal and defect of malicious insider in the organization.

(5) Accountability and Litigation: Depth of organization strategy and techniques to deal with possible litigation from breach is unsure.

(6) Unknown Risk Profile: Features and functionality of cloud services are well explained unlike security procedures, configuration, patching, logging, auditing, vendor sharing profile etc., leaving customer with unknown risk profile (Table 1).

Cloud services dedicated around-the-clock resources to maintain the security of data stored on cloud infrastructure and responding to breaches. The additional security measure enterprises implement the most is data encryption, followed by identity access policies and regular audits which can make it more robust than normal on premises system. Recently Gartner introduced concept called Cloud Access Security Broker (CASB), used to describe a particular set of cloud security solutions focused primarily on security and compliance monitoring for SaaS applications centered around four main pillars: visibility, compliance, data security, and threat protection.

Perhaps the most compelling discussion points on the Internet's transformation of Cloud computing world is the sociopolitical practices and its implications overall. Earlier digital boundary less world, information flows solely from few to the many, with little interactivity. Now, information flows from the many to the

Table 1 Cloud customer concerns versus reality

Concern	Reality
Is personalized data is secure in a cloud server?	Though at present cloud service providers (CSPs) have better physical security but many factors like frequent audit, alter mechanism can confirm servers holding personalized data is secure
Is Cloud Service Provider is responsible for securing data?	Origination of the data lost needs to be investigated and concluded before the onus goes to cloud service provider. Reason being since access is restricted to individuals within an Organization; hence they should own the end-responsibility for data lost
Is different users/competitors in the same cloud can attack each other or vulnerable to data leakages?	Current set of technology makes little difficult for one virtual machine to attack another virtual machine though in same infrastructure. Based on typical technology used by most providers all layers of a virtual machine are isolated from each other
Is data in the cloud can be located anywhere as long as it's encrypted?	Encrypted data is still subject to data protection law if the encryption can be reversed and the data personally identifiable. Current situation calls to understand which country and data protection legislation is applicable to concerned data, whether encrypted or not due to data legislation across country/boundary

many, multimodal and interact beyond one country/continent each second. By disintermediation government and corporate control of communication, horizontal communication networks have created a new landscape of social and political change, where jurisdiction should be clear from legal point of view. Jurisdiction describes bounds and limits of legal authority sometimes refers to a geographical or political territory, in general refer to both, out of which jurisdictions are constituted, it describes the way the power or control may be exercised. Jurisdiction of a country is the legal system of that territory, cloud providers have their own legal authority and they do business in different territory having different law and enforcement. For example legal age is different, stake holders are different, legal dealing are also different. So user of cloud provider needs a secure, aware platform where apart from deploying services legal issues should be clarified and deals properly.

4 Critical Issues and Jurisdiction

Jurisdiction refers to a particular geographical area, following a well-defined legal authority. On the contrary, concept of cloud is borderless. User needs a smart device-like smart phone, laptop, smart watch, and internet connection to access cloud resources, transactions is easy but prone to disagreement and alignment challenges. In cloud platform location of user, provider, and datacenters may have different geographical regions having different territory and following different jurisdiction. Usage of cloud computing cannot be refrained but at the same time cross-border conflict and related threat needs regulatory framework too. Presently all most every enterprise are using cloud, include a wide range of items such as book, clothing, software, vacation packages, insurance, luxury jewelry, etc. Strong regulatory framework will encourage more customers to use cloud components. Till date most consumer cost of legal penance by legal feud is not proportionate to claim amount. Concrete amendment of policies between cloud provider and law maker will facilitate the paradigm shift to cloud computing.

5 ODR Mechanism

Many enterprises are using cloud resources for their business as IaaS, PaaS, and SaaS, then any kind of dispute between parties is more likely to be solved by Online Dispute Resolution (ODR) [15]. Dispute resolution done online is ODR. ODR is a version of ADR (Alternative Dispute Resolution), where parties having disagreement are handled by using ICT (Information and Telecommunication Technology). ODR is a vast field, includes interpersonal disagreement like consumer to consumer (C2C), business to business (B2B), and business to consumer (B2C). It is a vital medium to solve cross-border disputes in an economical way. ODR mechanisms for cloud consumer disputes: Cross border transactions faces difference in jurisdiction, standard, linguistic terms, and many more. ODR provides an appropriate forum for both parties in terms of convenience and accessibility. ODR is introduced in 1996/97 in USA, Canada, and strongly vouched by European Commission. Course of its actions is: (1) Arbitration: Documents-only arbitration has been used to solve consumer disputes in ODR. Most European jurisdictions standard arbitration clauses bind only on business and optional on consumers. Unlike in the US, arbitration clauses are enforceable on business and consumer. (2) Automated Settlement System: It is highly innovative, cost-effective, software driven form of ODR suitable for monetary claims. In this process both parties does successive blind bids. The procedure involves subsequent blind bidding by the both parties. Software is used to wrap the final settlement when bidding reaches within the range for both parties. Web-based platforms back the process of settlement. (3) Complaints Assistance: It acts as a kind of online law center. It is communication effective to register a complaint and call for compensation. (4) Mediation: It is one of the flexible and voluntary ODR methods. Compensation basically deals with small monetary claims with minimal fees.

ODR Service providers: organization which provides ODR services are categorized into three forms: (1) Generally accessible, provided by independent provider. It is an open access forum for any claimant of any jurisdiction can use them to seek redress. Two main disadvantage are expensive and enforcement. If respondent user or business is not an ODR member then it is difficult to enforce the decision. (2) Trustmark ODR provider: Much government, military, health, education, etc., domain has moved to cloud wholly of partially. Trust factor plays a major role in consumer confidence using cloud resources across different demographics and redress mechanism of cloud user. Therefore some federations-like consumer association, government, trade, few private sectors too joined trustmark schemes. So members of trustmark adhere to the dispute resolution scheme, which boosts user of cloud and e-commerce. (3) ODR services provided in the context of a marketplace: Theses are basic portals which provide rating systems and feedback from users. Consumer trust factor can be achieved to a large extent by transparent feedbacks. Many market places like Yahoo, Amazon, E-bay provides 'money back guarantee' which avoid disputes in the first place and reinforce consumer confidence on cloud providers and internet purchases.

ODR problem-solving technique in Cloud platform: We are trying to identify the requirement to handle legal battle in cloud platform and second suggest some critical processes to apply consumer ODR.

(1) ODR service provider should be independent and impartial. Funding and board structure of components of ODR should be aligned to each other and strictly neutral, unlike the existing schemes [1].

(2) ODR should allow secrecy and confidentiality at the same time should promote public interest. Disputes between different parties following different jurisdictions should have alignment in regulation, gap needed to be identified and rules needed to be documented which addresses most scenarios. Decisions made by arbitration under Uniform Domain Name Dispute Resolution

Procedure (UDRP) till dates are published on ICANN website. Unfortunately it suffers from system authority.

(3) Communication being very important, most ODR services offer only English as a medium of communication. More different linguistic services should be provided.

(4) Cost of dispute resolution should be proportional to the amount at stake. A standard procedure should be maintained and when dispute is higher as well as complex then ODR need to be scaled according. CNIL (*Commission Nationale de l'Informatique et des Libertés*) *of France* recently fined Google 100, 000 euro's for its street view activity, which is quiet a peanut amount to raise eyebrow. Penalties need to be substantially higher than they are lately [1].

6 Needed Regulatory Framework

To facilitate policy integration between the multi-domain cloud environment, customer and law enforcement, a trust-based regulatory framework is quintessential. Many platforms are there like ODR, SCAP (Security Control Automation Protocol), FISMA. SCAP is platform for expressing and organizing as well as measuring security information in standardized way, even identify vulnerabilities [16]. Adoption to SCAP will embellish security posture of any organization. Since cloud computing is an ongoing process so law and enforcement needed for it is still not concrete, self-security drive needed from cloud to boost customer confidence. Adhering to FISMA, cloud could provide secure governance to both industry and law makers. In Table 2, cloud features along with related issues are discussed.

Table 2 Cloud computing features and required regulatory framework

Cloud feature	Related issue	Needed Regulatory Framework	Cloud contribution
Multitenancy	• Improper isolation • Private and confidentiality breach of data, • Improper handling of data	• Each provider of all capacity and domain should register under an organization and • Data confidentiality should be the topmost priority	• Ensure data confidentiality by improvised encrypting tools • Declaring its legal obligation
Complex, dynamically changing environment and data	• Overrated trust factor • Lack of transparency • Accountability issues	• Iterative 360 degree audit of "trust factor" of each provider to maintain the standard	• Trust factor need to the metric of business of all capacity

(continued)

Table 2 (continued)

Cloud feature	Related issue	Needed Regulatory Framework	Cloud contribution
Data duplication and proliferation; unknown geographical location	• Severe transborder data flow compliance issues • Linguistic issue • Breach in privacy	• Accountability of concern organization • Backed up by severe consequences when guilty	• Facilitate multilingualism in services
Convenient and enhanced data access from multiple locations	• User incomprehension • Gap in law and enforcement in different jurisdiction • Discrepancy between organizational policy and territorial policy	• Risk management of conflict between policies • Facilitate transparency of data handling procedure among stake holders	• Make tool to provide smart user interface to get in and out of their service request

Cloud applications are vast and when Internet of thing is already in vision, regulatory framework needed to strengthen. User accountability, transparency, and handling procedures will not only boost user's faith and confidence in cloud computing, but also will streamline the quintessential needed law and enforcement in cloud security.

7 Conclusion

ODR is definitely the much needed platform to solve dispute of all ranges across different jurisdiction for users of cloud computing. Improvised schemes and standards like Trustmark will enforce fair resolution among different parties. Security management can be further fine-tuned by adhering to act like FISMA and SCAP. Trust factor of consumers on cloud usage needed to be strengthened. Clearly most domain are moving to cloud, so reinforcing the privacy/security and body of law supporting cloud will make the shift concrete and consolidate. Privacy solutions and interoperable specifications needed to be critically reassessed with appropriate jurisdiction. Prospect of cloud computing is still in its infancy, as it emerges, more regulatory framework needed to be included over time, which will widespread its adoption.

References

1. Kuan W. H., Hörnle, J., and Millard, C.: Data protection jurisdiction and cloud computing–when are cloud users and providers subject to EU data protection law? The cloud of unknowing. International Review of Law, Computers & Technology,26(3), 129–164 (2012).
2. Kandukuri, B.R., Rakshit, A.: Cloud security issues, Services Computing. In: IEEE International Conference on Services Computing, pp. 517–520 (2009).
3. Kaufman L.M.: Data security in the world of cloud computing. IEEE Security & Privacy. 7 (4), 61–64 (2009).
4. Hassan Takabi, H., James Joshi, B.D., and Ahn, G. J.: Security and privacy challenges in cloud computing environments, IEEE Security & Privacy, 8(6), 24–31(2010).
5. Jaatun, M.G., Pearson, S., Gittler, F. and Leenes, R.: Towards Strong Accountability for Cloud Service Providers. In: CloudCom, pp. 1001–1006 (2014).
6. Benghabrit, W., Grall, H., Jean-Claude Royer, J.C., Sellami, M., Azraoui, M., Elkhiyaoui, K., Melek Önen, Anderson Santana De Oliveira, and Bernsmed, B.: A Cloud Accountability Policy Representation Framework. In: CLOSER, 4th International Conference on Cloud Computing and Services Science, pp. 489–498 (2014).
7. Xiao, Z., and Yang, X.: Security and privacy in cloud computing. IEEE Communications Surveys & Tutorials. 15(2), 843–859 (2013).
8. Sundareswaran, S., Squicciarini, A. C., and Lin, D.: Ensuring distributed accountability for data sharing in the cloud. IEEE Transactions on Dependable and Secure Computing. 9(4), 556–568 (2012).
9. Padhy R. P., Patra, M. R. and Satapathy, S. C.: Cloud computing: security issues and research challenges. International Journal of Computer Science and Information Technology & Security (IJCSITS), 136–146(2011).
10. Ullah, K. and Khan, M. N. A.: Security and Privacy Issues in Cloud Computing Environment: A Survey Paper. International Journal of Grid and Distributed Computing, 7(2), 89–98(2014).
11. Subashini, S., and Kavitha, V.: A survey on security issues in service delivery models of cloud computing. Journal of network and computer applications, 34(1), 1–11(2011).
12. Zissis, D., and Lekkas, D.: Addressing cloud computing security issues. Future Generation computer systems, 28(3), 583–592(2012).
13. Jena, T., Mohanty, J. R., Sahoo, R.: Paradigm Shift to Green Cloud Computing. Journal of Theoretical & Applied Information Technology, 77(3), 394–402 (2015).
14. Flaherty, P. D., & Ruscio, G.: Stormy Weather: Jurisdiction over Privacy and Data Protection in the Cloud, Internet and E Commerce Law in Canada, (2013).
15. Hörnle, J.: Online dispute resolution in business to consumer e-commerce transactions. Journal of Information, Law and Technology, 2, (2002).
16. National Institute of Standards and Technology, https://scap.nist.gov/publications/index.html

A Modified Genetic Algorithm Based FCM Clustering Algorithm for Magnetic Resonance Image Segmentation

Sunanda Das and Sourav De

Abstract In this article, we have devised modified genetic algorithm (MfGA) based fuzzy *C*-means algorithm, which segment magnetic resonance (MR) images. In FCM, local minimum point can be easily derived for not selecting the centroids correctly. The proposed MfGA improves the population initialization and crossover parts of GA and generate the optimized class levels of the multilevel MR images. After that, the derived optimized class levels are applied as the initial input in FCM. An extensive performance comparison of the proposed method with the conventional FCM on two MR images establishes the superiority of the proposed approach.

Keywords Segmentation · Clustering algorithm · Fuzzy *C*-Means algorithm · Genetic algorithm

1 Introduction

In medical image processing, segmentation is widely used as it is considered as a complex, challenging task due to the intrinsic property of medical images. Accurate segmentation is needed to analyze the affected area in medical imaging so that computer-aided segmentation techniques can be effectively applied. An attempt has been made here to detect the affected portions in MRI imaging technology using

S. Das (✉)
Department of Computer Science & Engineering,
University Institute of Technology, The University of Burdwan,
Burdwan 713104, West Bengal, India
e-mail: sunanda0301@gmail.com

S. De
Department of Computer Science & Engineering,
Cooch Behar Government Engineering College,
Cooch Behar, West Bengal, India
e-mail: sourav.de79@gmail.com

© Springer Nature Singapore Pte Ltd. 2017   435
S.C. Satapathy et al. (eds.), *Proceedings of the 5th International Conference on Frontiers in Intelligent Computing: Theory and Applications*, Advances in Intelligent Systems and Computing 515, DOI 10.1007/978-981-10-3153-3_43

such segmentation algorithms, which will facilitate the easy and exact detection of regions of interest using computer systems eliminating the need to identify manually.

Out of the many such methods, we are considering the fuzzy C-means (FCM) [1], a soft clustering procedure. FCM allows pixels to have relation with multiple clusters with a varying membership degree and thus more reasonable in real applications. But FCM has also some demerits. First, it does not consider the spatial information thereby is highly sensitive to noise and imaging artifacts. Ahmed et al. [2], Chuang et al. [3], Yang et al. [4], Adhikari et al. [5] have developed different algorithm to incorporate the spatial information into conventional FCM for better segmentation of images. Second, during acquisition of MR Images irregular biasing of intensity values of pixels generally occurs which leads to noise incorporation in the images. In [6] a nonlinear enhancement function in wavelet domain is implemented and also an iterative enhancement algorithm using morphological filtering is proposed for further enhancing the edges. Manjón et al. proposed in [7] two new methods for denoising MR images, which exploits the sparseness and self-similarity properties of image. In [8] a modified anisotropic diffusion algorithm is derived to improve the estimation of the diffusion constant to facilitate better edge detection. Third problem is that a local optimal solution due to poor initialization of cluster centres is generated and it often converges to local minima. Many soft computing approaches like genetic algorithm [9, 10], particle swarm optimization [11] have been applied to improve this problem. In Jhansi and Subashini [12], a GA based FCM algorithm is proposed.

Here we have also proposed a modified GA based FCM method to overcome the above described third problem of FCM. In the population generation stage, the chromosomes are generated on the basis of weighted mean approach. The crossover probability is varied throughout the generation stages. The resultant optimized class levels, derived by the MfGA, are employed as the input of the FCM algorithm. The proposed MfGA based FCM algorithm is compared with the traditional FCM algorithm [1]. Two standard MRI images and two standard fitness functions are used in regard. It is found that the proposed MfGA based FCM algorithm gives better result than the standard FCM algorithm to segment the multilevel MR images.

2 Modified Genetic Algorithm (MfGA)

Like conventional GA, MfGA provides a near-optimal solution in a large, complex, and multimodal problem space. MfGA is the modified version of GA where a small change is applied in population initialization and crossover part. Steps of MfGA are discussed below:

2.1 Population Initialization

In GA, we randomly choose n number of class levels to produce n number of clusters that is a demerit. To overcome the problem of little difference between two classes, which may happen in GA, here, we have chosen $n + 1$ number of temporary class levels at the first time within the range of maximum and minimum gray levels of the test image represented as $R_1, R_2, R_3, \ldots, R_{n+1}$ randomly. Then from these $n + 1$ class levels we have to produce n number of classes. The original class levels viz. L_i are generated after taking the weighted mean between the temporary class levels of R_i and R_{i+1}. The weighted mean is calculated using the following formula:

$$L_i = \frac{\sum_{j=R_i}^{R_{i+1}} f_j * I_j}{\sum_{j=R_i}^{R_{i+1}} f_j} \qquad (1)$$

where f_j is the frequency of the jth pixel and I_j shows the intensity value of the jth pixel. Now the weighted mean value of the class levels are taken to generate the original chromosome. Thus, we get the actual class levels of the chromosomes, i.e., $L_1, L_2, \ldots, L_i$. This helps in reducing the percentage of misclassification and increasing the accuracy in the segmentation.

2.2 Selection

Selection operator determines which individuals are chosen for mating purpose based on their fitness value. Here, we apply Roulette Wheel selection procedure, a proportional selection algorithm where the number of copies of a chromosome that goes into the mating pool for subsequent operations is proportional to its fitness.

2.3 Crossover

Crossover is performed on more than one parent solutions to produce a child solution from them. Crossover probability is to indicate a ratio of how many couples will be picked for mating. Unlike GA, here, a new crossover probability which is inversely proportional to the number of iteration is introduced. In other words, we can say that the crossover probability will decrease as the number of iteration is increased. So that the better chromosomes are remain unchanged and go to the next generation of population. This can be done with the given equation:

$$C_p = C_{\max} - \left(\frac{C_{\max} - C_{\min}}{IT_{\max} - IT_{\text{cur}}} \right) \qquad (2)$$

where C_p is the crossover probability, $C_{\max}$ is the maximum crossover probability (which is taken here 0.8), $C_{\min}$ is the minimum crossover probability (which is taken here 0.5). $IT_{\max}$ is the maximum number of iteration to be done on the genetic algorithm and IT_{cur} is the current iteration that is going on.

2.4 Mutation

After crossover, mutation is done based on mutation probability (0.01), which is basically a measure of the likeness that random elements of considered chromosome will be flipped into something else.

3 Proposed Methodology

In this paper, we want to introduce our proposed method in which we have improved the drawback of FCM like: it will easily converge to the local minima point and the clustering will be affected if the initial centroids values are not correctly initialized. To overcome this drawback, we first need to optimize the initial centroids values of FCM. To solve this problem we have incorporated modified GA with FCM. The advantage of using MfGA is that it gives us a global optimum value. In this proposed method, we have used the output of the MfGA as the input of FCM. Algorithm of proposed method is given below:

3.1 Algorithm 1: Modified Genetic Algorithm Based FCM

Step 1: Initialize the parameters—the number of clusters c, population size N, maximum crossover probability (0.8), and mutation probability (0.01) and error $\varepsilon(0.1 \times 10^{-6})$.

Step 2: Initialize the population: N numbers of chromosomes are created where each chromosome contains $l + 1$ class level. From $l + 1$ class level generate l class level using weighted mean formula Eq. (1).

Step 3: Fitness function of each individual is now calculated.

Step 4: Best-fitted chromosomes are selected through Roulette wheel selection method and taken part to crossover and mutation against a crossover and mutation probability (described in Sect. 2) to produce new off spring.

Step 5: Judge the termination condition of evolution, if it is satisfied then the evolution stops, otherwise go to step 3.

Step 6: The best individual generated by modified genetic algorithm is decoded and got the global optimized cluster centres.

Step 7: Now this cluster centres have input as the initial value of FCM algorithm. And then get the final segmented results by FCM.

4 Experimental Result

We have performed the above proposed method on two standard MR images [13, 14] and analyze the quality of the segmented images based on two standard quantitative measures: correlation coefficient (ρ) [15] and empirical measure $(\mathbf{Q})$ [16]. The class levels generated by the MfGA based FCM and conventional FCM on the basis of the two fitness function $(\rho$ and Q) are tabulated in Tables 1 and 2. Each type of experiments is done for 50 times against 8 clusters but only three good results of each method are tabulated here. The best result obtained by each process is boldfaced. These tables emphasis that the values of correlation coefficient and empirical measure of our proposed method are far better than the conventional FCM, thus it proves the superiority of the proposed method. Mean and standard deviations and also the run time of FCM method are evaluated for the two algorithms using different fitness functions for two MR image and these results are reported on Tables 3 and 4. From Tables 3 and 4 it is clear that not only the quality of the segmented image are enhanced using our proposed method but also using it FCM takes less time than the conventional FCM.

Table 1 Performance of FCM and MfGA based FCM of MR Image 1

Fitness function	Method	Class levels	Performance
Correlation coefficient (ρ)	FCM	0, 21, 55, 72, 87, 125, 176, 187	0.9623
		0, 16, 44, 63, 74, 82,110, 147	0.9752
		1, 27, 64, 83, 111, 146, 188, 242	**0.9793**
	MfGA based FCM	0, 15, 54, 76, 102, 135, 174, 232	0.9948
		0, 17, 63, 85, 120, 157, 200, 247	**0.9953**
		0,15, 57, 76, 102, 139, 186, 241	0.9950
Empirical measure (Q)	FCM	0, 27, 64, 84, 112, 146, 187, 240	52325.50
		0, 14, 33, 57, 73, 95, 140, 175	34228.94
		0, 22, 54, 69, 82, 94, 171, 178	**24581.86**
	MfGA based FCM	0, 11, 46, 70, 90, 121, 162, 229	6344.95
		0, 12, 38, 60, 81, 121, 169, 235	**3519.88**
		0, 12, 52, 70, 86, 111, 150, 221	5700.98

Table 2 Performance of FCM and MfGA based FCM of MR Image 2

Fitness function	Method	Class levels	Performance
Correlation coefficient (ρ)	FCM	0, 13, 29, 44, 61, 84, 109, 229	0.8731
		0, 21, 42, 60, 83, 107, 173, 245	0.9130
		0, 19, 39, 80, 104, 127, 160, 242	**0.9613**
	MfGA based FCM	**0, 20, 41, 60, 83, 107, 173, 245**	**0.9913**
		0, 18, 38, 55, 75, 98, 125, 237	0.9910
		0, 14, 29, 43, 58, 81, 107, 232	0.9907
Empirical measure (Q)	FCM	**0, 18, 39, 56, 79, 102, 141, 238**	**25296.40**
		1, 21, 42, 60, 83,107, 173, 245	56753.60
		0, 16, 35, 52, 71, 92, 115, 236	63698.00
	MfGA based FCM	**0, 13, 28, 42, 58, 80, 107, 231**	**12536.58**
		0, 15, 32, 47, 63, 84, 109, 231	14195.46
		0, 14, 31, 46, 61, 84, 109, 234	13963.00

Table 3 Mean and standard deviation using different fitness function and mean time for different algorithm for MR Image 1

Method	Correlation coefficient		Empirical measure	
	Mean ± Std. Div	Mean time (s)	Mean ± Std. Div	Mean time (s)
FCM	0.9789 ± 0.0091	22	20661.27 ± 16059.13	14
MfGA based FCM	0.9947 ± 0.0005	4	9507.81 ± 3154.19	3

Table 4 Mean and standard deviation using different fitness function and mean time for different algorithm for MR Image 2

Method	Correlation coefficient		Empirical measure	
	Mean ± Std. Div	Mean time (s)	Mean ± Std. Div	Mean time (s)
FCM	0.9401 ± 0.0431	31	31518.88 ± 37483.18	27
MfGA based FCM	0.9904 ± 0.0017	8	29485.48 ± 22856.10	10

In Figs. 1 and 2, the values of the evaluation functions are plotted in graphs using 15 good results out of 50 results. The segmented MR output images obtained using the proposed approach and also the conventional FCM algorithm is

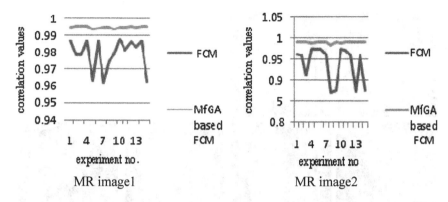

Fig. 1 Comparison between correlation coefficient values of FCM and MfGA based FCM

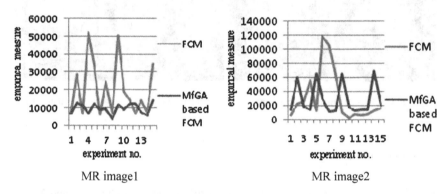

Fig. 2 Comparison between empirical measures of FCM and MfGA based FCM

demonstrated afterwards. The original and segmented MR test images derived by the FCM algorithm and the MfGA based FCM algorithm are depicted in Fig. 3. The original MR images are presented in Fig. 3a–b; the segmented images by the FCM algorithm are presented in Fig. 3c–f; the segmented image derived by MfGA based FCM are presented in Fig. 3g–j. It is observed from all the images in Fig. 3 that the proposed approach gives better segmented outputs than the same derived by conventional FCM. At the end, it can be concluded that the proposed MfGA algorithm gives better results than the FCM algorithm in respect of quantitatively and qualitatively.

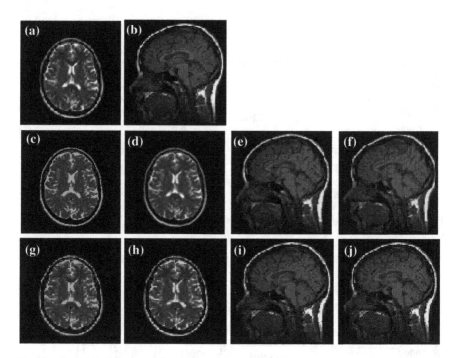

Fig. 3 a–b: Original image; **c–f**: segmented image using FCM; **g–j**: segmented image using MfGA based FCM

5   Conclusion

FCM is an algorithm that entertains with a result of local search ability and MfGA produces a result with global search ability. So, we have merged these two algorithms and generate an algorithm, which produces a good segmentation result with optimized cluster centres. In this paper, we compared our proposed method with pure FCM and it is proved that our proposed method gives better performance quantitatively and also qualitatively.

References

1. Bezdek,J.C.: Pattern recognition with fuzzy objective function algorithms. New York, NY: Plenum (1981)
2. Ahmed, M.N., Yamany, S.M., Mohamed, N., Farag, A.A., Moriarty, T.: A modified fuzzy C-means algorithm for bias field estimation and segmentation of MRI data. IEEE Transactions on Medical Imaging 21 (3), 193–199 (2002)
3. Chuang, K.S., Tzeng, H.L., Chen, S., Wu, J., Chen, T.J.: Fuzzy C-means clustering with spatial information for image segmentation. Computerized Medical Imaging and Graphics 30 (1), 9–15 (2006)

4. Yang, Z., Chung, F.L., Shitong, W.: Robust fuzzy clustering-based image segmentation. Applied Soft Computing 9 (1), 80–84 (2009)
5. Adhikari, S. K., Sing, J. K., Basu, D. K., Nasipuri, M.: Conditional spatial fuzzy C-means clustering algorithm for segmentation of MRI images, Applied Soft Computing 34,758–769 (2015)
6. Srivastava, A., Alankrita, Raj, A., Bhateja, V.:Combination of Wavelet Transform and Morphological Filtering for Enhancement of Magnetic Resonance Images, Proc. of International Conference on Digital Information Processing and Communications (ICDIPC 2011), Part-I, Ostrava, Czech Republic, CCIS-188. 460–474 (2011)
7. Manjón, J. V., Coupé, P., Buades, A., Collins, D. L., Robles, M.:New Methods for MRIdenoising based on sparseness and self-similarity. Medical image analysis, 16(1), 18–27 (2012)
8. Srivastava, A., Bhateja, V., Tiwari, H.: Modified Anisotropic Diffusion Filtering Algorithm for MRI, Proc. (IEEE) 2nd International Conference on Computing for Sustainable Global Development (INDIACom-2015). 1885–1890 (2015)
9. Nie, S., Zhang, Y., Li, W., Chen, Z.: A fast and automatic segmentation method of MR brain images based on genetic fuzzy clustering algorithm, Proc. of International Conference on Engineering in Medicine and Biology Society.5628–5633 (2007)
10. Hall, L. O., Ozyurt,I.B., Bezdek, J. C.: Clustering with a genetically optimized approach, IEEE Trans. Evol. Comput. 3 (2),103–112 (1999)
11. Li,L., Liu, X., Xu,M.: A novel fuzzy clustering based on particle swarm optimization. First IEEE International Symposium on Information Technologies and Applications in Education. 88–90 (2007)
12. Jansi, S., Subashini, P.: Modified FCM using Genetic Algorithm for Segmentation of MRI Brain Images. 2014 IEEE International Conference on Computational Intelligence and Computing Research. 1–5 (2014)
13. http://www.imaios.com/en/e-Anatomy/Head-and-Neck/Brain-MRI-in-axial-slices
14. http://www.imaios.com/en/e-Anatomy/Head-and-Neck/Brain-MRI-3D
15. De, S., Bhattacharyya, S., Dutta, P.: Efficient grey-level image segmentation using anoptimised MUSIG (OptiMUSIG) activation function. International Journal of Parallel, Emergent and Distributed Systems, 26(1), 1–39 (2010)
16. Borsotti, M., Campadelli, P., Schettini, R.: Quantitative evaluation of color image segmentation results. Pattern Recognition Letters 19(8), 741–747 (1998)

Skill Set Development Model and Deficiency Diagnosis Measurement Using Fuzzy Logic

Smita Banerjee and Rajeev Chatterjee

Abstract A skill set is the ability of performing a particular job. Skill set is acquired by improving the psychomotor domain of the human being. Deficiencies in skills need to be measured and addressed. This may improve the level of skill and reduce deficiency. Deficiency diagnosis is a process of identification of the skills that are lacking in any learner. In this research work, authors have proposed a model that identifies the various deficiencies of a learner.

Keywords E-learning · Skill development · Deficiency diagnosis

1 Introduction

In today's scenario, e-learning [1] has taken an important role in education system. E-learning has become most popular learning and skill development [2] method in recent days. This has become possible as learner is the center of learning in this process. E-learning is a learning methodology using electronic technology to help in acquisition and comprehension of knowledge and the progress can be measured from the behavior pattern of a learner. To improve and emphasize the skill and knowledge of a learner, the only way is assessment [3]. Deficiency measurement [4] is one of the important steps in assessment. In this paper, the authors proposed a model that will find the deficiency of skill set that needs to overcome during the learning process.

The paper is organized as follows. Section 2 describes the review of the existing works. Section 3 explains the proposed skill set development model. Learner assessment mechanism is presented in Sect. 4. Section 5 describes the result and analysis. Conclusion and future scope is given in Sect. 6.

S. Banerjee (✉) · R. Chatterjee
National Institute of Technical Teachers' Training and Research, Kolkata, India
e-mail: smitabanerjee69@gmail.com

R. Chatterjee
e-mail: chatterjee.rajeev@gmail.com

© Springer Nature Singapore Pte Ltd. 2017
S.C. Satapathy et al. (eds.), *Proceedings of the 5th International Conference on Frontiers in Intelligent Computing: Theory and Applications*, Advances in Intelligent Systems and Computing 515, DOI 10.1007/978-981-10-3153-3_44

2   Review of Existing Works

2.1   On Skill Hierarchy and Type of Skills

Kokcharov [5] gave a suggestive hierarchy for skill. It includes five levels: Know, Play, Work, Solve and Invent. Another classification includes roles, i.e., Student, Apprentice, Specialist, Expert and Craftsman. Figure 1 illustrates the hierarchy of skills.

Ghirardini [6] proposes that skills are of three different types such as cognitive skills, interpersonal skills and psychomotor skills. E-learning can be used mostly to develop cognitive and interpersonal skills. However, with certain tools and techniques in e-learning, now-a-days development of skill set related to psychomotor may also be enhanced, such as training on simulators.

2.2   On Fuzzy Logic

Fuzzy logic is based on the fuzzy set theory introduced by Zadeh [7] in 1965. The triangular function [8] shown in Fig. 2 as well as the mathematical formulae are described below.

$$\mu_A(x) = \begin{array}{ll} 0, & \text{if} \quad x \le a \\ \frac{x-a}{m-a}, & \text{if} \quad a \le x \le m \\ \frac{b-x}{b-m}, & \text{if} \quad m \le x \le b \\ 0, & \text{if} \quad x \ge b \end{array}$$

A membership function [9] $\mu_A(x)$ for a fuzzy set A on the universe of discourse x is defined as $\mu_A: x \rightarrow [0, 1]$, where each element of x is mapped to a value between

Fig. 1 Hierarchy of skills

Fig. 2 Shape of triangular function

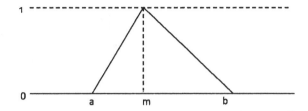

0 and 1. This is called membership value or degree of membership. The X axis represents the universe of discourse, whereas the Y axis represents the degrees of membership in the [0, 1] interval. Triangular function $\mu_A(x)$ defined by a lower limit a, an upper limit b and a value m, where $a < m < b$.

2.3 On Students' Group and Individual Assessment

Michael has proposed [10, 11] a mathematical model where he has used two alternative fuzzy measures: one, Shannon's formula for measuring a system's probabilistic uncertainty and another is the centre of mass method which is represented by a fuzzy set. Students' individual knowledge and skill was measured using fuzzy logic [11].

2.4 On Academic Performance Evaluation

Yadav et al. have proposed [12] assessment technique that does not focus on categorical evaluation of the subject and learner is diagnosed on the entire subject in a semester wise manner.

3 Proposed Methodology

Our proposed skill set development model has two phases: (i) assessment and (ii) deficiency diagnosis [4]. Figure 3 illustrate our proposed skill set development model.

Assessment will be taken in two parts, i.e., preliminary and main. If any student achieve minimum qualifying marks in the preliminary then he/she will go to the second part of the assessment, i.e., main. In the main part, there will be three different categories such as knowledge of the subject matter, problem solving abilities and analogical reasoning [11]. Marks will be generated by adding both preliminary and main parts of the assessment.

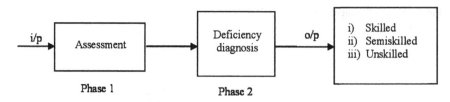

Fig. 3 Skill set development model

Assessment will be taken as:

1. Preliminary
2. Main

- Knowledge of the subject matter
- Problem solving abilities
- Analogical reasoning

3.1 Proposed Algorithm

PrelimExamProgram-Function The input dataset is stored in a file as a sequence of <key,value> pair, each of which represents a record in the dataset. The key is the students' name and value is the preliminary marks of each student. For each student, check that the preliminary mark is greater than or equal to the threshold mark. Only those candidates will qualify for the main exam. The pseudo code of PrelimExamProgram function is shown in Algorithm 1.

```
Algorithm 1. PrelimExamProgram (HM_Input, HM_Output)
Input: Randomly select a small group of students' name
and their preliminary exam marks from an unlabeled sample
pool
Output: <key', value'> Hash map pair, where key' contains
the name of the student and value' contains the
preliminary marks of the pass students
1. Initialize one Hash map to store the students' name
and their preliminary exam marks as a <key, value> pair;
2. Pre_th = Preliminary exam threshold mark;
3. FOR i = 1 to HM_Input.length do
4.    IF HM_Input[i].value(Premark ≥ Pre_th )
5.         MainExamProgram(HM_Input [i], HM_Output);
6.    END IF
7. END FOR
8. Output <key', value'> pair;
9. END.
```

MainExamProgram-Function After each Preliminary exam, dataset is stored in a file as a sequence of <key,value> pair in three different categories. After that we apply a combiner function to combine each student to pick the same key and append the values. The pseudocode of MainExamProgram function is shown in Algorithm 2.

```
Algorithm 2. mainExamProgram (Final_input, Final_Output)
Input: Collect the new training set in Final_input which
is same as HM_Output and assign a new label to each of
them i.e., X = {i₁, i₂, i₃...iₙ }
Declaration: S₁ = Knowledge of the subject matter, S₂ =
Problem solving skills, S₃ = Analogical reasoning
abilities.
Output: <key', value₁', value₂', value₃'> pair, where key'
contains the name of the student and value₁, value₂ and
value₃ contains the main Examination marks of S₁, S₂ and S₃
respectively.
1. S₁ U S₂ U S₃ ⊆ X;

2. Initialize three Hash map to store all the students'
name and their main exam marks as a <key, value₁>, <key,
value₂> and <key, value₃> Hash map pair respectively in
three different categories;
3. Use Combiner_Output for each student to pick the same
key and append the values;
4. Output <key', value₁', value₂', value₃'> pair;
5. END.
```

3.2 The Process of Deficiency Measurement

Deficiency of a student can be measured by the ratio of assessment result of that particular student and the maximum resultant value from the same assessment. The deficiency ratio of student i will be:

Stud_Def$_i$ = {(S_1, particular student's result/maximum result), (S_2, particular student's result/maximum result), (S_3, particular student's result/maximum result)}

For any particular student, if deficiency (Stud_Def) value equals to 1 then it signifies the performance value of that student's characteristic is maximum. So there is no need to define the skill category for that particular student. When comparing the range of deficiency, it should always consider the lower limit of the range (Table 1).

Table 1 The ranges between the performances of success

Deficiency	Range	Type of skills
Stud_Def=	0–0.33	Unskilled
	0.33–0.67	Semiskilled
	0.67–1	Skilled

4 Learner Assessment Mechanism

In Table 2, marks of five students (Stud$_1$.Stud$_5$) are given and the assessment follows the proposed algorithm. In this case, the proposed threshold mark in preliminary part is taken as 40.

According to algorithm 1, each student data will be validated. If student Stud$_i$ $1 \leq i \leq 5$ mark is more than or equal to threshold value, i.e., ≥ 40 in this case proceed for algorithm 2.

According to algorithm 2, qualified students from preliminary part are subject to proposed three categories in main part, i.e., S_1, S_2 and S_3. Marks of individual students based on three categories are evaluated.

5 Result and Analysis

In this section, the authors have taken the results and analysis has been made. Section 5.1 describes the evaluation model. Section 5.2 illustrates the Mamdani system using fuzzy logic.

5.1 The Evaluation Model

We are assessing a group of five students and students' individual performances after assessment are shown in below form:

Stud$_1$ = {(S_1, 0.25), (S_2, 0.5), (S_3, 0)}
Stud$_2$ = {(S_1, 0.5), (S_2, 0), (S_3, 0.75)}
Stud$_3$ = {(S_1, 0.75), (S_2, 0.5), (S_3, 0)}
Stud$_4$ = {(S_1, 0), (S_2, 0.25), (S_3, 0.5)}
Stud$_5$ = {(S_1, 0.5), (S_2, 0.75), (S_3, 0.25)}

Here we have calculated the deficiency ratio by measuring particular student's marks with respect to the highest result in same test:

Table 2 The mark of the Students

Sl. no	Name of the student	Preliminary exam marks	Main exam marks		
			S_1	S_2	S_3
1	Stud$_1$	70	20	60	59
2	Stud$_2$	50	40	37	60
3	Stud$_3$	80	20	80	79
4	Stud$_4$	34	–	–	–
5	Stud$_5$	25	–	–	–

Stud_Def$_1$ = {(S_1, 0.33), (S_2, 0.67), (S_3, 0)}
Stud_Def$_2$ = {(S_1, 0.67), (S_2, 0), (S_3, 1)}
Stud_Def$_3$ = {(S_1, 1), (S_2, 0.67), (S_3, 0)}
Stud_Def$_4$ = {(S_1, 0), (S_2, 0.33), (S_3, 0.67)}
Stud_Def$_5$ = {(S_1, 0.67), (S_2, 1), (S_3, 0.33)}

On comparing the above shown students' deficiency diagnosis, for student Stud_Def$_1$, in the section of knowledge of the subject matter, student Stud_Def$_1$ get 0.33, i.e., matched with the range (0.33–0.67). So we can conclude that student Stud_Def$_1$ is a semiskilled person in knowledge of the subject matter. Similarly, on problem solving ability section, we can conclude that student Stud_Def$_1$ is a skilled person. The performance of five students and their respective skill categories are shown in Table 3.

5.2 Mamdani System Using Fuzzy Logic Toolbox

The technique to identify the skill level of a learner is simulated in MATLAB [13] using the Mamdani type fuzzy inference system (FIS). Three input variables (knowledge of the subject matter, problem solving abilities and analogical reasoning) were loaded each having membership function value as unskilled, semiskilled, and skilled. The output variable (skill level) has seven membership functions namely extremely bad, very bad, bad, fair, good, very good and extremely good. The fuzzy rules governing the operation of the fuzzy logic controller are specified in Fig. 4.

From this rule editor S_1 = 0.235 (unskilled), S_2 = 0.416 (semiskilled), S_3 = 0.724 (skilled) and output (skill level) is 0.569, i.e., fair. The rule editor is given in Fig. 4.

Rules given in Fig. 4 are applied to the inputs and the output of the Mamdani type fuzzy inference system-based controller and displayed during simulators

Table 3 The categories of each student as per their assessment result

Sl. no.	Name of the student	S_1	S_2	S_3
1	Stud$_1$	Semiskilled	Skilled	Unskilled
2	Stud$_2$	Skilled	Unskilled	–
3	Stud$_3$	–	Skilled	Unskilled
4	Stud$_4$	Unskilled	Semiskilled	Skilled
5	Stud$_5$	Skilled	–	Semiskilled

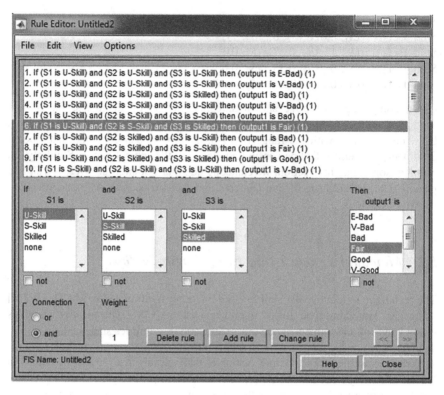

Fig. 4 Rule editor

shown in Fig. 5. By varying individual rules in the rule viewer, we conclude which rules are active or how these rules influence the results.

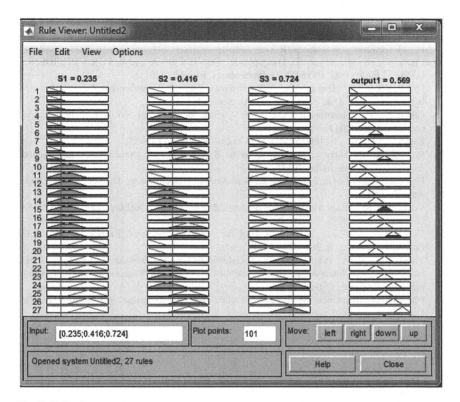

Fig. 5 Rule viewer

6 Conclusion and Future Scope

The proposed model tries to identify the deficiency of learners in the job environment. The model is working using fuzzy inference system. The proposed work is based upon certain threshold assumptions. In future, the authors of this paper may propose a system that may have mechanism for adjustment of the threshold properties intelligently. However, it is beyond the scope of the existing research work.

References

1. Mayer, R.: Elements of a Science of E-Learning. Journal of Educational Computing Research. 29, 297–313 (2003).
2. Lahwal, F., Amaimin, M., Al-Ajlan, A.: Perception Cultural Impacts: Principles for Trainer's skills for E- Learning. 986–993 (2009).

3. Hettiarachchi, E., Huertas, M., Mor, E.: E-assessment in high-level cognitive courses: Improving student engagement and results. 2014 14th International Conference on Advances in ICT for Emerging Regions (ICTer). (2014).
4. Chatterjee, R., Mukherjee, S., Dasgupta, R.: DESIGN OF AN LMS FOR CONFIDENCE BASED LEARNING. INTED2011 Proceedings. 619–626 (2011).
5. Igor Kokcharov, P.: Hierarchy of Skills, http://www.slideshare.net/igorkokcharov/kokcharov-skillpyramid2015, Date of Access (DOA): September 10, 2015.
6. Ghirardini, B.: E-learning methodologies A guide for designing and developing e-learning courses., Rome (2011).
7. Zadeh, L.: Fuzzy sets. Information and Control. 8, 338–353 (1965).
8. Roy, S., Chakraborty, U.: Introduction to Soft Computing: Neuro-Fuzzy and Genetic Algorithm. Pearson, India (2013).
9. Hajek, P.: Fuzzy Logic, http://plato.stanford.edu/entries/logic-fuzzy/, Date of Access (DOA): April 7, 2016.
10. Voskoglou, M.: Fuzzy Measures For Students` Mathematical Modelling Skills. IJFLS. 2, 13–26 (2012).
11. Voskoglou, M.: Fuzzy Logic as a Tool for Assessing Students' Knowledge and Skills. Education Sciences. 3, 208–221 (2013).
12. Yadav, R., Soni, A., Pal, S.: A study of academic performance evaluation using Fuzzy Logic techniques. 2014 International Conference on Computing for Sustainable Global Development (INDIACom). (2014).
13. Performance Evaluation by Fuzzy Inference Technique. International Journal of Soft Computing and Engineering. 3, (2013).

A Study on Various Training Programmes and Their Effects Offered by the IT Firms

Pattnaik Manjula and Pattanaik Balachandra

Abstract The competency of an employee is evaluated or identified through some techniques. The success of the Training Program can be evaluated only after it meets the need for which it was called for. This research will help the organization to scale up their training tools and methodology for any kind of training within the organization. An attempt is being made to identify the awareness on satisfaction level of the employees of an IT firm. The nature and behavior of employees are described by descriptive research design. Both the data collection methods have been used and selection of study area includes IT firms, it was selected for conducting the survey based on the judgment of the employees of all cadres. The sample size is 50. Analysis and interpretation of data being done by using statistical tools as percentage method and Chi-square method.

Keywords Training programs · Chi-square test · Descriptive research

1 Introduction

In current trend, Training Programmes are highly essential to improve the productivity. They have to be well organized. The following steps are considered necessary in any Training Programmes run by an organization: Identifying the Training needs, Roll out the Training Calendar, Send an Invite to the Participants, skill testing, Evaluation of output, Followed by giving rewards and taking Feedback. The software industry is the most important source to improve the finance

P. Manjula (✉)
Department of Accounting & Finance, Faculty of Business and Economics,
Mettu University, Mettu, Ethiopia
e-mail: drmanjula23@gmail.com

P. Balachandra
Department of Electrical and Computer Engineering, Faculty of Engineering
and Technology, Mettu University, Mettu, Ethiopia
e-mail: balapk1971@gmail.com

© Springer Nature Singapore Pte Ltd. 2017
S.C. Satapathy et al. (eds.), *Proceedings of the 5th International Conference on Frontiers in Intelligent Computing: Theory and Applications*, Advances in Intelligent Systems and Computing 515, DOI 10.1007/978-981-10-3153-3_45

position of the countries by creating jobs to millions of people throughout the world. In the past few years, the Indian software industry has been growing by 50% compounded annually. The average training budget for large companies was $17.4 million, while midsize companies allocated an average of $1.5 million, and small companies dedicated an average of $338,709 [1]. IT industry is knowledge-based industry, so the training programmers and the trainers who are the knowledge workers have to provide and keep themselves filled with latest and emerging technologies, which help the industry to work effectively [2]. Training programmes are very essential for the industry to update its workforce regularly to compete the demand of global market. Infosys Company achieved many awards from American Society for Training and Development (ASTD) as the world's best company for providing best training development opportunities to its employees which leaves a remarkable recognition in world. Its profitability was revised by running its global business foundation school which provides various programmes for all fresher's to prepare them for the technical environment opportunities available in market. Similarly IBM has a separate and distinct IBM educational department which was especially established to train and development employees. IBM is known for the use of e-learning programs to address its learning solutions by interaction, simulation, games, collaborative learning, etc. IBM connection is the most popular programme of IBM which helps the company to enhance its young skilled workers and improve their productivity as well as the organizations [3].

1.1 Objectives

General Objective: To study the various training programmes and their effects offered by the IT firms to their employees.
  Specific Objectives:

- To study the awareness level of the employees regarding the various training programmes.
- To analyze the employees level of satisfaction with the existing training programmes.
- To find out the real training need of the employees.

2 Literature Review

Indian IT firms are focused to bring maximum growth in economy. They are well experienced in managing various human resource issues and challenges. The finance of IT professionals increases by 25–30% annually. IT firms of China, Philippines, and Singapore are competitors to India. According to Michael J. Jucios, "training is any process by which the attitudes, skills and abilities of employees to perform specific jobs are improved" [4]. At present, in India IT firms have been working on the lower end of the value chain and keep on enforcing to move up for retaining the competition. This implies the firms working for technological advancement and product development. In turn, these initiatives require availability of experienced and more competent software professionals for sufficiently long period of time. There is a short supply of IT professions in USA and Europe and literally throughout the world. There is growing demand of well-qualified and skilled Indian IT professionals USA and other developed countries. So there are a large number of Indian software professionals who have been moving to USA and Europe. The development activities are designed for a systematic planning for the future organizational requirement and to improve the existing performance [5].

Steps in training programs:

Training programmes are very useful and play an important role for upgrading the IT industry. Therefore, they have to be performed very carefully. The following are the required steps of an effective training program:

Identifying the Training needs, Roll out the Training Calendar, Send an Invite to the Participants, skillful performance and acquiring knowledge, Evaluation of output, encouragement through Rewards and collecting Feedback.

Projects of different platforms have to be handled by Indian IT firms using latest technology. It is leading to a situation whereby software organizations exclusively working on lower levels of value chain increasingly find it difficult to attract and retain competent software professionals [6, 7].

So, the success of Indian software organization is depending upon the ability and efficiency by resolving conflicts emerging from the requirements of the market. It is in this context that this study explores and strives to explain human issues and challenges experienced by the Indian software industry.

3 Research Methodology

Descriptive research design is used for this research. The sampling method adopted for this study is convenience sampling under non-probability sampling [8]. Both primary and secondary data were used for this study. Primary data collected by distributing questionnaires to different employees of software companies, apart from that interviews were conducted with selected persons of the companies using judgmental sampling. The various sources of secondary data are books, periodicals, journals, directories, magazines, statistical data sources, etc. The secondary source used for this study is company profile, scope, need, review of literature. Due to improper response of the employees for collection of data due to their busy scheduled work, unwillingness, and unawarance, convenience sampling is used and the sample size is 50. The units selected may be each person who comes across the investigator [9].

4 Data Analysis and Interpretation

Inference: Table 1 clearly shows that all the respondents agreed that the training given by the company is need-based.

Inference: From Table 2 and Fig. 1, it was found that 62% of the respondents felt that the training programmes help them to update their knowledge, whereas 32% of the respondents felt that it would improve their knowledge and 6% of the respondent say that it is compulsory to meet the challenges.

Inference: In Table 3 it is clear that 88% of the respondents feel that their understanding about the concept and the theory is good, whereas 12% of the respondents feel that their understanding is excellent.

Inference: From the Table 4 and Fig. 2, it is clear that 94% of the respondents agree that they are satisfied, whereas 6% of the respondents strongly agree that they are satisfied with the result of the training programmes organized by the company.

Inference: Table 5 and Fig. 3, clearly shows that 90% of the respondents feel that the training programmes organized by the company will increase their productivity.

Table 1 Analysis on need based training program

S. No	Particulars	No of respondents	Percentage
1	Yes	50	100
2	No	0	0
	Total	50	100

Table 2 Opinion about training programmes

Parameters	No. of respondents	Percentage
Improve performance	16	32
Useful to update	31	62
Compulsory	3	6
Any	0	0
Total	50	100

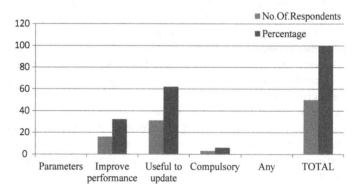

Fig. 1 Opinion about training programmes

Table 3 Analysis on the result of training programmes

Parameters	No. of respondents	Percentage
Excellent	6	12
Good	44	88
Average	0	0
Poor	0	0
Total	50	100

Table 4 Satisfactions level of the result of the training programmes

Parameters	No. of respondents	Percentage
Strongly agree	3	6
Agree	47	94
Disagree	0	0
Strongly disagree	0	0
Total	50	100

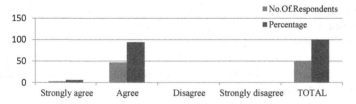

Fig. 2 Satisfactions level of the result of the training program

Table 5 Analysis on increase in productivity

Parameters	No.of.respondents	Percentage
Yes	45	90
No	5	10
Total	50	100

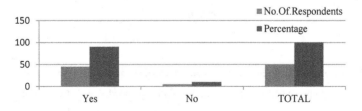

Fig. 3 Analysis on increase in productivity

5 Statistical Tools

5.1 Chi-Square Analysis

In sampling analysis for comparing a variance to a theoretical variance, chi-square test is used. Chi-square test is a significant test in all other statistical test available. It is an important nonparametric test. It can be used to determine categorical data, whether it may be dependant or independent. It is used to compare between theoretical population and actual data [10]. it is a method to judge the significance relationship between two attributes. We require only the degrees of freedom for using this test. Chi-square test is significant and can be ascertained by looking at the tabulated values of chi-square for given degrees of freedom at a certain level of significance.

H_0 = There is no significant relationship between number of training attended and the employees opinion (Table 6).

H_1 = There is a significant relationship between number of training attended and the employees opinion.

Formula: $Chi - square = (O - E)^2 / E$

where O = observed frequency

E = expected frequency

If two distributions (observed and theoretical) are exactly alike, the degree of freedom is worked out as follows:

Degree of freedom $= (c - 1)(r - 1)$

where 'c' means the number of columns which are 3 and 'r' means the number of rows which is maximum up to 4 can be considered [11] (Table 7).

$$\text{Level of Freedom} = 0.05,$$
$$\text{Degree of Freedom} = (r - 1) * (c - 1)$$
$$= (4 - 1) * (3 - 1)$$
$$= 3 * 2$$
$$= 6$$
$$\text{Table Value} = 12.6$$

Table 6 Relationship between training programs attended and the employees opinion

Parameters	Improve performance	Update skill and knowledge	Compulsory	Total
1–3	1	17	0	18
3–5	14	14	3	31
5–7	1	0	0	1
7–10	0	0	0	0
None	0	0	0	0
Total	16	31	3	50

Table 7 Calculation of Chi-square test

O	E	$(O - E)$	$(O - E)^2$	$(O - E)^2/E$
1	6	−5	25	4.16
17	11	6	36	3.27
0	1	−1	1	1
14	10	4	16	1.6
19	14	5	25	1.78
3	2	1	1	0.5
1	0.5	0.5	0.25	0.5
				12.81

Inference: Since the calculated value is greater than the table value, H_0 is rejected. There is significant relationship between the frequency of training programmes attended and the employee's opinion.

Test-2: Relationship between the Efficiency of The Trainer And The Quality Of The Training Programmes

H_0: There is no significant relationship between Efficiency of the trainer and Quality of the training programmes.

H_1: There is a significant relationship between Efficiency of the trainer and Quality of the training programmes

Formula: Chi-square $= (O - E)^2/E$

Where O = observed frequency

E = expected frequency

If two distributions (observed and theoretical) are exactly alike, the degree of freedom is worked out as follows:

Degree of freedom $= (c - 1)(r - 1)$

where 'c' means the number of columns which are 3 and 'r' means the number of rows which is maximum up to 4 can be considered (Table 8).

$$\textbf{Level of Significance} = 0.05,$$
$$\textbf{Degree of freedom} = (r - 1) * (c - 1)$$
$$= (4 - 1) * (3 - 1)$$
$$= 3 * 2$$
$$= 6$$
$$\textbf{Table Value} = 12.6$$

Table 8 Calculation of Chi-square test for test-2

O	E	$O - E$	$(O - E)^2$	$(O - E)^2/E$
5	3	2	4	1.33
20	17	3	9	0.53
0	5	−5	25	5
0	0	0	0	0
1	3	−2	4	1.33
14	17	−3	9	0.53
10	5	5	25	5
				13.72

Inference: Since the calculated value is greater than the table value, $H0$ is rejected. There is a significant relationship between the efficiency of the trainer and quality of the training programmes.

6 Findings and Suggestions

6.1 Findings

- All the respondents have idea about various training programmes.
- Majority of the respondents are happy with the time management in the training programmes.
- All the employees are fully satisfied with the trainers and the materials used in the training programmes.
- All the employees are allowed to raise question during the training section.
- All the employees are satisfied with the aids used in the training programmes.
- It is found that majority of the employees feel that their knowledge and skills, and productivity are increased because of the training programmes provided by the firms.
- Majority of the respondents prefer to have their training programmes inside the premises itself.
- It has been found that the training programmes are need-based and it has to be conducted at regular basis to provide adequate service for all the employees.
- All the employees are satisfied with the ambience in which the training programmes are conducted.

Majority of the employees feel that their feedbacks are considered by the management.

6.2 Suggestions

- Trainers should adopt innovative techniques, so that they can update their skill and knowledge much more.
- Training program needs to be conducted on the regular basis.
- More on the job training is needed which increases productivity.
- Separate training cell can be made to make training effective and to provide whenever it is necessary.
- Training officer has to focus more on developing employee's skills and ability.

7 Conclusion

From the study it is concluded that as for IT firms in concerned, their training programmes are well recognized, accepted, and welcomed by the employees. The training department is really taking sincere effort to improve the effectiveness through the training programmes. But still if they use new methods and techniques accordingly with the changing market demand, it will further enhance the employees to increase their skills and knowledge and not only that it will definitely increase their morale.

References

1. 'Training report November- December 2014", pp no. 16–29, www.trainingmag.com
2. "The Indian Information Technology Training Industry' International journal of Business Research, volume 7, Issue 2, March2007.
3. "Training and Development – Issue in the Indian Context" Global Journal of Finance and Management, @ Research India publication, volume 6 Number 7, PP no. 599–608, 2014.
4. Subba Rao 'Human Resource Management and Industrial Relations (Text and Cases)" New Edition Himalaya Publication, New Delhi, 2011.
5. 'The importance of Training And Development programs in Hotel Industry "International Journal of Business and Administration Research Review, Vol.1 Issue-5, PP no.50–56,. April-June 2014.
6. 'Employee Training and Development-A Case Study of State Bank of Hyderabad, Gulbarga City" by Dr.M.Surat kumara & Mallareddy Tatareddy.
7. Mirza's saiyadain Human Resource Management, Second Edition Tata Mc Graw Hill publishing company Ltd.2002.
8. Khanka S S Human Resource Management (Text and Cases) Third Edition, S Chand and Co. Ltd New Delhi 2010.
9. Gary Dessier, Human Resource Management, Eight Edition Pearsons Education. 2002.
10. Milkovich, Human Resource Management, Times mirror higher Education group.1997.
11. C.R Kothari, Research Methodology, Second edition, Wishova prakashan. 2002.

Analysis of Trustworthiness and Link Budget Power Under Free Space Propagation Path Loss in Secured Cognitive Radio Ad hoc Network

Ashima Rout, Anurupa Kar and Srinivas Sethi

Abstract The security issues of cognitive radio network have taken more attention recently due to the new challenges in wireless communication. Confronts related to such a debate seem more predominant in presence of the malicious secondary users when the transmission range of licensed users is shorter compared to the network size. In this paper, a model is introduced which verifies the distance both geographically and through the measuring distance obtained by received power of cognitive user. This is achieved by minimizing the interference to the primary licensed user and upon the faithful operation of the secondary user by calculating the trustworthiness of all users irrespective of their priority. This necessitates focus upon the free space propagation path loss of the transmitted signal. Thus, analysis of the trustworthiness becomes essential followed by calculation of link budget power that ensures the designing cost without extra overhead during a secured communication with the received power and free space propagation path loss.

Keywords Cognitive radio network (CRN) · WLAN · Trustworthiness · Path loss · Link budget

1 Introduction

Cognitive radio (CR) is a smart radio that can be configured dynamically and is planned in such a way so as to use the fittest wireless transmission channel in its region. This detects the available spectrum in an ad hoc wireless scenario auto-

A. Rout (✉) · A. Kar
Electronics and Telecommunication Engineering, IGIT Sarang, Khalapal, Odisha, India
e-mail: ashimarout@gmail.com

A. Kar
e-mail: anurupa_mar28@yahoo.in

S. Sethi
Computer Science Engineering & Applications, IGIT Sarang, Khalapal, Odisha, India
e-mail: srinivas_sethi@igitsarang.ac.in

© Springer Nature Singapore Pte Ltd. 2017
S.C. Satapathy et al. (eds.), *Proceedings of the 5th International Conference on Frontiers in Intelligent Computing: Theory and Applications*, Advances in Intelligent Systems and Computing 515, DOI 10.1007/978-981-10-3153-3_46

matically thereby changing its transmission and reception parameters in order to perform the communication processes accordingly. Alternatively termed as an intelligent radio in a wireless mesh network, it changes the path of the messages dynamically taken between two users vis-a-vis changes the frequency band used by the message dynamically in the process of communication.

A CR is completely a reconfigurable radio which can sense and adapt its communication parameters availing secondary usage of the spectrum. The parameters may be visualized as carrier frequency, transmission power, modulation type, bandwidth, symbol rate for the proficient usage of the spectrum. The network needs to make certain that the intervention of secondary user (SU) does not make any interference to the legitimate user, enabling the primary user (PU) to hold the rights of the spectrum access.

In infrastructure-less architecture which is stated as ad hoc architecture [1], there is no infrastructural support. If a mobile station (MS) is able to identify about the presence of other MSs in nearby location and can be connected through a wireless communication standard or protocol, then a link is set up which enables formation of an ad hoc network. Interlinking of two CR terminals is possible by using certain existing communication protocols (e.g., WiFi, Bluetooth) or unused spectrum. In an ad hoc infrastructure, a CR user can communicate with other CR users by utilizing both licensed and unlicensed frequency bands [1].

According to present environmental status security has become a major challenge to the wireless communication. Hence, in order to nullify the security problems, trustworthiness of a user is necessary and it depends upon the received power of users. The practical path loss estimation technique [2], i.e., link budget calculation is needed to improve the security of wireless network.

The sketch of the paper described about the problem definitions in Sect. 2 followed by path loss models in wireless scenario which stated a brief description about different types of path loss models and their significance in Sect. 3. Section 4 illustrates the designing concept of the proposed model. Simulation results and analysis are presented in Sect. 5. The expression of conclusion has been made in Sect. 6.

2   Problem Definitions

In [3], the authors have mentioned about a trust-based algorithm which is based on secure spectrum access in CRN. The given technique estimated the distance of a user from the base station in terms of both the received power strength and geographical positioning of the user. The trust values are evaluated by verifying the distance values between the users and base station. Here, the trusted SUs are categorized into three groups as per their trust value and occupied channel rate. It has also mentioned about transmission technologies of cognitive radio and explained that upon receiving requests from users belonging to same priority group, a token is generated and given to the user that has high power and energy.

Shrestha et al. [4], authors have described about an effective procedure which could verify the source of the occupied spectrum information from an authentic licensed user (LU) by enhancing the spectrum utilization efficiency effectively and reducing the interference effect to that LU. Here, the authors have also mentioned about the spectrum sharing scenarios, security threats, misbehaving user detection, and implementation of trust matrices. It has also stated about relative trustworthiness which is calculated by two distances such as, distance calculated by the received power and the coordinate axis.

In [5], the authors have mentioned about a framework which represented the ad hoc concept in cognitive radio network by using the trustworthy collaboration of spectrum sensing. This framework provides the protection to the dependent users without causing any obstacle to the primary users. The hierarchical architecture concept in [6] mentioned as to how the data fusion centers and cluster heads manage the trust value of a cognitive user and also about the misbehaving users being punished by the fusion center for neglecting their trust values. The authors mentioned about the determination of trustworthiness and comparison of path loss models in [3–7]. The above calculations are carried out in various ways using several models like Okumura model, Hata model, and Lee model by varying user station antenna height, base station antenna height and transmitter–receiver antenna height taking into account the system frequency of 900 MHz.

None of these papers have discussed about the secured cognitive radio ad hoc network using the propagation path loss. In this paper, the propagation path loss factor has been taken into consideration in order to calculate the relative trustworthiness of the user and link budget power calculation to analyze the secured cognitive radio ad hoc network.

3 Path Loss Models in Wireless Scenario

When a signal undergoes a transmission process, it experiences some propagation path loss which is the variations between the transmitted power levels as well as received power levels represented by the attenuation of signal caused by free space propagation path loss. This free space propagation path loss is dependent upon four propagation mechanisms, i.e., reflection, refraction, diffraction, and scattering. There are two aspects of path loss model present in a wireless environment such as [2], large-scale and small-scale path loss.

The mean signal strength of the distance between the transmitter and receiver is calculated by propagation model. This is necessary in determining the coverage area of a transmitter, and is addressed as large-scale path loss model. Large-scale path loss models can again be classified in two ways [2], as outdoor propagation model and indoor propagation model. Outdoor propagation model is used in case of an irregular terrain. These models are used to predict the signal strength at a particular receiving point in a specific area [2]. In indoor propagation model [2], path loss occurs within the buildings which are in a small area of coverage.

Small-scale path loss occurs [2], when fast fluctuation of the amplitudes, frequency, phases, or multipath delays of propagated signal for a very short period of time or distance traveled by the signal. Small-scale path loss models are of two types [2], depending upon Doppler spread as well as multipath time delay. Doppler spread happens due to the effects of fast fading and slow fading [2]. The later one depends upon flat fading and frequency selective fading [2].

Let us assume the propagation path loss for large-scale model in the proposed design. The free space path loss in dB is given by [2],

$$P_l = 10 \log \frac{P_t}{P_r} \tag{1}$$

4 Proposed Model

To provide better security in cognitive radio ad hoc network environments, two important parameters have been considered in this paper such as, trustworthiness and link budget power. By calculating free space propagation path loss value and incorporating this in trustworthiness and link budget power calculation, the results will be produced.

4.1 Trustworthiness

When a primary user is attacked by a harmful user or a malicious user, then the level of false alarm and miss detection increases, which leads to the insecurity of the data or signal contained by primary users. So, in order to improve and maintain the security of a network, calculation of trustworthiness, which serves as a faithful feature, seems desirable. An example of cognitive radio ad hoc network using the concept of free space propagation path loss and other enviable parameters has been reflected below.

As per the coordinate system

Let, (x, y) = Coordinate of the transmitting antenna

(x_1, y_1) = Coordinate of the receiver

The distance between transmitting antenna and the receiver is calculated by [4], this can be also considered as the distance calculated by the geographical point of view.

$$D = \sqrt{(x - x_1)^2 + (y - y_1)^2} \tag{2}$$

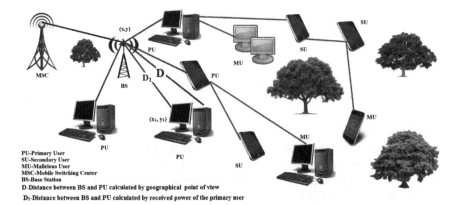

PU-Primary User
SU-Secondary User
MU-Malicious User
MSC-Mobile Switching Center
BS-Base Station
D-Distance between BS and PU calculated by geographical point of view
D₁-Distance between BS and PU calculated by received power of the primary user

Fig. 1 Example of cognitive radio ad hoc network

The received power of a radio wave is given by [4] (Fig. 1)

$$P_r = \frac{P_t G_t G_r (H_t)^2 (H_r)^2}{D_1^4 L} \tag{3}$$

where,

P_r = Received power of a user/receiver
P_t = Transmitting power of the transmitting antenna
G_t = Gain of the transmitting antenna
G_r = Gain of the receiver
H_t = Height of the transmitting antenna
H_r = Height of the receiver
D_1 = Distance measured from transmitting antenna to the receiver calculated by the receiver power of the primary user (PU)
L = Losses of the system
Considering, $G_t = G_r = H_t = H_r = L =$ Unity

The relationship between Received power of a user/receiver and distance measured from the transmitting antenna to the receiver is given [4] in Eq. (4),

$$P_r = \frac{1}{D_1^4} \tag{4}$$

$$\Rightarrow D_1^4 \alpha \frac{1}{P_r} \tag{5}$$

Again, the propagation path loss of a radio wave is directly proportional to the distance between the regulating antenna and the user. Thus, a relation can be established as follows,

$$D_1^4 \alpha P_l \tag{6}$$

Again P_l can be defined as by a formula [8],

$$P_l = P_t - P_r = P_{l0} + 10\gamma \log_{10} \frac{D_1}{D_0} + X_g \tag{7}$$

where [8],

P_{l0} = Path loss of a signal from the reference distance D_0
D_0 = It is the reference distance usually taken as 1 km
γ = It is the path loss exponent, for vacuum, infinite space it is 2.0
X_g = It is the normal (or Gaussian) variable having zero mean caused by flat fading which reflects the attenuation in decibel. In case of no fading this value is zero.
P_t = Consider transmitted power as 46.9 dB

Let's assume P_{l0} in the range of 10–50 dB
By taking into consideration Eqs. (5) and (6) it is concluded that,

$$D^4 \alpha \frac{P_l}{P_r} \tag{8}$$

$$\Rightarrow D^4 = \frac{KP_l}{P_r} \tag{9}$$

where K [2] is the constant of proportionality that is assumed as path loss exponent in this proposed model which is varied from 2 to 4 in urban cellular system.

$$\Rightarrow D = \sqrt[4]{\frac{KP_l}{P_r}} \tag{10}$$

If the distance calculated by the coordinate system matches with the distance calculated by the received power [4], then the system or the user can be considered as "Trustworthy User."

So, by considering the above statement the relative trustworthiness "RTw" can be calculated by the following equation [4],

$$RTw = \min\left(\frac{D}{D_1}, \frac{D_1}{D}\right) \tag{11}$$

4.2 Link Budget Design Using Path Loss

Link budget can be generalized and defined as the calculation of the total gains and losses in a transmission system. It determines the received power of a user at the receiving station. This also includes some parameters like, transmitted power of the regulating antenna, gains of both transmitting and receiving antenna, and path loss. Consequently, by calculating the link budget it is possible to design a network which can work properly without causing extra charge or cost. Link budget is often used in satellite system. Mathematically, the link budget can be stated as [9],

$$P_r(dBm) = P_t(dBm) + \text{Gains}(dB) - P_l(dB) \tag{12}$$

where

P_r (dBm) = Received power in dBm, that is $P_r(dBm) = 10\log_{10}\left(\frac{P_r}{1\,\text{mW}}\right)$

P_t (dBm) = Transmitted power in dBm, that is $P_t(dBm) = 10\log_{10}\left(\frac{P_t}{1\,\text{mW}}\right)$

Gains (dB) = Antenna gains in dB

P_l (dB) = Path loss in dB

Substituting Eq. (7) in Eq. (12) and considering Gains (dB) = Unity, the link budget received power may be calculated as,

$$P_r(dBm) = 10\log_{10}\left(\frac{P_t}{1\,\text{mW}}\right) + 1 - \left(P_{l0} + 10\gamma\log_{10}\frac{D_1}{D_0} + X_g\right) \tag{13}$$

5 Simulation Results and Analysis

In this context, MATLAB 2013 is used to produce all the simulation results. The parameters that have been considered are given below,

Transmitted power = 46.9 dB
Cell radius = 500 m
Cell diameter = 1 km
No. of users in a cell = 100
Distance between the users = 10–100 meters
$\gamma = 2$ [Path loss exponent for vacuum, infinite space]
$X_g = 0$, it is the normal (or Gaussian) variable having zero mean caused by flat fading which reflects the attenuation in decibel. In case of no fading this value is zero [8] (Figs. 2 and 3).

Hence, upon performing the above simulation some relative trustworthiness (RTw) values are produced. There is no specified threshold value for relative trustworthiness. By performing the simulation, it has been seen that the increasing exponential curves are becoming saturated that is becoming constant after 0.5. However, the threshold value for RTw may be assumed here as 0.5, as there is no specified threshold value for it, which indicates if the RTw value of a user is obtained above the threshold value then it is to be considered as trustworthy user

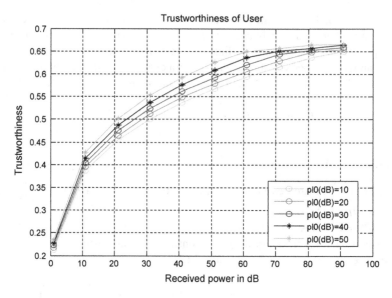

Fig. 2 Trustworthiness of the user

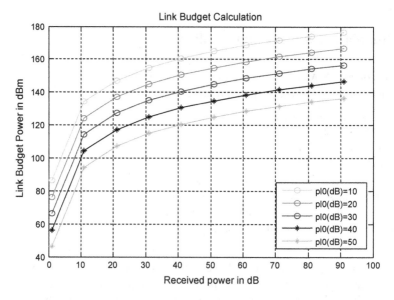

Fig. 3 Link budget calculation using free space path loss value

else a malicious user. It has been also seen that the distance measured by the received power is gradually decreasing with the increase received power of the user. The best result can be seen when the received power is greater than 30 dB. Here, the path loss increases with the increase in separation distance between the users.

Accordingly, it may be concluded that the relative trustworthiness increases with increases in received power.

Again, by calculating the link budget we can make an inference that, the link budget power is increasing with the increase in received power. The best results can be seen at path loss of a signal of a reference distance D_0 taken as 10 dB.

Thus, it is visualized that if the received power is increased then the trustworthiness and subsequently the link budget power will be increased simultaneously. The best results can be seen when the path loss is minimum. By calculating the link budget in minimum free space path loss, an enhancement in the link budget power will be resulted, which ultimately would capitalize higher trustworthiness value for the designed network.

6 Conclusions

Cognitive radio improves the spectrum efficiency effectively without causing any interference to the authentic user. It should make sure that the secondary user must get a message of occupancy of spectrum from the primary user. By allowing changes in the frequency of a temporarily unused licensed spectrum and adjusting some operating parameters for the area variations, CR provides future generation wireless devices for rapid growth in data transmission. So, to enhance the communication process in future generation devices cognitive radio plays a major role. Though the communication medium is wireless there are various security challenges occur. So in order detect and nullify these security issues, a model has been designed. In this proposed model, trustworthiness and link budget power of users are calculated, by taking free space propagation path loss into account. For minimum path loss value, the best results can be seen for both trustworthiness and link budget power.

References

1. Sazia Parvin, Farook Khadree Husain, Omar Khadree Husain, SongHan, Biming Tian, and Elizabeth Chang: Cognitive Radio Network Security: A Survey. Elsevier Journal of Network and Computer Application. 1084–8045 (2012).
2. Theodore S. Rappaport: Wireless Communications Principals and Practice", Second edition published by PEARSON Publication, Indian edition (2009).
3. Prasanna Venkatesan K. J. and Vijayarangan: Trust Based Algorithm for Secure Spectrum Access in Cognitive Radio Networks. K. R. Venugopal and L. M. Patnaik (Eds.) ICCN 2013, pp. 50–62 Elsevier Publications (2013).
4. Shrestha, Junu, Sunkara, Avinash, Thirunavukkarasu and Balaji: Security in Cognitive Radio. A project report to The Faculty of the Department of General Engineering of San Jose State University (2010).

5. Muhammad Faisal Amjad, Baber Aslam Afraa Attiah, and Cliff C. Zou: Towards trustworthy collaboration in spectrum sensing for ad hoc Cognitive Radio Networks. DOI:10.1007/s11276-015-1004-2 Springer Science & Business Media New York (2015).
6. www.utexas.edu/research/mopro/papercopy/chapter11.pdf (Retrieved as on 26.04.2016).
7. Jianwu Li, Zebing Feng, Zhiqing Wei, Zhiyong Feng, and Ping Zhang: Security Management Based on Trust Determination in Cognitive Radio Networks. DOI:10.1186/1687-6180-2014-48 , EURASIP Journal on Advances in Signal Processing a Springer Open Journal (2014).
8. Julius Goldhirsh and Wolfhard J. Vogel: Handbook of Propagation Effects for Vehicular and Personal Mobile Satellite Systems (1998).
9. M. A. Alim, M. M. Rahman, M. M. Hossain and A. Al-Nahid: Analysis of Large-Scale Propagation Models for Mobile Communications in Urban Area. ISSN: 1947-5500 Vol.7 No.1, International Journal of Computer Science and Information Security (IJCSIS) (2010).

A Secure and Lightweight Protocol for Mobile DRM Based on DRM Community Cloud (DCC)

Hisham M. Alsaghier, Shaik Shakeel Ahamad, Siba K. Udgata and L. S.S. Reddy

Abstract DRM provides a secure solution for the illegal distribution of digital content through communication networks. We propose a Secure and lightweight protocol for mobile DRM based on DRM community cloud (DCC) and banking community cloud (BCC). Non-repudiation property is a very important property that needs to be ensured for DRM. Non-repudiation property in this protocol is achieved using wireless public key infrastructure (WPKI), universal integrated circuit card (UICC) at the client side, DRM community cloud (DCC) at the cloud provider (CP) and banking community cloud (BCC) at the Issuing Bank. BCC and DCC are a Cloud of Secure Elements (CSE). Our proposed protocol achieves end-to-end security from the client to DCC and BCC.

Keywords Digital · Rights management (DRM) · Banking community cloud (BCC) · DRM community cloud (DCC) · Mobile agent · Content provider (CP) · Universal integrated circuit card (UICC) · Cloud of secure elements (CSE) · Wireless public key infrastructure (WPKI)

H.M. Alsaghier (✉) · S. Shakeel Ahamad
College of Computer and Information Sciences, Majmaah University,
Al Majmaah, Kingdom of Saudi Arabia
e-mail: h.alsaghier@mu.edu.sa

S. Shakeel Ahamad
e-mail: ahamadss786@gmail.com; s.ahamad@mu.edu.sa

S.K. Udgata
School of Computer and Information Sciences, University of Hyderabad,
Hyderabad, India
e-mail: udgatacs@uohyd.ernet.in

L.S.S. Reddy
Department of CSE, KL University, Guntur, India
e-mail: vc@kluniversity.in

© Springer Nature Singapore Pte Ltd. 2017
S.C. Satapathy et al. (eds.), *Proceedings of the 5th International Conference on Frontiers in Intelligent Computing: Theory and Applications*, Advances in Intelligent Systems and Computing 515, DOI 10.1007/978-981-10-3153-3_47

1 Introduction

Digital rights management (DRM) allows content and content providers to securely distribute digital content and to control its access and use [1]. DRM comprises the complete process of managing and controlling the access and consumption of protected content [1]. The exponential development of information and communication technology (ICT) with the evolution of the Internet in the early 1990 has disturbed balance between the interests of content providers and consumers. Consumers got an opportunity to make perfect copies of digital content and distribute it around the world without any cost. DRM provides a secure solution for the illegal distribution of digital content through communication networks. Digital content, such as copyright-protected music or books and medical data from unauthorized peers must be accessed by paid users only [2]. In order to provide protection and control the access of digital content we need a secure and lightweight framework that ensures end to end security from the client to the content provider (CP). Existing solutions based on DRM [3, 4] does not achieve non-repudiation property at both the ends, i.e., at the client side (UICC) and at CP as digital signatures are not generated in secure elements so they are not considered as signatures and moreover the communication and computational cost of the proposed protocols is more. There is no clarity in [3, 4] how the content is protected between CP and User (U) during the transit. We propose a Secure and Lightweight Protocol for Mobile DRM based on DRM DCC and BCC. Non-repudiation property is a very important property that needs to be ensured for DRM. Non-repudiation property in this protocol is achieved using WPKI, Universal Integrated circuit card (UICC) at the client side, DCC at the CP and BCC at the Issuing Bank. BCC and DCC are a Cloud of Secure Elements (CSE). Our proposed protocol achieves end to end security from the client to DCC and BCC.
  Note:

(a) We adopt DSMR proposed in [5] in the proposed protocol.
(b) *Trusted Execution Environment (TEE)*: The Secure Element of User (U) (i.e., UICC), Cloud of Secure Elements (CSE) in BCC and DCC provides TEE environment mobile agent will get executed at CSE of BCC and DCC safely.

2 Related Work and Contributions Made

Related Work

(a) Existing non-repudiation solutions on DRM such as [3, 4] does not ensure non-repudiation property at both the ends, i.e., at the client side (UICC) and at CP as digital signatures are not generated in secure elements so they are not considered as signatures.

(b) Authors of [4] paper did not show how to protect the Assets, such as DRM keys and license files, temporary keys, and decrypted, compressed content.
(c) The communication and computational cost of the proposed protocol in [4] is more.
(d) There is no clarity in [4] where the encrypted, integrity-protected database is stored in the device permanent storage of CP.
(e) There is no clarity in [4], where the cryptographic execution is done at the (CP side and U side.
(f) There is no clarity in [4] how the content is protected between U and CP, U, and RI during the transit.
(g) TEE (Trusted Execution Environment) was not used neither at CP, RI, nor at U so digital signatures generated are not considered as Qualified Electronic Signatures.

Contributions Made

a) Proposes a Secure and Lightweight Protocol for Mobile DRM based on DCC and BCC.
(b) Non-repudiation property in this protocol is achieved using WPKI, UICC at the client side, DCC at the CP, and BCC at the Issuing Bank.
(c) The communication and computational cost of the proposed protocol is very less.
(d) SLMDDCC achieves end to end security.
(e) Our proposed protocol (SLMDDCC) withstands well-known attacks.

3 Proposed Protocol (SLMDDCC)

Our proposed Secure and Lightweight Protocol for Mobile DRM based on DRM Community Cloud (SLMDDCC) protocol for DRM based on DCC and BCC. Following are the participants in SLMDDCC ecosystem

3.1 Participants

(a) **Trusted Service Manager (TSM)**: Certification Authority (CA) issues certificates, binds public keys, and revokes certificates. It also acts as a Trusted Service Manager (TSM) which acts as a catalyst between CP/RI and MNO's worlds from a technical perspective.
(b) **User (U)**: User is an entity which is directly involved in our proposed protocol. She/he possesses a mobile phone with UICC as a Secure Element with GPRS/4G/5G networks.

(c) **DRM Community Cloud (DCC)**: Our proposed DCC contains CP and RI servers. Our proposed DCC consists of CI and RI. DCC is cloud of Secure Elements.

  a. **Content Provider (CP)**: Content Provider delivers content.
  b. **Rights Issuer (RI)**: RI's functions include to delegate permissions, restrains DRM Content and generation of Rights Objects [6].

(d) **Banking Community Cloud (BCC)**: BCC is established by the central bank of the country for the banking community. BCC employs payment gateway (PG) which acts as an adjudicator in resolving disputes. In our proposed BCC each bank will have its own hardware security module (HSM) which is personalized by the respective bank. So in this proposed protocol issuing bank (IB), acquirer bank, (AB) and payment gateway (PG) are the entities in BCC. Banks host their applications on its own HSM in BCC. The connectivity between BCC and Banks would be through MPLS or Lease lines with required IPSEC/GRE, IPS, IDS, etc., in place. The connectivity between BCC and Customers are through 4G/5G networks which are protected with communication security (using TLS) and application security (using ECDSA and AES algorithms).

  a. **Issuing Bank (IB)**: IB is the financial institution of the User (U) and is trusted by the User (U).
  b. **Acquiring Bank (AB)**: AB is the financial institution of the beneficiary or payee and is trusted by the payee (in this protocol CI is the beneficiary or payee).
  c. **Payment Gateway (PG)**: It functions as an adjudicator.

(e) **Mobile Network Operator (MNO)**: MNO is an entity which provides network connection for the User (U).

(f) **Wireless PKI (WPKI)**: Our proposed protocol adopts WPKI for achieving end to end security. WPKI is adopted for resource constrained devices such as mobile phones and PDAs.

3.2 Proposed SLMDDCC Protocol

SLMDDCC Protocol has two phases they are

a. Negotiation Phase
b. Payment Phase

Our proposed protocol employs two mobile agents for performing two different functions: (i) Negotiation Agent (PAg1) for gathering information and negotiating and (ii) Payment Agent (PAg2) for making payments (Fig. 1).

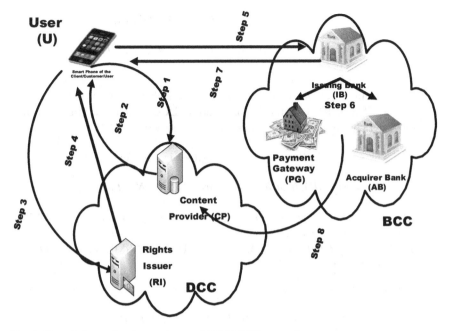

Fig. 1 Negotiation and payment phases of SLMDDCC protocol

(a) **Negotiation Phase**

PAg1 is collects information and negotiates about the content and CP information. Following are the steps involved in this phase

$$Step1: U \rightarrow {}_{PAg1}CP: DSMR_{U_{CP}}(MS1), PubKey_U$$

$$MS1 = UID, ItemNo, T_U, N_U, URLItemNo$$

In step 1 User (U) sends a Payment Agent 1 (PAg1) with *MS1* to Content Provider (CP), PAg1 carries this information in the form of DSMR to CP. PAg1 gets executed at the trusted execution environment (TEE) of CP (CP and RI are a part of DCC which is a grid of Secure Elements). CP recovers message MS1 from the DSMR.

$$Step2: CP \rightarrow {}_{PAg1}U: DSMR_{CP_U}(MS2), PubKey_{CP}$$

$$MS2 = UID, ItemNo, T_{CP}, N_U, N_{CP}, Price, TID$$

In step 2CP sends a message *MS2* to U through PAg1 in the form of DSMR containing *UID, ItemNo, T_{CP}, N_U, N_{CP}, Price, TID*.

$$Step3: U \rightarrow {}_{PAg1}RI: DSMR_{U_{RI}}(MS3), PubKey_U$$

$$MS3 = UID, ItemNo, URLItemNo, T_U, N_U, Price, TID$$

After receiving MS2 from CP, U sends MS3 to RI through PAg1 in the form of DSMR containing $UID, ItemNo, URLItemNo, T_U, N_U, Price, TID$. PAg1 gets executed at the TEE of RI (CP and RI are a part of DCC which is a grid of Secure Elements). RI checks the DSMR it has received and accepts MS3.

$$Step4: RI \rightarrow {}_{PAg1}U: DSMR_{RI_U}(MS4), PubKey_{RI}$$

$$MS4 = RO, UID, ItemNo, URLItemNo, T_{RI}, N_U, N_{RI}, Price, TID$$

In step 4 RI sends a message *MS4* to U through PAg1 in the form of DSMR containing $RO, UID, ItemNo, URLItemNo, T_{RI}, N_U, N_{RI}, Price, TID$.

Note: /* $U \rightarrow {}_{PAg1}CP: DSMR_{U_{CP}}(MS1)$ means U sends DSMR message with the Payment Agent1 to CP */

(b) **Payment Phase**

$$Step5: U \rightarrow {}_{PAg2}IB: DSMR_{U_{IB}}(MS5), PubKey_U$$

$$MS5 = \{RO, UID, ItemNo, URLItemNo, T_U, N_U, Price, TID, AccCP, PIN_U, (PI)_{SYYKEY_{UIB}}\}$$

PAg2 is generated by the User (U) for carrying MS5 to the Issuing Bank (IB), PAg2 carries this information in the form of DSMR to IB. PAg2 gets executed at the Trusted Execution Environment (TEE) of IB (IB and AB are a part of BCC which is a grid of Secure Elements). IB checks the DSMR it has received and accepts MS5.

$$Step6: IB \rightarrow AB\&PG: (MS6)$$

$$MS6 = \{Success, RO, UID, ItemNo, URLItemNo, T_{IB}, N_{IB}, Price, TID, AccCP\}$$

IB, PG, and AB are an integral part of BCC which is a grid of Secure Elements. In step 6IB sends message containing $Success, RO, UID, ItemNo$ $commaURLItemNo$, $T_{IB}, N_{IB}, Price, TID, AccCP$ to both AB and PG without mobile agents through leased lines in private banking network.

$$Step7: IB \rightarrow {}_{PAg2}U: DSMR_{IB_U}(MS7)$$

$$MS7 = \{Success, ItemNo, T_{IB}, N_{IB}, Price, TID, AccCP\}$$

IB sends $Success, ItemNo, T_{IB}, N_{IB}, Price, TID, AccCP$ to U.

$$Step8: AB \rightarrow CP: (MS8)$$

$$MS8 = \{Success, ItemNo, T_{AB}, N_{AB}, Price, TID, AccCP\}$$

IB sends $Success, ItemNo, T_{AB}, N_{AB}, Price, TID, AccCP$ to CP.

Notations:
See Table 1.

4 Security Analysis

A. *End to End Security and Forward Secrecy*

Proposed SLMDDCC ensures end-to-end security and forward secrecy using DSMR which was proposed in [5].

B. *Secrecy of Payment*

Secrecy of Payment is ensured by encrypting the PI using shared symmetric key between U and the IB. CP and RI fail in retrieving information from PI.

C. *Key pair generation and storage at the User side in secure element*

UICC is used at (U) which is a secure element. UICC is used for generating and storing client's credentials.

Table 1 Notations

U	User	**TID**	Transaction identifier
CP	Content provider	**ItemNo**	Item number of content
RI	Rights issuer	**URLItemNo**	URL of item number of content
BCC	Banking community cloud	**AccCP**	Account information of CP
DCC	DRM community cloud	$DSMR_{X_Y}(MS)$	DSMR generated by participant 'X' and which should be verified by 'Y' on MS
IB	Issuing bank	**MS**	Message
AB	Acquirer bank	N_X	Participant 'X' generated nonce
MNO	Mobile network operator	T_X	Participant 'X' generated timestamp
TSM	Trusted service manager	**Success**	Success
CA	Certifying authority	**PI**	Payment information
UID	User identity	**RO**	Rights objects
Price	Price	**WPKI**	Wireless public key infrastructure
DSMR	Digital signature with message recovery	$SYYKEY_{XY}$	Symmetric shared between 'X' & 'Y'

D. *Mobile Agent Execution at DCC and BCC*

Both the mobile agents will get executed at their respective TEE's of the DCC and BCC

E. *Identity protection from CP, RI*

User enrolls for anonymous identity with CA and IB both CA and IB know original identity of U. So both CP and RI will not be able to know the real identity of User.

F. *Fewer consumption of resources*

Proposed protocol uses mobile agents and DSMR mechanism so certificate validation is not required as it is achieved while recovering the message from the DSMR. So our proposed protocol consumes fewer resources.

G. *Prevents Double spending and Over spending*

Suppose U tries to double spend PI, IB will come to know this from timestamps and nonce so User will be unsuccessful in double spending PI. If the User or CP tries to overspend, IB can avoid U and CP as it U's balance for every transaction.

H. *With stands well-known attacks*

Timestamps and nonce which are included in the DSMR avoid replay attacks in our protocol. An intruder (In) cannot impersonate as U to CA and IB because intruder (In) is not in possession of U's private key, so impersonation attack is not possible in our protocol. Intruder (In) is not in possession of receiver's private key so man in the middle attack is not possible in our protocol.

Table 2 Comparative Analysis with Related Works

Features	Protocols		
	[4]	[3]	SLMDDCC
End to end security	No	No	Yes
Forward secrecy	No	No	Yes
Key pair generation and storage at the user side in secure element	No	No	Yes
Digital signatures generated in tamper resistant devices	No	No	Yes
Secure execution of mobile agent at CP's TEE	No	No	Yes
Identity protection from CP & RI	No	No	Yes
Privacy of transaction from eavesdropper	No	No	Yes
Withstands well-known attacks	Yes	No	Yes
Secrecy of payment	No	No	Yes
Fewer consumption of resources	No	No	Yes
Prevents double spending and over spending	No	No	Yes

5 Comparative Analysis of SLMDDCC Protocol with Related Works

See Table 2.

6 Conclusions and Future Work

We propose a SLMDDCC Protocol based on DCC and BCC. Proposed protocol (SLMDDCC) is secure and lightweight as the communication and computational cost is very less since our proposed protocol uses mobile agent and DSMR and all the entities involved in the protocol are in the same cloud. Our proposed protocol achieves end-to-end security and is consumes fewer resources. Our future work is to verify the proposed protocol using BAN logic and Scyther tool.

References

1. Mercè Serra Joan, Bert Greevenbosch, Harald Fuchs, Anja Becker and Stefan Krägeloh. Overview of OMA Digital Rights Management. Handbook of Research on Secure Multimedia Distribution (pp. 55–70). http://www.irma-international.org/viewtitle/21307/
2. Onieva, J., Lopez, J., Roman, R., Zhou, J., &Gritzalis, S. (2007). Integration of nonrepudiation services in mobile DRM scenarios. Telecommunication Systems, 35, 161–1765.
3. Chin-Ling Chen (2008). A secure and traceable E-DRM system based on mobile device. Expert Systems with Applications 35 (2008) 878–886.
4. Chung-Ming Ou, C.R. Ou (2011). Adaptation of agent-based non-repudiation protocol to mobile digital right management (DRM). Expert Systems with Applications 38 (2011) 11048–11054.
5. Tseng,Y., J. Jan, H. Chien, (2003). Digital signature with message recovery using self-certified public keys and its variants. Applied Mathematics and Computation 136 (2–3) (2003) 203–214.
6. DRM Architecture, Open Mobile Alliance, OMA-AD-DRM-V2_1–20081014-A, URL:http://technical.openmobilealliance.org/Technical/release_program/docs/DRM/V2_1-20081106-A/OMA-AD-DRM-V2_1-20081014-A.pdf

Cloud Based Malware Detection Technique

Sagar Shaw, Manish Kumar Gupta and Sanjay Chakraborty

Abstract Security is one of the major concerns in cloud computing now-a-days. Malicious code deployment is the main cause of threat in today's cloud paradigm. Antivirus software unable to detect many modern malware threats which causes serious impacts in basic cloud operations. This paper counsels a new model for malware detection on cloud architecture. This model enables identification of malicious and unwanted software by amalgamation of multiple detection engines. This paper follows DNA sequence detection process, symbolic detection process, and behavioural detection process to detect various threats. The proposed approach (PMDM) can be deployed on a VMM which remains fully transparent to guest VM and to cloud users. However, PMDM prevents the malicious code running in one VM (infected VM) to spread into another noninfected VM with help of hosted VMM. After detecting malicious code by PMDM technique, it warns the other guest VMs about it. In this paper, a prototype of PMDM is partially implemented on one popular open-source cloud architecture—Eucalyptus.

Keywords Malware · Eucalyptus · Antivirus · Security · Cloud computing · DNA sequence · Symbolic detection and behavioural detection · Sandbox

S. Shaw (✉) · M.K. Gupta · S. Chakraborty
Department of Computer Science & Engineering, Institute
of Engineering & Management, Kolkata, India
e-mail: shaw.sagar09@gmail.com

M.K. Gupta
e-mail: gupta.manish414@gmail.com

S. Chakraborty
e-mail: sanjay.chakraborty@iemcal.com

© Springer Nature Singapore Pte Ltd. 2017
S.C. Satapathy et al. (eds.), *Proceedings of the 5th International Conference on Frontiers in Intelligent Computing: Theory and Applications*, Advances in Intelligent Systems and Computing 515, DOI 10.1007/978-981-10-3153-3_48

1 Introduction

Detecting malicious file is a complicated work. The big amount of new malware files are growing at a shocking rate. Microsoft receives over 150 thousand new unknown files each day to be analyzed. Antivirus software is one of the most widely used tools for detecting. In this paper, we suggest a new model where a file mainly undergoes these processes to detect malicious behavior.

1.1 DNA Sequence Detection Process

DNA sequencing is the process of determining the precise order of nucleotides within a DNA molecule to identify regions of local or global similarity.

1.2 Symbolic Detection Process

In symbolic detection process we cluster the files and use symbol to detect malware.

1.3 Behavioural Detection Process

Analyzing behavior of the file is one of the best ways to detect malicious file. In Behavioural detection process we use Anubis sandbox to detect new malicious file. This proposed malware detection model is deploying into cloud architecture which gives the resultant as cloud deployment model (CDM) with the help of Eucalyptus.

2 Background

2.1 Cloud Computing

Cloud computing is a common term. According service model it is divides in 3 types: [1]. Software-as-a-service, Platform-as-a-service, Infrastructure-as-a-service [2].

2.2 Security in the Cloud Computing

The Security in the Cloud is provided by many companies to detect malware with industry-leading detection rates. "Cipher Cloud" is company which provides service. We are providing a basic infrastructure of cloud malware detection see Fig. 1 [3, 4].

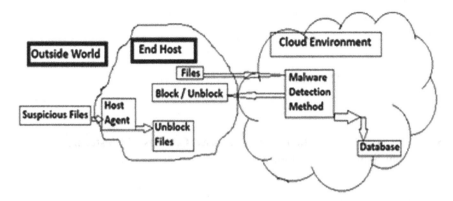

Fig. 1 Cloud malware detection technique

Fig. 2 Traditional antivirus detection method

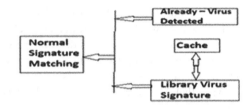

2.3 Related Work

The traditional antivirus software's can detect only those malware whose signatures are already present in the databases. This approach is based on the anomaly (Fig. 2) [3].

3 Proposed System

Paper proposes a new malware detection system built on cloud environment. Initially, we will divide the system architecture into two main sections according to the mechanism of action of each part. First part, explains the PMDM and the second part, explains CDM.

PART—I

3.1 Proposed Malware Detection Model

The proposal is to find the optimal solutions to the problems of antiviruses and improve performance and find possible alternatives for a better working environment without problems with high efficiency and flexibility.

In this malware detection model, total three process are used to explain the mechanism,

3.1.1   Process 1: DNA Sequence Detection Process

3.1.2   Process 2: Symbolic Detection Process Consists of Clustering and Symbolic Detection

3.1.3   Process 3: Behavioural Detection Process Using Sandbox Testing

All of them are explained below in detail see Fig. 3.

In Process 1 we go for DNA Sequence checking, the initial step is the extraction of DNA sequence from a file is done by converting the file into its binary form and change each two corresponding bits into a DNA sequence character by using Tables 1 and 4. The conversion is completely reversible.

After that Malware_Sequence_Database is created using these steps shown in Fig. 4.

The various input files are converted into binary files then converted into FASTA sequence and merged to create a Malware_Sequence_Database (Fig. 5).

Now when new files come, it first convert to the DNA or fast a sequence then check with the Malware_Sequence_Database using Blast online software. The result of this comparison is a BLAST report [6], determine that the file is malicious or not see Fig. 6.

In Process 2 those files are undetected in process 1 are comes where we first cluster the files according to their file format, by checking the file format see Fig. 7.

Then, we use symbolic detection technique in which files are converting to symbol using symbol database table. Then after we match symbol file with the existing symbol database which contain symbol of conventional malware signatures, see Table 2.

If the file symbols are not matched with the existing symbol database then the file may be a new malicious file. Otherwise, the files are detected in process 1 and 2 and blocked for the third process. In Process 3, the files which are passed through the second process are only go for the third process. In this process we detect malicious files using a virtual machine that extensively used for this type of analysis by testing and running the file into a sandbox gives an optimal result to detect malware. For this purpose, we use Anubis sandbox [7] which is free available. Anubis interact with file using API call and check the behavior of the file to identify whether it contain malware or not.

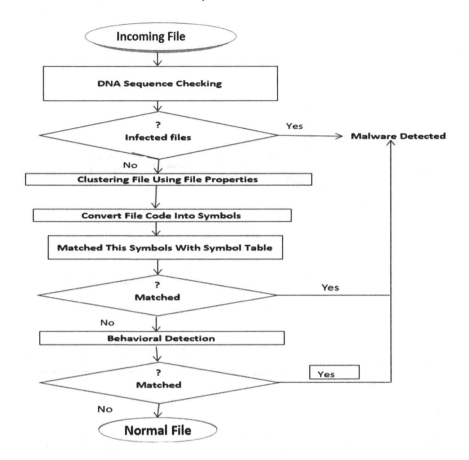

Fig. 3 Flowchart for detection method

Table 1 DNA sequence mapping table [5]

Binary bits	DNA character	Binary bits	DNA character
00	T	10	C
01	G	11	A

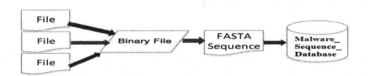

Fig. 4 Creating BLAST database

>c: /user/name.txt
GTAGGGCCCGTTTGGCCAAAAATTTTTTTT

Fig. 5 Example of DNAs

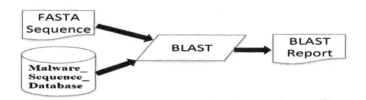

Fig. 6 Comparing FASTA sequence with Malware_Sequence_Database

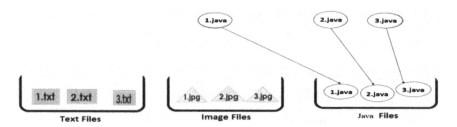

Fig. 7 Clustering file formats

Table 2 Symbol database table

SL No.	String	Symbol	SL No.	String	Symbol
1.	:A	◀	2.	explorer	‼
3.	Start	↕	4.	shutdown	¿
5.	goto	§	6.	%random%	‰

PART—II

3.2 Cloud Deployment Model

The PMDM discuss in Part-I is deploy into cloud architecture, i.e., CDM by using a free open-source computer software Eucalyptus. The purpose of this CDM is to implement PMDM in real cloud environment. In this experiment, we determine the fitness of the proposal into the Eucalyptus architecture. PMDM is valuate against known and unknown malicious attacks. Here, the PMDM system is partially

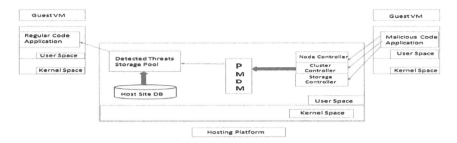

Fig. 8 Proposed Malware Detection Method (PMDM) embedded in cloud deployment model (CDM)

implemented in the Eucalyptus architecture due to the architectural (infrastructure) limitations. The PMDM is mainly used to detect the malicious code based on the above discussed processes. And keep them as a set of warnings in a dedicated thread storage pool and block the malicious file to enter into a Guest VM or Guest Operation System see Fig. 8 [1, 8, 9].

4 Result Analysis

4.1 Result of DNA Sequence Process

Several experiments were designed to evaluate the usefulness of this process including document gathering, modification of DNA sequences, database creation, and software identification.

4.1.1 Document Gathering

Total 1020 files were collected, types, and counts of file were recorded in Table 3.

4.1.2 Modify DNA Sequence

Converting the file into its binary form and change each two corresponding bits into a DNA sequence using by a java program character see Table 1.

Table 3 File counts

186 Text files	152 Image Files	194 HTML Files
169 Java Files	99 Binary Files	

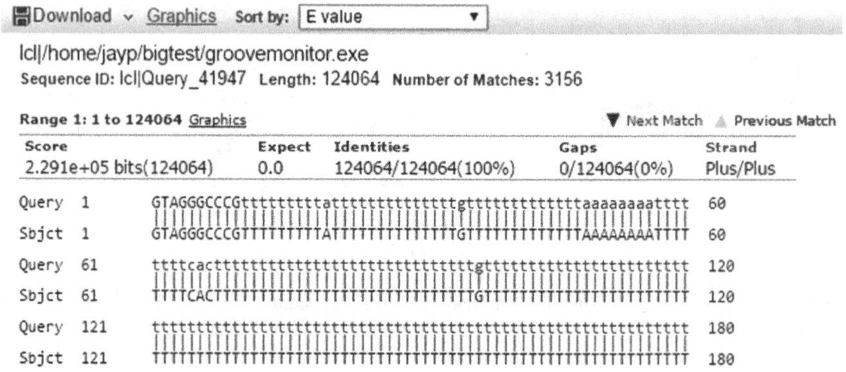

📁Download ⌄ Graphics Sort by: | E value   ▼ |

lcl|/home/jayp/bigtest/groovemonitor.exe
Sequence ID: lcl|Query_41947 Length: 124064 Number of Matches: 3156

Range 1: 1 to 124064 Graphics   ▼ Next Match ⚠ Previous Match

Score	Expect	Identities	Gaps	Strand
2.291e+05 bits(124064)	0.0	124064/124064(100%)	0/124064(0%)	Plus/Plus

```
Query  1    GTAGGGCCCGtttttttttattttttttttttttttgtttttttttttttttaaaaaaaatttt  60
            ||||||||||||||||||||||||||||||||||||||||||||||||||||||||||||||||
Sbjct  1    GTAGGGCCCGTTTTTTTTTATTTTTTTTTTTTTTTGTTTTTTTTTTTTTTTAAAAAAAATTTT   60

Query  61   ttttcactttttttttttttttttttttttttttttttttgttttttttttttttttttttttt  120
            |||||||||||||||||||||||||||||||||||||||||||||||||||||||||||||||
Sbjct  61   TTTTCACTTTTTTTTTTTTTTTTTTTTTTTTTTTTTTTTTGTTTTTTTTTTTTTTTTTTTTTTT  120

Query  121  ttttttttttttttttttttttttttttttttttttttttttttttttttttttttttttttt  180
            |||||||||||||||||||||||||||||||||||||||||||||||||||||||||||||||
Sbjct  121  TTTTTTTTTTTTTTTTTTTTTTTTTTTTTTTTTTTTTTTTTTTTTTTTTTTTTTTTTTTTTTT  180
```

Fig. 9 Descriptive analysis result of groovemonitor.exe

Fig. 10 Pseudocode for matching symbol with symbol table database

Take symbolfile name as Input
Scan symbolfile
While not EOF do
    If symbolfile match with virus symbol Then
        Print "File contain virus"
    Else
        Print "File does contain virus"
    Endif
End While

4.1.3 Database and Software

The result which comes out from Process 1 is obtained by using BLAST software. We analyze the above files and observe the result one of the result output is given below: Groovemonitor.exe malware on BLAST and found a 100% identities matched, i.e., it is a malware, see results in Fig. 9.

Same Analysis process is applied on an image file which is malware free and it result is below 70% as expected. From the above various files results the groovemonitor.exe file is blocked and the Image File1.bin and other files are pass for the second process (Figs. 10, 11).

4.2 Result of Symbolic Detection Process

For symbolic detection there is also several operations were designed to evaluate the usefulness of this process including file clustering, converting characters into symbols and matching symbol with symbol table database.

Fig. 11 Symbolic detection result of symbolvsample.txt

4.2.1 File Clustering

In file clustering different types were cluster according to their file format see Fig. 7.

4.2.2 Converting File Characters into Symbols

The conversion of file characters into symbol is done by using Table 2.

4.2.3 Matching Symbol with Symbol Table Database

The matching of symbols with symbol table is done by using the following pseudo code.

Files match are found in the symbol table, as example here symbolvsample.txt file and it is blocked after that and other files that not matched are pass for the third process.

4.3 Result of Behavioural Detection Process

The files which pass the second process are only entertain in this process like from the above result all the files are entertained excluding groovemonitor.exe and symbolvsample.txt files. All the files are the input to the Anubis sandbox after that Anubis will check the files using API call and check the behavior of the file to identify whether it contain malware or not.

Table 4 Comparison between traditional malware detection versus proposed malware detection

Features	Traditional malware detection method	Proposed malware detection method
Effected by security attacks	More	Less
Process time	More	Less
Implementation	Expensive	Cheap
For large number of files	Decrease performance	Increase performance
Attackers intrusion	Easy	Difficult
Cloud solution	May or may not be available	Available

4.4 Comparison Between Traditional Malware Detection Versus Proposed Malware Detection

See Table 4.

4.5 Advantages

4.5.1 Advantage of DNA Sequencing

Using DNA Sequencing detection method we can detect malware without opening the file, it will save time because we need not to see whole file content to detect malware.

4.5.2 Advantage of Symbolic Detection Process

We cluster the file which gives a benefit of postinfection protection, i.e., we actually know which portion of file content we have to see exactly for malware detection. Then matching the converted file with symbol database, will increase time efficiency as we not required to see whole malware signature matching only a small part of traditional malware signature symbol is sufficient to detect malware [10].

4.5.3 Advantage of Behavioural Detection Process

In behavioural malware detection solve the problem of cannot cope with malware variants, i.e., malware can change their code and compiler setting to bypass the detection or zero day protection problem, i.e.. once a new malware is produced its signature or symbol is unknown, so by testing the malicious file into a sandbox we can say that it is malicious or not. The overall advantage of the system is that it increases the efficiency and effectiveness for detection of malwares [10].

5 Conclusion and Future Work

To conclude, it proposes an effective and advanced cloud security method that can detect the different malware attacks during cloud communication. It is also partially implemented on a popular architecture of Eucalyptus with modified. This paper discusses basic techniques in brief needed for the development of PMDM as it is actually cheap, requires less processing time and provides good performance for large numbers of files compare to other traditional malware detection systems. It is totally transparent to the user. We used the both optimal traditional detection methods and modern era methods to detect malwares in this paper. The proposal of this work is to find the best solutions to the problems of anti-malwares and improve performance and find possible alternatives for a better working environment without problems with high efficiency and flexibility. In future, we will see an increase in the dependence of cloud computing. The advantages of this approach include an increase in the number of clients that can be served for every physical server.

References

1. Marinescu D.C., "Cloud Computing: Theory and Practice", MK Publication (2013).
2. Dahl G.E., Stokes J.W. et al., "Large-scale malware classification using random projections and neural networks", IEEE International Conference
3. Hatem S.S., wafy M.H., et al., "Malware Detection in Cloud Computing", International Journal of Advanced Computer Science and Applications (IJACSA), Vol 5, Science and Information (2014).
4. Graham M., "Behaviour of Botnets and Other Malware in Virtual Environments", The Open Web Application Security Project (2014).
5. Oberheide J., Cooke E. et al., "CloudAV: N-Version Antivirus in the Network Cloud", 17th conference on Security symposium, pp- 91–106 (2008).
6. Pedersen j., Bastola D., et al., "BLAST Your Way through Malware Malware Analysis Assisted by Bioinformatics Tools", International Conference on Security and Management (2012).
7. Mandl T., Bayer U.et al., "ANUBIS ANalyzing Unknown BInarieS The automatic Way", VIRUS Bulletin Conference, v 1.0.02 (2009).
8. Johnson D, Murari K. et al., "Eucalyptus Beginner's Guide- UEC Edition", v1.0 (2010).
9. Parmar H., Champaneria T., "Comparative Study of Open Nebula, Eucalyptus, Open Stack and Cloud Stack", International Journal of Advanced Research in Computer Science and Software Engineering, Vol.4, No. 2, pp 714–721 (2014).
10. https://www.youtube.com/watch?v=fV5kED7nryw.

Abnormal Network Traffic Detection Using Support Vector Data Description

Jyostna Devi Bodapati and N. Veeranjaneyulu

Abstract Outlier detection also popularly known as anomaly detection is the process of recognizing whether the given data is normal or abnormal. Some of the applications of this outlier detection are: network intrusion detection, fraud detection, database cleaning, etc.; In most situations, there is scarcity of abnormal data where as plenty of normal data is available. This is the biggest challenge of novelty detection. The characteristics of abnormal or outlier data are often unknown beforehand. Density estimation methods can be used for novelty detection tasks. These methods work only when the assumed data distribution is same as the underlying data distribution which may not be known in advance. C-SVDD and ν-SVDD are used for novelty detection tasks in our experiments. Experiments are performed on a toy data set of bivariate and overlapping classes and real-time multivariate data. Different kernels are also used for experimental studies. All experiments shows that RBF (Gaussian) kernel gives better performance than the other types of kernels. Experimental results on both artificial and real-world data are reported to illustrate the promising performance of outlier data detection.

Keywords Abnormal traffic detection · Novelty detection · Density estimation · One-class classification · C-SVDD and ν-SVDD

1 Introduction

Pattern recognition is the process of identifying patterns in the data and has gained tremendous focus of the researchers in the recent past. There are many branches in pattern recognition: Classification, Clustering and regression are the most useful

J.D. Bodapati (✉)
Department of CSE, Vignan's University, Vadlamudi, AP, India
e-mail: jyostna.bodapati82@gmail.com

N. Veeranjaneyulu
Department of IT, Vignan's University, Vadlamudi, AP, India
e-mail: veeru2006n@gmail.com

© Springer Nature Singapore Pte Ltd. 2017
S.C. Satapathy et al. (eds.), *Proceedings of the 5th International Conference on Frontiers in Intelligent Computing: Theory and Applications*, Advances in Intelligent Systems and Computing 515, DOI 10.1007/978-981-10-3153-3_49

tasks for many real-time applications. These tasks come under the broad category of supervised or unsupervised learning. If class labels are used along with the data for training process then that type of learning is known as supervised learning. Classification and regression fall under this category. If class labels are not used along with the data in training process then that type of learning is known as unsupervised learning. Clustering is an unsupervised learning as labels are not used in training. Classification is the task of assigning a class to the given test data sample. Regression is a function approximation task in which a continuous value is to be assigned to a test sample. Classification is a special case of regression where the output of the data we approximate is discrete. On the other hand output of the function approximation task is Continuous. For classification task we can use models like simple KNN, Gaussian mixture models (GMM)-based Baye's classification, artificial neural networks (ANN)-based Multilayer feedforward neural networks (MLFFNN), support vector machine (SVM)-based classification. For function approximation task, we can use models like linear model for regression, Generalized radial basis function (RBF), SVM-based regression. For clustering one can use models like K-means clustering, GMM-based clustering, Agglomerative hierarchical clustering, decision trees.

In addition to these tasks outlier detection has been gaining focus in the recent literature of machine learning and pattern recognition as it's applications are plenty. In this task, one has to make a decision whether a test observation comes from the ordinary or abnormal class. The given observation is decided as ordinary if it falls in the same distribution of the existing observations (it is an *inlier*). The given observation is decided as abnormal if it does not fall in the same distribution of the existing observations (it is an outlier). A tremendous interest in this outlier recognition task has been noticed in the recent past as there are plenty of real-time applications of this method. Some applications of this task are:

- Network Intrusion Detection—Detect whether someone is trying to access unauthorized user accounts, downloading tons of pdf files, plenty of MP3 s, or doing anything else unusual on the network.
- Fraud Detection—Someone is trying to crack pin numbers of Credit Cards, Telephone Bills, Medical Records etc.;
- Database Cleaning—We want to check whether someone stored fraudulent details in a database, mislabeled digits, and abusive photographs in a social network album.

This approach is helpful in applications to detect presence or absence of breast cancer, automatic mass detection in mammograms, diabetes positive or negative, hepatitis positive or not, filter spam mails from normal mails, in sonar mines, presence of thyroid or not, network traffic is normal or unusual and several others.

Major challenge in this abnormality detection task is availability of the data. Collection of abnormal data is a tedious job as there is lack of availability in such data. The reasons for unavailability of this data are: Abnormalities are usually very rare, or the data that describe the fault conditions may not be available. For these reasons this of task of Outlier or novelty detection cannot use conventional classification methods. One proposed solution in the literature is to model data of the normal class using a similarity or distance metric and use a threshold to measure abnormality.

As the training examples of unusual data are not available, this task is popularly known as one-class classification (OCC), abnormality recognition, novelty recognition, or outlier detection.

In literature many approaches to novelty detection are proposed. Each approach has it's own advantages and disadvantages.

One simple method could be use a distance measure. Find the mean of the class and to decide the test sample find its distance from the class mean if the distance crosses certain threshold decide it as abnormal.

Statistical-based approaches: In case of classification in statistical approaches data is modeled for each class separately. An unseen or test example is assigned to a class that with maximum a posterior. In this outlier detection task as we do not have the abnormal data following steps are being followed to detect whether a sample is normal or not.

Step1: Collect the data of the normal class.
Step2: Model the data of the normal class alone.
Step3: When a test sample comes find its likelihood (l) to the normal class data.
Step4: if l is less than predetermined threshold assign it the normal class otherwise assign it to the abnormal class.

Advantage of this approach is, it is the simplest approach that follows construction of a density function for data of a known class. The down side of this method is that reaching a proper threshold is a challenge. Another drawback is selection of the type of distribution. We do not which type of distribution best fits the data especially when the data is high in dimension. Two popularly used distributions are Gaussian and multinomial distribution. One better way of using statistical modeling techniques is use Gaussian mixture model that can better represent the data.

Another way of categorizing these novelty detection approaches is based on whether they need parameter estimation is required or not.

a. Parametric approaches [1]: In Parametric approaches, we have to estimate the parameters. We assume that data fits a type of distribution. The type of distribution could be mixture of Gaussians or it could be a multinomial distribution. With the assumption of mixture of Gaussian distribution we have to estimate mean and covariance along with the mixture coefficient. A test example is classified based on the decision rule of winner takes all. We have to estimate the parameters that maximize the likelihood of the given data. Usually, EM algorithm is used for estimation of the parameters. It is an iterative approach and there is a possibility of reaching the local minimum.

Challenge with these parametric approaches is reaching the local minimum, and no one knows which type of distribution better fits the data. For these reasons these parametric models became outdated in the literature. To overcome these issues nonparametric methods have been proposed in the literature.

b. Nonparametric approaches [2]: In this category, data need not to follow any statistical properties or assumptions. Popular approaches in this category are: KNN-based approach, Parzen window-based method, and cluster assumption approaches.

K-Nearest Neighbor-based approach- K-nearest neighbor method is one of the conventional ways to classify the data which can also be used for novelty detection problem. When a test sample is given, identify K-nearest samples to the test sample and assign it to the normal class, if majority of those examples belong to the normal class otherwise assign it to the abnormal class. Major challenge in this method is selecting appropriate value for 'K.'

Parzen window based approach- This is another conventional method for data classification which can also be adapted to novelty detection problem. Selection of proper width parameter is challenging.

2 Support Vector Data Description (SVDD) [3]

The goal is to come up with the minimal enclosing hyper sphere for the data of normal class, as only data of normal class is available. To test whether an unseen example belongs to normal class or abnormal class, check whether that data point falls within or outside the hyper sphere. If the data point is within the hyper sphere then it belongs to the normal class otherwise it belongs to the abnormal class [4].

Mathematically, this problem of finding the hyper sphere around the data can be formalized as given below. Consider the data set $\{x_i\}$, $i = 1, 2,..., N$, N is the total number of samples and x_i is a real vector of d dimensions, i.e., $x_i \in R^d$. Assume that the hyper sphere is described by the center 'a' and the radius 'R,' then the training objective is to find the minimal enclosing hyper sphere that is finding a hyper sphere with minimum radius (R). The constraint is that the hyper sphere has to span the entire surface of the normal class data. In simple words, all the normal class data points are expected to be inside the hyper sphere and all the abnormal class data points are expected to be outside the hyper sphere. Such hyper sphere is called as minimal hyper sphere as it is the hyper sphere that covers all the data of normal class with minimum possible radius. The down side of this objective function is: if only few examples are in the training set, a hyper sphere with large R would be obtained and obviously which cannot represent the data properly.

To overcome that we allow (slack) few normal class samples to lie outside the hyper sphere. This hyper sphere is called as soft minimal hyper sphere as slack on the data is allowed. To obtain such a soft minimal hyper sphere, the following constrained optimization problem has to be solved:

$$F(R, a, \xi_i) = R^2 + C * \sum \xi_i$$

In the above objective function, 'C' gives the amount of slack we allow on the training data that is the number of normal class examples that are allowed to lie outside the hyper sphere. The above objective function is subject to the following constraints.

$$(x_i - a)^T (x_i - a) \leq R^2 + \xi_i; \xi_i \geq 0$$

Lagrangian for the soft minimal hyper sphere is:

$$L(R, a, \alpha_i, \xi_i) = R^2 + C \sum \xi_i - \sum \alpha_i \{R^2 + \xi_i - (x_i - a)^2\} - \sum \gamma_i \xi_i$$

with all Lagrange multipliers being nonnegative. By taking the partial derivative and setting that to 0, new constraints are obtained:

$$\sum \alpha_i = 1; \quad a = (\sum \alpha_i x_i) / (\sum \alpha_i) = \sum \alpha_i x_i; \quad C - \alpha_i - \gamma_i = 0$$

Dual form of the above lagrangian is:

$$L = (\sum \alpha_i (x_i . x_i)) = \sum_{ij} \alpha_i \alpha_j (x_i . x_j)$$

With constraints: $0 \leq \alpha_i \leq C; \sum \alpha_i = 1$.

The center of the soft minimal hyper sphere is represented as a linear combination of the support vectors. Set of training examples for which α_i are nonzero contribute to the computation of the centre 'a' and are called as the support vectors. The radius R of the sphere can be obtained by computing the distance from the centre of the sphere to a support vector with a weight smaller than C.

Decision logic: To test an unseen example its distance from centre is computed if the distance is greater than R implies the example is abnormal else it is normal.

If the hyper sphere covers all the training data without any error then it is known as minimal hyper sphere. Hyper sphere with slack is known as soft minimal hyper sphere.

3 Experimental Results

In this paper, all the experiments are done using two different datasets. One is a bivariate toy data set and the other one is a network traffic data of 18 dimensional data. Various types of kernels like linear, polynomial, and Gaussian kernels are used for our experiments. It is observed that on both data sets Gaussian kernel results in better results than other types of kernels. Both C-SVDD and ν-SVDD are used for the experiments.

3.1 Bivariate Data: C-SVDD Model

An appropriate value of sigma is chosen based on the validation data (Fig. 1).

Observations: Accuracy values are found in between 90–100. Maximum accuracy for validation data is at sigma = 1/90. To build the C-SVDD model, experiments were done with the cost parameter. On introducing the cost parameter, accuracy has improved to the range 95–100. Based on validation data, cost is taken to be 0.22.

In Fig. 2, yellow points correspond to training data points, black points correspond to test data points belonging to abnormal class. The red star points correspond to the test data points that have been classified as normal class. The square red points are bounded support vectors. The large red circles are the support vectors. The blue points are the test data points belonging to normal class. The cyan region is the decision region.

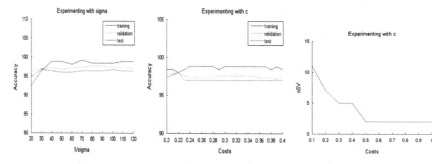

Fig. 1 Choosing sigma, cost parameters, number of support vectors versus cost

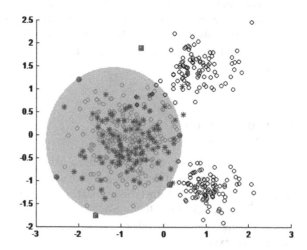

Fig. 2 Hyper sphere formed using C-SVDD on bivariate data

Table 1 Performance of C-SVDD on bivariate data

	True positive (%)	False alarm (%)
Training	100	0
Validation	98.99	1.96
Test	98.50	3

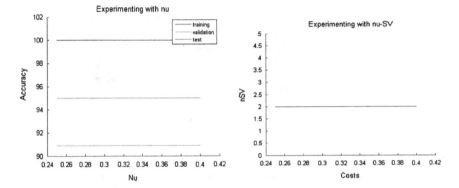

Fig. 3 Accuracy versus ν and number of support vectors versu ν

Observations: Very few points are misclassified. This misclassification is due to the overlapping nature of the data. Three bounded and three unbounded support vector points are found (Table 1).

Results: Accuracy on test: 98%

Bivariate data: ν-SVDD

No significant change was observed as ν was varied and 0.5 is chosen for our experiments. We did not notice much difference in the accuracy values compared to C-SVDD. However, the number of support vectors reduced (Fig. 3).

Following figure shows the details of bounded and unbounded support vectors (Fig. 4).

Results: Accuracy on test: 95%

See Table 2.

3.2 Multivariate Data

This section gives the performance of SVDD on network data. The data is pre-processed and converted into numeric form. An appropriate value of sigma is chosen based on the validation data (Fig. 5).

Observations: Accuracy values are in the range 99.92–100. Based on highest accuracy for validation data, sigma is chosen to be 2.1277e-066. Experiments are done to determine the cost parameter in C-SVDD. Accuracy values are found to be very close to 100% for all the data. Introduction of cost parameter, improves

Fig. 4 Hyper sphere along
with support vectors on data
using ν-SVDD

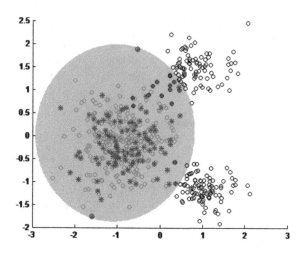

Table 2 Performance of
ν-SVDD on bivariate data

	True positive (%)	False alarm (%)
Training	100	0
Validation	100	21.47
Test	100	13.04

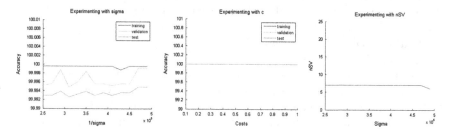

Fig. 5 Choosing sigma, cost and number of support vectors versus 1/sigma

performance. Based on the number of support vectors, cost value is chosen as 1.
The number of support vectors is found to decrease as we increase the value of cost.

Results: Accuracy: 99.9949%

See Table 3.

ν-SVDD: The value of ν is taken to be 0.5 because of similar results to the
bivariate case (Fig. 6).

Table 3 Performance of
C-SVDD on multivariate data

	True positive (%)	False alarm (%)
Training	99.99	0
Validation	99.99	0
Test	99.99	0

Fig. 6 Choosing ν value

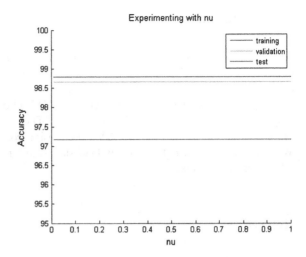

Fig. 7 Number of SVs
versus ν

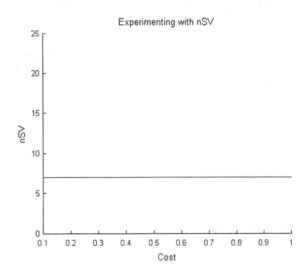

Table 4 Performance of
ν-SVDD on multivariate data

	True positive (%)	False alarm (%)
Training	99.99	0
Validation	99.99	0
Test	99.99	0

Observations: There is no change in accuracy value. There is an increase in the
number of support vectors (Fig. 7).

Results: Accuracy: 99.9949%

See Table 4.

4   Conclusions

Based on the experimental results we conclude that for abnormal network flow detection, ν-SVDD gives the best performance for both bivariate and multivariate data. ν-SVDD gives best true positive and false alarm values in the case of bivariate data. Advantage of using ν-SVDD over C-SVDD is choosing the cost parameter is easy as it is within the range of 0 to 1. In C-SVDD choosing the cost parameter is a big challenge but in the case of ν-SVDD choosing cost parameter is very simple. We observed that on both data sets Gaussian kernel results in better results than other types of kernels.

References

1. Chao He, Mark Girolami, "Novelty detection employing an L2 optimal non-parametric density estimator", Pattern Recognition Letters 25 (2004) 1389–1397.
2. David M.J. Tax, Robert P.W. Duin, "Support vector domain description", Pattern Recognition Letters 20, (1999), 1191–1199.
3. Mingrui Wu, Jieping Ye, "A Small Sphere and Large Margin Approach for Novelty Detection Using Training Data with Outliers", IEEE Transactions on Pattern Analysis and Machine Intelligence, vol. 31, NO. 11, November 2009.
4. David M J Tax, Robert P W Duin, "Data Domain description using Support Vectors", ESANN 1999, ISBN 2-600049-9-X, pp 251–256.

Real-Time Automotive Engine Fault Detection and Analysis Using BigData Platforms

Yedu C. Nair, Sachin Kumar and K.P. Soman

Abstract This paper is aimed at diagnosing automotive engine fault in real-time utilizing BigData framework called spark. An automobile in the present day world is equipped with millions of sensors which are under the command of a central unit the ECU (Electronic Control Unit). ECU holds all information about the engine. A network of ECUs connected across the globe is a source tap of BigData. Leveraging the new sources of BigData by automotive giants boost vehicle performance, enhance loco driver experience, accelerated product designs. A piezoelectric transducer coupled to the ECU captures the vibration signals from the engine. The engine fault is detected by carving the problem into a pattern classification problem under machine learning after extracting cyclostationary features from the vibration signal. Spark-streaming framework, the most versatile BigData framework available today with immense computational capabilities is employed for engine fault detection and analysis.

Keywords Bigdata · Vibration signal · Cyclostationary · Spark streaming · Machine learning

1 Introduction

Automobile industry has witnessed a drastic revolutionary growth in the past decade. Hence engine's and associated component performance, maintenance, etc., is of utmost importance today. This has created the necessity to design and develop an accurate engine condition monitoring [1] and fault detection system to reduce maintenance and increase the life span of the machine. Detection of faults at early stages can eliminate unwanted, component replacements, thereby cutting the cost

Y.C. Nair (✉) · S. Kumar · K.P. Soman
Centre for Computational Engineering and Networking (CEN)
Amrita School of Engineering, Amrita University,
Amrita Vishwa Vidyapeetham, Coimbatore, India
e-mail: yeducnair777@gmail.com

S.C. Satapathy et al. (eds.), *Proceedings of the 5th International Conference on Frontiers in Intelligent Computing: Theory and Applications*, Advances in Intelligent Systems and Computing 515, DOI 10.1007/978-981-10-3153-3_50

factor and saving the resources. Condition monitoring and fault detection may be accomplished by performing analysis on engine vibration signal or effective acoustic emission signal [2].

A healthy engine generates vibration pulses in a specific pattern. When a fault is induced in the engine this pattern changes and are classified as faulty signals [3]. Automobile Engine fault may be mainly attributed to fatigue or excessive wear of specific engine parts, faults linked to electrical system and control units, faults of fuel system as a result of poor fuel quality, etc. [4]. Potential sources of vibration are mainly: piston-crank mechanism movement, inputs resulting from the work of the fittings of the engine (that is alternator, compressor), burning pressure, inputs transmitted from the motor-car body [5], and the drive transmission system. Acoustic emission wave on the other hand is airborne and can be captured by using a simple microphone unlike the vibration signal that employs complex sensors. In an engine in running condition, stress wave traverses through the mechanical parts which are the consequences of sudden release of strain energy. This stress wave is called an acoustic emission wave [6]. Extracting the feature out of the acoustic wave is more complex when compared to that of feature extraction on vibration signal.

Here the analytics is computed on vibration data in real time over spark-streaming framework. The signal is captured by means of piezoelectric accelerometer [7] coupled to an IC engine, which is passed through a signal conditioning unit by the ECU. Cyclostationary feature extraction [8] approach is implemented after which data is streamed into sparks distributed framework. Here we applied real-time streaming classification algorithms (Data Mining) for fault identification [9]. A model for real-time driver warning system is also proposed.

Real-time streaming is implemented using the spark-streaming frame work [10] and data classification is accomplished by utilizing spark-MLlib, which is a machine learning library [11] that is developed and supported by Apache Spark. Spark Streaming, an extension of the core Spark (API) enables high-throughput; scalable, fault-tolerant stream processing of live data streams. Data can be ingested onto spark framework from sources like Flume, Twitter, Kafka, TCP sockets, ZeroMQ or Kinesis and complex algorithms can be used to process the streamed data with high-level functions like map, reduce, join, window, etc. The streamed data is subjected to processing for fault detection and can be stored to BigData databases like Cassandra/Hbase in HDFS (Hadoop Distributed File System) for future analysis [12].

2 Methodology

Implementation part comprises of vibration signal capture, signal conditioning, cyclostationary feature extraction, and data classification using spark streaming (using spark-MLlib).

2.1 Vibration Signal Capture

The vibration signal is collected by means of a piezoelectric accelerometer fixed to an aluminum plate mounted on the testing engine frame. The signal captured at about 3000 rpm engine speed is fed to a signal conditioning unit for noise removal and filtering. Mainly four classes of data are generated here based on different conditions of the bearings.

2.2 Cyclostationary Feature Extraction

A cyclostationary process is defined as a nonstationary process which has a periodic time variation (about mean and autocorrelation) in some of its statistics and that can be characterized in terms of the order of its periodicity [13]. Vibration signals exhibit cyclostationary behavior and this peculiar characteristic is employed for feature extraction and classification. Cyclostationary process can be interpreted to be comprised of multiple interleaved stationary processes [14]. An added advantage of cyclostationary feature is that they are not susceptible to noise.

Spectral Correlation Function (SCF), which are the actual cyclostationary features are extracted after taking the Fourier Transform of cyclic autocorrelation function and is given as

$$S_x^\alpha(f) = \int_{-\alpha}^{\alpha} R_x^\alpha(\tau)e^{-i2\pi f\tau}d\tau, \tag{1}$$

$$\text{where} \quad R_x^\alpha(t,\tau) = \sum_\alpha R_x^\alpha(\tau)e^{i2\pi \alpha t} \tag{2}$$

i.e., the Fourier transform of the. periodic autocorrelation function $R_x(t,\tau)$.

Cyclic Frequency Domain Profile (CDP) is computed from a maximum of the normalized SCF over all cyclic frequencies and is defined as

$$C_x^\alpha(f) \triangleq \frac{S_x^\alpha(f)}{[S(f+\frac{\alpha}{2})S(f-\frac{\alpha}{2})]^{\frac{1}{2}}} \tag{3}$$

Classification is achieved from a unique CDP pattern of vibration schemes.

2.3 Spark Streaming Using MLlib

The feature extracted data is fed into spark-streaming framework by means of TCP socket streaming. To maintain data integrity we can use Kafka if needed as a messaging queue. The streaming data is discretized into tiny microbatches by the spark-streaming [15] framework unlike other streaming frameworks where data is processed as one record at a time. The data is accepted in parallel and buffered in the memory of spark's worker nodes by the streaming receivers. The latency-optimized framework runs the tasks, processing the batches and generating the results. The tasks for the scheduled jobs in spark are dynamically assigned to the worker nodes based on data locality and resources available. This contributes to proper load balancing and speedy fault recovery. On this streamed data, we apply machine learning algorithms and perform data classification by building and updating the model (Fig. 1).

Spark streaming allows arbitrary RDD (Resilient Distributed Dataset) [16] computations under the abstraction of Dstreams. Dstreams are sequence of RDDs. The system is fault tolerant that is it keeps track of parent RDDs. Master or Driver saves the state of Dstreams to a check point file. In case the master fails it can be restarted using the check point file. The framework periodically saves the DAG (Directed Acyclic Graphs) of Dstreams to fault-tolerant storage, probably a check point directory that should be configured initially. Automatic restart is possible. Spark along with shark and spark streaming forms a unistack that can solve all your data analytics. Spark has been optimized to process batches in milliseconds thus achieving low latency [17]. Spark streaming is built on top of spark combining both batch jobs and streams. Advantage is that we need to design algorithm once and run it on spark standalone mode as well as on spark streaming.

In this case we rely on spark's streaming machine learning algorithms (streaming linear regression) [18] that can simultaneously learn from streaming data and also apply the model on real-time streaming data. Larger class of machine learning problems is dealt by creating a model offline and then applying the same online on real-time data. When new data creeps in model parameters are continually updated. Spark-MLlib [19] actually implements a distributed version of stochastic gradient descent along with various parameters (stepsize, numIterations) and regularizations such as L1, L2, etc. The cluster setup (Testing environment spark) includes a master machine and three worker nodes each with core i7 processor, 16 GB RAM, and 3 TB hard disk storage (Fig. 2).

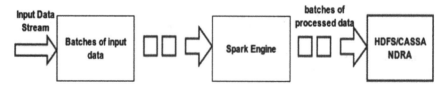

Fig. 1 Spark-streaming framework

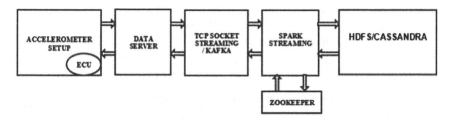

Fig. 2 System Overview

2.4 Data Set Description and Output Performance Statistics

Each record in the data set comprises of 8190 points from which 50 cyclostationary features are extracted. Mainly 4 class data is considered here out of which class one corresponds to data with healthy or normal working conditions and others corresponding to different fault conditions. Real-time streaming linear regression written in Scala is used for classification. We can build the model in offline or

CONFUSION MATRIX (60/40 DATA)						CONFUSION MATRIX (90/10 DATA)				
		CLASS LABELS						CLASS LABELS		
		0	1	2	3			0	1	2	3
CLASS LABELS	0	37	2	0	0	CLASS LABELS	0	13	0	0	0
	1	1	30	7	0		1	1	11	0	0
	2	1	5	27	0		2	0	0	6	0
	3	1	0	0	38		3	0	0	0	17

CONFUSION MATRIX (70/30 DATA)						CONFUSION MATRIX (80/20 DATA)				
		CLASS LABELS						CLASS LABELS		
		0	1	2	3			0	1	2	3
CLASS LABELS	0	29	1	0	0	CLASS LABELS	0	16	0	0	0
	1	4	16	12	0		1	4	11	9	0
	2	0	2	24	0		2	0	1	14	0
	3	2	0	0	32		3	1	0	0	26

Fig. 3 Confusion Matrix

Summary Statistics				
	60/40 DATA	**90/10 DATA**	**70/30 DATA**	**80/20 DATA**
Precision	0.8859	0.9791	0.8278	0.8170
Recall	0.8859	0.9791	0.8278	0.8170
F1 Score	0.8859	0.9791	0.8278	0.8170

Fig. 4 Summary statistics

online mode. Initially we create two folders for storing training data and testing data. The feature extracted data is formatted by using Scala code as $(x, [y1, y2, y3])$, where x is the label and $y1$, $y2$, $y3$ are the features. Whenever a new text read comes in the training directory, the model update is done and whenever a record is placed in testing directory we get the predictions. As the data feed to the training directory increases, predictions are likely to get better. The confusion matrix and other parameters like precision, recall, F1 score computed on a sample set of data for 100 iterations and for different test-train data percentage combinations is depicted in the table below. For example 60/40 DATA refers to 60% training data and 40% test data (Fig. 3).

The diagonal elements in the confusion matrix show how many are correctly classified by applying the pattern classification algorithms on the feature extracted data. The summary statistics obtained is shown in the table below (Fig. 4).

The weighted parameters obtained are entered into the Fig. 5.

Data classification and fault detection were achieved with better statistical parameters at exceptionally faster rate using spark-MLlib. Compared to other conventional single node classification modes, better performance is rendered by the parallel processing algorithms and the highly efficient distributed architecture of spark.

Whenever a fault signal is detected a response is send back to the data server via TCP socket streams and this in future can be send wirelessly to the respective automobiles. Data coming from different automobiles can be identified with a tag.

Weighted Parameters				
	60/40 DATA	**90/10 DATA**	**70/30 DATA**	**80/20 DATA**
Weighted precision	0.8865	0.9806	0.8453	0.8575
Weighted recall	0.8859	0.9791	0.8278	0.8170
Weighted F1 score	0.8860	0.9790	0.8192	0.8054
Weighted false positive rate:	0.0365	0.0077	0.0514	0.0444

Fig. 5 Weighted Parameters

3 Conclusion

In this paper, we have developed a software model that is based on BigData platform and is capable of handling, classifying, and identifying faults in automotive engine vibration data in real time. Faults detected at an early stage can prolong the lifespan of engine and its component parts. This can be proved beneficial to both the consumer and manufacturer. Proper tuning after catastrophic fault detections can contribute to overall efficiency of the engine. Also faults detected at an earlier stage can reduce or avoid unnecessary frequent part replacements, thereby preventing the wastage of resources.

Acknowledgements The authors would like to thank all at Centre for Computational Engineering and Networking for their support.

References

1. Randall and Robert Bond, Vibration-based condition monitoring: industrial, aerospace and automotive applications. John Wiley and Sons, 2011.
2. Gu F., Li, W., Ball, A. D. and Leung, A. Y. T, "The condition monitoring of diesel engines using acoustic measurements, part 1: acoustic characteristics of the engine and representation of the acoustic signals", SAE 2000 World Congress, Noise & Vibration, Detroit, USA, SAE Paper 2000- 01-0730, pp. 51–57.
3. De Silva and Clarence W, "Vibration: fundamentals and practice", CRC press, 2006.
4. M. R. Parate & S. N. Dandare, "IC Engine Fault diagnosis Using ROC', International Journal of Advancements in Technology, ISSN 0976-4860, pp 68–78.
5. J. J. Gertler, M. Costin, X. Fang, R. Hira, Z. Kowalczuk and Q. Luo, "Model-based on-board fault detection and diagnosis for automotive engines", Control Engineering Practice, vol. 1, pp. 3–17, 1993.
6. B Eftekharnejad and D Mba, 'Seeded fault detection on helical gears with acoustic emission', Applied Acoustics, Vol 70, No 4, pp 547–555, April 2009.
7. A. Albarbar, S. Mekid, A. Starr, and R. Pietruszkiewicz, "Suitability of MEMS accelerometers for condition monitoring: An experimental study," Sensors (Basel), PMCID: PMC3672998, vol. 8, no. 2, pp. 784–799, 2008.
8. Antoni JAl'rAt'me, FrAl'dAl'ric Bonnardot, A. Raad, and Mohamed El Badaoui. "Cyclostationary modelling of rotating machine vibration signals. Mechanical systems and signal processing", 18(6):1285–1314, 2004.
9. Fugate, M. L., Sohn, H. & Farrar, C. R., "Vibration-Based Damage Detection Using Statistical Process Control", Mechanical Systems and Signal Processing, Vol. 15, No.4, 2001, pp. 707–721.
10. Spark-streaming, https://spark.apache.org/docs/latest/streaming-programming-guide.html.
11. Spark-MLlib, http://spark.apache.org/mllib/.
12. Konstantin Shvachko, et al., "The Hadoop Distributed File System," Mass Storage Systems and Technologies (MSST), IEEE 26th Symposium on IEEE, 2010, http://storageconference.org/2010/Papers/MSST/Shvachko.pdf.
13. Edgar Estupiñan, Paul White, César San Martin A Cyclostationary Analysis Applied to Detection and Diagnosis of Faults in Helicopter Gearboxes.

14. Sachin Kumar S, Neethu Mohan, Prabaharan Poornachandran, Soman K.P, "Condition Monitoring in Roller Bearings using Cyclostationary Features" WCI '15 Proceedings of the Third International Symposium on Women in Computing and Informatics

15. Discretized Streams: An Efficient and Fault-Tolerant Model for Stream Processing on Large Clusters. Matei Zaharia, Tathagata Das, Haoyuan Li, Scott Shenker, Ion Stoica. HotCloud 2012. June 2012.

16. M. Zaharia, M. Chowdhury, T. Das, A. Dave, J. Ma, M. McCauley, M. Franklin, S. Shenker, and I. Stoica. Resilient distributed datasets: A fault-tolerant abstraction for in-memory cluster computing. In NSDI, 2012.

17. Matei Zaharia, Mosharaf Chowdhury, Michael J. Franklin, Scott Shenker, Ion Stoica. "Spark: Cluster Computing with Working Sets". Hotcloud 2010. June 2010.

18. Spark-MLlib, https://spark.apache.org/docs/latest/mllib-frequent-pattern-mining.html.

19. Spark-framework, https://spark.apache.org/research.html.

Load Flow Analysis of Distribution System Using Artificial Neural Networks

M. Suresh, T.S. Sirish, T.V. Subhashini and T. Daniel Prasanth

Abstract In distribution system to determine static states at each node or bus and operating conditions, the load flow studies are very crucial. The load flow studies are very important, not only in finding static states but also during distribution system planning and its extension. In this paper, the load flow problem has been solved by artificial neural networks and these networks are efficient to describe the relation involved within the raw data. Two types neural networks are proposed to solve load flow problem of a distribution system, first one is Radial Basis Function Neural Network (RBFN) and other one is Multilayer Feedforward Neural Network with Backpropagation Algorithm (MFFN with BPA). The mathematical model of distribution load flow comprises a set of nonlinear algebraic equations that are solved using network topology-based distribution load flow which is usurped as reference off-line load flow. A series of training data is generated using off-line load flow, which is used to train the neural networks. The training data consists of different loading conditions and voltages corresponding to each and every node in the distribution system. The neural networks are trained with series of training data and tested with a loading which is not present in training data. Results obtained from two neural networks closely agrees with the reference off-line load flow result of same loading. The results of neural networks are compared together and computational time of two neural networks is considerably small.

M. Suresh (✉) · T. Daniel Prasanth
Vignan's Institute of Information Technology, Visakhapatnam, India
e-mail: msushe56@gmail.com

T. Daniel Prasanth
e-mail: tdanielprasanth@gmail.com

T.S. Sirish
Gayatri Vidya Parishad College of Engineering (A), Visakhapatnam, India
e-mail: srinivas.sirish@gmail.com

T.V. Subhashini
Anil Neerukonda Institute of Technology and Sciences, Visakhapatnam, India
e-mail: tvsubhashini@gmail.com

© Springer Nature Singapore Pte Ltd. 2017   515
S.C. Satapathy et al. (eds.), *Proceedings of the 5th International Conference on Frontiers in Intelligent Computing: Theory and Applications*, Advances in Intelligent Systems and Computing 515, DOI 10.1007/978-981-10-3153-3_51

Keywords Backpropagation algorithm (BPA) · Load flow solution · RBFN · Mean square error (MSE) · Multilayer feedforward neural network with backpropagation algorithm (MFFN with BPA)

1 Introduction

Distribution system is part of power system and is directly connected to the consumers. The consumers always need quality of power supply, i.e. voltage within permissible limits, less power number of power interruptions and more reliability. Distribution load flow is more important for the analysis of various distribution system parameters under different conditions [1]. To cognize the variation of voltage with respect to load and distribution system response during fault conditions, there is need of intelligent load flow. In distribution systems, it is very difficult to simulate atmospheric conditions and abnormal conditions. Sometimes the conventional load flows do not converge with real-time data on the distribution system because of high R/X ratio [2, 3]. The neural networks are more efficient and intelligent in processing the real-time data on distribution system. By using artificial neural networks, it is easy to do contingency analysis and simulate abnormal conditions, ther is no need to solve nonlinear equations of the distribution system [4]. Mostly, two types of neural networks RBFN and MFFN with BPA are used in power system applications like load forecasting, fault diagnosis and state estimation [5]. A network topology-based distribution load flow is used to generate training data which consists of a set of load patterns with voltages [6]. The training data is used to train the neural networks with the help of supervised learning rule, i.e. the input to neural network is the variation of load patterns and the corresponding voltages are the targets [2, 7]. The neural networks are also most suitable for network security assessment. The feedforward neural networks with error propagation algorithm are more efficient in analysis of real-time applications like load forecasting and fault diagnosis [8].

2 Power Flow Solution

It is an iterative method for finding network parameters voltage, load angle, active power and reactive power at each and every node of the distribution system is called distribution load flow solution. Here the system is assumed to be a three-phase balanced system. Topology-based distribution load flow is considered as reference off-line load flow.

2.1 Three-Phase Topology-Based Distribution Load Flow

Off-line distribution load flow algorithm in stepwise [1]

Step 1 From the network topological data, read both line data and bus data of distribution system

Step 2 Frame the Bus Injection Branch Current (BIBC) and (BCBV) matrix

$$[B] = [BIBC] * [I] \quad [\Delta V] = [BCBV] * [B] \tag{1}$$

Step 3 Obtain the distribution load flow matrix as

$$[\Delta V] = [DLF] * [I] \quad \text{Where } [DLF] = [BIBC] * [BCBV] \tag{2}$$

Step 4 For first iteration, set k = 0
Step 5 Increment the value of 'k', i.e. k = k + 1
Step 6 Update voltages by following eq., as

$$I_i^k = I_i^r(V_i^k) + jI_i^k(V_i^k) = \left(\frac{P_i + Q_i}{V_i}\right)^*$$

$$[\Delta V^{k+1}] = [DLF][I^k] \text{ and } [V^{k+1}] = [V^0] + [\Delta V^{k+1}] \tag{3}$$

Step 7 If max $\left(\left|I_i^{k+1}\right| - \left|I_i^k\right|\right) > tolerance$ go to step 6
Step 8 After obtaining the final voltages, calculate current flow in the branches and losses in the distribution system
Step 9 Print the bus voltages, losses, line flow or currents in the branches
Step 10 Stop.

3 Proposed Methodology

In this proposed methodology all parameters existing at nodes are active, reactive powers at nodes and resistance, reactance of each branch of distribution system. With help of supervised learning, the inputs to the neural network are variation of active, reactive powers at the nodes and corresponding voltages are the targets. The resistance and reactance are constants for a distribution system; they are not given to the neural network.

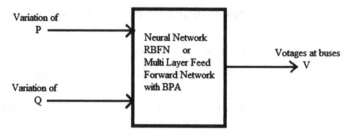

Fig. 1 Block diagram of the proposed methodology

3.1 Block Diagram of Proposed Methodology

In the block diagram, apart from inputs and outputs going inside; the architecture of the module consists of neural networks which could be either RBFN or MFFN with BPA. The input neurons and output neurons of inside neural networks are selected based on the bus number of distribution system (Fig. 1).

4 Radial Basis Function Neural Network

This method becomes more popular because of simple structure and easy to train the network [3].

The input given to the radial basis function neural network is feed forwarded to hidden layer and the output at the hidden layer jth unit is computed as follows:

$$h_j = \frac{\varphi(\|x - c_j\|)}{\delta_j} \tag{4}$$

where ϕ is the nonlinear radial basis function, X is the input vector,
C_j, δ_j are centre and centre spread parameter respectively.

5 Multilayer Feedforward Neural Network with Error Backpropagation Algorithm

MFFN with BPA has been employed to solve load flow problem of distribution system. The same training data is used for both RBFN and MFFN with BPA.

The generalized delta rule and backpropagation algorithm both are referred as same. This backpropagation algorithm consists of two steps; first one is forward pass and second one is backward pass. This algorithm gives gradient of error

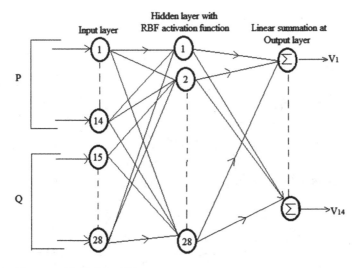

Fig. 2 Architecture of RBFN for 15-bus system

function through differentiation. The first step forward pass involves the training data forward propagated layer by layer through the neural network similar to perceptron. The training data is processed by activation function at each and every layer. After forward computation, the error will be calculated at the output layer with known target values. The second step backward pass involves the output error propagated back layer by layer for calculating error at hidden layer, since target is unknown. Repetition of this procedure will be done until a relatively small rate of change of error is achieved.

For 15-bus system first one is slack bus and the variation of active and reactive powers at remaining 14 buses are given input to the neural network shown in Figs. 2 and 3.

5.1 Backpropagation Algorithm in Steps

Step 1 The performance of neural networks will be enhanced when inputs and outputs of neural network lie in the range of 0–1. Hence, input and outputs are normalized with respect their maximum values. Assume there are 'L' inputs given by $\{I\}_{l \times 1}$ 'n' outputs $\{o\}_{n \times 1}$ normalized form 'm' is the number of hidden layer units

Step 2 Define the size of [v] and [w] that stand for the weights that connecting input to hidden and hidden to output respectively

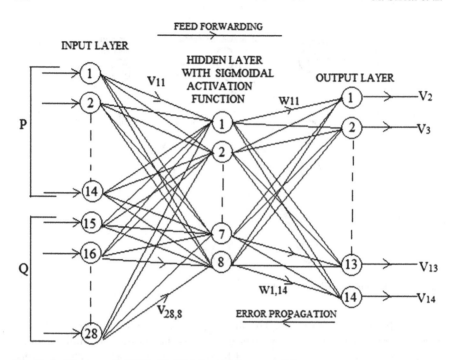

Fig. 3 Architecture of MFFN with BPA for 15-bus system [8]

Step 3 In general 'λ' assumed as threshold value can be zero

$$[v]^0 = [random weights] [w]^0 = [random weights] \; [\Delta v]^0 = [\Delta w]^0 = [o]$$

Step 4 training data includes a set of inputs and outputs

$$\{o\}_{I \atop l \times 1} = \{I\}_{I \atop l \times 1} \tag{5}$$

Step 5 Calculate input to the hidden layer by

$$\{I\}_{H \atop m \times 1} = [v]^T_{m \times l} \{o\}_{I \atop l \times 1} \tag{6}$$

Step 6 evaluate output of the hidden layer units by using sigmoid function as

$$\{o\}_H = \left\{ \begin{array}{c} \cdot \\ \frac{1}{(1 + e^{-I_{Hi}})} \\ \cdot \end{array} \right\}_{m \times 1} \tag{7}$$

Step 7 inputs of the output layer are obtained as follows:

$$\{I\}_{o \atop n \times 1} = \{w\}^T_{n \times m} \{o\}_{H \atop m \times 1} \tag{8}$$

Step 8 let the output layer units evaluate the output using sigmoid function as

$$\{o\}_o = \left\{ \begin{array}{c} \cdot \\ \frac{1}{(1 + e^{-I_{oj}})} \\ \cdot \end{array} \right\} \tag{9}$$

Step 9 For the jth training set, mean square of the error between the actual output and obtained output is obtained as

$$E^p = \frac{\sum (T_j - O_{oj})^2}{n} \tag{10}$$

Step 10 find $\{d\}$ as

$$\{d\}_H = \left\{ \begin{array}{c} \cdot \\ (T_k - O_{ok})O_{ok}(1 - O_{ok}) \\ \cdot \end{array} \right\}_{n \times 1} \tag{11}$$

Step 11 find [y] matrix as

$$[y]_{m \times n} = \{o\}_{H \atop m \times 1} <d>_{1 \times n} \tag{12}$$

Step 12 find the change in weights

$$[\Delta w]^{t+1}_{m \times n} = \alpha [\Delta w]^t_{m \times n} + \eta [y]_{m \times n} \tag{13}$$

Step 13 Find

$$\{e\}_{m \times 1} = [w]_{m \times n} \{d\}_{n \times 1} \tag{14}$$

$$\{d^*\} = \left\{ \begin{array}{c} \cdot \\ e_i (O_{Hi})_{m \times 1} (1 - O_{Hi})_{m \times 1} \\ \cdot \end{array} \right\} \tag{15}$$

Then find [x] as $[x]_{1 \times m} = \{o\}_{I \atop 1 \times 1} <d^*> = \{I\}_I <d^*>_{1 \times m}$ \quad (16)

Step 14 find or update change in weights from input layer to hidden layer

$$[\Delta v]_{1 \times m}^{t+1} = \alpha[\Delta v]_{1 \times m}^{t} + \eta[x]_{1 \times m} \tag{17}$$

Step 15 update weights from input to hidden layer as well as hidden layer to output layer

$$[v]^{t+1} = [v]^{t} + [\Delta v]^{t+1} \quad [w]^{t+1} = [w]^{t} + [\Delta w]^{t+1} \tag{18}$$

Step 16 find error rate as $errorrate = \frac{\sum E^p}{nset}$ and repeat steps 5–16 until the convergence in the error rate is less than the tolerance value.

6 MATLAB/Simulation Results

The simulation studies are carried out using MATLAB. The proposed neural networks tested on two case studies 15-bus system and 33-bus system. The single line diagram of 15-bus system is shown in Fig. 4.

Case study 1: first network topology-based distribution load flow conducted on 15-bus system and a series of data generated for training of neural networks. The training data consists of two types of load patterns. First bus assumed as slack bus.

Type-1 load patterns: the active and reactive power varied from 10 to 160% in steps of 10% at each and every node simultaneously, i.e. 16-load patterns.

Type-2 load patterns: the active power and reactive power varied from 10 to 150% in steps of 10% at a single bus by keeping active and reactive powers at the remaining buses are constant and repeat the same for all 14 buses, i.e. 15-load patterns for single bus*14 buses gives 220 load patterns. The training data of 15-bus system containing 226 load patterns and corresponding voltages.

Fig. 4 Single Line Diagram of 15-Bus System

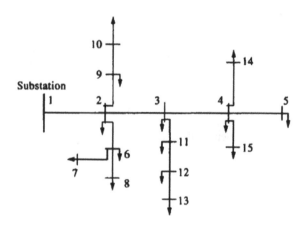

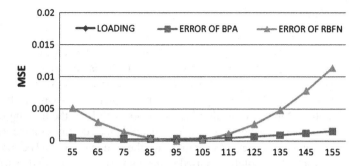

Fig. 5 RBFN error versus MFFN with BPA error

Case study 2: similarly for 33-bus system 16-type-1 load patterns and type-2 load patterns are 15-load patterns for single bus*32 buses gives 480 load patterns. The training data of 33-bus system containing 496 load patterns and corresponding voltages. The training data is used to train the neural networks, the output of neural networks is weights. The obtained weights are used for testing the neural networks. The testing data is out off training data. After testing the results of neural networks are compared with each other to find accuracy between proposed neural networks with help of MSE as shown in Fig. 5. Then the more accurate neural network results are compared with reference off-line load flow, i.e. topology-based distribution load flow as shown in Table 1.

Table 1 Load Flow Result of MFFN with BPA for 15-bus system

Serial number	Bus voltages	Voltages from off-line load flow base case(target)	Voltages from MFFN with BPA
1	V_2	0.9713	0.9712
2	V_3	0.9567	0.9564
3	V_4	0.9509	0.9506
4	V_5	0.9499	0.9495
5	V_6	0.9581	0.9579
6	V_7	0.9559	0.9555
7	V_8	0.9568	0.9565
8	V_9	0.9679	0.9678
9	V_{10}	0.9669	0.9668
10	V_{11}	0.9499	0.9495
11	V_{12}	0.9457	0.9452
12	V_{13}	0.9444	0.9440
13	V_{14}	0.9486	0.9482
14	V_{15}	0.9484	0.9480

7 Conclusion

In this work, two modules of neural networks have been developed to solve load flow solution to radial distribution networks. First one is Radial Basis Function neural network and second one is Multilayer Feedforward network with Backpropagation Algorithm. Radial basis function neural network is easy to train and simple in structure. Multilayer feedforward network with BPA the training process is slow, and its ability to generalize a pattern mapping task depends on the learning rate and the number of neurons present in the hidden layer. The proposed methodology or load flow solution is used to test on two standard radial distribution test systems 15-bus system and 33-bus system. Radial Basis function neural network shows an excellent performance while testing with loading levels which are present in training data. It gives zero error for training data. Radial basis function neural network gives more error while testing with different loading conditions other than training data or not present in training data. In practical loading conditions are not similar to training set. Multilayer feedforward neural network shows a very good performance irrespective of loading condition weather it is present in trained data or not. It gives very less error (MSE) in acceptable range. By comparing the results of both RBFN and MFFN with BPA, it is concluded that MFFN with BPA is best suitable for load flow solution and having less computational time and best suitable to process real-time distribution system data.

References

1. Jen-Hao Teng, "A Network-Topology-based Three-Phase Load Flow for Distribution Systems", Proc. Natl. Sci. Counc. ROC(A) Vol. 24, No. 4, 2000. pp. 259–264.
2. A. Rathinam, S. Padmini, V. Ravikumar, Conference on Information and Communication Technology in Electrical Sciences (ICTES 2007), "Application Of Supervised Learning Artificial Neural Networks [CPNN, BPNN] For Solving Power Flow Problem", IET-UK International Dr. M.G.R. University, Chennai, Dec. 20–22, 2007. pp. 156–160.
3. Mohammad Khazaei, Shahram Jadid, "Contingency Ranking Using Neural Networks by Radial Basis Function Method", 978-1-4244-1904, ©2008 IEEE.
4. D Das, D P Kothari, A Kalam, "Simple and efficient method for load flow solution of radial distribution networks", Electrical Power & Energy Systems, Vol. 17, No. 5, pp. 335–346, 1995.
5. D. Das, H.S. Nagi and D.P. Kothari, "Novel Method for solving radial distribution networks," Proceedings IEE Part C (GTD), vol.141, no.4, pp. 291–298, 1991.
6. S. Ghosh and D. Das, "Method for Load−Flow Solution of Radial Distribution Networks," Proceedings IEE Part C (GTD), vol. 146, no. 6, pp. 641–648, 1999.
7. R. Ranjan and, D. Das, "Simple and Efficient Computer Algorithm to Solve Radial Distribution Networks," International Journal of ElectricPower Components and Systems, vol.31, no. 1,: pp. 95–107, 2003.
8. Thomas Wei Kwang Lee, "Artificial Neural Network Security Assessment", Ph.D. Dissertation, Dept of Engineering Physical Science & Architecture, University of Queensland, 2003.

An Enhanced Bug Mining for Identifying Frequent Bug Pattern Using Word Tokenizer and FP-Growth

K. Divyavarma, M. Remya and G. Deepa

Abstract Nowadays bugs are the commonly occurring problems in many types of software. In order to prevent from these issues, a detailed study of bugs is an essential thing. Bugs are classified based on their severity in corresponding bug repositories. Some of the bug repositories are Mozilla, Android, Google Chromium, etc. So finding the most frequently occurring bugs is the right solution for the software malfunctioning. Thus it can help developers to prevent those bugs in the next release of the software. In this paper, our main aim is the mining of bugs from the bug summary data in the bug repositories by applying FP-Growth, one of the best techniques for finding frequently occurring pattern using WEKA.

Keywords Association rules · Bug summary · FP-growth · Stemming · Stop-word · Tokenization

1 Introduction

The integral part of a software organization is to provide quality assured products to their clients. In order to make customers happy, they should build quality products. Some of the open source bug tracking systems such as Bugzilla, Red mine, etc., provide a platform where clients report their issues directly to the repositories. Based on the severity of the issue, the quality assurance team manager assigns the bugs to the developer in their team by setting the milestones and thereby ensuring trust for the end users.

K. Divyavarma · M. Remya · G. Deepa (✉)
Department of Computer Science & IT, Amrita School of Arts and Sciences,
Amrita Vishwa Vidyapeetham, Amrita University, Kochi, India
e-mail: deepsgopi@gmail.com

K. Divyavarma
e-mail: divyavarma375@gmail.com

M. Remya
e-mail: rem_92@live.com

© Springer Nature Singapore Pte Ltd. 2017
S.C. Satapathy et al. (eds.), *Proceedings of the 5th International Conference on Frontiers in Intelligent Computing: Theory and Applications*, Advances in Intelligent Systems and Computing 515, DOI 10.1007/978-981-10-3153-3_52

The fundamental component of a bug repository is their bug summary data, which is in the form of text. We have to apply text mining techniques to mine those data for finding the commonly occurring patterns of bugs. Our goal is to propose a text mining method to extract bugs from the bug summary and find its matching patterns using FP-growth algorithm in the machine learning tool WEKA. Our paper is arranged as follows: Sect. 2 includes the background of the proposed work, Sect. 3 deals with proposed work, Sect. 4 contains experimentation and results, Sect. 5 describes other research papers related to this work, Sect. 6 describes Conclusion.

2 Background

In this section, we briefly discuss about bug summary and different text mining techniques and association algorithm, which we used for our proposed work done in WEKA.

2.1 Bug Summary

A typical bug report contains so many attributes such as Reporter name, Product, OS, Summary, Description, etc. When you report a bug in the bug repository, its most important part is the bug summary data. Using this data, the developers can check the importance of bug and at the same time they can check for duplication also.

2.2 Text Mining

Text mining is the process of extraction of meaningful information from the textual data for further data mining purposes [1, 2]. The various techniques applied for text mining are:

Tokenization

Tokenization is the process of splitting up large strings into tokens. These tokens are in the form of either words or digits. It filters punctuations as well as escape characters and symbols appeared in a string. In WEKA, there are different tokenization methods such as n-gram tokenizer, alphabetic tokenizer and word tokenizer.

E.g.: Issue in phone volume

After tokenization, the tokens created are:

Issue, in, phone, volume

Stemming

Stemming is a process of minimizing each word in a text to its root form. For e.g. "crashing" diminish to its root form "crash" after stemming. There are various stemming algorithms available in WEKA such as Porter-Stemmer, Snowball, Lovins, etc.

Stop-words removal

Stop-words are the words which will not cause any effect on the textual data. The words such as is, was, were, etc., are the examples of stop-words. These words should be removed unless the number of attributes will be more and thus it will affect the performance of association algorithms.

2.3 FP-Growth

FP-Growth is an association algorithm for generating frequent patterns from the given dataset [3]. Unlike other pattern matching algorithms such as A priori, FP-Growth will not generate candidate. So when compared to others, it has high performance [4]. There are mainly 2 steps for this algorithm. First, construct a Frequent Pattern Tree from the given dataset. After that, generate frequent patterns from the Frequent Pattern Tree that is already constructed.

2.4 Association Rules

The association rule is of the form A → B, where A ∩ B is a null set [5, 6]. The support and confidence are the two terms which are used to measure the strength of the association rules [7].

For a rule A → B,

$$\text{Support (coverage)} = \frac{\sum(A \cup B)}{N} \qquad (1)$$

where N is the total number of transactions

$$\text{Confidence (accuracy)} = \frac{\sum(A \cup B)}{\sum A} \qquad (2)$$

3 Related Work

Currently, we are aware of one research paper which is based on the bug mining from bug repositories. Kiran Kumar B, JayadevGyani and Narasimha G [8] proposed a method for mining frequent patterns from the bug summary available in the open bug repositories by applying text mining techniques. They took bug summary data from the Mozilla and Eclipse bug repositories for performing their work. Their work consist of applying text mining techniques such as tokenization, stemming and stop-words removal for mining the bug summary and after that they applied Apriori algorithm for finding the frequent patterns.

JaweriaKanwal and OnaizaMaqbool [9] provide an approach for automatically prioritize the newly arrived bug report by using Support Vector Machines.

4 Proposed Work

Bug repository is a place, where we can view and store the information of bugs. A bug repository contains so many attributes such as, ID, Status, Summary, Priority, etc. The Summary attributes indicate the associated issues. We propose a method for mining bugs from the bug repository. The proposed method is depicted in the Fig. 1.

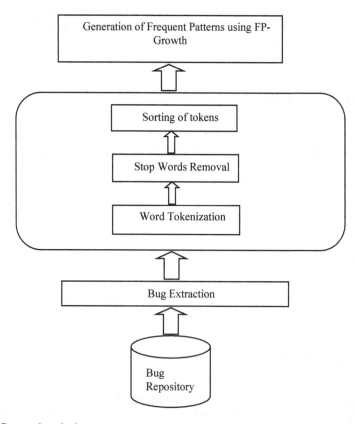

Fig. 1 Proposed method

Here we choose Word Tokenizer as the tokenization method. Compared to other tokenization methods, Word tokenizer will select words as well as digits for its processing whereas other tokenization methods will take only words. Both words and digits are equally important while considering a bug (Nexus7, iPhone6, etc.). So while considering a bug, Word Tokenizer is the best tokenization method.

Generally, while mining textual data stemming is considered as a part of pre-processing. In the case of simple text mining, stemming will not cause that much trouble. But in the case of bug mining, technical terms are associated with the bugs. So by applying, stemming will change the exact meaningful information. For example, "exception in Iphone6" is bug. After applying Stemming, the word "exception" will become "except". Both except and exception is entirely different in technical terms. So in this proposed method, we avoid stemming process.

Stop-words are words that are not much significant in a bug. So we have to remove it. Removing stop-words from a bug will not change its meaning. For example, "Samsung S5 is crashed". Here 'is' is a stop word. After removing it, does not cause any effect on its meaning.

That is {Samsung, S5, crashed}

After applying Word tokenization, and stop-words removal to the bug summary data, a set of words are obtained which are known as Tokens. So next step is to sort all these tokens.

For generating frequent patterns of bugs, FP-Growth algorithm is chosen. A bug repository contains large number of bug reports. Thus, it generates large number of attributes too. While compared to other association algorithms which takes only less than 20 attributes whereas FP-Growth can handle more than 5000 attributes. So, FP-Growth is the best way to find pattern matching of bugs.

5 Experimentation and Outcomes

Since Android is the largely used operating system in the world, we decided to take bug datasets from the Android's open bug repository. From the android bug repository, we choose 'Issues on nexus devices', 'Issues on Samsung devices' and 'Issues on signals in Android devices' for conducting our experiment. So pre-processing of the bugs should be necessary before finding frequent patterns. Generally, pre-processing methods of textual data consist of Tokenization, Stop-Words removal and Stemming. The Table 1 shows the result we obtained after applying pre-processing without stemming and the number of rules generated by applying FP-Growth algorithm with a confidence level 0.9 in WEKA.

Generally, we have to obtain the summary of generated rules. Consider an example of three rules:

1. $\{x\} \rightarrow \{z\}$, includes itself and 3
2. $\{y\} \rightarrow \{z\}$, includes itself and 3
3. $\{x,y\} \rightarrow \{z\}$, includes itself

Table 1 After preprocessing and FP-growth

	Issues on Nexus devices	Issues on Samsung devices	Issues on signals
Unigram Tokenizer			
Total instances	316	19	33
Total attributes	965	164	194
Total association rules	Nil	133	5
Bigram Tokenizer			
Total instances	316	19	33
Total attributes	3108	391	477
Total association rules	2	24	35
Trigram Tokenizer			
Total instances	316	19	33
Total attributes	5431	617	753
Total association rules	2	122	145
Word Tokenizer			
Total instances	316	19	33
Total attributes	1012	180	215
Total association rules	11	271	28
Alphabetic Tokenizer			
Total instances	316	19	33
Total attributes	964	162	194
Total association rules	Nil	168	5

Therefore, 1 and 2 forms the summary of these three rules. They are also known as Direction Setting (DS) rules [10]. That means, with the help of this, one can go for further details.

The results shown in Table 1 prove that Bugs using Word Tokenizer generate large number of association rules using FP-Growth. But in the case of Issues on signals, both the Bigram and Trigram Tokenizers generate more number of rules yet Word Tokenizer is chosen as the best tokenization method. From Table 2, we can summarize that 'there is a bug in Nexus 5's lollipop version'. Similarly from Tables 3 and 4, we can summarize that 'Galaxy S5 driver is broken' and 'There is

Table 2 Rules of issues on Nexus devices

Top 5 Rules of Word Tokenizer	
1.	[lollipop=1]: 74 ==> [nexus=1]: 74 <conf:(1)> lift:(1.01) lev:(0) conv:(0.47)
2.	[4=1]: 70 ==> [nexus=1]: 70 <conf:(1)> lift:(1.01) lev:(0) conv:(0.44)
3.	[5=1, on=1]: 113 ==> [nexus=1]: 113 <conf:(1)> lift:(1.01) lev:(0) conv:(0.72)
4.	[5=1, lollipop=1]: 68 ==> [nexus=1]: 68 <conf:(1)> lift:(1.01) lev:(0) conv:(0.43)
5.	[5=1]: 253 ==> [nexus=1]: 252 <conf:(1)> lift:(1) lev:(0) conv:(0.8)

Table 3 Rules of issues in Samsung

Top 5 Rules of Word Tokenizer	
1.	[s5=1]: 2 ==> [galaxy=1]: 2 <conf:(1)> lift:(2.11) lev:(0.06) conv:(1.05)
2.	[s=1]: 2 ==> [galaxy=1]: 2 <conf:(1)> lift:(2.11) lev:(0.06) conv:(1.05)
3.	[driver=1]: 2 ==> [galaxy=1]: 2 <conf:(1)> lift:(2.11) lev:(0.06) conv:(1.05)
4.	[broken=1]: 2 ==> [galaxy=1]: 2 <conf:(1)> lift:(2.11) lev:(0.06) conv(1.05)
5.	[as=1]: 2 ==> [galaxy=1]: 2 <conf:(1)> lift:(2.11) lev:(0.06) conv:(1.05)

Table 4 Rules of issues of signals in Android devices

Top 5 Rules of Word Tokenizer	
1.	[sigsegv=1]: 8 ==> [fatal=1]: 8 <conf:(1)> lift:(2.75) lev:(0.15) conv:(5.09)
2.	[code=1]: 4 ==> [fatal=1]: 4 <conf:(1)> lift:(2.75) lev:(0.08) conv:(2.55)
3.	[sigsegv=1]: 8 ==> [11=1]: 8 <conf:(1)> lift:(3.67) lev:(0.18) conv:(5.82)
4.	[nexus=1]: 7 ==> [5=1]: 7 <conf:(1)> lift:(4.13) lev:(0.16) conv:(5.3)
5.	[fatal=1, 11=1]: 8 ==> [sigsegv=1]: 8 <conf:(1)> lift:(4.13) lev:(0.18) conv:(6.06)

an error in fatal code sigsegv 11 in android devices' respectively. Therefore, when taking top 5 rules, word tokenizer provides more accurate summary while compared to others.

6 Conclusion

A software bug is a fault that can affect the performance of the system. Most of these bugs are due to human intervention. If an error occurred in software, it will become more expensive in terms of developmental cost, maintenance cost and recovery cost. Bugs can either be reported by the users or developers with the help of bug reports, which are then added to the bug repository. It is the duty of the Quality Assurance (QA) manager to find these reported bugs. After that, based on the priority of these bugs, the QA will assign it to skillful/capable developers.

By knowing the frequent pattern generated from the bug repository, help the developer to take preventive measures against its chance of occurring in future. In this paper, we presented a way for identifying bugs using, Word Tokenizer without stemming. The stemming techniques available till now are not efficient ways for bug mining. By finding stemming algorithms especially for bugs in future may be useful for mining issues.

References

1. Y, Zhou., Y, Tong., Gu, Ruihang., H, Gall.: Combining Text Mining and Data Mining for Bug Report Classification. *ICSME*. 311–320 (2014).
2. V, Neelima., N, Annapurna., V, Alekhya., B, Vidyavathi.: Bug Detection through Text Data Mining. *IJARCSSE*. 564–570 (2013).
3. shweta, M.S, Drgarg,.K.: Mining Efficient Association Rules Through Apriori Algorithm using Attributes and Comparative Analysis of Various Association Rule Algorithms. *IJARCSSE*. 306–312. (2013).
4. Rashmi, S., Prof nitin, S.: An Improved Association Rule Mining With Fp Tree Using Positive And Negative Integration. *JGRCS*. 46–51. (2012).
5. Yasmeen, S.: Software Bug Detection Algorithm using Data mining Techniques. *IJIRAE*. 105–108. (2014).
6. Drkanak, S., Rajpoot,..D..S..: A Way to Understand Various Patterns of Data Mining Techniques for Selected Domains. *IJCSIS*. 186–191. (2009).
7. marukatat, R.: On the Selection of Meaningful Association Rules. In Julio, P & adem, K (Eds), *Data mining and knowledge discovery in real life applications* pp. 75–88. (2009).
8. hahsler, M, chelluboina, S.: Visualising Association Rules : Introduction to the R-extension Package arulesViz.
9. kumar, K., Jayadev, .G, Narsimha, G.: Mining Frequent Patterns from Bug Repositories. *IJARCSSE*. 698–704. (2014).
10. Kanwal, J, Maqbool, O.: Managing Open Bug Repositories through Bug Report Prioritization Using SVMs. *ICOSST*. pp. 22–24. (2010).

Implementing Anti-Malware as Security-as-a-Service from Cloud

Deepak H. Sharma, C.A. Dhote and Manish M. Potey

Abstract In Security-as-a-service model the objective is to provide security as one of the cloud services. In this model the security is provided from the cloud in place of traditional on-premise implementation. The objective of this initiative is to provide Anti-Malware functionality as a cloud service. This paper provides implementation framework for Anti-Malware system from the cloud as a service. The framework uses several existing file scanning web-based anti-malware engines. The Anti-Malware SecaaS offers all the benefits provided by Security-as-a-Service (SecaaS) model. The proof-of-concept (POC) prototype of Anti-Malware AM-SecaaS is implemented and evaluated successfully. An innovative approach is used to integrate this POC with other SecaaS options so that various SecaaS options are provided to users intelligently and transparently.

Keywords Cloud computing · Cloud security · Security-as-a-service · Anti-malware security-as-a-service

1 Introduction

In Security-as-a-Service model security is provided as one of the cloud services. In this model the security is provided as a commodity instead of on-premise implementation. The capability of traditional on-premise security implementations can be enhanced by security-as-a-service model. The services provided by SecaaS model

D.H. Sharma (✉) · M.M. Potey
Department of Computer Engineering, K. J. Somaiya College of Engineering,
Mumbai, India
e-mail: deepaksharma@somaiya.edu

M.M. Potey
e-mail: manishpotey@somaiya.edu

C.A. Dhote
Department of Information Technology, PRMIT&R, Amravati, India
e-mail: vikasdhote@rediffmail.com

© Springer Nature Singapore Pte Ltd. 2017
S.C. Satapathy et al. (eds.), *Proceedings of the 5th International Conference on Frontiers in Intelligent Computing: Theory and Applications*, Advances in Intelligent Systems and Computing 515, DOI 10.1007/978-981-10-3153-3_53

can work in hybrid way in-premise security solutions. The traditional Anti-Malware solutions have already matured. However, due to the growth of cloud computing, virtualization, and multi-tenant resource sharing there are several new targets for malware. The major problem with on-premise anti-malware solutions is that virus/malware signatures have to be updated regularly, and eventually the size of database of signatures becomes too large. The increasing size of malware/virus signature makes the process of malware detection very slow. Several families of new malware bring more threats to the organizations. A necessity is being felt to move anti-malware protection from user level desktop machines to the cloud. This solution from the cloud can be collaborative and more effective to counter threats from several new types of malwares. The cloud service provider can use several anti-malware engines collaboratively thus providing zero day attack countermeasures as well.

The primary purpose of anti-malware solution is to monitor the traffic flowing to clients' organization to prevent any malicious software to cause any interruption, interception, modification of data, applications, and systems. There are several cloud-based anti-malware systems that exist. Several of these are freely available and can be used over the web in the form of web interface or public API. The POC prototype implementation specifically uses Virus Total (https://www.virustotal.com) and Meta defender (https://www.metadefender.com) cloud APIs. In actual practice several engines can be collaboratively used for more effective solution.

This paper proposes an Anti-Malware Security-as-a-service (AM-SecaaS) framework. This AM-SecaaS will be available on-demand; it will be portable, and available on pay-per-use costing basis. The paper discusses issues related to Anti-Malware functionality delivered as cloud service. This paper discusses the Anti-Malware SecaaS option in following sections. In Sect. 2, related work is discussed. In Sect. 3, the scope and POC framework of AM-SecaaS in public cloud is discussed. In Sect. 4, evaluation of AM-SecaaS prototype is discussed. Finally, conclusion and future scope of work is discussed in Sect. 5.

2 Anti-Malware Security Related Work

In ThinAV [1], the authors have introduced an anti-malware system for Android using preexisting web-based file scanning engines. The system provides anti-malware security for smart phones. The evaluation of performance in wide area network has also been discussed.

In CloudSEC architecture [2] the authors have proposed a notion of Collaborative Security against the distributed attacks that originate from malware. The paper proposes architecture for automated malware detection and containment based on collaborative approach.

The authors of CAS [3] have proposed a fast and efficient technique for detection of malware families based on correlation signatures. The model uses Advanced

Persistent Threats correlation. The model also proposes to move the anti-virus functionality to the cloud instead of user desktops.

The SECaaS [4] proposes a user centric approach to give users more control of over their security in the cloud for their cloud-based applications. It proposes several use-cases, one of them is anti-malware cloud to monitor VMs and detect spyware, viruses, and worms.

In paper [5], the authors have discussed the improvements in malware detection based on file relations. The paper proposes a file verdict system built on a semi-parametric classifier model to combine file contents and file relations together for malware detection.

The authors of [6] have discussed several options available for Security-as-a-service but Anti-Malware service is not described by them.

3 POC Anti-Malware-SecaaS Implementation

Anti-Malware as a service has been implemented as a combination of various mechanisms from different vendors. It is provided as cloud service by accessing it through the browser. The users need not install any software on their desktops or devices so there is no question of regular signature updates. All the software is cloud-based; it resides on the cloud, and users can use multiple anti-malware engines simultaneously. Various elements of AM-SecaaS architecture are as shown in Fig. 1.

The architecture of Anti-Malware Security-as-a-service (AM-SecaaS) mainly involves protection of users' organization from any kind of malicious software. In POC implementation, file level cloud-based anti-malware engines from various vendors like Virus Total and Meta Defender have been used. As shown in Fig. 1, before delivering to the organization the incoming stream is cleaned and all policies are enforced. The file is uploaded to anti-malware cloud, and it is scanned before it can be sent out on the public network. Similarly any incoming file will be scanned before being entering the users' organization.

The Windows Azure public cloud has been used to test proof-of-concept (POC) prototype of AM-SecaaS. This AM-SecaaS framework can be implemented in all types of cloud environments. Virtual Machines (VMs) are used to build all AM-SecaaS components in a public cloud environment. The important characteristics of AM-SecaaS are that it is available on-demand, elastic, and portable. It can make use of security solutions from various vendors simultaneously. All these features are possible as the VM instances can be started and stopped dynamically on the go based on need.

The emphasis here is on Anti-Malware scan for any incoming or outgoing file from different types of users, viz, remote, desktop, or mobile devices. Anti-Malware functionality is provided as a cloud service. Figure 1 shows the system architecture

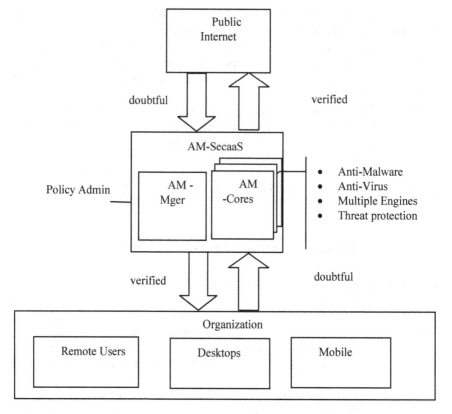

Fig. 1 Implementation of POC AM-SecaaS

of POC; the main components of the system are AM-SecaaS core and AM-SecaaS manager. The core functions are implemented in AM-SecaaS core module. AM-SecaaS manager implements the managerial functions like policy enforcement, and Intelligence. AM-Secaas is placed between the public internet and client organization assets. When any traffic comes from the public internet it is cleaned by applying all policies, and then it is delivered to the Client Organization as demonstrated in the figure. A client organization can have several types of resources which are to be protected.

The AM-SecaaS core module has other functionalities like anti-malware, anti-virus, and threat protection using multiple engines from various vendors.

The implementation is done in form of Virtual machines in Windows Azure cloud setup. The AM-SecaaS Core and AM-SecaaS Manager reside and operate from separate Virtual machines. In actual scenario the anti-malware systems from multiple vendors can be used for practical purpose. This type of solution will be more effective than using anti-malware from single vendor in an on-premise solution.

The Anti-Malware functionality has been implemented in the POC. It is delivered as a Cloud Service. Various types of clients on different devices (Desktops, Laptops, and Handheld devices, etc.) can access it by simply using browser.

4 Evaluation of AM-SecaaS

To evaluate the effectiveness of our proof-of-concept prototype several experiments have been conducted. The objective is to compare an on-premise solution vis-à-vis AM-SecaaS in public cloud. The POC is evaluated on the basis of discussion in article [7] and evaluation criterion discussed in paper [8, 9]. The evaluation has been done based on following criteria:

- Reliability: the service is highly reliable and available owing to redundancy provided because of multiple web servers implemented in the cloud environment. The testing of POC was done in the form of web servers implemented in cloud environment for providing uninterrupted services to the clients.
- Effectiveness: the services can be made more effective by designing the core module to use anti-malware engines from various vendors. There is no need of any regular updates on any of the client devices. The entire signature database resides on cloud. Different types of anti-malware techniques can also be integrated together in this form of cloud service.
- Performance: the performance was tested by comparing the average time taken by on-premise anti-malware detection with respect to AM-SecaaS mechanism. The testing was done by running the AM-SecaaS under anti-malware engines in various cloud environments. The traffic in the public cloud also affects the overall overhead. In POC evaluation the overhead does not increase by 15–20% (refer Fig. 2). It is not very high as compared to its advantages offered over traditional systems.

Fig. 2 Comparison of Anti-Malware AAS System

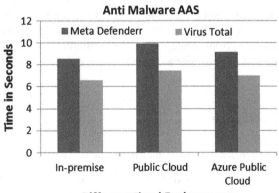

- Flexibility: the solution can be made to work with existing traditional systems also. The POC implementation has used file-based anti-malware detection from Virus Total and Meta defender but in actual systems the anti-malware systems from multiple vendors can be actually used to run on VMs. The clients can be provided with more flexibility by allowing them to select different functionalities of security as per their need.
- Control: the client can exercise the use of service from several devices, viz, desktops, laptops, and handheld devices, etc. The management from users' point of view can be very easy through a central console.
- Privacy and Security: before any doubtful inbound traffic enters clients' organization it is filtered by AM-SecaaS. This filtering enforces all policies defined. This ensures the privacy and security of data in a client organization.
- Cost of ownership: it works on the pay-per-use costing model. There is no upfront investment from client, and client does not install anything in on-premise implementation. AM-SecaaS is available to client as Operational Expenses (OPEX) model.

5 Conclusion and Future Scope

The paper introduced delivery of Anti-Malware as Security-as-a-Service (AM-SecaaS) option. It was presented in the form of a framework which can be used by cloud service provider to deliver Anti-Malware as a cloud service. Thus AM-SecaaS has various important cloud features like portability, elasticity, and pay-per-use service. The AM-SecaaS was realized as a combination of anti-malware engines in public cloud to work in the cloud environment. This solution can also be used in a hybrid manner to work best with existing on-premise traditional implementations. Their security capabilities can also be thus enhanced. With AM-SecaaS, users can protect their resources by creating a virtual private area with in cloud space. An innovative approach of integrating POC with several other Security-as-a-Service options was followed. All these security services have been provided through a common portal of various security services.

In future the system effectiveness can be enhanced by using anti-malware engines from multiple vendors simultaneously. The additional functionalities related to anti-malware management can be added to service more useful.

References

1. Chris Jarabek, David Barrera, and John Aycock. 2012. ThinAV: truly lightweight mobile cloud-based anti-malware. In *Proceedings of the 28th Annual Computer Security Applications Conference* (ACSAC '12). ACM, New York, NY, USA, 209–218.

2. J. Xu, J. Yan, L. He, P. Su and D. Feng, "CloudSEC: A Cloud Architecture for Composing Collaborative Security Services," *Cloud Computing Technology and Science (CloudCom), 2010 IEEE Second International Conference on*, Indianapolis, IN, 2010, pp. 703–711.
3. W. Yan, "CAS: A framework of online detecting advance malware families for cloud-based security," *2012 1st IEEE International Conference on Communications in China (ICCC)*, Beijing, 2012, pp. 220–225.
4. Mohammed Hussain and Hanady Abdulsalam. 2011. SECaaS: security as a service for cloud-based applications. In *Proceedings of the Second Kuwait Conference on e-Services and e-Systems*(KCESS '11). ACM, New York, NY, USA,, Article 8, 4 pages.
5. Yanfang Ye, Tao Li, Shenghuo Zhu, Weiwei Zhuang, Egemen Tas, Umesh Gupta, and Melih Abdulhayoglu. 2011. Combining file content and file relations for cloud based malware detection. In Proceedings of the 17th ACM SIGKDD international conference on Knowledge discovery and data mining (KDD '11). ACM, New York, NY, USA, 222–230.
6. Tim Mather, Subra Kumaraswamy, and Shahed Latif, 2009, Cloud Security and Privacy. An Enterprise Perspective on Risks and Compliance, O'Reilly Media, 336.
7. Websense white paper, Seven Criteria for Evaluating Security-as-a- Service Solutions, 2010.
8. Deepak Sharma, Dr. C A. Dhote, Manish Potey, 'Security-as-a-Service from Clouds: A comprehensive Analysis', IJCA Volume 67-Number 3, April 2013.
9. Deepak Sharma, Dr. C A. Dhote, Manish Potey, 'Security-as-a-Service from Clouds : A survey' IIJC Vol 1 Issue 4, October 2011.

Recognize Online Handwritten Bangla Characters Using Hausdorff Distance-Based Feature

Shibaprasad Sen, Ram Sarkar, Kaushik Roy and Naoto Hori

Abstract In this paper, an effort has been made to emphasize the usefulness of Hausdorff Distance (HD) and Directed Hausdorff Distance (DHD) based features for the recognition of online handwritten Bangla basic characters. Every character sample is divided into N number of rectangular zones and then HD- and DHD-based features have been computed from every zone to every other zone. These distance measurements are served as feature values for the present work. Experiment has been done on a set of 10,000 character dataset. Multilayer Perceptron (MLP) produces the best result with an accuracy of 95.57% when sample character is divided into 16 rectangular zones and DHD-based procedure has been considered.

Keywords Online handwriting recognition · Hausdorff distance · Directed hausdorff distance · Bangla script

1 Introduction

Due to easy accessibility of handheld devices such as A4 Take Note, iPad, smartphones, etc., at an affordable price, these devices are gaining popularity day by day which, in turn, makes Online Handwriting Recognition (OHR) an upcoming

S. Sen (✉)
Future Institute of Engineering & Management, Kolkata, India
e-mail: shibubiet@gmail.com

R. Sarkar
Jadavpur University, Kolkata, India
e-mail: raamsarkar@gmail.com

K. Roy
West Bengal State University, Barasat, India
e-mail: kaushik.mrg@gmail.com

N. Hori
University of Texas, Austin, USA
e-mail: hori.naoto@gmail.com

© Springer Nature Singapore Pte Ltd. 2017   541
S.C. Satapathy et al. (eds.), *Proceedings of the 5th International Conference on Frontiers in Intelligent Computing: Theory and Applications*, Advances in Intelligent Systems and Computing 515, DOI 10.1007/978-981-10-3153-3_54

research domain. On those devices people can write information freely in their normal style and written data can be saved in the form of online information. In this way, people not only save extra time but also this procedure minimizes the chance of mistyping that may arise when writing with a keyboard. Though some good research works are found in the literature for Devenagari script [1–8], the same cannot be said for the Bangla script. The presence of limited research materials in the literature proves the truthfulness of the statement. S. Bag et al., in [9], have presented a novel technique for handwritten Bangla characters irrespective of writing direction in 2D plane. Here, authors have considered structural shape of a character as different skeletal convexity of its constituent strokes. Another dimension is highlighted by the authors in [10], where they have extracted component strokes at character level. Stroke level sequential and dynamic information then served as feature values for stroke recognition. Recognized strokes are then used to form the character by matching the stroke sequences in the database. In [11], S.K. Parui et al. have manually grouped constituent strokes of characters into 54 classes based on the shape similarity at graphemes level. For recognition of strokes one HMM for each stroke class has been constructed. In the later stage, characters have been identified from stroke classification results with the help of 50 lookup tables. U. Bhattacharya et al., in [12], have explained a procedure to recognize Bangla basic characters by applying direction code-based features. According to the authors in [13] structure- or shape-based representation can be formed at stroke level where each stroke is represented by a string of shape features. An unknown stroke is then recognized by comparing it with a previously built stroke database using DTW (Dynamic Time Warping) technique. In [14], R. Ghosh has discussed about designing of a novel feature vector by taking into account writing direction, curvature, slope, curliness, and standard deviation of x and y coordinates in stroke level. Here, author has divided all constituent strokes into nine local zones to generate the feature vector and fed to the Support Vector Machine (SVM) classifier for classification purpose. S.P. Sen et al., in [15], have mentioned the accuracy achieved while combining some online (point-based and structural features) and offline (quad-tree-based longest run and convex hull) features in order to recognize handwritten Bangla characters. In [16], authors have highlighted the efficiency achieved by implementing customized version of popularly known distance-based feature extraction technique for online handwritten Bangla character recognition. Here, characters are divided into N segments and then distances are computed from each segment point to every other point. Individual power of global information and local information has been explored in paper [17]. Global information has been used to estimate shape statistics and some important local information of Bangla character have been extracted through local information-based procedure. They have also highlighted the recognition accuracy for combination of global and local information-based approaches.

Feature extraction plays a significant role to design a state-of-the-art character recognition system for any language. In this current work, two almost similar feature extraction procedures have been explored which are termed as HD and DHD based features, respectively. In HD-based approach, the maximum of forward and backward hausdorff distances between any pair of zones is considered as feature component, whereas, DHD-based procedure computes both forward and backward hausdorff distances between any paired zones/set of points and treats them as feature values to recognize handwritten online Bangla characters. The detailed discussion about both the feature extraction procedures has been reported in later section of the paper. A brief description of Bangla script has been discussed in [16].

2 Database Preparation and Preprocessing

200 handwritten samples for each Bangla character have been collected from 100 different individuals belonging to diverse sections of the society with varying age groups, educational background, gender, etc. Considering the 50 distinct character symbols of Bangla script, size of the current database becomes 10,000. No strict burden enforced on the writers during data collection process apart from the issue that the writers were requested to write constituent strokes of characters as part of basic stroke database [10]. In this experiment, preprocessing follows the same steps as mentioned in [16] except characters have been normalized into 90 points rather than 64.

3 Feature Extraction

3.1 Hausdorff Distance (HD) and Directed Hausdorff Distance (DHD) Based Features

In the current work, an expedient shape-based feature extraction approach based on HD has been adopted to recognize online handwritten Bangla basic characters. Generally, forward HD $h(X, Y)$, from set X to set Y is computed using maxmin function as defined in Eq. (1).

$$h(X, \ Y) = \frac{max}{x \in X} \left\{ \frac{min}{y \in Y} \{d(x, y)\} \right\}, \tag{1}$$

where a and b represent the points of set X and set Y, respectively, and $d(x, y)$ is any metric in between the points. In the current work, $d(x, y)$ is taken as the Euclidian distance between x and y. The same procedure is applied to calculate the distance $h(Y, X)$ from set Y to X, which is known as backward HD. The calculation of forward or backward HD between any two sets X and Y having different number of points is explained in Algorithm 1.

```
1.   k = 0
2.   for every point xᵢ of X,
2.1  dist = Infinity ;
2.2  for every point yⱼ of Y
     dᵢⱼ = d (xᵢ , yⱼ)
     if dᵢⱼ< dist then
     dist = dᵢⱼ
2.3  if dist > k then
     k = dist
```

Algorithm 1. Forward or backward HD calculation between any two sets/zones

For HD-based feature calculation, computation of HD values between zones/sets reflects the asymmetric property as introduced by maxmin function in Eq. (1). Due to this asymmetry, values of h(X, Y) and h(Y, X) may not be same always. HD measurement for set X and Y, takes maximum value from h(X, Y) and h(Y, X) as shown in Eq. (2). Algorithm 2 describes the way of finding HD values among different zones when sample character is divided into N rectangular zones. If we closely observe Algorithm 2, then it can be seen that this HD-based procedure basically tries to match the points of two zones/sets, or more specifically, it attempts to find the similarities of shapes present in different zones. In the present work, as described in Algorithm 2, the sample character is divided into N rectangular zones (see Fig. 1, where N = 16, and black-colored points describe the contributed pixels in the respective zones). It can be observed that some of the zones may not have any data pixel; it generally depends on shape of the sample character. Please note that for this distance calculation strategy, only maximum of h(X, Y) and h(Y, X) acts as feature component. From Algorithm 2, this can be noticed that for the entire character sample N*(N-1)/2 [i.e., (N-1) + (N-2) + ... + 1] a number of distance values have been produced as features. Therefore, for HD-based feature computation a length of 6, 36, 120, and 300 element feature vectors have been produced when character images are divided into 4, 9, 16, and 25 rectangular zones, respectively.

$$h(X, Y) = \max \{h(X, Y), h(Y, X)\} \tag{2}$$

Fig. 1 Sample character when broken into 16 rectangular zones

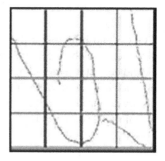

```
1. Divide the character sample into N rectangular zones
2. for p=1 to N-1 do
     2.1 for q=p to N do
         if p!=q do
         find HD between zone_p and zone_q
         end for
     end for
```

Algorithm 2. HD calculation for any character sample segmented into N number of rectangular zones

```
1. Divide the character sample into N rectangular zones
2. for p=1 to N-1 do
     2.1 for q=1 to N do
         if p!=q do
         find DHD between zone_p and zone_q
         end for
     end for
```

Algorithm 3. DHD calculation for any character sample segmented into N rectangular zones

In contrast, for DHD-based feature calculation, the character sample is divided into N rectangular zones same as described in HD-based technique. Then DHD values are computed from every zone to all other zones. Algorithm 3 specifies the DHD-based feature calculation procedure when any character sample is divided into N number of zones. It is worth mentioning that the distance values of h(X, Y) and h(Y, X) of Eq. (1), also known as forward and backward HDs, respectively, are not equal most of the time. Hence, in this work DHD values has been computed from each zone to all other zones to get the exclusivity of the directional distances. As spread of data pixels for different characters are dissimilar in different zones/sets because of their shape structures, this feature extraction approach can be assumed to work well to identify the shapes which are similar/dissimilar in nature. While considering a particular zone and following Algorithm 3, N-1 number of DHD measurements has been generated. As a result a total of N*(N-1) number of DHD values has been produced considering the entire character image which are serving as feature values for the present work. In the current experiment, different values of N such as 4, 9, 16, and 25 are taken to profoundly achieve the discriminatory local information. Hence, a length of 12, 72, 240, 600 element feature vectors have been produced corresponding to the values of N which is of double length compared to HD-based feature calculation. Thus feature produced in HD-based feature calculation is basically a subset of the features produced when DHD-based procedure is applied.

4 Result and Discussion

To test the effectiveness of the said feature vectors produced by HD- and DHD-based procedures for the recognition of handwritten Bangla characters, some well-known classifiers such as MLP, SVM, BayesNet, Simple Logistic have been applied. We have used fivefold cross-validation scheme on the total dataset. Table 1 reflects the recognition rates of the said classifiers for both the feature extraction strategies when sample character is divided into 4, 9, 16, 25 rectangular zones. By analyzing the data as recoded in Table 1, it can be said that for all the classifiers success rate increases as division of character sample increases from 4 to 16 for both the feature extraction procedures. Dividing the character into more number of zones means more number of components constituting the character have been produced; thus more discriminative feature set can be obtained for the recognition of the character under consideration. In contrast, when specimen sample is divided beyond 16 zones then success rate starts falling downward, as observed from Table 1. This happens because when character is divided into say, 25 rectangular zones then number of components gets increased and thus size/length of the components gets decreased as well and it becomes less informative; as a result overall recognition accuracy declines.

Here, MLP produces best recognition of 95.57% when DHD-based features are used and sample character is divided into 16 rectangular zones. It has also been observed that irrespective of the zoning scheme and the classifiers applied, DHD-based feature extraction procedure outperforms HD-based technique. This is because feature set produced by the DHD procedure can be considered as the superset of the feature set produced by HD technique; as the former one considers both forward and backward distances as feature component whereas, HD-based procedure considers only maximum of forward and backward HD as feature value. The graphical behavior of all the classifiers, used here, for both the feature extraction approaches has been shown in Fig. 2. Here, blue, red, green, and violet

Table 1 Success rates of different classifiers for both DHD- and HD-based feature estimation procedures considering different zoning scheme (*bold styles* data indicate maximum accuracy for a particular zoning scheme and *shaded cell* indicates maximum accuracy achieved by the present technique irrespective of the zoning scheme or feature extraction procedure)

Classifier	HD based features				DHD based features			
	Number of Zones				Number of Zones			
	4	9	16	25	4	9	16	25
MLP	**55.9**	83.67	91.97	91.75	73.97	89.59	**95.57**	93.8
BayesNet	46.09	79.78	88.31	88.1	67.15	82.25	88.61	88.24
SVM	52.29	**92.16**	92.8	92.4	85.7	**93.64**	94.43	93.16
SimpleLogistic	41.18	83.19	91.97	91.88	66.99	90.81	95.36	**94.85**

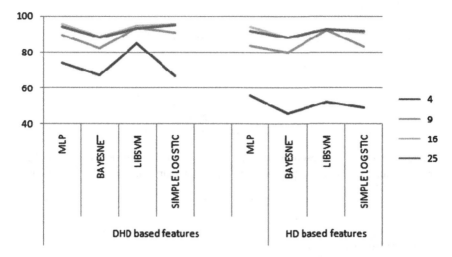

Fig. 2 Graphical behavior of the said classifiers for DHD- and HD-based feature estimation procedures considering different zoning schemes

lines represent the nature of classifiers when sample character is divided into 4, 9, 16, and 25 rectangular zones, respectively. From Fig. 2, it can be easily observed that green line representing 16-rectangular zones, always lies on top of all the other lines irrespective of the feature extraction approaches. It has also been observed from this graph that recognition rates of all the said classifiers are higher for DHD-based feature calculation than HD-based computation.

A comparative analysis has been reported in Table 2 that contains some of the past works in this domain, to highlight the effectiveness of the proposed feature extraction strategies.

Though DHD-based feature estimation works satisfactorily still certain misclassifications have been observed. Table 3 shows some of the misrecognized

Table 2 A comparative assessment of the present technique with some recently reported works

Method	Features used	Accuracy (in %)
Roy et al. [10]	Structural + Point-Based Feature	88.23
Parui et al. [11]	Shape-based feature (Shape and size of stroke)	87.7
Sen et al. [15]	Combination of some online and offline Feature (point-Based feature + quad-tree-based feature)	83.92
Bhattacharya et al. [12]	Directional code feature	83.61
R. Ghosh [14]	Structural and directional feature	87.48
Proposed Technique	HD-based feature	92.8
	DHD-based feature	**95.57**

Table 3 Most confusing character pairs

Original character (200 samples)	Misclassified as (Samples out of 200)
ভ	উ(8)
ন	ল(8)
থ	ঘ(6)
ঙ	উ(4) উ(4)
ঘ	য(4) ম(4)

character pairs by the present technique. After a vigilant analysis it can be said that strong structural similarity between the character pairs misleads the classification procedure.

5 Conclusion

In the current work, HD as well as DHD-based feature extraction techniques are applied to recognize online handwritten Bangla characters. Basic maxmin function has been used to estimate shape similarities between different constituent parts of the character sample when the same is divided into N number of rectangular zones. Though few characters are misclassified still it can be concluded that this approach can be fit into the character recognition of other language also, because this technique efficiently handles the complexity level of the Bangla script. In future, a nonsymmetric zoning approach could be employed to cope up with the structurally similar characters misclassified by the system. Another plan is to apply this technique for the recognition of strokes in stroke-based character recognition procedure.

References

1. Connell, S.D., Sinha, R.M.K., Jain, A.K.: Recognition of Unconstrained Online Devenagari Characters. In: 15th International Conference on Pattern Recognition, pp. 368–371 (2000)
2. Joshi, N., Sita, G., Ramakrishnan, A.G., Deepu, V.: Machine Recognition of Online Handwritten Devanagari Characters. In: Proceedings of International Conference on Document Analysis and Recognition, pp. 1156–1160 (2005)
3. Swethalakshmi, H., Jayaraman, A., Chakravarthy, V.S., Sekhar, C.C.: On-line Handwritten Character Recognition for Devanagari and Telugu Scripts Using Support Vector Machines. In Proceedings of International Workshop on Frontiers in Handwriting Recognition, pp. 367–372 (2006)
4. Swethalakshmi, H., Sekhar, C.C., Chakravarthy, V.S.: Spatiostructural Features for Recognition of Online Handwritten Characters in Devanagari and Tamil scripts. In: Proceedings of International Conference on Artificial Neural Networks, vol. 2, pp. 230–239 (2007)
5. Kumar, A., Bhattacharya, S.: Online Devanagari Isolated Character Recognition for the iPhone Using Hidden Markov Models. In International Conference on Students Technology Symposium, pp. 300–304 (2010)

6. Tripathi, A., Paul, S.S., Pandey, V.K.: Standardization of Stroke Order for Online Isolated Devanagari Character Recognition for iPhone. In IEEE International Conference on Technology Enhanced Education, pp. 1–5 (2012)
7. Kubatur, S., Sid-Ahmed, M., Ahmadi, M.: A Neural Network Approach to Online Devanagari Handwritten Character Recognition. In international conference on High Performance Computing and Simulation, DOI:10.1109/HPCSim.2012.6266913 (2012)
8. Lajish, V.L., Kopparapu, S.K.: Online Handwritten Devanagari Stroke Recognition Using Extended Directional Features. In IEEE 8th International Conference on Signal Processing and Communication System, DOI:10.1109/ICSPCS.2014.7021063 (2014)
9. Bag, S., Bhowmick, P., Harit, G.: Recognition of Bengali Handwritten Characters Using Skeletal Convexity and Dynamic Programming. In International Conference on Emerging Application of Information Technology, pp. 265–268 (2011)
10. Roy, R.: Stroke-Database Design for Online Handwriting Recognition in Bangla. In International Journal of Modern Engineering Research, pp. 2534–2540 (2012)
11. Parui, S.K., Guin, K., Bhattacharya, U., Chaudhuri, B.B.: Online Handwritten Bangla Character Recognition using HMM. In International Conference on Pattern Recognition, pp. 1–4 (2008)
12. Bhattacharya, U., Gupta, B.K., Parui, S.K.: Direction code based features for recognition of online Handwritten characters of Bangla. In International Conference on Document Analysis and Recognition, pp. 58–62 (2007)
13. Bandyopadhyay, A., Chakraborty, B.: Development of online handwriting recognition system: a case study with handwritten Bangla character. In World Congress on Nature and Biologically Inspired Computing, pp. 514–519 (2009)
14. Ghosh, R.: A novel feature extraction approach for online Bengali and devenagari character recognition. In International Conference on Signal Processing and Integrated Networks, pp. 483–488 (2015)
15. Sen, S.P., Paul, S.S., Sarkar, R., Roy, K., Das, N.: Analysis of different classifiers for On-line Bangla Character Recognition by Combining both Online and Offline Information. In 2nd International Doctoral Symposium on applied computation and security Systems (2015)
16. Sen, S.P., Sarkar, R., Roy, K.: A Simple and Effective Technique for Online Handwritten Bangla Character Recognition. In 4th International Conference on Frontiers in Intelligent Computing: Theory and Application, pp. 201–209 (2015)
17. Sen, S.P., Bhattacharyya, A., Das, A., Sarkar, R., Roy, K.: Design of Novel Feature Vector for Recognition of Online Handwritten Bangla Basic Characters. In 1st International Conference on Intelligent Computing & Communication (2016)

A Framework for Dynamic Malware Analysis Based on Behavior Artifacts

T.G. Gregory Paul and T. Gireesh Kumar

Abstract Malware stands for malicious software. Any file that causes damage to the computer or network can be termed as malicious. For malware analysis, there are two fundamental approaches: static analysis and dynamic analysis. The static analysis focuses on analyzing the file without executing, whereas dynamic analysis means analyzing or observing its behavior while it is being executed. While performing malware analysis, we have to classify malware samples. The different types of malware include worm, virus, rootkit, trojan horse, back door, botnet, ransomware, spyware, adware, and logic bombs. In this paper, our objective is to have a breakdown of techniques used for malware analysis and a comparative study of various malware detection/classification systems.

Keywords Dynamic malware analysis · Cuckoo sandbox · Features extraction · Machine learning

1 Introduction

Malware analysis is analogous to cat and mouse game. As new malware analysis systems are developed, malware writers write in such a way that they can thwart analyze. Static analysis mainly composed of reverse engineering, the malware sample with the help of disassembler, and analyzes what the program does. Example for such disassembler is IDA pro [1]. The main drawback of static analysis is it will thwart the analysis result due to obfuscation and packing. But dynamic analysis shows you exactly what the malware does [2]. Some of the tools used for dynamic analysis are procmon, process explorer, and regshot [1]. For running

T.G. Gregory Paul (✉) · T. Gireesh Kumar
TIFAC CORE in Cyber Security, Amrita School of Engineering, Amrita Vishwa Vidyapeetham, Amrita University, Coimbatore, India
e-mail: gregorypaultg@gmail.com

T. Gireesh Kumar
e-mail: gireeshkumart@gmail.com

© Springer Nature Singapore Pte Ltd. 2017   551
S.C. Satapathy et al. (eds.), *Proceedings of the 5th International Conference on Frontiers in Intelligent Computing: Theory and Applications*, Advances in Intelligent Systems and Computing 515, DOI 10.1007/978-981-10-3153-3_55

malware deliberately and monitoring the results, we require a safe environment. This safe environment will ensure that malware is not spreading to production machines.

For malware detection, there are two approaches: signature-based malware detection and behavior-based malware detection [3]. In signature-based malware detection, some of the categories employed are hash signatures, byte signature, and heuristics signature. In signature-based approach, it does not require the malware sample to run, so the results obtained will be very fast. The hash signature involves a hash value, which is created by a hash function (one-way function). The hash function will convert a large amount of data into a single value and this will be extremely accurate. The most commonly used hash functions are MD5 and SHA 1 [1]. The byte signature is a signature method, which finds the sequence of file bytes that are present in a file or data stream. The heuristic-based signature involves techniques used to detect malware by their behavior which includes API logger and rule based of APIs. Some of the tools used to create signature are ClamAV, yara, ssdeep, and titan engine [4]. The drawbacks of the signature-based approaches are it produces inaccurate results, it requires prior knowledge including the set of known signature and also the chance of missing a zero-day malware is higher. In behavior-based approaches, it utilizes artifacts the malware sample generates during execution. This kind of system produces higher accuracy rate when compared to signature-based approach.

The rest of the draft arranged systematically as follows: Sect. 2 describes related works focuses on malware research and comparison of various malware detection/classification methods. Section 3 describes dynamic malware analysis framework cuckoo sand box, and in Sect. 4 we finally conclude the draft.

2 Related Research Works, Discussion, and Comparison

In this section, we are surveying novel approaches in malware analysis, and a comparative study of various malware detection/classification methods. The comparative study includes different parameters such as the type of analysis, analysis environment, file formats supported, software used, features extracted for classification/detection of malware, representation of features, machine learning algorithm used, and evaluation parameter which shows in Tables 1 and 2.

Pirscoveanu [5] in 2015 suggested a system for classification of a large amount of malware in an automated manner. For that, they developed a malware testing environment where malware samples are executed in an instrumented environment and trace their behavioral data. The system architecture consists of main components as follows customized version of cuckoo sandbox where dynamic analysis of malware is done, InetSim [5]—an Internet simulator which replies to requests from malware while executing in instrumented environment, SSH (Secure shell)—all commands between the control unit and host is established using secure shell to improve security, and MongoDB—all collected behavior data is stored in data

Table 1 Comparison of various malware detection/classification technique

Study	Detection/classification	Analysis type	Analysis environment	File format supported	Software used
Pirscovean, Radu S in 2015	Classification	Dynamic	VM	DLL, EXE, DOC, PDF, PPT, HTML, XSL and URL	Customized cuckoo sandbox, InetSim, MongoDB, WEKA
Mohaisen, Aziz, Omar Alrawi, and Manar Mohaisen in 2015	Classification	Dynamic	VM	DLL, EXE, DOC, PDF, PPT, HTML, XSL and URL	AutoMal, MaLabel
Shijo, P. V., and A. Salim in 2015	Detection	Static and dynamic	VM	Portable Executable	Cuckoo sandbox, WEKA
Kawaguchi, Naoto, and Kazumasa Omote in 2015	Classification	FFRI data set	–	–	–
Ozsoy, Meltem in 2015	Detection	Dynamic	VM	Portable executable	PIN tool

Table 2 Comparison of various feature representation and evaluation parameter

Study	Features	Representation	Machine learning algorithm used	Evaluation parameter
Pirscoveanu, Radu S in 2015	API calls, DNS request, mutexes, registry keys	Frequency of API calls, Modified sequence of distinct API calls, Count of distinct files, mutexes, registry keys	Random forest (10 k fold cross validation)	Precision-0.9 F measure—0.898
Mohaisen, Aziz, Omar Alrawi, and Manar Mohaisen in 2015	File system, Registry, Network	Map feature value between 0 and 1	SVM, Log Regression, k nearest neighbor, Perceptron	Precision—99.5% Recall—99.6%
Shijo, P. V., and A. Salim in 2015	PSI, API call sequence	Frequency representation of API call gram	SVM, Random forest	Accuracy—98.7%
Kawaguchi, Naoto, and Kazumasa Omote in 2015	API calls	Binary representation	SVM, K nearest neighbor, Naive bayes, Random forest, C4.5	Accuracy—83.4%
Ozsoy, Meltem in 2015	Architectural, Memory, Instruction events	Binary representation	Logistic regression, Neural network	Accuracy—90%

management system. In this article, 151 different API calls are chosen as the significant features which are represented in different formats. The first one is frequency representation where the count of each API call is computed and represented in matrix form. The matrix representation means each column gives specific feature and each row represents a malware file. The second representation is sequence representation where first 200 API call sequence is modified to recover unseen similarities. The sequence is customized in such a way that repetition of the same API call continuously is discarded, without considering the frequency of API calls. The third representation is counter representation. It consists count of different parameters represented in columns such as DNS request, accessed files, mutexes, and registry keys. For classification, a decision tree-based algorithm—Random forest is used to represent four different class namely Trojan, potentially unwanted program, adware, and rootkit.

Mohaisen et al. [3] in 2015 developed a system architecture AMAL mainly focuses to analyze malware sample automatically and classify the malware samples into different categories based on their behavior artifacts. AMAL composed of two subcomponent: AutoMal [3] and MaLabel [3]. AutoMal [3] consists of a tool to collect behavioral artifacts produced during the execution of malware in a virtualized environment. While MaLabel [3] uses artifacts produced during dynamic analysis of malware to create representative features. These features are used for building classifiers to classify malware samples efficiently based on their behavior. The main features of AutoMal can be characterized as it allows malware sample to process on a priority basis, instrument the guest OS via script and allows multiple input file formats also. The behavior artifacts generated are used by MaLabel for classification and clustering of malware. The features used for classification can be grouped into three categories. The first one is file system features—the frequency for unique file extensions created in specified locations. Example, locations like app data, temp, program files, and other common paths. The next one is registry features —the frequency for registries modified, created, and deleted. The third one is network features—the network features can be grouped into three categories. The first group includes the frequency of distinct IP addresses, frequency of connections initiated at specified port numbers. The next group includes HTTP features like count of GET, HEAD, and POST request, count for response codes namely 500, 400, 300, and 200. The third group comprises DNS features like count for CNAME, PTR, A and MX record lookups [3]. After extracting different features, map feature values in the range of 0 and 1. For classification of malware, different machine learning algorithms are used.

Shijo and Salim [6] in 2015 designed a malware detection system comprising of both static and dynamic approach. The system architecture consists of cuckoo sandbox for dynamic analysis and string utility for static analysis. In static analysis, printable string information (PSI) is used as a feature. Due to code obfuscation techniques, many unwanted PSI will be inserted to the malicious file. To remove unwanted PSI frequency-based approach is used. In this technique, extracted PSIs are sorted according to the frequency of occurrence and applied a threshold to eliminate less relevant PSIs. In dynamic analysis, gram-based method [7] is used to analyze the API call sequence. The set of 3 API call gram and 4 API call gram for each are generated from the logs of the cuckoo sandbox and each API call gram is sorted according to the frequency of occurrence. From the sorted list of each, API call gram, grams below a particular threshold are eliminated so as to choose most relevant API call gram. If a particular API call gram is present, we will set the attribute value as true ('1') otherwise false ('0'). The machine learning tool WEKA [8] is used for classification. The designed system which comprises of both static and dynamic approach shows it is capable to detect malware efficiently with high accuracy rate.

Naoto and Omote [9] in 2015 introduced a new methodology to classify malware functions efficiently using initial behavior of APIs. Most of the antivirus company mainly tries to detect malware not to classify efficiently. Thus, the classification accuracy of malware is still low. In this draft, they used FFRI dataset 2014 which was collected from the analysis logs during the dynamic analysis of malware. For the creation of dataset, they use the cuckoo sandbox. In addition, to that, it includes analysis logs gathered by FFRI yarai analyzer Professional [9]. The malware functions are defined by referring symantec security response (SSR) information and the malware functions described in this paper are the backdoor, downloader, send information, key logger, copy itself, send spam, display ad, remote control [9]. The proposed method includes three stages: first one is API extraction, the second one is learning stage, and the third one is classification stage. In API extraction stage, it will extract APIs used by malware and creates a database of APIs. The feature vectors and malware functions are represented in binary format means if a particular API or malware function in the present value of the attribute will be one otherwise zero. While on extracting phase, APIs which occur two or more time is discarded. After extracting feature, labeling and classification of the malware are done.

Meltem [10] in 2015 suggested new architecture to differentiate malware from legitimate programs. In this article, they propose MAP (Malware Aware Processors [10])—hardware-based detector which is capable of detecting malware in real time. The MAP architecture consists of three units: Feature Collection, Prediction Unit, and Online detection. The feature collection module gathers features used for classification. The features are directly collected from the processor pipeline. Mainly features related to three events are used. They are architectural, memory addresses, instructions. The features related to architectural events include count of memory write/read, taken and immediate branches and uneven memory calls. The features related to memory address patterns include count of memory address distance histogram and memory address distance histogram mix [10]. The features related to executed instructions includes count of instruction categories, the count of opcodes with the major variation, the presence of categories, and the presence of opcodes. The input to the prediction unit is these features. It implements the classifier that provides binary detection with state '1' signifying malware sample and state '0' representing normal program. The output of the prediction unit is a time sequence of the decisions taken overtime. The input to the third module is the consecutive decisions of the classifier and will provide a real-time result on presently executed programs. The detection algorithm must filter out false positives and quickly detect true malicious behavior. To make a decision about the process they are using exponentially weighted moving average algorithm.

3 Dynamic Malware Analysis Framework

As mentioned earlier for dynamic analysis requires a safe environment-sandbox. The malware sandbox is a software or hardware appliance that receives suspicious files and returns an overview of their functionality. The goal of malware sandbox is to run an unknown and untrusted application or file inside an isolated or controlled environment and get the information what it does. One of the common malware sandbox which is able to run the malicious file in an instrumented virtual machine environment is a cuckoo sandbox [11]. The cuckoo sandbox is a free open source analysis framework developed in python comprising modular and customizable property. It is a standalone automated malware analysis framework which shows all the behavior artifacts created during the execution of the malware. Some of the benefits of automated malware analysis system are automation of specific tasks, can be integrated with the defense mechanism, process high volumes, and useful in digital forensics/incident response. The main features of cuckoo sandbox are as follows: (1) The analyst will be able to assign run time for each malware sample in the instrumented environment. (2) Run a concurrent analysis of malware. (3) Analyst can control the operating system via python script interface before running the malicious binary. (4) It can support multiple file formats like DLL, EXE, DOC, PDF, PPT, HTML, XSL, and URL. (5) With VM cloaking mechanism, we are able to create dozens of VM so that analyst can choose which OS malware to be analyzed. (6) Able to take the snapshot of the machine state so that after malware analysis we can restore back to the previous clean state. (7) Built in anti-sandboxing techniques like the emulation of human interaction 0. (8) Cuckoo generates Java Script Object Notation (JSON) report, Hyper Text Markup Language report and also it provides real-time reporting of data—MongoDB. (9) It supports packages like tcpdump (for network capture), pydeep (for calculating the hash of files), PyMongo (for storing the result in MongoDB), YARA (for matching yara signature), and volatility (forensic analysis on memory).

3.1 Experimental Setup and Result of Cuckoo Malware Analysis Framework

The experimental environment is set up on an Ubuntu 14.04 LTS as host OS and created a VM [12] with windowsXPSP3 as guest OS for the execution of malware in a controlled environment. Then install all required packages like python framework, ssdeep, pydeep, tcpdump for analysis. The configuration of each guest OS is applied via agent.py (a python script inside agent folder). The agent.py file should be copied to the guest OS startup folder so that while on malware is invoked, it executes the analyzer contained in it. While executing cuckoo agent which is a basically an XMLRPC server listens to the socket 0.0.0.0.8000 and loads the analyzer module. Before analysis of malware, we have to take the snapshot of the system state and also make sure that guest OS XP is pinging Ubuntu host OS. And then edit the configuration files in cuckoo (cuckoo.conf, auxillary.conf, kvm.conf, processing.conf, and virtualbox.conf [11]). After launching cuckoo submit the malware to the guest OS, then analysis of binary will start. After the analysis, to show the analysis report we have to start the web service. The report of the malware analysis includes various categories such as info, date of analysis, how much time it took for each sample, guest OS version, file size, file type, and hash of the file. The analysis result of cuckoo sand box includes trace of API calls, files (deleted, created and modified), file metadata, registry (deleted, created and modified), network traffic (PCAP file), memory dump, screen shots of VM during the execution of malware, signature, process memory dump, and virus total scan result. The behavior summary includes files, registries which are deleted, created and modified which shows in Fig. 1. The count of files and registries can be used as a feature to classify malware sample.

The process dump comprises of process id, parent process id, timestamp, thread, the sequence of API calls, flags, base address, passed API calls, and failed API calls which show in Fig. 2. The sequence of API calls is one of the significant features for the classification of malware.

The virus total scan comprises the result of 55 antivirus software. It will show a detection rate which means the number of the antivirus software that detected the particular sample as malicious. While inspecting the scan result, we can conclude that there is no efficient mechanism to categorize malware samples since one sample was named differently by different antivirus software's.

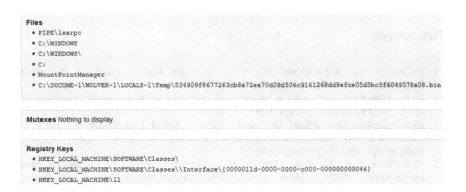

Fig. 1 Behavior summary

registry | filesystem | process | services | network | synchronization

534909f8677263cb8e72ee70d09d506c9161268dd9efce05d5bc5f6049578a08.bin PID: 1972, Parent PID: 208

Timestamp	Thread	Function	Arguments
21:53:04,946	168	LdrLoadDll	Flags => 1244816 BaseAddress => 0x77dd0000 FileName => advapi32
21:53:04,946	168	LdrGetProcedureAddress	Ordinal => 0 FunctionName => RegOpenKeyA FunctionAddress => 0x77ddefb8 ModuleHandle => 0x77dd0000
21:53:04,946	168	RegOpenKeyExA	Handle => 0x00000080 Registry => 0x80000002 SubKey => SOFTWARE\Classes\

Fig. 2 Process dump

4 Conclusion and Future Enhancement

In this work, we have presented a dynamic malware analysis framework which shows the behavior artifacts created during the execution of the malware. Due to code obfuscation and packing techniques static analysis will have a less performance when compared to dynamic analysis where malware is executing in an instrumented environment. From the studies shows API calls is one of the key features that describe malicious behavior. There are various approaches to representing API calls like binary, frequency and n-gram representation. And also, the commonly used machine learning algorithms for malware classification/detection are support vector machine (SVM) and random forest which shows high predictive nature and accuracy. In the future works, we will extract features from the behavior artifacts created during execution of malware and propose a highly efficient malware detection and classification system.

References

1. Sikorski, Michael, and Andrew Honig. *Practical Malware Analysis: The Hands-On Guide to Dissecting Malicious Software*. No Starch Press, 2012.
2. Egele, Manuel, et al. "A survey on automated dynamic malware-analysis techniques and tools." *ACM Computing Surveys (CSUR)* 44.2 (2012): 6.
3. Mohaisen, Aziz, Omar Alrawi, and Manar Mohaisen. "Amal: High-fidelity, behavior-based automated malware analysis and classification." *Computers & Security* (2015).
4. Malware tips, https://malwaretips.com.
5. Pirscoveanu, Radu S., et al. "Analysis of Malware behavior: Type classification using machine learning." *Cyber Situational Awareness, Data Analytics and Assessment (CyberSA), 2015 International Conference on*. IEEE, 2015.
6. Shijo, P. V., and A. Salim. "Integrated Static and Dynamic Analysis for Malware Detection." *Procedia Computer Science* 46 (2015): 804–811.

7. Naval, Smita, et al. "Employing Program Semantics for Malware Detection." *Information Forensics and Security, IEEE Transactions on* 10.12 (2015): 2591–2604.
8. University of Waikato, http://www.cs.waikato.ac.nz.
9. Kawaguchi, Naoto, and Kazumasa Omote. "Malware Function Classification Using APIs in Initial Behavior." *Information Security (AsiaJCIS), 2015 10th Asia Joint Conference on.* IEEE, 2015.
10. Ozsoy, Meltem, et al. "Malware-aware processors: A framework for efficient online malware detection." *High Performance Computer Architecture (HPCA), 2015 IEEE 21st International Symposium on.* IEEE, 2015.
11. Cuckoo Sandbox, http://www.cuckoosandbox.org.
12. Jiang, Xuxian, Xinyuan Wang, and Dongyan Xu. "Stealthy malware detection through vmm-based out-of-the-box semantic view reconstruction." *Proceedings of the 14th ACM conference on Computer and communications security.* ACM, 2007.

Evaluation of Machine Learning Approaches for Change-Proneness Prediction Using Code Smells

Kamaldeep Kaur and Shilpi Jain

Abstract In the field of technology, software is an essential driver of business and industry. Software undergoes changes due to maintenance activities initiated by bug fixing, improved documentation, and new requirements of users. In software, code smells are indicators of a system which may give maintenance problem in future. This paper evaluates six types of machine learning algorithms to predict change-proneness using code smells as predictors for various versions of four Java-coded applications. Two approaches are used: method 1-random undersampling is done before Feature selection; method 2-feature selection is done prior to random undersampling. This paper concludes that gene expression programming (GEP) gives maximum AUC value, whereas cascade correlation network (CCR), treeboost, and PNN\GRNN algorithms are among top algorithms to predict F-measure, precision, recall, and accuracy. Also, GOD and L_M code smells are good predictors of software change-proneness. Results show that method 1 outperforms method 2.

Keywords Machine learning algorithms · Undersampling · Feature subset selection (FSS) · Code smells · Software change-proneness

1 Introduction

After software is developed and delivered to the customer, it enters maintenance phase. During maintenance, the software is corrected and new functionalities may also be added. A study [1] has indicated that 80% of the maintenance work is non-corrective. Software maintainability is an important software quality attribute.

K. Kaur (✉) · S. Jain
University School of Information and Communication Technology (U.S.I.C.T), Guru Gobind
Singh Indraprastha University (G.G.S.I.P.U), New Delhi, India
e-mail: kdkaur99@gmail.com

S. Jain
e-mail: shilpijain0203@gmail.com

© Springer Nature Singapore Pte Ltd. 2017
S.C. Satapathy et al. (eds.), *Proceedings of the 5th International Conference on Frontiers in Intelligent Computing: Theory and Applications*, Advances in Intelligent Systems and Computing 515, DOI 10.1007/978-981-10-3153-3_56

Change-proneness (CP) is defined as the probability that a given software module will undergo change during maintenance period. Change-prone modules multiply maintenance effort and cost, which makes early change-proneness prediction an important task. It is hypothesized that code smells are indicators of change-proneness [2]. In this paper, six machine learning algorithms, [3] probabilistic and general regression neural network (P\G), group method of data handling polynomial network (GM), cascade correlation network (CCR), treeboost, gene expression programming (GEP), and discriminant analysis (DA), are applied to identify change-prone classes of various versions of four open source Java applications For our case study, we have taken code smells as independent variables and change-proneness as dependent variable. This paper examines 10 software code smells. Code smells investigated are [4, 5, 6, 7]: feature envy (F_E), long method (L_M), God class (GOD), empty catch block (ECB), unprotected main program (UMP), dummy handler (DH), nested try statement (NTS), careless cleanup (CC), exceptions thrown from finally block (ETFFB) and overlogging (OL). We have applied correlation-based feature selection (CFS) to select important code smells which are significant predictors of change-proneness.

Two research questions addressed in this paper are:

- Which code smells are good predictors of Change-Proneness?
- Which machine learning algorithm is best in predicting software change-proneness?

To the best of our knowledge, no previous work has been done to investigate the relationship between software code smells on change-proneness using these machine learning algorithms. The rest of this paper is organized in the following manner: Sect. 2 presents literature review. Section 3 gives empirical data collection. Section 4 presents research methodology. Section 5 presents result analysis and Sect. 6 gives conclusion and future work.

2   Literature Review

Kreimer [8] proposed an adaptive learning approach based on machine learning algorithms to detect software code smells. These detection methods were developed so that they are easily adaptable in multiple software maintenance scenarios. A prototype tool was also developed for detecting the software code smells. Yamashita and Counsell [9] investigated whether code smells are indicators of system level maintainability. They concluded that code smells were more promising indicators of maintainability problems than static code metrics-based approach. Li and Shatnawi [10] studied relationship between code smells and class error probability in three releases of Eclipse project and found that code smells were positively associated with error-prone classes. Mantyla et al. [11] studied how different code smells were corelated with each other. Emden and Moonen [12] presented an approach for detecting code smells automatically and discussed that their approach can be integrated into design of software inspection tool. Rao and Reddy [13] proposed a quantitative method that made use of change propagation probability matrix to detect important code smells. Olbrich et al. [14] investigated two code smells using historical data of three open source systems that were being developed from past 7–10 years. They concluded that God and brain classes were more change-prone and defect-prone than other kinds of classes. D'Ambros et al. [15] investigated the impact of software design flaws on software defects. Using six case studies the authors [15] have compared different design flaws and analyzed the relationship between design flaws and software defects. They have concluded that there exists no such design flaw, which could be considered as a threat to software defects. Barstad et al. [16] classified source code into well written or badly written source code. Their analysis was based on static source code metrics. The input data was collected using student hand-ins, peer reviews, and public data sets. They concluded that machine learning algorithms can easily identify codes, which are well written as compared to the code which is badly written.

Code smells can also be detected by the symbolic logic representation and analytic learning technique as proposed by SakornMekruksavanich [17]. They investigated six types of code smells namely: lazy class, data clump, long parameter list, switch statement, refuse bequest, and temporary field. It was concluded that the proposed design flaw detection method was efficient as compared to other methods.

Khomh et al. [2] investigated the impact of software code smells on software change-proneness. They studied 29 types of code smells on 9 releases of Azeuras and 13 releases of Eclipse. They found out that software code classes having code smells are more likely to be change-prone as compared to classes without code smells. However, most software quality datasets have an imbalanced or skewed distribution due to presence of large number of software modules that are not change-prone as compared to modules that are change-prone [18].

To the best of our knowledge, no study till date has considered the impact of imbalanced datasets in code smells based change-proneness prediction. Gao and Khoshgoftaar [18] conducted an experiment on six real-world software quality datasets and found that if Random undersampling is performed before Feature selection, then prediction performance increases. However, their study is based on static code metrics.

3 Empirical Data Collection

This section discusses the procedure followed to extract data for the analysis of our case study. For investigating the dependency of software code smells on change-proneness, we have applied machine learning algorithms on large open source Java based applications. The datasets used in this study are listed in Table 1. The reason for selecting these datasets is to check the validity of the model on source codes of different sizes. The four datasets namely: Mobac (1.9.1, 1.9.7,

Table 1 List of datasets

Software version	Date (Last date modified)	Downloaded from	Repository to find change
Mobac 1.9.1	2011-09-09	https://sourceforge.net/projects/ mobac/?source=directory	https://github.com/ larroy/mobac
Mobac 1.9.7	2012-05-17		
Mobac 1.9.16 (till v2 alpha 3)	2014-02-06		
Jajuk1.10 (till v1.10.7)	2012-8-12	https://sourceforge.net/projects/ gogui/?source=directory	https://github.com/ jajuk-team/jajuk
Gogui1.1	2008-7-10	https://sourceforge.net/projects/ gogui/?source=directory	https://github.com/ icehong/gogui
Gogui1.2 (till v1.4)	2010-2-28		
Openrocket12.3	2012-3-18	https://sourceforge.net/projects/ openrocket/?source=directory	https://github.com/ openrocket/ openrocket
Openrocket13.5	2013-5-4		
Openrocket13.9.1 (till v14.03)	2013-10-6		

1.9.16), Gogui (1.1, 1.2), Jajuk (1.10), and Openrocket (12.3, 13.5, 13.9.1) can be downloaded from sourceforge.net as listed in Table 1. Figure 1 gives step-by-step procedure to extract data empirically and apply machine learning methodologies using two techniques.

The following steps explain the procedure followed to extract data empirically (method 1):

Step 1: Smells collection

Smells are important as they determine the quality of the code written by the developer. They can be used as indicators to predict change-proneness of software [2]. For our investigation, we have extracted 3 non-exception handling smells namely [4]: Feature Envy (F_E), long method (L_M), and God class (GOD) using eclipse plugin called Jdeodorant [6]. Exception handling smells occur when an unwanted situation is not handled properly. These situations are unwanted because they affect the predefined flow of the program. Such situations are called exceptions. Seven exception handling smells [5] collected using Robusta tool [7] are: empty catch block (ECB), unprotected main program (UMP), dummy handler (DH), nested try statement (NTS), careless cleanup (CC), overlogging (OL), and exception thrown from finally block (ETFFB). Robusta is an Eclipse plugin.

Step 2: Change calculation

Change between two versions of dataset is calculated by analyzing commits on GITHUB. GITHUB has many repositories, which keep records of changes made in software code. For our experiment, we have used different repositories to find change variable for each version of the source codes as listed in Table 1. For two software versions, if the commit has recorded change in the Java file, then the change variable is made TRUE otherwise FALSE.

Fig. 1 Step-by-step
procedure of model

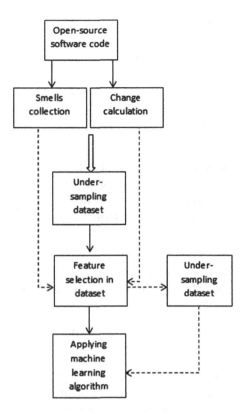

Step 3: Under-sampling of datasets

Most real-world software quality data sets suffer from class-imbalance problems. Class-imbalance problem for a software quality data set means the number of software modules with quality issues such as change-proneness (CP) or defect proneness (DP) are far less in number than number of non-change-prone (nCP) or non-defect-prone (nDP) modules. Due to this reason, machine learning algorithms may find it difficult to distinguish between software modules with quality issues (such as CP) from those without quality issues. The classes that are in abundance are called majority classes while the classes which are limited are called minority classes. The data sets are balanced using spread subsample filter in Weka (3.6.13) [19] at distribution spread 1.0. This filter uses undersampling technique to balance skewed data distribution. It is a supervised instance-based filter supported in Weka.

Step 4: Feature subset selection (FSS) in dataset

Feature selection is a technique of identifying the main representative features on which model depends. FSS on software quality datasets not only helps in identifying relevant code smells for predicting change-proneness but also reduces learning time [18]. In addition to this, FSS improves predictive accuracy of various

Table 2 Attributes selected by CFS

Data version	Smells selected by method 1	Smells selected by method 2
Mobac 1.9.1	NTS, CC, **GOD, L_M**	DH, OL, **GOD, L_M**
Mobac 1.9.7	ECB, UMP, **GOD, L_M**	ECB, DH, UMP, CC, **GOD, L_M**
Mobac 1.9.16	DH, UMP, ETFFB, **GOD, L_M**	DH, UMP, NTS, ETFFB, **GOD, L_M**
Gogui 1.1	NTS, **GOD, L_M**	ECB, DH, UMP, NTS, ETFFB, **GOD, L_M**
Gogui 1.2	UMP, ETFFB, **GOD, L_M**	UMP, **GOD, L_M**
Jajuk 1.10	NTS, **GOD, L_M**	NTS, **GOD, L_M**
Openrocket 12.3	ECB, DH, **GOD, L_M**	ECB, **GOD, L_M**
Openrocket 13.5	ECB, CC, **GOD, L_M**	ECB, UMP, NTS, **GOD, L_M**, F_E
Openrocket 13.9.1	DH, **GOD, L_M**	ECB, NTS, **GOD, L_M**, F_E

software quality models. In this work, correlation-based feature selection (CFS) is applied using Weka [19]. It is based on the following formula [20]:

$$\text{Merit}_S = \frac{n\overline{\text{Rcf}}}{\sqrt{n + n(n-1)\overline{\text{Rff}}}} \tag{1}$$

where n is the number of features in a subset of features S. $\overline{\text{Rcf}}$ is the mean of the correlations between each predictor and response variable. $\overline{\text{Rff}}$ is the mean of pairwise correlations between two predictors. Table 2 shows the features selected by CFS for the datasets using methods 1 and 2.

Step 5: Application of machine learning algorithms

Machine learning algorithms make predictions based on predictor variables. For our analysis, we have applied six machine learning algorithms to test the datasets at 10 cross-validation using DTREG tool [3]. The performance of each algorithm is studied and compared using ranking. For all the performance parameters, average ranks are calculated for all the datasets. Average ranks for both the techniques are calculated separately.

For method 2, Steps 3 and 4 are interchanged.

4 Research Methodology

In this section, a brief overview of the methods and algorithms [3] used to analyze the dependency of change variable on code smells is given. These algorithms are evaluated on Intel core i-5 processor supporting 64 bit Windows 7 operating system.

4.1 Random Undersampling (RUS)

Random undersampling is a preprocessing technique used to balance the data. Datasets can be imbalanced. An imbalanced data set has unequal numbers of majority class and minority class instances. When the classes which are in majority are removed to balance the skewed data distribution then it is called undersampling. Our model applies undersampling using spread subsample filter at distribution spread 1.0. This filter is supported in Weka [19].

4.2 Correlation-Based Feature Selection (CFS)

Classification models can be created by identifying set of representative attributes. From a large set of attributes, a subset of features is identified. These attributes help the learning algorithms to focus on important aspects of data. We have applied CFS using best first search methodology. It selects those code smells which have low correlation with each other but high correlation with change-proneness. Best first search method selects a node by applying specified rules to select a path closer to the solution. For method 1 (M1), undersampling is performed first and then feature selection because recently Gao and Khoshgoftaar [18] have found that this data processing approach gives best prediction accuracy. This paper checks ML algorithms for method 2 (M2) also where undersampling is done after feature selection.

4.3 Machine Learning Methods

4.3.1 Probabilistic and General Regression Neural Network (P\G)

Probabilistic neural network method is a classification algorithm where the target variable is categorical in nature. A variable is said to be categorical when it can take value from a fixed set of values. Concept-wise probabilistic neural network works similar to K-nearest neighbor algorithm. General regression neural network takes the target variable to be continuous variable. A continuous variable is a variable that can obtain its value from infinitely many numbers of values.

4.3.2 Group Method of Data Handling Polynomial Network (GM)

It is an inductive model used for mining data. A model is regarded as inductive if it sorts out the solutions to predict the best result based on external criteria. This model establishes nonlinear relations between multidimensional data automatically.

4.3.3 Cascade Correlation Network (CCR)

It is a neuron-based network where in the first step hidden units are added only once. Second, new units are created using learning algorithm. At each step of the learning algorithm, the data is trained until error is minimized. Candidate units are created by connecting input units with hidden units. No weights are assigned between candidate units and output units. The training is stopped till a maximum value of correlation is obtained.

4.3.4 Treeboost (TB)

It is a stochastic gradient boosting method. Here error is minimized by applying predictive function in series repeatedly. Every time output is generated each function is weighted to generate new output. Treeboost creates decision trees in series using randomization function. This type of technique improves the accuracy of the system.

4.3.5 Gene Expression Programming (GEP)

It is an evolutionary technique that models the data like tree structures. They model by varying the shapes and sizes of tree structures. They are encoded using chromosomes. Chromosomes are linear structures. Chromosomes help in creating data for next generations as they are mutable.

4.3.6 Discriminant Analysis (DA)

Discriminant analysis generalizes linear discriminant given by Fisher. This model is a classification model and not a regression model. In this technique new transformed target is created based on two predictors. This transformed target predicts more accurately as compared to when only a single predictor is used.

5 Result Analysis

This section analyzes the experimental results. To evaluate and compare machine learning algorithms for predicting change-proneness based on code smells, we have employed six learning algorithms on our datasets. To balance the datasets, random undersampling (RUS 1.0) technique is used. CFS is used to select features. For each dataset, smells are selected as representative attributes for predicting change-proneness. The following performance measures are analyzed for each learning algorithm of our model:

- AUC measure: The receiver operating characteristics curve depicts sensitivity and specificity. It determines the accuracy of the model by observing area under the curve.
- Recall: It determines the fraction of actually useful attributes that are obtained.

$$Recall = True\ Positive/(True\ Positive\ +\ False\ Negative) \tag{2}$$

- Precision: It determines the fraction of the attributes that are obtained which are actually useful.

$$Precision\ =\ True\ Positive/(True\ Positive\ +\ False\ Positive) \tag{3}$$

- F-measure: It is a combination of precision and recall. It combines the two by taking harmonic mean.

$$F-measure\ =2\left(Precision\ {}^{*}Recall\right)/(Precision\ +\ Recall) \tag{4}$$

- Accuracy: It determines how well the classifier learns the data.

$$Accuracy = (True\ positive\ +\ True\ Negative)/(True\ positive\ +$$
$$True\ Negative\ +\ False\ Positive\ +\ False\ Negative) \tag{5}$$

5.1 Analysis of Results After Ranking

Ranking the performance measures AUC, F-measure, recall, precision, and accuracy for all the nine datasets helps in comparing machine learning algorithms for techniques 1 and 2. Average of ranks is determined by taking mean of all the ranks for all algorithms for each dataset. The average ranks found for method 1 (M1) and method 2 (M2) are shown in Table 3. From Table 3, it can be observed that for AUC, GEP with average rank 1.11 for M1 and 1.05 for M2 is the best method with Rank 1, whereas DA comes second (R2). P\G performs worst with average ranks 5.44 (M1) and 5.77 (M2). CCR with average ranks 2.33 (M1) and 2.5 (M2) outperforms all the other algorithms to calculate F-measure. TB comes second whereas GM is the worst, whereas CCR, P\G and TB are among the top algorithms to predict precision, recall, and accuracy, whereas GM and DA are the bottom two machine learning algorithms. Also, avg. ranks R1 for M1 is less than M2 for most of the cases.

Table 3 Ranking of Machine Learning Algorithms based on average ranks

Rank	AUC		F-Measure		Precision		Recall		Accuracy	
	M1	M2	M1	M2	M1	M2	M1	M2	M1	M2
R1	**GEP (1.11)**	**GEP (1.05)**	**CCR (2.33)**	**CCR (2.5)**	**CCR (2.55)**	**CCR (2.61)**	**CCR, TB (2.94)**	**PG (3)**	**CCR (2.33)**	**CCR (2.5)**
R2	DA (2.88)	DA (2.94)	TB (2.61)	TB (2.66)	PG, TB (2.83)	PG (2.7)	GEP (3.16)	CCR (3.05)	TB (2.61)	PG (2.83)
R3	CCR (3.44)	GM (3.11)	GEP (2.77)	GEP (2.88)	GEP (3)	GEP (3.05)	DA (3.44)	TB (3.11)	GEP (2.83)	TB (2.88)
R4	GM (3.88)	CCR (3.55)	P/G (3.16)	P/G (2.94)	GM (4.72)	TB (3.22)	PG (3.77)	GEP (3.5)	PG (3.16)	GEP (2.94)
R5	TB (4.33)	TB (4.55)	DA (4.72)	DA, GM (5)	DA (5.05)	GM (4.22)	GM (4.72)	DA (3.8)	DA (4.88)	GM (4.72)
R6	PG (5.44)	PG (5.77)	GM (5.38)			DA (5.16)		GM (4.5)	GM (5.16)	DA (5.11)

5.2   Discussion of Results

Software change-proneness prediction identifies the classes, which are more change dependent using code smells as indicators. After applying this model on multiple versions of open source codes, among all six machine learning algorithms used GEP obtains rank 1 (R1) for AUC for both the methods. Whereas CCR, P\G and TB are among the top ranks to predict F-measure, recall, precision, and accuracy. GM and DA give worst performances for the above three performance parameters. From Table 2, we conclude that GOD and L_M smells are better predictors of change-proneness using both the techniques because GOD and L_M software code smells are selected by CFS for all the datasets. Also Table 3 shows that M1 predicts better than M2.

6   Conclusion and Future Work

The main motive of this paper was to compare different types of machine learning algorithms on the basis of AUC, F-measure, accuracy, recall, and precision by ranking them based on their performance. This paper concludes that:

- Among all 10 types of smells God class and long method smells are better predictors of change-proneness using both the methods as they are selected by CFS for all the datasets (Table 2).
- Gene expression programming (GEP) gives the maximum AUC value for all the nine datasets using both techniques (Table 3).
- CCR, P\G and TB are among the top methods to predict F-measure, recall, precision, and accuracy (Table 3).
- Method (M1) outperformed method (M2) as average ranks (R1) for M1 is less than average ranks for M2 for most of the performance parameters (Table 3).

This work can be taken further by comparing more learning algorithms on large software systems to verify the nature of this model.

References

1. Pigoski, T. M.: Practical software maintenance: Best practices for managing your software investment. Wiley Computer Pub., New York (1996)
2. Khomh, F., Di Penta, M., Gueheneuc, Y. G.: An exploratory study of the impact of code smells on software change-proneness. In: 16th IEEE Working Conf. on Reverse Engineering (WCRE), pp. 75–84(2009)
3. DTREG, https://www.dtreg.com/
4. Fowler, M.: Refactoring: improving the design of existing code. Pearson Education, India (1999)

5. Chen, C. T., Cheng, Y. C., Hseih, C. Y., WU, I. L.: Exception handling refactorings: Directed by goals and driven by bug fixing. J.Systems and Software, vol. 82(2), pp. 333–345 (2009)
6. Tsantalis, N., Fokaefs, M., Chatzigeorgiou, A.: JDeodorant: Identification and removal of feature envy bad smells. IEEE Int'l Conf. on Software Maintenance (ICSM), pp. 519–520 (2007)
7. Robusta, http://pl.csie.ntut.edu.tw/project/Robusta/Robusta_User_Instruction.pdf
8. Kreimer, J.: Adaptive detection of design flaws. Electronic Notes in Theoretical Computer Science, vol. 141 (4), pp. 117–13(2005)
9. Yamashita, A., Counsell, S.: Code smells as system-level indicators of maintainability: An empirical study. J. Systems and Software, vol. 86 (10), pp. 2639–2653 (2013)
10. Li, W., Shatnawi, R.: An empirical study of the bad smells and class error probability in the post-release object-oriented system evolution. J. Systems and Software, vol. 80(7), pp. 1120–1128 (2007)
11. Mäntylä, M., Vanhanen, J., Lassenius, C.: A taxonomy and an initial empirical study of bad smells in code. IEEE Int'l Conf. on Software Maintenance (ICSM), pp. 381–384 (2003)
12. Emden, E., Moonen, L.: Java quality assurance by detecting code smells. In: 9th IEEE Working Conf. on Reverse Engineering, pp. 97–106 (2002)
13. Rao, A. A., Reddy, K. N.: Detecting bad smells in object oriented design using design change propagation probability matrix. Int'l MultiConf. on Engineers and Computer Scientists. vol. I, pp. 19–21, Hong Kong (2008)
14. Olbrich, S. M.,Cruze, D. S., Sjøberg,D. IK.: Are all code smells harmful? A study of God Classes and Brain Classes in the evolution of three open source systems. IEEE Int'l. Conf. on Software Maintenance (ICSM),pp. 1–10 (2010)
15. Ambros, M. D., Bacchelli, A., Lanza,M.: On the impact of design flaws on software defects. In: 10th IEEE Int'l. Conf. on Quality Software (QSIC), pp. 23–31 (2010)
16. Barstad, V., Goodwin, M., Gjøsæter, T.: Predicting Source Code Quality with Static Analysis and Machine Learning.NorskInformatikkonferanse (NIK) (2014)
17. Mekruksavanich, S.: Design Flaws Detection in Object-Oriented Software with Analytical Learning Method. Int'l. J. e-Education, e-Business, e-Management and e-Learning, vol. 1(3), p. 210 (2011)
18. Gao, K., Khoshgoftaar, T. M., Napolitano, A.: Combining feature subset selection and data sampling for coping with highly imbalanced software data. Proc. 27th Int'l. Conf.on Software Engineering and Knowledge Engineering (2015)
19. WEKA, http://www.cs.waikato.ac.nz/ml/weka/
20. Hall, M.A.: Correlation-based feature selection for machine learning. PhD Diss. The University of Waikato (1999)

Snort Rule Detection for Countering in Network Attacks

Venkateswarlu Somu, D.B.K. Kamesh, J.K.R. Sastry
and S.N.M. Sitara

Abstract Phones are turning into the surely understood method for relationship; strategies helping adaptability connote a genuine asset of issues in light of the fact that their preparatory style did not execute effective assurance. A novel structure work of turn imperceptible framework strikes, known as versatility-based avoidance, where an adversary partitions an unsafe payload in a manner that no part can be recognized by ebb and flow ensuring strategies, for example, the most cutting edge framework assault acknowledgment procedures working in condition full method. Snort is a free Network Intrusion Detection System blending several benefits provided by trademark, strategy, and variation from the norm focused examination and is respected to be the most regularly executed IDS/IPS mechanical advancement globally. This report recommends various changes for improving Snort Security Platform and different gathering is suggested to strengthen the measure of rays which can be inspected, and Snort's multi-threading open doors are scrutinized.

Keywords Mobile · RFID · NID · TCP

1 Introduction

An unavoidable effect of the immense accomplishment of client flexibility is an expanding perceivability of cellular telephones and frameworks to a broad scope of strikes. Notwithstanding listening stealthily on wireless signals, break, GSM mimic,

V. Somu (✉)
Department of CSE, KL University, Vaddeswaram, India
e-mail: somu23@kluniversity.in

D.B.K. Kamesh · J.K.R. Sastry
Departmentof ECM, KL University, Vaddeswaram, India
e-mail: kameshdbk@kluniversity.in

S.N.M. Sitara
Department of ECE, KL University, Vaddeswaram, India

© Springer Nature Singapore Pte Ltd. 2017
S.C. Satapathy et al. (eds.), *Proceedings of the 5th International Conference on Frontiers in Intelligent Computing: Theory and Applications*, Advances in Intelligent Systems and Computing 515, DOI 10.1007/978-981-10-3153-3_57

social mechanical development, we show a novel type of strikes known as cell avoidance that can be utilized to cell techniques, for example, Mobile IPv4, Mobile IPv6, and Wireless. Portable avoidance controls the understood weaknesses of cell systems helping clear flexibility where meandering occasions do not upset built-up connections. This is a compulsory element for all projects requesting a consistent relationship, yet it uncovers cell hubs and related frameworks to so known as "stealth" framework strikes. They do not control shortcomings in system executions, however, the impacts of cell hub relocations on diverting.

Society has gotten to be dependent on a broad scope of cell cellular telephones. For instance, most Visa swipes at eating spots are directed with cell cellular telephones. RFID is broad in stock administration. To lower infrastructural expenses and to satisfy their representatives, organizations are trying to join purported "Bring Your Own Device" (BYOD) which are the rules that are used to permit workers so as to increase the oversaw access for the internal framework assets with the phone cell telephones (for the most part smart phones telephones).

At first gives a model of the cell avoidance assault that can be utilized to any surely understood cell technique. At that point, we prescribe an advanced cure that controls a novel route for NIDS joint effort as shown in Fig. 1 with respect to packet transfer. The proposed arrangement permits examining of inward state data among a few NIDSs actualized in distinctive frameworks or framework segments. The general cure is consolidated into a model which grows Snort, however, it can be effortlessly customized to whatever other NIDS on the grounds that the execution is relying upon a light and compact specialists and an arrangement of modules overseeing diverse systems. This flip outline guarantees awesome adaptability with respect to execution and extend capacity. We affirm the viability and productivity of the recommended structure for diverse blends of movement rates and framework routines. We can verify that the recommended cure can distinguish versatility-based strikes in all analyzed genuine circumstances at a negligible expense with respect to proficiency.

"Conseil Européen pour la Recherche Nucléaire (CERN)" is an organization that operates largest networks with processing of different IDS/IPS in real time network applications. This document will discover how Snort, a free Program Attack Recognition Program, can be used to protect and observe such a very huge business network. An Attack Recognition Program (or IDS) is consisting of software and/or

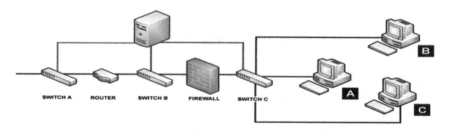

Fig. 1 Intrusion detection over network sending packet information in NIDS system

components developed to identify undesirable efforts of obtaining, adjusting, and/or limiting of PCs. An IDS has been used to identify several kinds of harmful actions that can bargain the protection and believe in of a computer techniques system. These risks are various, and consist of network strikes against insecure solutions, information motivated strikes on programs, variety centered strikes such as benefit escalation, illegal accesses, or viruses. Snort is a free NIDS application. Mixing the benefits of trademark, method and abnormality centered examination Snort is the most commonly implemented IDS/IPS technological innovation globally. It has able to execute advanced level analysis on the trace owing through indicator. Snort is available in two different variations. Snort Security System (Snort SP) is an expansion of Snort, using the same motor, but enabling much more choices, such as multiple threads. Snort SP is still in beginning try out stage and is very likely to enhance its efficiency before the final launch.

Mobile device ad hoc networks replace a main reference of data regarding the interference of observation devices for wireless domains. Considering a sample of Thamilarasu et al. [1] we suggest an interference for cross-layer Observation Device (IOS) In arrangement to reduce these attacks in ad hoc networks for the clash at the center, direction gets changed and the pack drops occurs. The coating plan is there to notice the interference of protocol layers which are not similar and is there to use the data from one layer of one network to a totally different layer Zhang et al. [2] predicts an issue collaborative IDS planning in which each node divide with its data riches. Remaining authors [1] note the point which is related to a hard reply to an interference by arranging an supple scheme which resists on the normal gravity of charge and the humiliation for the matrix show. Here we notice that all the schemes are absorbing center on ad hoc networks which are considerably away from the functionality are internet being used by mobile devices. Which are found on networks to this paper concerned. To a near cause, We separate from high using host IDS [3], from the algorithms which contains statistical profiles (e.g., [4–11]) which does not effect even with the mobile destruction.

It is not agreed for all the scheme using issues for the interference of observation devices which combine the information from sources which are not similar and send awareness to an aggregator which is used for examining and connecting all the available data. The results which are obtained are dependent on various IDS planning: graded, hierarchical, and independent fro cloud systems, peer to peer, for which the main aim is to move away from the losses generated by single points. These devices reduce human effort. To evaluate such unbiased work, it is associated along with NIDS planning such that the state of moving can be utilized. To consider an example, cooperation between parallel NIDS architectures is possible by relation position roaming from an NIDS to another. In [1] the authors illustrate few NIDS inside variables that are helpful in recognizing one or more inner variables of NIDS which are recognized in an agreement having a goal by which a network of worth shared between all other cooperative NIDS sensors issue between those dissimilar network links can happen. Such a same type of machine for minimum level cooperate analysis is helped to tool the NIDS cluster [1], while the technique to import/export the state information into/from a NIDS is provided by various framework. State hike is used to answer all these and to better production, while the theory proposed by you defines an original method from stopping an invader to get benefited from mobility to keep away from observation. Such is succeeded by providing help to a mechanism in which the state data collaborated mobile node that follows mobile node in this network. The work has been improved in several ways with respect to [9–22] in many ways. A common and modular framework is planned which maintains various mobile protocols. The snort software is enhanced and executed. Good number of results are modified that enabled us for fact finding. These outposted success in practical production.

2   Research Methodology

We show the versatile avoidance assault by taking the most developed stateful NIDS architectures into consideration, in view of the fact that stateless frameworks can be effectively circumvent by a few sorts of assaults and are currently censured. In NIDS, the data of a system parcel, i.e., significant to interruption, is used to make an overhaul inward state regarding the dynamic transport level associations. For every interaction, a preprocessor puts up data above the data and two requested arrangements of loads given by the terminal points. The location calculation is then resulted on the whole state data. Albeit no single parcel has enough data to identify an interruption, an NIDS recognizes it, by accessing data removed from diverse bundles.

The issues start when a situation is taken in view of permitting hub portability in which an assailant can pursue after avoiding versatility-based NIDS. Here, an attacker gets benefited from system portability and procedures avoiding in view of assault fracture keeping in mind the final goal to get away from recognition by advanced and powerful stateful NIDS frameworks.

In three different situations this assault can be managed. It is noticed that following situations are free of the portable hub meandering innovation, versatility based on NIDS avoidance can be completed.

The following reasonable conditions occur:

1. The assault is a destructive terminal load blaming remote weakness.
2. It is conceivable to partition the destructive terminal load in no less than 2 segments. Portion 1, Portion 2 such that neither Portion 1 nor Portion 2 are identified by NIDS marks.
3. The procedure is straight. Dynamic transport level associations are not obstructed by normal handover process.

The final problem is fulfilled by a few improvements and system conventions, for example, Mobile IP V 4, Mobile IP V6, and layer-2 conventions for handover crosswise over remote access focuses and arranges.

In this section we describe mobile attacker fixed with victim server specification in real time application progression (Fig. 2).

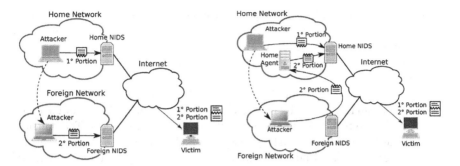

Fig. 2 Data Direct Transmission with Tunneling Process in Domain Name Service applications

The first situation is that in which two hubs, Victim and Attacker, can get into through the Internet. The assailant uses a portable hub that works in a home network first which permits hub portability. To evade after the risk, the assailant will jump to a Foreign Network at some good time. The home and the remote systems both are carefully observed by mark based NIDs. First, the aggressor holds a remote helplessness of the casualty hub along with the parts that compares payload in Portion 1 and Portion 2. As IP bundle fracture is infirmed in IP V6 and are recognized easily by present NIDSs as weird system movement, the assailant wraps the two bits in two undivided TCP sections with sequential grouped numbers. At such point, the assailant makes up TCP make up the casualty and moves Portion 1 from home system. The home NIDS breaks down Portion 1 [3], overhauls its state data, even though it has insufficient data to identify interruption. In the meanwhile, the aggressor comes outside network. After this relocation, bundles between the casualty and the assailant can be processed through two different plans: Tunneling or Direct communication in which the Portion 2 of its payload is directed through the remote system. Bit 2 is caught by the NIDS outside which has not got the past bundle and which do not have any required state data to show the assault. Such a state data is controlled by the home NIDS, even it is delicate as it do not get portion. Consequently, any of the two NIDs depicted in the home as well as in the outside cannot recognize the assault. Such a disappointment is not a consequence of a bug in NIDS executions. Hub versatility allows the vindictive portable aggressor in performing a stealth assault which is not because of any bug in NIDS executions. Fortuitous upon how this hub portability is executed, a stateful NIDS induced in the casualty hub's system may have the capacity to recognize the interruption endeavor.

2.1 Snort Rule Progression

Different parameters are used for defining Snort standard. A guideline is made out of 2 particular parts—the tenet header and the tenet choices. The tenet header will contain the guidelines activity, convention, IP address of source and IP address of destination and net masks and the port of the source, port of the destination data. The messages which are readily available will may standard choice segment and data on those parts of the bundle ought to be investigated to figure out whether the principle move ought to be made. Here is a specimen principle:

Caution tcp any – 10.0.0.0/ 24 80

(content:- |00 00 00|; profundity: 8; msg:- "Bad Bytes"; sid:- 1234)

An algorithm is triggered by these guidelines when 4 out of 8 bytes of the traffic that is pushed to port 80 of the system with the address 10.0.0.0/24. The principle exceptional ID:—1234, the ready message is "Bad Bytes." Guidelines are intense and number of potential outcomes are available. Searching for required bytes of data at specific location is conceivable. To check the quantity of occurrences before a match is found is preferable. It is likewise conceivable to utilize "Perl Compatible Regular Expressions" (PCREs) on information, and to confine the inquiry to some particular bytes.

Every standard (or set of principles) ought to be broke down for asset utilization. Execution is a basic component with high load. The element is likely being hard to assess given the sort/measure of traffic. Case of arrangements are like "Compare CPU load", "Compare rate of traffic investigated" and Usage of Snort Rule Profiling."

An assessment to give indisputable results, principles ought to be assessed with comparative sum/nature of traffic. For every tenet (or set of principles), the benefits ought to be assessed. On the off chance that the proportion of positives for a tenet is as well high, in such a case it is not that much helpful. Top to bottom investigation of the principle and some connection with the end-clients are expected to legiti-mately assess this. Guideline intricacy is basically in light of the quantity of bytes checked in the traffic; specific should as much as possible. Tenets checking not very many bytes are relied upon to produce a ton of counterfeit positives with a huge traffic. However, it is likewise relying upon bytes. Verifying for a lengthy and normal string that trigger more counterfeit positives compared with a couple of abnormal bytes.

This arrangement of guideline tries to gathering all standards demonstrating that an assault is in advancement. As whatever other enormous association or organi-zation, the fundamental issue with this set is that CERN is always under assault, and along these lines there are continually many cautions activated by Snort. Grunt's assault scope is wide. There are tenets gone for recognizing specific vulnerabilities, standards examining irregular utilization of a convention, tenets identifying animal power endeavors, guidelines distinguishing anomalous traffic, and so on.

The known assaults are taken into consideration. The outgoing assaults are significantly improve and could demonstrate. To create one such set, a project was composed to reverse rules, which will consider the CERN as the likely wellspring of assaults. Both assault sets (ordinary and switched) are deployed and the same procedure is repeated. Every ready assault was broke down and the source standard crippled if there should arise an occurrence of false positive or undesirable caution.

2.2 Snort Based IDs System

Bundle sniffing and logging capacities are basic components of Snort. Grunt catches crude bundles with lib and after that it interprets and preprocesses them preceding sending them to the identification motor.

The preprocessing incorporates premature parcel droppings, grouping, layer 3 IP section reassembly, layer 4 TCP session recreations etc. The discovery motor checks bundle headers and in addition payloads against a few a large number of tenets put away in a database of pre-characterized assault marks, as appeared in Fig. 3.

Caution TCP any 192.168.240.0

111(Content |00 01 86 a5|; msg "Mounted Access")

Grunt guidelines are separated into two sensible areas, the tenet header and the standard alternatives. The principle header contains guideline's activity, convention, source and destination IP addresses, net covers and the source and destination ports data. The tenet choice segment contains ready messages and data on which parts of the parcel ought to be investigated. This is done to figure out whether the standard move ought to be made. On a chance that one guideline coordinates, a move is made to rely upon the principle arrangement for the activity. Two of the most utilized activities are "alert" and "log." The alarming office will report any suspicious bundle that has been distinguished. The logging office exists to log full parcel data. Grunt is equipped for yielding "alert" and "log" information in a mixed bag of arrangements and routines.

When we read grunt precludes we find various assault classes for instance foreswearing of administration assaults, ping assault, character mocking assault and so on each sorts of assault contains different alarms identified with a specific

Fig. 3 Snort detection system verification for software components

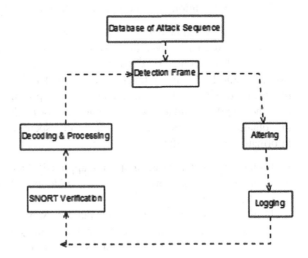

signature. It additionally recognizes the quantity of sources that create assaults and the quantity of destinations that gets those assaults. Each signature of assault will have one kind ID. From that ID one can know the full details.

3 Results and Analysis

In this segment we portray the system execution results when we are utilizing diverse standard structure for recognition of Denial-of-Service assaults and other system assaults in system correspondence process. For this procedure we are creating diverse Snort principle structures like DOS, DDOS, Web Attack, and SCAN. In our proposed approach we are creating distinctive grouping structure for every hub present in system, and after that they are ascertaining individual order time setting up association for distinguishing attacks. Consider an Internet bundle that contains a variety of a known assault, there ought to be some robotized approach to recognize the parcel as almost coordinating a NIDS assault signature. On the off chance that a specific explanation has an arrangement of conditions against it, a thing may coordinate a condition's portion. While Boolean rationale would give the worth false to the inquiry 'does this thing match the conditions', our rationale could permit the thing to match to a lesser degree as opposed to not in any manner. This standard can be connected when looking at an Internet parcel against an arrangement of conditions in a SNORT guideline. Our speculation is that if everything except one of the conditions are met, a caution with a lower need can be issued against the Internet bundle, as the parcel may contain a variety of a known assault. While usage, speculation on account of coordinating system bundles against guidelines, includes permitting a parcel to produce a caution if:

- The conditions in the standard do not all match, yet the greater part of them do;
- The main conditions that do not coordinate precisely about match.

At the point when actualizing summed up standards, the execution time was 1 s to process and change over the first 1,325 principles into an aggregate of 6,975 guidelines. The summed up Content execution time was 2 s to process and change over the same 1,325 unique tenets, into a sum of 18,265 principles. These execution times would effortlessly be adequate for most potential uses, for example, every time the SNORT standards were downloaded for mark overhauls. The increment in the quantity of tenets influenced the time spent preparing system movement information as takes after:

- Using the first standards, Snort took approx 100 s to prepare 1,635,267 bundles;
- Using the summed up (rearranged) guidelines, Snort took approx 400 s to handle the same parcels;
- Using the summed up substance standards, Snort took approx 1,000 s to handle the parcels. The adjustment in Snort's handling time is an increment of around four to ten times and generally in accordance with the increment in the quantity of principles (Fig. 4).

Fig. 4 Comparative data
analysis with respect to time
in both techniques

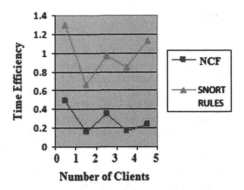

As appeared in the above figure, it recognizes the examination results between both existing and proposed methodologies created in our application. In our current methodology we need to create Collaborative procedure for location of Denial-of-Service, web assaults in system correspondence. In this system we are not giving any tenet structure procedure to recognition of those assaults present in the system correspondence. In this strategy we was created Intrusion discovery framework guidelines structure for creating system execution with equivalent need estimations of every hub present in network.

4 Conclusion

Presently, just the surface has been scratched with respect to summed up NIDS standard coordinating and it is hard to make any complete conclusions. On the other hand, a percentage of the more strange matches against summed up tenets have revealed insight into how speculation may help Snort, or to be sure any NIDS, in discovering indistinct assaults. The procedures examined, created and investigated have raised an extensive number of false alarms. When these alarms are dispensed with, some conceivably intriguing cautions radiate through. Further examination is required to focus completely how compelling speculation can be. For example, it is imperative to work out how to recognize all the more consequently false positives cautions from bona fide new alarms created by summed up guidelines. From the outcomes what's more, examination in this paper, it appears that specifically applying speculation to the substance and unisubstance Snort principle parameters ought to be examined further. One speculation in respect to why applying specu-lation to the unit content choice string seems more helpful is that URI (e.g. website page location) strings could without much of a stretch shift crosswise over assaults. An assault including a URI string may have the same impact if a somewhat dis-tinctive catalog name is utilized, particularly where standard index names may fluctuate crosswise over web server establishments.

References

1. Albin, N. Rowe, A realistic experimental comparison of the Suricata and Snort intrusion-detection systems, in: Enokido, T. (Ed.), Proc. 26th Int. Conf. Advanced Information Networking and Applications, WAINA'12. IEEE, Los Alamitos, CA, March 2012, pp. 122–127.
2. Butun, S.D. Morgera, R. Sankar, A survey of intrusion detection systems in wireless sensor networks, Commun. Surv. Tutorials 16 (1) (2014) 266–282.
3. A Collaborative Framework for Intrusion Detection in Mobile Networks" by Mauro Andreolini, Michele Colajanni, Mirco Marchetti, proceedings in Information Sciences 321 (2015) 179–192.
4. T. Alpcan, C. Bauckhage, A.D. Schmidt, A probabilistic diffusion scheme for anomaly detection on smartphones, in: P. Samarati, M. Tunstall, J. Posegga, K. Markantonakis, D. Sauveron (Eds.), Information Security Theory and Practices. Security and Privacy of Pervasive Systems and Smart Devices, Springer, Berlin, DE, 2010, pp. 31–46.
5. M. Colajanni, M. Marchetti, A parallel architecture for stateful intrusion detection in high traffic networks, in: G. Carle (Ed.), Proc. 1st Workshop on Monitoring, Attack Detection and Mitigation, MonAM'06, IEEE, Los Alamitos, CA, 2006, pp. 9–16.
6. M. Andreolini, S. Casolari, M. Colajanni, M. Marchetti, Dynamic load balancing for network intrusion detection systems based on distributed architectures, in: M. Wolf, F. Quaglia, D. Avresky (Eds.), Proc. 6th Int. Symp. Network Computing and Applications, NCA'07, IEEE, Los Alamitos, CA, 2007, pp. 153–160.
7. M. Becher, F. Freiling, J. Hoffmann, T. Holz, S. Uellenbeck, C. Wolf, Mobile security catching up? Revealing the nuts and bolts of the security of mobile devices, in: D. Frincke (Ed.), Proc. Int. Symp. Security and Privacy, SP'11, IEEE, Los Alamitos, CA, 2011, pp. 96–111.
8. L.D. Carli, R. Sommer, S. Jha, Beyond pattern matching: a concurrency model for stateful deep packet inspection, in: Proc. 21st Conf. Computer and Communications Security, SIGSAC'14, ACM, New York City, NY, 2014, pp. 1378–1390.
9. M. Colajanni, D. Gozzi, M. Marchetti, Enhancing interoperability and stateful analysis of cooperative network intrusion detection systems, in: R. Yavatkar, D. Grunwald, K. Ramakrishnan (Eds.), Proc. 3rd Int. Symp. Architectures for Networking and Communication Systems, ANCS'07, ACM, New York City, NY, 2007, pp. 165–174.
10. M. Colajanni, L.D. Zotto, M. Marchetti, M. Messori, Defeating NIDS evasion in mobile IPv6 networks, in: L. Bononi, A. Banchs (Eds.), Proc. 1st Int. Symp. World of Wireless Mobile and Multimedia Networks, WoWMoM'11, IEEE, Los Alamitos, CA, 2011, pp. 1–9.
11. M. Colajanni, L.D. Zotto, M. Marchetti, M. Messori, The problem of NIDS evasion in mobile networks, in: T.E. Ghazawi, L. Fratta (Eds.), Proc. 4th Int. Conf. New Technologies, Mobility and Security, NTMS'11, IEEE, Los Alamitos, CA, 2011, pp. 1–6.
12. M. Curti, A. Merlo, M. Migliardi, S. Schiappacasse, Towards energy-aware intrusion detection systems on mobile devices, in: Proc. 1st Int. Conf. High Performance Computing and Simulation, HPCS'13, IEEE, Los Alamitos, CA, 2013, pp. 289–296.
13. P. Garcia-Teodoro, J.E. Diaz-Verdejo, G. Macia-Fernandez, E. Vazquez, Anomaly-based network intrusion detection: techniques, systems and challenges, Comput. Secur. 28 (1) (2009) 18–28.
14. L. Etienne, \A short Snort rulesets analysis," tech. rep., CERN CERT, 2009.
15. D. Bon_glio, M. Mellia, M. Meo, D. Rossi, and P. Tofanelli, \Revealing skype tra_c: when randomness plays
16. with you," SIGCOMM Comput. Commun. Rev., vol. 37, no. 4, pp. 37{48, 2007.
17. S. A. Baset and H. G. Schulzrinne, \An analysis of the skype peer-to-peer internet telephony protocol," in INFOCOM 2006. 25th IEEE International Conference on Computer Communications. Proceedings, pp. 1{11,2006.

18. E. Freire, A. Ziviani, and R. Salles, \Detecting skype rows in web traffic," in Network Operations and Management Symposium, 2008. NOMS 2008. IEEE, pp. 89{96, April 2008.
19. F. D. P. Biondi, \Silver Needle in the Skype." Black Hat Europe'06, Amsterdam, the Netherlands, Mar. 2006.
20. D. B. Y. Kulbak, \The eMule Protocol Specification." DANSS, Hebrew University of Jerusalem, Jan. 2005.
21. Oinkmaster." http://oinkmaster.sourceforge.net, cited June 2009.
22. Dumbpig-Automated checking for Snort rulesets." http://leonward.wordpress.com/2009/06/07/dumbpig-automated-checking-for-snort-rulesets/, cited July 2009.

Trust and Energy-Efficient Routing for Internet of Things—Energy Evaluation Model

Carynthia Kharkongor, T. Chithralekha and Reena Varghese

Abstract The internet of thing (IOT) is an upheaval of traditional internet. It is new revolution that not only allows transmission and exchange of data but also communication between the physical objects in the real world. These heterogeneous devices are pervasive, ubiquitous in nature that changes dynamically and frequently. Mostly, these devices are low-powered devices and have less computation power and capacity. Traditional routing techniques in ad hoc network do not take security and energy into consideration. To extend the lifetime of the network, the energy supply and consumption of the node is an important aspect. The routing of the packet should be as such that even the low-powered devices have the ability to receive and transmit the packet. The presence of malicious node will make the network more susceptible to the different attacks and threat. To overcome this problem, an energy-efficient routing protocol with a centralized controller is integrated with IOT devices.

Keywords Internet of things · Ubiquitous · Pervasive · Smart environment · Centralized

1 Introduction

Internet of things refers to the interconnection and intercommunication of the devices in the real world. The word "thing" in IOT relates to any physical device or object which are equipped or embedded with software, sensors, actuators, RFID tags, etc., connecting to the internet. The word "Internet" is the interconnection of networks linking millions of devices for exchanging information resources and services. Internet of things is a new emerging technology in this era. The term 'Internet Of Things' was coined by Kevin Ashton in 1999. The IOT not only connects the variety of heterogeneous devices but also the various types of services,

C. Kharkongor (✉) · T. Chithralekha · R. Varghese
Pondicherry University, Pondicherry, India
e-mail: carynethia@gmail.com

© Springer Nature Singapore Pte Ltd. 2017
S.C. Satapathy et al. (eds.), *Proceedings of the 5th International Conference on Frontiers in Intelligent Computing: Theory and Applications*, Advances in Intelligent Systems and Computing 515, DOI 10.1007/978-981-10-3153-3_58

networks, and applications thereby forming an intelligent environment where device communicate and exchanges data. Nowadays, technology has immensely advanced ranging from mobile computing, cloud computing, real-time computing, and machine learning [1]. As these fields progress technology yields to a smart environment. The IOT has a perception of introducing new and better business standard, new innovations that help to create a secure and smart community changing the living standard of the people [2]. The IOT application can be used for any type of system either centralized system or decentralized systems. Some of them are health monitoring, industrial domain, manufacturing site, defense domain, smart environment domain, etc. The IOT smart environment relates to the concept of connecting "anything," "anyplace," "anywhere," at "anytime." Data can be access at any time and at any place via any application or cloud. IOT is broadly described into two categories: general-purpose platforms (GPPs) and dedicated-purpose devices (DPDs). The general-purpose platforms include phones, tablets smart phones, and net-book that can perform any kind of operation depending on the user's choice of application dedicated-purpose devices (DPDs) consist of mostly low-powered and light-weight devices such as RFID tags, sensors, and wireless monitors. The DPDs can vary ranging from different sizes and shapes each well equipped to operate a particular function and linked to cloud resources. These devices report their location and status to the application or to resources in the cloud updating the assets and data, which provide the overall societal awareness and control. Number of connected devices defines the growth and rate in IOT. As population increases every year, IOT expands because of the increase in the number of users. Researchers assumed that by the 2020, billions of devices will be linked to each other. The resources that support such devices also need to be scale: wired and wireless networks, cloud resources, servers, application server providers, and so forth. The merging of wireless communication networks such as sensor networks has lead to introduction of new products that conserve less energy and resources [3]. The same principle is used in IOT where devices have less computation power, limited power supply, limited battery, and low communication skill. These devices sense the environment and monitor the environment for transmission of the packet. This is the reason why most of the node's energy is wasted. This is a challenge especially for the low-powered devices as they initially have limited power. It is necessary in the IOT environment to effectively communicate using limited resources, low capacity, and reduced energy consumption. This signifies the importance of connectivity between the heterogeneous devices as well as transfer and exchange of data [1].

In this paper, Sect. 1 gives the brief introduction about the internet of things. Section 2 specify the related works and literature review in regard to IOT. Section 3 states the problem statement in the existing environment and Sect. 4 presents the solution to the problem. It also mentions the detail explanation and modeling of the proposed system. Section 5 gives the simulation for the proposed system and Sect. 6 shows the results of the proposed in comparison with the other three routing protocols.

2 Literature Survey

In [2] conducted a survey relating to security, privacy, and trust of data in IOT. This paper concluded that for IOT services, security, and privacy levels needs to be improved. The main part that is missing is a unified framework for the different heterogeneous devices regarding security requirements. The paper also lists out that research issue in IOT namely mobility of devices and integration of different communication technologies.

Sriram Sankaran et al. in [4] modeled energy consumed for routing in IOT. The routing has been modeled using Markov Chain. By execution of traces for a single node, the probability for state transition is being derived. The model shows that after evaluation the data predicted for energy consumed is more or less same as that in real observation by using NS-2 simulation.

Paul Lon Reun Chze et al. in [5] proposes a secure multi-hop routing for IOT communication. This proposed work for creating an IOT network, routing, and authentication is combined. Each device needs to be authenticated before joining IOT Network. This will assure that only legitimate nodes are present in the network. Hello messages are broadcasted at a regular timing for discovering the neighbors. Then the device will verify the neighbor against itself. If both matched, the device is able to communicate else it cannot communicate.

Oladayo Bello et al. in the paper [6] emphasizes on communication of devices taking in consideration the low-powered devices where communication is between device to device (D2D) rather than communicating via base station. The drawback of D2D communication is that it does not have automated security for authenticating the devices. Privacy and security of transmission of data became the utmost challenge for IOT. Many attacks have also been discussed such as interference attack where data patterns is being analyzed while transmission between devices, distributed denial of service which occurs due to jamming in the communication of D2D.

Sloochang et al. in [7] explains the requirements named 6A Connectivity that specifies how communication occurs in IOT architecture. He presented a novel approach for needs and requirements that are required for IOT architecture relating to routing. A novel component called autonomous system of things (ASOT), which signifies a device sharing the same routing services and policies. This paper also presented the challenge that takes occurs in regard to interoperability of ASOT with each other. This is due to huge variety of features relating to interoperability.

Hao Zhang et al. in [8] implemented a secure routing protocol for IOT. The security is enhanced by trimming the number of handshake in IOT. The paper also specifies that privacy can be protected by embedding protection such as authentication, password management, malicious node detection and other serious attacks.

In [9] this paper explains in detail the different solutions in regard to various attacks in WSN and IOT on the basis of security requirement. They also provide a study of the classification of the existing protocols required for creating a secure communication. These protocols are being analyzed to identify the drawbacks in the existing work in order to find the solution.

3 Problem Statement

With the advancement of technology billions of devices will be connected via wired or wireless communication. Due to this, security of data is being compromised thus exposing the data to various attacks in the environment. In addition to this, the end-to-end communication between the devices is not guaranteed. Furthermore, most of the energy supply of the node is used in sensing and monitoring the network, transmitting and receiving the packet, processing the operations and establishing the communication between the neighbors nodes [9]. The lifetime of the node is decreased due to these reasons. Sometimes the powered device is unable to transmit the data although it is trustworthy node because of low energy there by it will treated as a malicious node. The energy consumption of the node is an important issue in the Internet of Things.

4 Proposed System

To overcome these challenges and problems, a solution is conjure a centralized controller with internet of things. The architecture of the integration of centralized controller with IOT is given in Fig. 1.

This centralized controller is introduced which monitors the network topology, routing purposes, managing the lifetime of the network as well as scalability of the network. The proposed system consists of the controller and the network devices which are actually IOT devices. The lifetime of the network is maximized until all the nodes in the network break down. The energy of the node is necessary for maximizing the network functionality. The proposed system takes into consideration the energy of the node during transmission and reception of the packet. In this paper, the energy of the node is fixed with the minimum value of 10 J which considers the low-powered devices. This minimum energy is required for sensing, monitoring, receiving, and transmitting the packet. Depending on this energy consumption, the trustworthy of the node is also checked. If the node's energy is below the threshold value but it is trustworthy, the node is not treated as malicious node and will not be block from the network.

4.1 Overview of the Algorithm

Assumptions:

- All the nodes are assumed to be trustworthy and having energy value above the threshold value
- The controller is centralized monitoring every action and move of the nodes in the network.

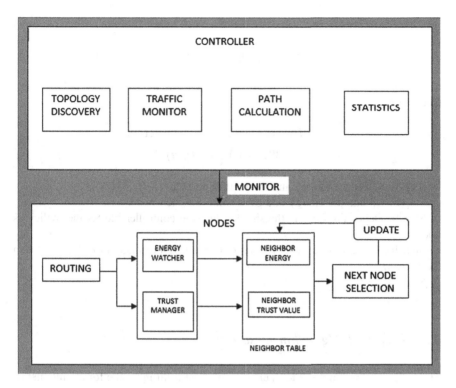

Fig. 1 Architecture of the centralized controller with IOT

Step 1: Deploy the nodes randomly in the network
Step 2: Initial energy for each node is 10 J and trust value is 5
Step 3: The nodes will register with the centralized controller
Step 4: Find the neighbor nodes using Euclidean distance

S = Select (Neighbors)

Step 5: Calculate the energy consumed by each node using the formula
The energy for transmitting the packet, Etd

$$\text{For long distance} = S^* \text{Etr} + \text{Eld}^* d^4, \text{ if } d \geq do \tag{1}$$

$$\text{For short distance} = S^* \text{Etr} + \text{Esd}^* d^2, \text{ if } d < do \tag{2}$$

$$\text{The energy for receiving the packet, Erd} = S^* \text{Etr} \tag{3}$$

The total energy from Eqs. (1), (2) and (3) is

$$Ett = Erd + Etd \tag{4}$$

$$\text{The remaining energy for each node, } Re = Rc - Ett \tag{5}$$

Step 6: Calculate the trust value by taking both the direct and indirect trust calculation

$$\text{Total Trust} = \text{Direct trust} + \text{Indirect trust}$$
$$= W1Tij^{d}(t) + W2Tij^{i}(t) \tag{6}$$

Step 7: Select the energy and trust fulfilling path
Step 8: Update the energy and trust value for each node.
Step 9: Depending on the threshold value, the controller blocks the malicious node.
Step 10: The controller will broadcast the blocked information to all the nodes in the network.
Step 11: End

4.2 Proposed Algorithm Description

This proposed system provides a energy and trust fulfilling route for the internet of things. The overall flowchart is given in Fig. 2.

The controller controls and monitors the overall activity of the nodes in the network. The functions of the controller are topology discovery, monitoring the traffic, and calculating the path based on the trust and energy. In this proposed system, there are two level of security: the first level and second level security. The first level is in the node level where the nodes will be forwarding the data based on the energy and the trust value. Initially, the nodes have an initial trust value of 5 and energy value of 10 J. The node will forward the data to the trustworthy node by checking an optimized path between the energy value and trust value. It is assumed that each node should possess minimum energy of 10 J for forwarding the data. The distance between the neighbor nodes $(x1, x2)$ and $(y1, y2)$ is calculated using the Euclidean distance [7]

$$\sqrt{(x1 - x2)^{2} + (y1 - y2)^{2}} \tag{7}$$

Depending on the distance, the energy spent to transmit will also vary. If the transmission range is for a large distance, the amount of energy spent by a node will also increase in comparison to shorter range. Furthermore, the amount of energy for amplification for long distance Eld and short distance Esd will also vary depending on the distance. A distance do is defined as the threshold distance, which act as a borderline between the short and long distance. The minimum energy for both

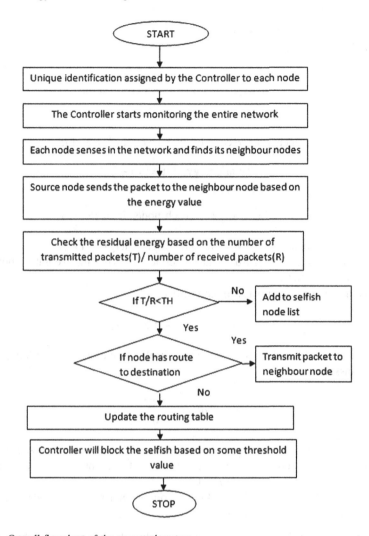

Fig. 2 Overall flowchart of the proposed system

transmitting and receiving the packet, Etr is mandatory for the nodes to enable transmission of data. The calculation for energy consumption by each IOT device is given below:

- The energy for transmitting the packet, Etd

$$\text{For long distance} = S^* \text{Etr} + \text{Eld}^* d^4, \text{ if } d \geq do \qquad (8)$$

$$\text{For short distance} = S^* \text{Etr} + \text{Esd}^* d^2, \text{ if } d < do \qquad (9)$$

where d = the distance between the two nodes
S = the size of the data

$$do = \sqrt{\frac{Eld}{Esd}} \text{ is the threshold distance} \qquad (10)$$

- The energy for receiving the packet, $Erd = S^* Etr$ $\qquad (11)$

$$\text{Total energy, } Ett = Erd + Etd \qquad (12)$$

- The remaining energy for each node, $Re = Rc - Ett$ $\qquad (13)$

where the remaining energy = Re and current energy = Ec [12]
The calculation for total trust value is done by taking the weight of both the direct and indirect trust.
Total Trust = Direct trust + Indirect trust

$$= W1\, Tij^d(t) + W2\, Tij^i(t) \qquad (14)$$

- Direct Trust is taken from the node itself depending on the number of transmission

$$Tij^d(t) = \frac{Sij(t)}{Rij(t)} \qquad (15)$$

where $Sij(t)$ is the number of the packets, which has been send successfully from node i to node j at time t and $Rij(t)$ is the number of packet that are received by node j from node i at time t.
- Indirect trust is calculated by taking recommended trust that node j receives from the neighbor node i about node k.

After evaluating the energy and trust, the values from both the attributes have been combined together to give an energy trust fulfilling path. Even though the malicious node is present in the network, the nodes will neglect the malicious node and finds another route that leads to the destination (Figs. 3, 4, 5 and 6).

The second level of security in this proposed system is by using the centralized controller. The controller will monitor the activity of the IOT devices and has all the information regarding the nodes such as packet forwarded, packet drop, statistics of the network, delay, congestion and accordingly can perform the required action in the network. Even if the node in the initial state behaves as a good mouthing node,

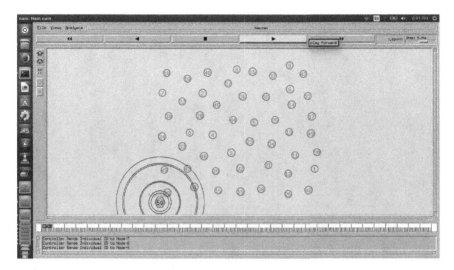

Fig. 3 Deploymnet of nodes and registration with the controller

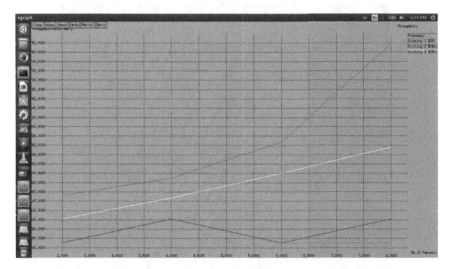

Fig. 4 Throughput

it can detect the malicious node activity. The controller has the ability of controlling the activity of node if found to be malicious. It can deny access to the malicious node into the network. It also calculates their energy value consumption and as well as the trust value. For checking the trust value, the controller will observe the packet delivery ratio of the two-third of the total interaction.

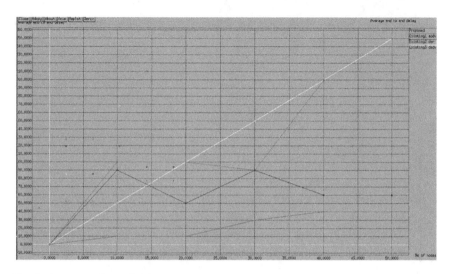

Fig. 5 Average end-to-end delay

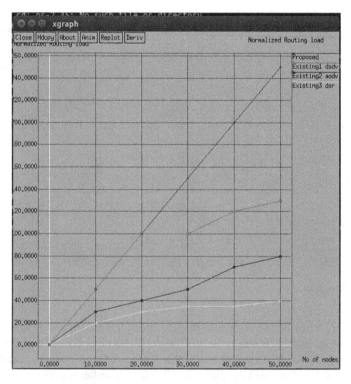

Fig. 6 Normalized routing overhead

Table 1 Simulation table

Simulation parameters	Values
Simulator	N-2.35
Number of nodes	50
Simulation time	3500
Pause time	0.01
MAC protocol	IEEE 802.11
Initial energy value	10 J
Initial trust value	5
Packet size	512

- $$\text{Total number of packet dropped } {}^{'}R^{'} = (d - h) \tag{16}$$

where $h = 2$ is the relaxation for congestion, delay and h is the number of packet dropped

- $$\text{Successfully forwarded packets, } S = s - (d - h) \tag{17}$$

where s is the number of packets that have been sent successfully

If $(S > R)$ then there will no change in trust value else decrement the trust value. If the trust value is below the threshold value, it will block the node and will be denied access to the network. The controller will broadcast this blocked information to all nodes in the network (Table 1).

5 Simulation Setup

The proposed system is simulated using NS-2 Simulation. Initially in the starting of the simulation, all the nodes are assumed to be trustworthy node with initial energy 10 J. The proposed system is compared with three other routing protocols AODV, DSR and DSDV using some parameters. The simulation table is given below.

6 Results and Discussion

The proposed system has been implemented using NS-2 Simulation. The proposed system is compared with other routing protocols using the following metrics:

- Throughput: is the ratio of the number of the packets that have been sent to the number of packets that have been received [10].

- Average end-to-end delay: it is the time for the packet to reach the destination from the source taking in consideration the delay, congestion, propagation, and transfer time [11].
- Normalized routing packet: is the number of routing packets send to the packets received at the destination [12].

From the graph, it is observed that the throughput increases using this proposed routing. The other three routing protocols AODV, DSDV, and DSR do not give good performance as compared with the proposed routing.

Using the proposed routing, the average end-to-end delay is less in comparison with the other three routing protocols AODV, DSR, and DSDV.

It is seen from the graph that the normalized routing overhead increases for the proposed routing as compared to AODV, DSR, and DSDV routing protocol.

7 Conclusion

The proposed system has been implemented using NS-2 Simulation. From the results, it has been observed that using the controller, the network efficiency improves. The throughput and normalized routing overhead increases where as the average end-to-end delay decreases using this proposed routing as in comparison to the other routing protocols.

References

1. Luigi Atzori, Antonio Ierab, Giacomo Morabito, The Internet of Things: A survey, Elsevier, Computer Networks 54 (2010) 2787–2805
2. S. Sicari, A. Rizzardi, L.A. Grieco, A. Coen-Porisini, Security, privacy and trust in Internet of Things: The road ahead, Elsevier, Computer Networks 76 (2015) 146–164.
3. Dr. Neeraj Sharma, Simarjot Kaur," Overview of Various Routing Protocols in Wireless Sensor Networks", Journal of Network Communications and Emerging Technologies (JNCET) Volume 2, Issue 2, June (2015)
4. Sriram Sankaran, Ramalingam Sridhar, Modeling and Analysis of Routing in IoT Networks, international Conference on Computing and Network Communications (CoCoNet'15), Dec. 16–19, 2015, Trivandrum, India
5. Paul Loh Ruen Chze, Kan Siew Leong, A Secure Multi-Hop Routing for IoT Communication, IEEE World Forum on Internet of Things (WF-IoT), 3014
6. Oladayo Bello, Sherali Zeadally, Intelligent Device-to-Device Communication in the Internet of Things, IEEE SYSTEMS JOURNAL, 2014
7. S. Park, N. Crespi, I. Mines-telecom, T. Sudparis, H. Park, and S. Kim, "IoT Routing Architecture with Autonomous Systems of Things," pp. 442–445, 2014.
8. Hao Zhang, Tingting Zhang, A Peer to Peer Security Protocol for the Internet of Things: Secure Communication for the Sensible Things Platform, International Conference on Intelligence in Next Generation Networks, 2015

9. Aakanksha Sharma1, Kamaldeep jangra2, A survey – Energy Efficient Routing protocol for Homogenous & Heterogeneous Networks, International Journal of Research in Computer and Communication Technology, Vol 4, Issue 5, May -2015

10. Dr.Neeraj Sharna,Simarjot Kaur, "Overview Of Various Routing Protocols in Wireless Sensor Networks", Journal of Network Communications and Emerging Technologies (JNCET) Volume 2, Issue 2, June (2015)

11. Harsha Mishra, Prof. Vaibhav Kumar, Prof. Sini Shibu, "Cluster based Energy Efficient Routing for Wireless Sensor Network", Engineering Universe for Scientific Research and Management, Vol. 7 Issue 1 January 2015

12. Mohamed I. Gaber, Imbaby I. Mahmoud, Osama Seddik, Abdelhalim Zekry, "Comparison of Routing Protocols in Wireless Sensor Networks for Monitoring Applications", International Journal Of Computer Applications (0975-8887), Volume 113-No, 12, March 2015.

READ—A Bangla Phoneme Recognition System

Himadri Mukherjee, Chayan Halder, Santanu Phadikar
and Kaushik Roy

Abstract Speech Recognition is a challenging task especially for a multilingual country like India as the speakers are habituated in using mixed language and accent. Bangla is a very popular language in East Asia and a fully functional Automated Speech Recognition System (ASR) for it is yet to be developed. Every language embodies a set of sounds called phoneme set, which is the building block for the words of that language. READ (Record Extract Approximate Distinguish) is a Bangla phoneme recognition system, proposed toward the development of a Bangla ASR. To start with, Mel Scale Cepstral Coefficient (MFCC) features have been used for testing on a database of 1400 Bangla vowel phonemes and an accuracy of 98.35% has been obtained.

Keywords READ · ASR · Phoneme · MFCC · Approximation · MLP

1 Introduction

To use the digital resources more and more in a convenient way, the user interface of the devices needs to be brought within the grasp of the rustics. The user interface should be designed in such a way so that the users feel that they are interacting with their peers and not a device. One way of achieving this is making the devices verbally

H. Mukherjee (✉) · C. Halder · K. Roy
Department of Computer Science, West Bengal State University, Kolkata, India
e-mail: himadrim027@gmail.com

C. Halder
e-mail: chayan.halderz@gmail.com

K. Roy
e-mail: kaushik.mrg@gmail.com

S. Phadikar
Department of Computer Science & Engineering, Maulana Abul Kalam Azad
University of Technology, Kolkata, India
e-mail: sphadikar@yahoo.com

© Springer Nature Singapore Pte Ltd. 2017
S.C. Satapathy et al. (eds.), *Proceedings of the 5th International Conference on Frontiers in Intelligent Computing: Theory and Applications*, Advances in Intelligent Systems and Computing 515, DOI 10.1007/978-981-10-3153-3_59

interactive as verbal interaction has been one of the primary and natural modes of communication since the earliest of times. Interaction with the commonly available digital devices in Indic languages such as Bangla, Hindi, Malayalam, etc., is difficult because of the presence of compound characters and absence of proper input devices in Indic languages. Thus, there is a pressing need for the development of ASRs for the Indic languages.

Speech Recognition is the technique of identification of spoken words from voice signals. Analysis of speech signals requires them to be segmented into very short windows as some of the essential characteristics for classification do not change much in such small durations. A large change of these characteristics is observed if the signal is analyzed as a whole, thereby posing difficulty in classification. One of the basic challenges in speech recognition is the task of coping up with various factors like variability of speech signals with slight change of the acoustic condition, different speakers with different accents and emotional states, etc. All the words of a particular language are formed by restricted permutation of elements from the phoneme set of that language. Vowel phonemes are an essential entity for the words of a language. There are a very few meaningful, pronounceable words which do not have a vowel phoneme, thereby making vowel phoneme identification important. In the proposed system, MFCC [1] features have been used to start with. The sound produced while talking is filtered and shaped by our vocal tract, tongue, and teeth. It is this shape or envelope which determines what we actually hear. MFCC is an artificial way of replicating these envelopes which they are presumed to be.

In the remainder of the paper, related work is discussed in Sect. 2 followed by the proposed work in Sect. 3. Section 4 casts light on result and discussion. Finally, the conclusion is drawn in Sect. 5.

2 Related Work

The attempt of building an automatic speech recognizer started way back in 1930 when Dudley [2, 3] proposed a system for analysis and synthesis of speech signals. Computerized speech recognition was first attempted by Forgie et al. in 1959 [4] to recognize English vowels. English speech recognition has come a long way since then. Desai et al. have presented some of the commonly used speech recognition and feature extraction techniques in [5].

A number of highly accurate speech recognizers such as Dragon Speech Recognition Software [6], Microsoft Windows Speech Recognition [7], Google Cloud Speech API [8], etc., are now commercially available for English and a few other languages. Though work in the field of Bangla speech recognition started way back in 1975 [9] but it is yet to receive proper and widespread attention from researchers. Some stray works are available in literature but a full fledged ASR for Bangla is still a distant dream. There is a pressing need for an automated speech recognizer in Bangla, being the seventh most popular language [10] in the world. Ali et al. in [11] presented four models to recognize Bangla words. The first one was based on

MFCC features with Dynamic Time Warping (DTW) which yielded an accuracy of 78%, the next one was based on Linear Predictive Coding and DTW which yielded an accuracy of 60%. The next one used MFCC and Gaussian Mixture Model along-with posterior probability function for classification which yielded an accuracy of 84% and finally using Linear Predictive Coding (LPC)-based MFCC features along-with DTW resulted in an accuracy of 50%. Firoze et al. [12] designed a Bangla isolated word recognition system by employing spectral analysis-based feature and fuzzy logic for classification obtaining accuracy of only 80% on a modest dataset of 50 words. Hasnat et al. [13] proposed a technique to recognize both isolated and continuous speech in Bangla using Hidden Markov Model (HMM) on a vocabulary of 100 words recorded by only 5 speakers. Their feature set consisted of 39 features comprising of 12 MFCC Features, 1 energy coefficient, 13 first-order derivatives, and 13 second-order derivatives. For isolated speech recognition the accuracy was found to be 90% in speaker-dependent mode and only 70% in case of speaker-independent mode. For continuous speech the accuracy was 80% for speaker-dependent mode and only 60% for speaker-independent mode. Hossain et al. [14] employed various classification techniques for Bangla phoneme identification using MFCC features. They worked with 6 Bangla vowel phonemes (/অ/, /আ/, /ই/, /উ/, /এ/, /ও/) and 4 Bangla consonant phonemes (/ক/, /ট/, /ম/, and /শ/). A small database of 300 phonemes (30 each) was used for the purpose. An accuracy of 93.66% was obtained for Euclidean Distance Measure but it dropped to 93.33% and 92% on applying hamming distance and Artificial Neural Network respectively. Hasanat et al. [15] characterized Bangla phonemes based on 13 reflection coefficients with 13 autocorrelations. An unknown phoneme was classified based on eucledian distance of its derived coefficients from the phonemes in the reference library with an accuracy of 80%. Kotwal et al. [16] classified Bangla phonemes using hybrid features (MFCC and phoneme probability derived from the MFCCs and acoustic features). A HMM classifier was trained using a speech corpus of 3000 sentences spoken by 30 male speakers. An accuracy of 58.53% was obtained on a database of 1000 sentences spoken by 10 speakers.

3 Proposed Method

A bird's eye view of READ is shown in Fig. 1. The steps are described in the subsequent paragraphs.

3.1 Data Collection

Data collection is an extremely important aspect of any experiment. It is always very essential to minimize the amount of error which might creep into the data due to various external factors during the collection of data. To the best of our knowledge,

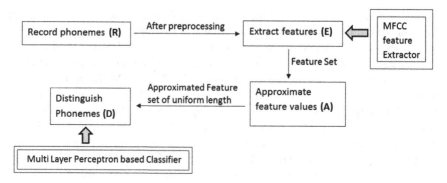

Fig. 1 Block diagram of READ

Table 1 Bangla Vowel Phonemes with their IPA symbol, alphabetic representation, pronunciation as in (example) and equivalent English pronunciation

IPA	Corresponding Bangla Representation	Pronunciation as in (example)	Equivalent English Pronunciation
ɔ	অ	অজগর	Distort
a	আ	আম	Metal
i	ই, ঈ	বিরাট	Cherry
u	উ, ঊ	ভূল	Tool
e	এ	বেল	Red
o	ও	ওজস্বী	Boat
æ	অ্যা	ত্যাগ	Tab

presently there is no standard Bangla phoneme database, so a database was built with the aid of volunteers comprising of 12 males and 8 females aged between 20 and 75. The phonemes along with their International Phonetic Alphabet (IPA) symbols, Bangla alphabet, Bangla pronunciation and English pronunciation is shown in Table 1. The recording of the phonemes was done in ordinary room condition with standard Frontech headphone (JIL-3442) using Audacity [17]. Volunteers were asked to pronounce the 7 vowel phonemes one after the other which were recorded in a single take. Each volunteer spoke 10 times. Twenty such volunteers contributed to the database. The result of successful separation using a semiautomatic amplitude-based separation technique along with an original track is shown in Fig. 2.

3.2 Feature Extraction and Approximation

3.2.1 Framing and Overlapping

Framing is done in order to capture the characteristics of audio signals which vary with time but tend to stay stable in short duration of time. 256 sample points were

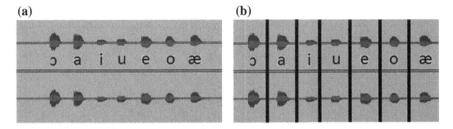

Fig. 2 Audio track of all the phonemes **a** before separation. **b** After separation at the *black vertical lines*

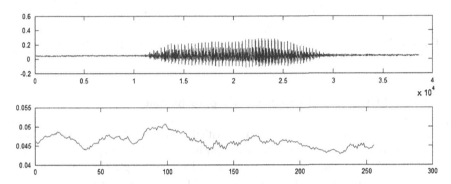

Fig. 3 First frame of phoneme ' ɔ ' in comparison to the entire signal

considered for every frame with an overlap of 100 sample points, because 256 is a perfect power of 2 which facilitates in obtaining Fourier transformation [1] of the frames later. The lower perfect power being 128 was very small and the higher power being 512 was very large for frame size. Figure 3 shows the first frame of the phoneme ' ɔ ' and the entire signal.

3.2.2 Windowing

A hamming window [1] was multiplied with every frame in order to keep the continuity in between the frames and remove jitters. Figure 4 shows the hamming window and result after multiplying the frame in Fig. 3 with the hamming window. The equation of Hamming Window is shown in (1).

$$w(n) = 0.54 - 0.46 \cos\left(\frac{2\pi n}{N - 1}\right) \tag{1}$$

$w(n)$ is the hamming window function. n ranges from start of the frame to end of the frame which is 0–255 here. N is the frame size which is 256 here.

(a)   (b)

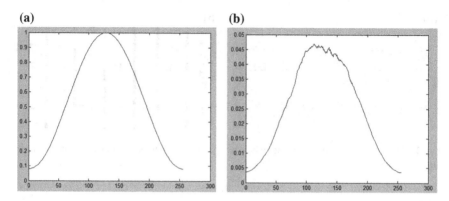

Fig. 4 **a** Structure of 256-point hamming window. **b** Result after multiplying the frame in Fig. 5 with the hamming window

3.2.3 MFCC Feature Extraction and Approximation

The Power Spectrum for each frame along with the generated Mel Filter bank [1] was used to extract nineteen MFCC feature from the phonemes using Discrete Cosine transformation [1] which are shown in Fig. 5. Different lengths of the audio clips resulted in variation in the number of columns of the MFCC features thereby making the task of classification difficult. To cope up with this, all the energy values of all the MFCC feature files were analyzed and the energy values were graded into 18 equally spaced classes. Each band of the MFCC features of the phonemes were then analyzed and the frequency of occurrence of energy values from each class was recorded. Following this, every frequency value of a class was assigned its percentage value based on the total number of values (number of frames). After doing this with all the bands of the MFCC values, a matrix of constant size (19×18) was obtained. Along with this another row was appended, containing the band values arranged in decreasing order of the sum of their energies. This matrix was converted into a single-dimensional array consisting of $(19 * 18) + 19 = 361$ values. Such 361 values were obtained for every phoneme. Feature values of a single instance of every phoneme is shown in Fig. 6.

4 Result and Discussion

Bangla vowel phonemes are one of the most important aspects of Bangla speech. As the size of the current data set was not so large, we used 5-Fold Cross Validation on our collected 200×7 (1400) data. A Multi Layer Perceptron (MLP)-based classifier from the well- known open-source tool WEKA [18] was used for the present work. As a total of 361 features were obtained for each phoneme, with 7 phoneme classes, so a 361-h-7 MLP classifer was designed with 1 hidden layer. The MLP was trained

(a) (b)

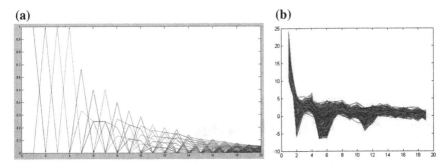

Fig. 5 a Mel Filter Bank. **b** 19 MFCC features in a single graph for the phoneme ' ɔ '

(a) (b) (c) (d)

(e) (f) (g)

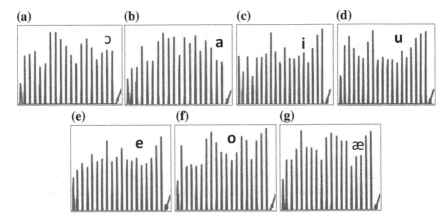

Fig. 6 a–g Showing approximated feature values of the Bangla vowel phonemes where x axis represents the features and y axis represents the magnitude of the feature values. Range of x axis is 1–361 and that of y axis is 0–100

using backpropagation method with a learning rate of 0.3 and a momentum of 0.2 with 1500 iterations. The number of neurons in hidden layer was chosen to be 50 based on trial run.

An accuracy of 98. 36% was obtained in the experiment which is very encouraging. Analysis of the result revealed the highest classification rate to be 100% for the phoneme ' ɔ '. In terms of successful recognition ' ɔ ' was followed by the phonemes 'e' and 'æ' with an accuracy of 99.5%. The third successful recognition rate of 99% was obtained for the phonemes 'i' and 'æ'. The recognition rate for the phonemes 'u' and 'o' were lacking in this context which were 95.5% and 96% respectively. The phoneme 'u' was confused as 'o' 7 times out of 200 instances and the vice versa occured 8 times, the phonemes 'u' and 'o' were confused with each other 15 times, which was the highest confused pair in our experiment. This is due to the fact that at times these two phonemes sound very close to each other which become indistinguishable by a lot of people. The closeness can be observed from Amplitude and

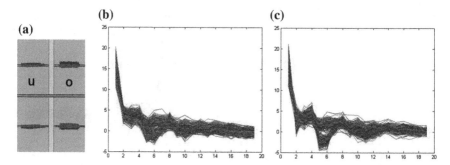

Fig. 7 **a** Amplitude plot of 'u' and 'o'. **b** Shows MFCC of 'u'. **c** Shows MFCC of 'o'

MFCC plots of Phonemes 'u' and 'o' as shown in Fig. 7. The mean absolute error was 0.0063, and the relative absolute error was 2.56% with a coverage of 99.5%.

5 Conclusion

The paper presents a system to classify Bangla vowel phonemes by approximating MFCC features. Since phonemes are the building blocks of speech in a particular language, so attempts of their identification can pave the way for an ASR. As mentioned earlier, due to the absence of a standard database, we could not compare our work. In future, the same technique will be applied on a larger Bangla consonant phonemes database to test its accuracy. The separated phonemes will be further processed to remove the leading and trailing silence parts and pre-emphasis of the phonemes will also be done to test its effect on the accuracy of the system.

References

1. L. Muda, M. Begam and I. Elamvazuthi, "Voice Recognition Algorithms using Mel Frequency Cepstral Coefficient (MFCC) and Dynamic Time Warping (DTW) Techniques", Journal of Computing, Vol. 2, Issue 3, pp 138–143, 2010
2. H. Dudley, "The Vocoder", Bell Labs Record, Vol. 17, pp. 122–126, 1939.
3. H. Dudley, R. R. Riesz and S. A. Watkins, "A Synthetic Speaker", J. Franklin Institute, Vol. 227, pp. 739–764, 1939.
4. J. W. Forgie and C. D. Forgie, "Results obtained from a Vowel Recognition Computer Program", The Journal of the Acoustical Society of America, Vol. 31, pp. 1480–1489, 1959.
5. N. Desai, K. Dhameliya and V. Desai, "Feature Extraction and Classification Techniques for Speech Recognition: A Review", International Journal of Emerging Technology and Advanced Engineering, Vol. 3, Issue 12, pp. 367–371, 2013
6. http://www.nuance.com/dragon/index.htm, visited on 25.04.2016
7. http://windows.microsoft.com/en-in/windows/set-speech-recognition#1TC=windows-7, visited on 25.04.2016

8. https://cloud.google.com/speech/, visited on 25.04.2016
9. M. Pramanik and K. Kido, "Bengali Speech : Formant Structures of Single Vowels And Initial Vowels of Words", In Proc. of ICASSP, Vol. 1, pp. 178–181, 1976.
10. Lewis, M. Paul, G. F. Simons, and C. D. Fennig, "Ethnologue: Languages of the World", Nineteenth edition. Dallas, Texas: SIL International, (eds.), 2016.
11. Md. A. Ali, M. Hossain and M. N. Bhuiyan, "Automatic Speech Recognition Technique for Bangla Words", International Journal of Advanced Science and Technology Vol. 50, pp. 51–59, 2013
12. A. Firoze, M. S. Arifin, R. Quadir and R. M. Rahman, "Bangla Isolated Word Speech Recognition", In Proc. of ICEIS, Vol. 2, pp. 73–82, 2011
13. M. A. Hasnat, J. Mowla and M. Khan, "Isolated and Continuous Bangla Speech Recognition: Implementation Performance and application perspective", In Proc. of SNLP, 2007.
14. K. K. Hossain, Md. J. Hossain, A. ferdousi and Md. F. Khan, "Comparative Study of Recognition Tools as Back-Ends for Bangla Phoneme Recognition", IJRCAR, Vol. 2, Issue 12, pp. 36–40, 2014
15. R. Karim, M. S. Rahman and M. Z Iqbal, "Recognition of spoken letters in Bangla", in Proc. of ICCIT, 2002.
16. M. R. A. Kotwal, Md. S. Hossain, F. Hassan, G. Muhammad, M. N. Huda and C. M. Rahman, "Bangla Phoneme Recognition Using Hybrid Features", In Proc. of ICECE, 2010
17. http://www.audacityteam.org/, visited on 25.04.2016
18. M. Hall, E. Frank, G. Holmes, B. Pfahringer, P. Reutemann and I. H. Witten, "The WEKA Data Mining Software: An Update", SIGKDD Explorations, Vol. 11, Issue 1, pp. 10–18, 2009

Information Fusion in Animal Biometric Identification

Gopal Chaudhary, Smriti Srivastava, Saurabh Bhardwaj
and Shefali Srivastava

Abstract This work presents the application of biometrics in animal identification, which is a highly researched topic in human recognition. Here, our analysis presents the identification of zebra in their natural habitat. All the techniques are tested on 824 Plains zebra images captured at Ol'Pejeta conservancy in Laikipia, Kenya. We have used coat strips as a biometric identifier which is unique in nature. To improve the performance of identification, information fusion of coat strips can be taken place from many points in zebra skin such as near legs, stomach and neck. Here two region near stomach (flake) and first limb (leg) is cropped from the textural pattern of strips of zebra is used in feature extraction. GMF, AAD, mean, and eigenface feature extraction methods are applied on flake and limb ROI of zebra. Then a novel image enhancement method: difference subplane adaptive histogram equalization is applied to improve the identification rate. Our technique is based on information fusion in fusing the score from stomach (flake) and first limb (leg) region. For this, sum, product, frank T-norm, and Hamacher T-norm rules are applied to validate the identification results. Information fusion improves the identification results from the previous reported results from eigenface, CO-1 algorithm, and stripecodes. The improvement in results verifies the success of our approach of information fusion using score level fusion.

Keywords Animal biometrics · Plains zebra · Coat strips

G. Chaudhary (✉) · S. Srivastava · S. Srivastava
Netaji Subhas Institute of Technology, Delhi, India
e-mail: gopal.chaudhary88@gmail.com

S. Srivastava
e-mail: smriti.nsit@gmail.com

S. Srivastava
e-mail: shefali9625@gmail.com

S. Bhardwaj
Thapar University, Patiala, India
e-mail: bsaurabh2078@gmail.com

© Springer Nature Singapore Pte Ltd. 2017
S.C. Satapathy et al. (eds.), *Proceedings of the 5th International Conference on Frontiers in Intelligent Computing: Theory and Applications*, Advances in Intelligent Systems and Computing 515, DOI 10.1007/978-981-10-3153-3_60

609

1 Introduction

Physiological traits like human iris, palmprint, fingerprints, face, and veins provide ample amount of research in human biometrics during last many decades. All these biometric modalities are used in human identification. But these system require artificially controlled acquisition conditions with normalized illumination and equal distance, cooperative user behavior [8]. These types of biometric approaches can be applied to animal identification in wildlife control and management systems and provide a number of research opportunities in this field. In animal, there are many types of markings, skin patterns, color patterns, which are permanent camouflage markings on their coats [10]. These patterns are highly stable and unique that mainly includes stripes and spots, which are species dependent and are important in animal behavior [15]. For example, eye spots or color codes of butterflies, stripes on zebras, patches of giraffe, tiger lines, etc., are skin coat pattern of animal are unique and stores the identity of individual. It is very easy to identify the different species on the basis of their pattern and lots of research have been done. To identify the inter-species variation is a topic of research now.

Accuracy in the data involving position and movement of individual animal plays a crucial role in conducting research on them. In past, tags and transmitters attached to the captured animals provides such information. But they suffer from several drawbacks like cost, physically invasive; require proximity to unwilling subjects, etc. Digital cameras that are widespread available provides an inexpensive alternate approach to the existing method. For data acquisition, animal biometrics differs from the human biometrics. The problem of natural habitat and those animals is not specific, model trained for data acquisition is a major hurdle. Videos and taking pictures are best means to acquire animal data. The data that generates from the preprocessing procedure can be used as test or train samples for feature extraction. Before the steps of feature extraction and matching, videos are processed as 2-D or 3-D images.

Number of techniques have been developed for animal identification [2, 14]. Stripe spotter [9] technique is based on the features known as stripe codes, binary values representing two-dimensional strings are designed to acquire the zebras stripe patterns. Modified edit distance dynamic-programming algorithm measure the similarity between stripe codes. Queries are run by calculating the similarity to each database image individually and returning the top matches. A median correct rank of 4 is achieved by stripe spotter which involves database of 85 plains zebras. Wild-ID [1] uses the original SIFT features and descriptors. It scores the query image against each database image separately. On a database of 100 Wildebeest images, Wild-ID achieved a false positive rate of 8.1×10^4, with a false rejection rate ranging from 0.06 to 0.08.

In our approach of information fusion, few rectangular area of animal coat is cropped from the main image and these cropped images will be used in typical score level fusion. For fusion, two regions are selected one at stomach, which is referred as flake side and other one is at first limb. Four score level fusion rules are applied on skin coats. Also a novel image enhancement method is also proposed to improve the resolution of region of interest (ROI) which is independent of distance between the animal of photograph clicked. Information fusion improves the accuracy of our system and is better from previous reported results.

The paper is organized as follows. Section 2 describes the proposed approach, which includes preprocessing that is described in Sect. 2.1 and enhancement is explained in Sect. 2.2. Section 3 gives the ides of score level fusion. Section 4 provide the normalization method used in the proposed approach. Section 5 demonstrated simulations and result analysis. Last Sect. 6 concludes the suggested work.

2 Proposed Approach

2.1 Preprocessing

Before feature extraction, all the animal images must be position invariant for a valid matching at the classifier. But in case of animal, photographs are non-constrained, non-coordinated, sometimes grouped images. So zebra images are not similar. To sufficiently remove the problem of orientation with the removal of background and other animals, region of interest (ROI) of fixed dimension is to be cropped from the images. For fusion, two regions are selected one at stomach that is referred as flake side and other one is at first limb. To locate both the region, two rectangular windows are selected of size 200×500 and 150×200 at flake and limb region, respectively. For cropping this window, few key points are selected from the image. First, background is removed from the image using [6] to get a binarized image using Ostu algorithm. After this, boundary is traced using boundary tracing algorithm. Then the centroid of the binarized image is calculated which is aligned near flake side of zebra. By adjusting the centroid point in upward direction, a fixed size ROI of 200×500 is cropped. For calculating the limb ROI, negative rate of change of boundary is calculated and point is used to crop the limb ROI of size 150×200. The procedure of extracting ROI is presented in Fig. 1.

2.2 Enhancement

For enhancement of ROI, an adaptive histogram equalization technique can be applied. As the photographs taken at a variable distance, so there is a intersample difference. But further to improve the ROI, a novel method of difference subplane adaptive histogram equalization is applied which is given in algorithm below. This method equalizes the changes due to different distance of photograph by differencing

ROI-Flake

ROI -Limb

Fig. 1 ROI extraction

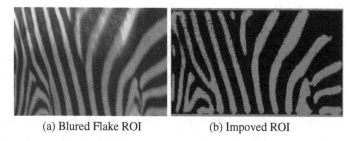

(a) Blured Flake ROI   (b) Impoved ROI

Fig. 2 ROI with difference subplane adaptive histogram equalization

the horizontal and vertical components of image. The improvement in the blurred ROI due to distance is shown in Fig. 2.

2.3 Feature Extraction

After cropping flake and limb ROI and enhancement, the ROI of images of limb and flake of size 150×200 and 200×500 are windowed in rectangular shape of size 15×20 and 20×50 each, respectively, thus creating totally 100 windows from each image. Then gaussian membership function (GMF) features a_i are obtained from i_{th} window and thus a feature vector of length 100 is obtained using Eqs. 1 and 2,

$$u_i = \frac{\exp -(x_k - \bar{x})^2}{2\sigma^2} \qquad (1)$$

Algorithm 1 Difference subplane adaptive histogram equalization

1: **procedure**
2:     $I \leftarrow rgb\ to\ gray\ (ROI)$
3:     $tt=50$
4:     $[m\ n] = Size\ of\ I$
5:     **if** $tt < m$ and i=1: tt: m **then**
6:     $thr \leftarrow gray\ thresholding\ of\ I(i\ :\ i+tt,\ n)$
7:     $Im_{horigontal} = thr \times I(i\ :\ i+tt,\ n)$
8:     $I \leftarrow conjugate\ I$
9:     **if** $tt < n$ and i=1: tt: n **then**
10:     $thr \leftarrow gray\ thresholding\ of\ I(i\ :\ i+tt,\ m)$
11:     $Im_{vertical} = thr \times I(i\ :\ i+tt,\ m)$
12:     $I_{enhanced} = Im_{vertical} - Im_{horigontal}$

$$a_i = \frac{1}{K}\Sigma_{i=0}^{K}x_i u_i \tag{2}$$

where x_k is the image value at k_{th} point of the window, $\bar{x}$ is mean image value, and σ is the standard deviation of the window, u_i is the membership function and a_i is the feature obtained from the i_{th} window [3]. The general AAD features, mean features, and eigenface features are also used for feature extraction for comparison.

3 Information Fusion

There are several fusion methods in literature that are applied in human biometrics [12, 13], while score level fusion is suggested to provide better performance in most of cases [5]. The score level fusion also called as confidence level fusion refers to combining the matching scores obtained from different classifiers. The block diagram of score level fusion is shown in Fig. 3.

3.1 Fusion Rules

Various score level fusion rules are reported in literature. Form all we have selected sum, product, Hamacher and frank T-norm for validation. Let R_i be the matching

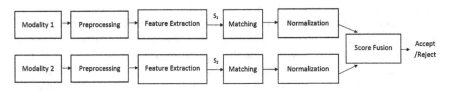

Fig. 3 Score level fusion

score obtained from i_{th} modality and R denotes the fused score or the combined score and N be the number of modalities.

1. Sum rule: $R = R_1 + R_2 + \cdots + R_N = \sum_{i=1}^{N} R_i$
2. Product rule: $R = R_1 * R_2 * \cdots * R_N = \prod_{i=1}^{N} R_i$
3. Hamacher t-norm: $R = \dfrac{R_1 R_2 R_3}{R_1 + R_2 + R_3 - R_1 R_2 - R_3 R_2 - R_1 R_3 + R_1 R_2 R_3}$
4. Frank t-norm: $R = \log_p \left(\dfrac{1 + (p^{R_1} - 1)(p^{R_2} - 1)(p^{R_3} - 1)}{p - 1} \right)$.

4 Score Normalization

For score level fusion, the similarity/dissimilarity scores of each modality must be ranged in a common level to make their fusion meaningful [7]. Here *Min-Max Normalization* method is used due to its simplicity. All the scores are shifted to a range of 0 and 1. Let s_k denote a set of matching scores, where k = 1, 2, ..., n and $s_{k'}$ denote normalized score. Then the normalized score is given as

$$s_{k'} = \frac{s_k - min}{max - min} \tag{3}$$

5 Experimental Results and Discussion

In simulations, the implementation of the suggested methods have been validated in identification and verification modes. In identification, system validates a zebra from all the enrolled zebra, i.e., 1:N mapping. While in verification, sample of zebra is compared with same zebra, that is, one versus one. K-nearest neighbor (KNN) classifier is used here to obtain the similarity/dissimilarity scores using with Euclidean distance with k-fold cross-validation. Receiver operating characteristic (ROC) curve is used to investigate the performance of the system, which is plotted between the genuine acceptance rate (GAR) and false acceptance rate (FAR).

ROC curve of flake ROI using euclidean distance with GMF-based features, AAD features, mean features, and eigenface are shown in Fig. 4. It is seen that GAR at FAR = 1 is 85.88%, 85.82%, 72.8%, and 60.38% and for FAR = 10, 100%, 97.93%, 94% and 75% for GMF, AAD, mean, and eigenface features, respectively. The recognition rate using KNN is calculated which is 92.4%, 89.1%, 86.6%, and 67.6% for GMF, AAD, mean, and eigenface features, respectively.

ROC curve of limb ROI using euclidean distance with GMF-based features, AAD features, mean features, and eigenface are shown in Fig. 5. It is seen that GAR at FAR = 1 is 66.72%, 63.26%, 62.81%, and 60.27% and for FAR = 10, 94.47%, 85.13%, 82.49%, and 75% for GMF, AAD, mean, and eigenface features, respectively. The recognition rate using KNN is calculated which is 88%, 82.5%, 80.3%, and 64.3% for GMF, AAD, mean, and eigenface features, respectively.

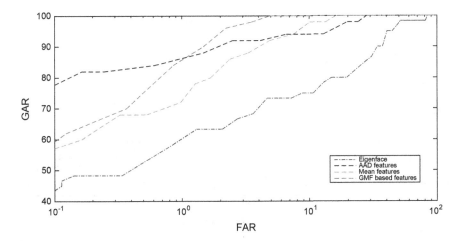

Fig. 4 Receiver operative characteristics of flake ROI using euclidean distance

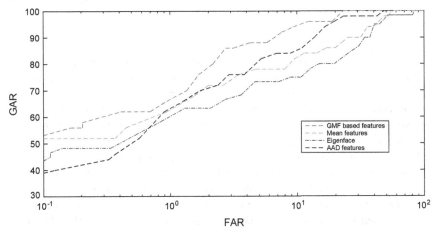

Fig. 5 Receiver operative characteristics of limb ROI using euclidean distance

To validate the information fusion using score level fusion, scores for limb ROI and flake ROI are calculated using euclidean distance. These scores are normalized using min-max normalization method. Then score level fusion is taken place using sum rule, product rule, Hamacher t-norm, and Frank t-norm rule. It is seen from Fig. 6, frank T-norm outperforms the other three rules and hit rate reaches its maximum value 1 at false alarm rate = 0.226. In the terms of area under the curve (AUC), it is also seen that Frank rule is better than other rules and verify most of cases where AUC = 0.9994. When compared to HotSpotter [4] where accuracy is 99%, CO-1 algorithm [11] where accuracy is 94%, StripeCodes [9] where accuracy is 96.6%, information fusion using frank T-norm gives better results and reaches to the 99.9% of queries for plain zebra's.

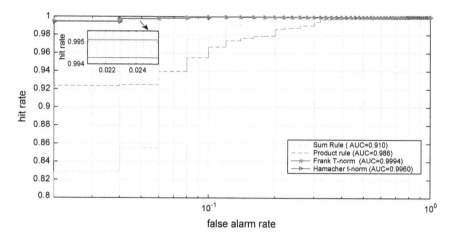

Fig. 6 Score level fusion of limb and flake side ROI

6 Conclusion

In this work, the applications of human biometrics techniques were applied in animal identification. Few skin markings and color patterns such as eye spots on butterflies and stripes on zebras can be used as biometric identifier and provide the unique information of animal. The identification of zebra in their natural habitat was suggested in this work through information fusion of coat pattern using score level fusion. All the techniques were tested on 824 Plains zebra images captured at Ol'Pejeta conservancy in Laikipia, Kenya. The textural pattern of strips of zebra was used in feature extraction using GMF, AAD, and mean, and eigenface feature extraction methods. GMF-based features gave the satisfactory performance on flake and limb ROI of zebra. To improve the performance of identification, information fusion of coat strips was taken place from flake and limb ROI of zebra. Our technique was based on information fusion in fusing the score from flake and limb ROI of zebra. For this, sum, product, frank T-norm, and Hamacher T-norm rules were applied to validate the identification results. Information fusion improved the identification results from the previous reported results from eigenface, CO-1 algorithm, and stripecodes. The improvements in results verify the success of our approach of Information fusion using score level fusion. Experimental results demonstrate that the proposed method can enhance the results effectively.

References

1. Bolger, D.T., Vance, B., Morrison, T.A., Farid, H.: Wild-id user guide: pattern extraction and matching software for computer-assisted photographic mark recapture analysis. Dartmouth College, Hanover, NH (2011)

2. Bradfield, K.S.: Photographic identification of individual Archey's frogs, Leiopelma archeyi, from natural markings. Department of Conservation Wellington, New Zealand (2004)
3. Chaudhary, G., Srivastava, S., Bhardwaj, S.: Multi-level fusion of palmprint and dorsal hand vein. In: Information Systems Design and Intelligent Applications, pp. 321–330. Springer (2016)
4. Crall, J.P., Stewart, C.V., Berger-Wolf, T.Y., Rubenstein, D.I., Sundaresan, S.R.: Hotspotter-patterned species instance recognition. In: Applications of Computer Vision (WACV), 2013 IEEE Workshop on. pp. 230–237. IEEE (2013)
5. Hanmandlu, M., Grover, J., Gureja, A., Gupta, H.M.: Score level fusion of multimodal biometrics using triangular norms. Pattern Recognition Letters 32(14), 1843–1850 (2011)
6. Hung, C.S., Ruan, S.J.: Efficient adaptive thresholding algorithm for in-homogeneous document background removal. Multimedia Tools and Applications 75(2), 1243–1259 (2016)
7. Jain, A., Nandakumar, K., Ross, A.: Score normalization in multimodal biometric systems. Pattern recognition 38(12), 2270–2285 (2005)
8. Jain, A.K., Pankanti, S., Prabhakar, S., Hong, L., Ross, A.: Biometrics: a grand challenge. In: Pattern Recognition, 2004. ICPR 2004. Proceedings of the 17th International Conference on. vol. 2, pp. 935–942. IEEE (2004)
9. Lahiri, M., Tantipathananandh, C., Warungu, R., Rubenstein, D.I., Berger-Wolf, T.Y.: Biometric animal databases from field photographs: identification of individual zebra in the wild. In: Proceedings of the 1st ACM international conference on multimedia retrieval. p. 6. ACM (2011)
10. Murray, J.: Mathematical biology: Spatial models and biomedical applications, volume ii (2003)
11. Ravela, S., Gamble, L.: On recognizing individual salamanders. In: Proceedings of Asian Conference on Computer Vision, Ki-Sang Hong and Zhengyou Zhang, Ed. Jeju, Korea. pp. 742–747 (2004)
12. Rodrigues, R.N., Ling, L.L., Govindaraju, V.: Robustness of multimodal biometric fusion methods against spoof attacks. Journal of Visual Languages & Computing 20(3), 169–179 (2009)
13. Ross, A., Jain, A.: Information fusion in biometrics. Pattern recognition letters 24(13), 2115–2125 (2003)
14. Shorrocks, B., Croft, D.P.: Necks and networks: a preliminary study of population structure in the reticulated giraffe (giraffa camelopardalis reticulata de winston). African Journal of Ecology 47(3), 374–381 (2009)
15. Turing, A.M.: The chemical basis of morphogenesis. Philosophical Transactions of the Royal Society of London B: Biological Sciences 237(641), 37–72 (1952)

Small World Network Formation and Characterization of Sports Network

Paramita Dey, Maitreyee Ganguly, Priya Sengupta and Sarbani Roy

Abstract The motivation of this paper is formation of sports network and characterization of the small world network phenomenon by analyzing the data of individual players of a team. Analysis of the network suggests that sports network can be considered as small world and inherits all characteristics of small world network. Making a quantitative measure for an individual performance in the team sports is important in respect to the fact that for team selection of International football matches, from a pool of best players, only 11 players can be selected for the team. The statistical record of each player is considered as a traditional way of quantifying the performance of a player. But other criteria like performing against a strong opponent or executing a brilliant performance against a strong team deserves more credit. In this paper, a method based on social networking is presented to quantify the quality of player's efficiency and is defined as the total matches played between each team members of individual teams and the members of different teams. The application of Social Network Analysis (SNA) is explored to measure performances and rank of the players. A bidirectional weighted network of players is generated using the information collected from English Premier League (2014–2015) and used for network formation.

Keywords Small world network · Social network analysis · Football · Small world coefficient · Clustering coefficient · Betweenness centrality.

P. Dey (✉) · M. Ganguly · P. Sengupta
Department of Information Technology, GCECT, Kolkata, India
e-mail: dey.paramita77@gmail.com

M. Ganguly
e-mail: maitreyee12aug@gmail.com

P. Sengupta
e-mail: priya.sengupta009@gmail.com

S. Roy
Department of Computer Science & Engineering, Jadavpur University, Kolkata, India
e-mail: sarbani.roy@cse.jdvu.ac.in; sarbani.roy@gmail.com

© Springer Nature Singapore Pte Ltd. 2017
S.C. Satapathy et al. (eds.), *Proceedings of the 5th International Conference on Frontiers in Intelligent Computing: Theory and Applications*, Advances in Intelligent Systems and Computing 515, DOI 10.1007/978-981-10-3153-3_61

619

1 Introduction

According to Watts and Strogatz model [1] of small world networks definition, it represents the set of networks that are highly clustered, like regular lattices, yet have small characteristic path lengths, like random graphs. Social network analysis (SNA) procures information about interrelationship of players within a network and success of a team depending on that factor. This paper focuses on the relationship between all players within the networks and how the small individual networks are connected to each other [2, 3]. The focus of this paper is to apply the social network analysis tool for the players of football and try to quantify the individual player as a team member instead of their individual performances. Advantage of social network-based analysis is that it incorporates quality of the players as a team member along with their individual performances which is very much important for team sports.

For the database of our study, we used total matches between players in English Premier League (EPL 2014–2015). Records of 16 teams namely Arsenal, Manchester city, Chelsea, Liverpool, Manchester United, Tottenhum Hotspur, Southampton, Stoke City, Crystal Palace, Everton, Aston villa, Sunderland, Hull city, Burnley, Westham—United, Norwich city are considered. The number of matches common to two players are considered as the edge between them. In recent research papers, there are trends to quantify the qualitative measures like team spirit, team belongingness of individual performance along with their individual performances in team sports. Analysis has been done in football [4, 5], baseball [6], basketball [7], and other ball sports for quantifying the time-dependent performance but in a manner of statistical analysis [8]. But the analysis using network is a recent approach [9]. For team selection strategy mainly machine learning, neural network techniques were used to predict the performance of individual player's based on their past performance [8]. Again, a model-free approach is used for finding the outcome of American soccer [10]. Neural network is also applied for the solution [9]. Performances of water polo also quantified through network-based approach. At present scenario network-based approach was developed for gradation of football teams, tennis [11], and cricket [12]. Though these papers focussed on formation of network, small world phenomenon or the application has yet to be explored.

In this paper, we explore different important parameters that can be derived from social network using existing algorithms. In Sect. 2, different social network parameters important for small world network and centrality measures are discussed. In Sect. 3 we are discussing about the sport network formation. Sections 4 and 5 are based on the result and statistical analysis of the data. Section 6 is concluded with the summarization of the paper.

2 Small World Network Characterisation

For characterization of a social network, the following parameters are important [1] for small world phenomenon. Though most of the nodes are not connected in a small world, but the distance between two nodes is significantly less.

2.1 Average Path

Average path can be expressed as the average number of connecting paths of all the shortest paths that can exist between all pairs of vertices. Shortest path length signifies the maximum connectivity between two players. It can be expressed as

$$L = \frac{2}{N(N-1)} \sum_{p,q} d_{p,q}$$

Where $d_{p,q}$ is the shortest path length between vertices p and q.

2.2 Clustering Coefficient

Clustering coefficient denotes the number of nodes to which a node is connected to other nodes which are also connected to each other. Clustering coefficient of sports network signifies characteristics for the players forming local clusters that is numbers of players those are influenced by that particular player. The dense local clusters signifies that players has great influence to other players.

In social networks, specially in small world network, generally all nodes are highly connected and the clustering coefficient is quiet high valued than the average clustering coefficient of random network. The local clustering coefficient of a node is defined as the number of complete graph (clique) that can be formed using the neighbor of that node. This property was first introduced by Watts and Strogatz [1] for defining small world coefficient. The Clustering Coefficient varies between 0 and 1, i.e., 0 when there is not at all clustering, and 1 for maximal clustering [13].

One common procedure for measuring clustering coefficient is to find existing triangles, i.e., to check that when two edges share a node, then in a network with high clustering, it is highly probable that a third edge exist to form a triangles [14].

2.3 Betweenness Centrality

Betweenness centrality, the most crucial and widely popular centrality measure, can be defined as proportionality of total shortest paths passing through that vertices to all possible shortest paths present in network [15]. A player with high value of centrality measure has large influence on the other players. Betweenness centrality signifies the influence of a particular node in the network, i.e., players with higher betweenness centrality play important roles within the network.

2.4 Random Graph

Random graph is a type of graph where vertices are connected through edges in a random way. Erdos and Reyni proved graph with large number of edges exhibit the property of random graphs unless certain rules are considered for the formation of graph [16]. For a nonregular degree sequence, there is no efficient algorithm known for random graph generation. We generate a random graph simply by the probability distribution, or by a random process to generate the network. Using Gephi software, an equivalent random graph was generated based on the assumption that a link for node pair connectivity have an uniform probability without perturbation of original node degree distribution.

3 Formation of the Sports Network

In this paper, we have investigated the small world phenomenon in the area of football network. The following criteria have been studied (1) Study the small world properties of these networks, (2) identifying the players in the networks that have high betweenness which signifies the key players of the teams.

3.1 Extraction of Data for EPL Network

Each team member is defined as a node in the graph, and connection between them as the edges of the graph. Matches played between them are defined as a weight of each edge. An adjacency matrix was formed, which defines the number of matches played between players. From these matrices we connect the players in Gephi, a free and open-source software for graph analysis. The EPL football team's squad member database was collected from the English premier league's formal website. We have collected the record of individual players for each team <Name, team, position, numbers of matches played, goals, yellow card, red card, total matches played> as shown in Fig. 1.

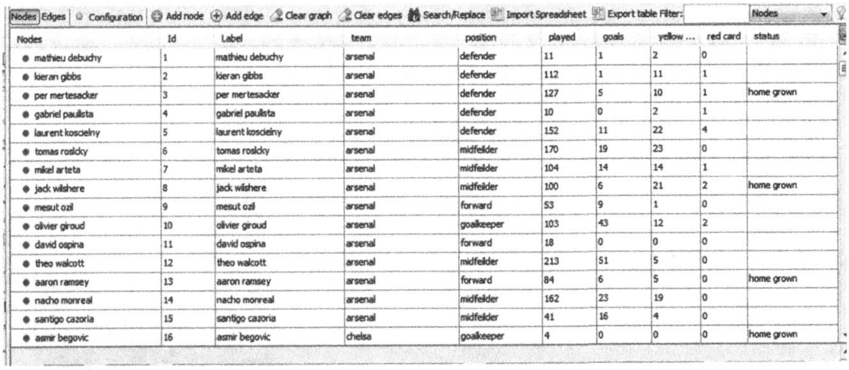

Nodes	Id	Label	team	position	played	goals	yellow ...	red card	status
mathieu debuchy	1	mathieu debuchy	arsenal	defender	11	1	2	0	
kieran gibbs	2	kieran gibbs	arsenal	defender	112	1	11	1	
per mertesacker	3	per mertesacker	arsenal	defender	127	5	10	1	home grown
gabriel paulista	4	gabriel paulista	arsenal	defender	10	0	2	1	
laurent koscielny	5	laurent koscielny	arsenal	defender	152	11	22	4	
tomas rosicky	6	tomas rosicky	arsenal	midfelder	170	19	23	0	
mikel arteta	7	mikel arteta	arsenal	midfeilder	104	14	14	1	
jack wilshere	8	jack wilshere	arsenal	midfeilder	100	6	21	2	home grown
mesut ozil	9	mesut ozil	arsenal	forward	53	9	1	0	
olivier giroud	10	olivier giroud	arsenal	goalkeeper	103	43	12	2	
david ospina	11	david ospina	arsenal	forward	18	0	0	0	
theo walcott	12	theo walcott	arsenal	midfeilder	213	51	5	0	
aaron ramsey	13	aaron ramsey	arsenal	forward	84	6	5	0	home grown
nacho monreal	14	nacho monreal	arsenal	midfeilder	162	23	19	0	
santigo cazoria	15	santigo cazoria	arsenal	midfeilder	41	16	4	0	
asmir begovic	16	asmir begovic	chelsa	goalkeeper	4	0	0	0	home grown

Fig. 1 Database in Gephi tabular form

4 Result and Discussion

4.1 Small World Characteristics of EPL Network

For a small world network, small world coefficient σ can be defined as

$\sigma = (C_{Actual} / C_{Random})/(L_{Actual} / L_{Random})$, should be greater than 1 for a network for being a small world network [1].

At first we have generated the actual network for EPL Football Network. Clustering coefficients and average path length for the network are calculated. Clustering coefficients for EPL network is 0.912 and average path length is derived as 1.572. Then we have generated random networks for both the networks using random rewiring probability. Clustering coefficients for random network is derived as 0.142 and average path length is derived as 1.153.

The small world coefficient is calculated as 4.71 which is not only greater than 1, but shows higher value than small world coefficients of karate (1.65) and the Internet networks (2.38) [3]. Table 1 represents a comparative chart of the parameters. This table clearly depicts that the football network inherits small world phenomenon.

Table 1 Comparison of various networks

Types of network	C_{Actual}/C_{Random}	L_{Actual}/L_{Random}	Clustering coefficient
Karate	1.774	1.075	1.65
Internet	2.555	1.072	2.38
EPL football	6.428	1.363	4.71

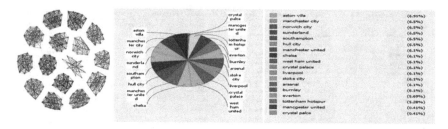

Fig. 2 Team wise distribution

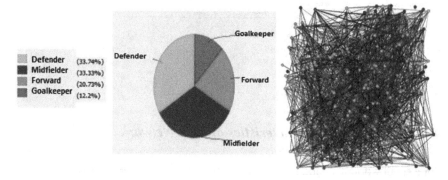

Fig. 3 Role wise distribution

4.2 Network for Individual Team

Figure 2 represents the the team wise distribution of sports small world network. We are working on 16 teams of EPL. Different color indicates different teams in the figure.

From these figures we get a clear knowledge of how players of different teams are connected to each other and defines various centrality measures. Each football team has forward, defender, midfielder, goalkeeper as different roles. Figure 3 show the percentage of individual player having specific role or position in the network. This feature can be very useful at the time of community detection.

5 Statistical Analysis

5.1 Clustering Coefficient Distribution

Figure 4 shows the clustering coefficient distribution of EPL Football network as value of clustering coefficient versus number of player counts. Small world networks have the characteristics of highly clustered nodes. From the analysis it can be stated that both the networks are highly clustered as maximum players have high clustering

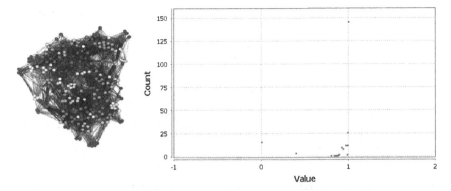

Fig. 4 Clustering coefficient distribution

coefficient value. Light-colored nodes define the low clustering value, whereas dark-colored nodes define the high clustering value.

Average clustering coefficient for EPL Football is derived as 0.9, by calculating a total of 6706 triangles.

5.2 Centrality Distribution

In small world network nodes with high betweenness centrality have a large influence through the network and considered as key nodes.

In EPL football network number of shortest paths is 4502, calculated for 246 players.

Serial No.	Name of the Player	Country	Betweenness Centrality	Total Matches Played
1	Adrian Mariapp	Crystal Palace	225	37
2	Ryan Shawcross	Stroke City	224	11
3	Maya Yoshida	Southampton	195.25	34
4	Jan Vertonghen	Tottenham Hotspur	192.22	103
5	Joe Hart	Manchester City	126	243
6	Jose Enrique	Liverpool	112.07	76
7	Asmir Begovic	Chelsea	13.75	4
8	Bran Saviavnovic	Chelsea	13.75	221
9	Cese Fabregas	Chelsea	13.75	40
10	Ramires	Chelsea	13.75	149

Fig. 5 Top ten players with high betweeness of EPL football network

Serial No.	Name of the Player	Country	Weighted Degree	Total Matches Played
1	Kieran Richardson	Aston Villa	420	28
2	Alan Hutton	Aston Villa	419	42
3	Steven Whittaker	Norwich City	411	41
4	Jose Enrique	Liverpool	407	76
5	Philippe Senderos	Aston Villa	400	8
6	Sebastien Bassong	Norwich City	399	76
7	Lee Cattermole	Sunderland	397	31
8	Gary Gardner	Aston Villa	396	27
9	Peter Odemwingle	Stoke City	391	24
10	Jordan Henderson	Liverpool	389	2

Fig. 6 Top ten players with high node degree of EPL football network

Figures 5 and 6 represents the top ten list of players with high betweenness and high node degree centrality respectively.

6 Conclusion

The focus of this paper is the verification and characterization of the small world phenomenon in EPL football network by collection of the data of players and matches. In this paper, two centrality measurements are evaluated for this network. The experimental results suggest that football network exhibits the property of small world network. Few key players may influence many other players. To summarize, we quantified the performance of players from the history of previous year data by studying the network structure of players. It can be observed from these networks that node connectivity is also influenced by the friends of friends phenomenon, and some players evolved as key players.

References

1. D.J. Watts., S.H. Strogatz, *Collective dynamics of small-world networks*, Nature, Volume 393, pp. 440–442, 1998.
2. M. E. J. Newman, *Random graphs as models of networks*, Brain Network, pp. 2–3, 2005.
3. Qawi K. Telesford, Karen E. Joyce, Satoru Hayasaka, Jonathan H. Burdette, and Paul J. Laurienti, *The Ubiquity of Small-World Networks*, Brain Connectivity, Volume 1, pp. 1–5, 2011.
4. E. Bittner, A. Nussbaumer, W. Janke, M. Weigel, *Football fever: goal distributions and non-Gaussian statistics*, European Physical Journal Vol. B, pp. 67–459, 2009.
5. E. Ben-Naim, S. Redner, F. Vazquez, *Scaling in Tournaments*, Europhysics Letters, pp. 124, 2007.
6. C. Sire, S. Redner, *Understanding baseball team standings and streaks*, European Physical Journal, pp. 473–481, 2009.

7. B. Skinner, *The price of anarchy in basketball*, Journal of Quantitative Analysis in Sports, pp. 3–6, 2010.
8. A. Heuer, C. Müller, O. Rubner, *Soccer: Is scoring goals a predictable Poissonian process?*, Europhysics Letters, pp. 89, 2010.
9. S.R. Iyer, R. Sharda, *Prediction of athletes performance using neural networks: An application in cricket team selection*, Expert Systems with Applications, Elsivier, pp. 36, 2009.
10. Y. Yamamoto, K. Yokoyama, *Common and Unique Network Dynamics in Football Games*, PLOS One, pp. 6–12, 2011.
11. D. Lusher, G. Robins, P. Kremer, *Measurement in Physical Education and Exercise Science, Social Network Analysis*, Sport in Globalised Societies. Changes and Challenges pp. volume-14, pp. 211–224, 2010.
12. S. Mukherjee, *Quantifying individual performance in Cricket? A network analysis of Batsmen and Bowlers*, Physica A 393, pp. 624–637, 2012.
13. Shiu-Wan Hung, An-Pang Wang, *A Small World in the Patent Citation Network*, IEEE International Conference on Industrial Engineering and Engineering Management, pp. 2–4, 2008.
14. Matthieu Latapy, *Main-memory Triangle Computations for Very Large (Sparse (Power-Law)) Graphs*, Theoretical Computer Science (TCS) 407 (1-3), pp. 458–473, 2008.
15. Ulrik Brandes, *A Faster Algorithm for Betweenness Centrality*, Journal of Mathematical Sociology 25(2), pp. 163–177, 2001.
16. Erdos, P. and Rnyi, *On the evolution of random graphs*, Publication of the Mathematical Institute of the Hungarian Academy of Sciences, pp. 17–61, 1960.

UWB BPF with Notch Band for Satellite Communication Using Pseudo-Interdigital Structure

Yatindra Gaurav and R.K. Chauhan

Abstract An ultra wideband bandpass filter with a notch band for satellite communication is stated in the paper using pseudo-interdigital structure. The structure is planar and there is no use of via or defected ground structure that makes the structure less complex and easy to fabricate. The insertion loss of the proposed filter is less than 0.8 dB and return loss more than 16.7 dB. The notch band centered at 8 GHz has insertion loss of 12.8 dB. The small size of the filter is 0.265 λg × 0.071 λg. The filter is designed and simulated in ADS software.

Keywords Ultra wide band · Bandpass filter · Psuedo interdigital structure and notch band

1 Introduction

An ultra wideband systems came into influenced when unlicensed use of spectrum 3.1–10.6 GHz is permitted by FCC commission in 2002 [1]. Since then ultra wideband filters as basic component of ultra wideband system and devices are continuously developing. Efforts are still concentrated on miniaturization of ultra wideband bandpass filters with improved performance for compact design of ultra wide band systems. Early the effort was to have the ultra wide band passband but lately the effort was shifted on eliminating the interferences with other channels within the passband with miniaturization and improved performance [2–5].

Y. Gaurav (✉) · R.K. Chauhan
Department of Electronics and Communication Engineering, Madan Mohan Malaviya
University of Technology, Gorakhpur, Uttar Pradesh, India
e-mail: ygaurav2000@gmail.com

R.K. Chauhan
e-mail: rkchauhan27@gmail.com

© Springer Nature Singapore Pte Ltd. 2017
S.C. Satapathy et al. (eds.), *Proceedings of the 5th International Conference on Frontiers in Intelligent Computing: Theory and Applications*, Advances in Intelligent Systems and Computing 515, DOI 10.1007/978-981-10-3153-3_62

629

Many techniques and design topology was used to introduced notch band to remove the interferences within passband. In some of the design a coplanar waveguide structure and stubs loaded various type of multimode resonators are used for notch band within passband [6–10]. A interdigital structure with different types of resonators are used to have the improved performance of passband with notch achieving small size of filters [11, 12]. Researches are still a main concentration for miniaturization and better passband performance with notch.

In this paper, an ultra wide band bandpass filter with a notch for satellite communication is designed and improved performance is achieved by using a pseudo-interdigital structure. The filter is designed on 1.6 mm thick FR4 substrate with dielectric constant as 4.4.

2 Design of Ultra Wide Band Bandpass Filter

To achieve the miniaturization and improved performance of passband the interdigital structure is modified. A pseudo-interdigital structure with six finger lines coupled to each other is proposed, see Fig. 1. The passband characteristics can be define by these coupled finger lines. The higher and lower cutoff frequency as well as notch centre frequency is govern by the finger length L1, L2 and L3. Decreasing the length L1 it is seen that lower and higher cut of frequency remains almost constant but the notch centre frequency increases, see Fig. 2. By decreasing the

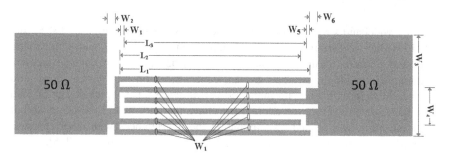

Fig. 1 Proposed UWB BPF

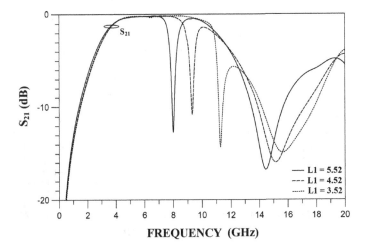

Fig. 2 Variation of insertion loss when L1 is varied

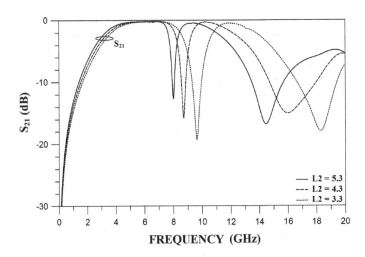

Fig. 3 Variation of insertion loss when L2 is varied

length L2 higher cut off frequency and notch centre frequency increases, whereas slight increase in lower cut off frequency is seen, see Fig. 3. On decreasing length L3 both lower and higher cut off frequency increases but the notch centre frequency remains constant, see Fig. 4. The coupling coefficient and selectivity is achieved by width of the finger lines and the slots between them.

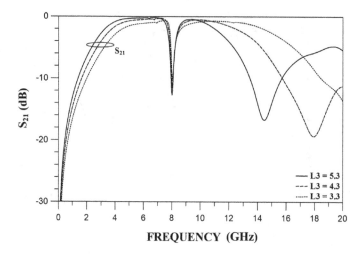

Fig. 4 Variation of insertion loss when L3 is varied

3   Parameters and Simulation

Proposed Filter Parameters are L1 = 5.52 mm, L2 = L3 = 5.3 mm, W1 = 0.15 mm, W2 = 0.2 mm, W3 = 2.87 mm and W4 = 1.05 mm, W5 = 0.1 mm, W6 = 0.28 mm, see Fig. 1. Simulated lower and higher cutoff frequency are 2.9 and 11.6 GHz where as insertion loss and return loss is found to be less than 0.8 dB and more than 16.7 dB. The attenuation centered at 8 GHz is 12.8 dB. The 10 dB rejection fraction bandwidth is found to be 2.3%, see Fig. 5. A comparative study of the presented design with the past design is done and shown in Table 1.

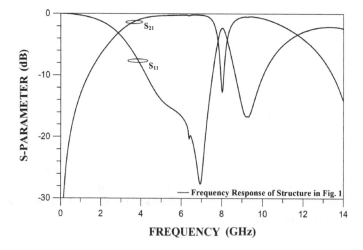

Fig. 5 Simulated S-parameter of proposed UWB BPF

Table 1 Comparison of present design with previous design

Ref. no.	Relative dielectric constant/thickness of dielectric(mm)	Insertion loss (dB)	Return loss (dB)	3-dB fractional bandwidth	Size (λg × λg)	Notch capability
6	2.2/0.508	0.9	11.6	113.5	0.3 × 0.17	Yes
7	6.15/0.635	1.33	8	102.8	1.05 × 0.51	Yes
8	2.2/0.127	1.1	10	125	0.27 × 0.22	Yes
9	3.38/0.81	0.94	12	123	0.94 × 0.14	Yes
10	2.55/0.8	1.5	10	122	0.75 × 0.48	No
11	2.65/1.0	<2	>20	118	1.012 × 0.54	Yes
12	3.5/0.508	1	13	118	0.58 × 0.12	Yes
This work	**4.4/1.6**	**<0.8**	**>16.7**	**120**	**0.265 × 0.071**	**Yes**

4 Conclusion

A planar and compact ultra wideband bandpass filter is designed with a notch for satellite communication. The comparative study with previous filter is done and shown in Table 1. The proposed filter is of small size of 0.265 λg × 0.071 λg (where λg is guided wavelength of 50 O microstrip at 6.85 GHz) having 3 dB fraction bandwidth of 120% with centre frequency of 6.85 GHz. The 3 dB rejection fraction bandwidth is 7.6%.

References

1. Revision of Part 15 of the Commission's Rules regarding ultrawideband transmission system, FCC, Washington, DC, ET-Docket, pp. 98–153 (2002).
2. Hong, J.S. and Shaman, H.: An optium ultra-wideband microstrip filter, Microwave Opt. Technol. Lett. 47, pp. 230–233 (2005).
3. Chen, H. and Zhang, Y.X.: A novel microstrip UWB bandpass filter with CPW resonators, Microwave Opt. Technol. Lett. 51, pp. 24–26 (2009).
4. Wu, X.H., Chu, Q.X., Tian, X.K. and Ouyang, X.: Quintuple-mode UWB bandpass filter with sharp roll-off and super-wide upper stopband, IEEE Microwave Wireless Compon. Lett. 21, pp. 661–663 (2011).
5. Wang, H., Zheng, Y., Kang, W., Miao, C. and Wu, W.: UWB bandpass filter with novel structure and super compact size, Electron Lett. 48, pp. 1068–1069 (2012).
6. Luo, X., Ma, J.G., Ma, K.X. and Yeo, K.S.: Compact UWB bandpass filter with ultra narrow notched band, IEEE Microwave Wireless Compon. Lett. 20, pp. 145–147 (2010).
7. Kim, C.H. and Chang, K.: Ultra-wideband (UWB) ring resonator bandpass filter with a notched band, IEEE Microwave Wireless Compon. Lett. 21, pp. 206–208 (2011).
8. Song, K., Pan, T. and Xue, Q.: Compact ultra-wideband notch-band bandpass filters using multiple slotline resonators, Microwave Opt. Technol. Lett. 54, pp. 1132–1135 (2012).
9. Mirzaee, M. and Virdee, B.S.: UWB bandpass filter with notch-band based on transversal signal-interaction concepts, Electron Lett. 49, pp. 339–401 (2013).

10. Zhu, H. and Chu, Q.X.: Compact ultra-wideband (UWB) bandpass filter using dual-stub-loaded resonator (DSLR), IEEE Microwave Wireless Compon. Lett. 23, 527–529 (2013).
11. Song, Y., Yang, G.M. and Geyi, W.: Compact UWB Bandpass Filter With Dual Notched Bands Using Defected Ground Structures, IEEE Microwave Wireless Compon. Lett. 24, No. 4, PP. 230–232 (2014).
12. Xu, K., Zhang, Y., Lewei, J., Joines, W.T. and Huo, Q.: Miniaturized Notch-Band UWB Bandpass Filters Using Interdigital Coupled Feed-Line Structure, Microwave Opt. Technol. Lett. 56, No. 10, pp. 2215–2217 (2014).

Finding Clusters of Data: Cluster Analysis in R

Tulika Narang

Abstract The paper discusses an essential data mining task, clustering. Clustering groups similar instances and results in classes of similar instances. In this paper, clustering methods k-means, SOM clustering, and hierarchical method of clustering are discussed and implemented in R. Before the application of clustering algorithms cluster tendency is evaluated to determine whether the data set is appropriate for clustering or not. Cluster tendency is also discussed in the paper.

Keywords Clustering · Data mining · R

1 Introduction

Clustering is a method that allocates data instances to groups. It results in cohesive groups called clusters. Clustering algorithms create classes or clusters such that the objects in a group are analogous and associated. The better the similarity the enhanced and distinct is cluster analysis [1].

Clustering is a classification method for placing similar instances in a group. It aims at high inter cluster similarity and also low intra-cluster similarity. It is unsupervised method of classification as class labels are not predetermined. It is an optimization problem. It is the minimization or maximization of a function subject to a set of limitations [2, 3].

The aim of clustering can be defined as, for a given dataset D = {d1, d2, ..., dn}, the desired number of clusters and a function Fn that computes the quality of clustering a mapping is evaluated as

$$M: \{1, 2, \ldots., n\} \rightarrow \{1, 2, \ldots., k\}$$

that minimizes the function Fn subject to some constraints [3, 4].

T. Narang (✉)
University of Allahabad, Allahabad, India
e-mail: n.tulika@gmail.com

© Springer Nature Singapore Pte Ltd. 2017
S.C. Satapathy et al. (eds.), *Proceedings of the 5th International Conference on Frontiers in Intelligent Computing: Theory and Applications*, Advances in Intelligent Systems and Computing 515, DOI 10.1007/978-981-10-3153-3_63

The function Fn is a similarity function. It determines the clustering quality and is stated in terms of similarity between objects. The similarity measure is an input to clustering algorithms.

Clustering differs from classification. In clustering the class labels are previously unknown and is supervised learning. In clustering groups and structures in the data are in some form similar without using known structures. In classification, the class labels are previously known. It is supervised learning [1, 5].

2 Preliminaries

2.1 Data Mining

Data mining is the process of finding hidden, previously not known, valuable information from data. It is nontrivial mining of inherent, previously unknown and useful information from data. It is the exploration and analysis of meaningful patterns from data. In simple words, it is mining knowledge from data. It includes various techniques and methods to find knowledge from the associated data. It is one of the essential activities of knowledge discovery in databases. The various essential data mining tasks are clustering, classification, association mining, prediction, correlation analysis, and outlier analysis.

The objective of data mining is to seek "nuggets" of information among the mass of data [6–9].

2.2 R Environment

R integrated environment provides functionality of computation, manipulation, and display of data. It is free software under the General Public Licence. R has an extensive range of statistical and graphical techniques [10].

3 Clustering Tendency

The evaluation of cluster tendency is an essential primary step in cluster analysis. It is important to determine whether clusters are present in a data set before applying clustering method.

3.1 Evaluating Clustering Tendency on IRIS Dataset

The library ("seriation") can be used for the evaluation of clustering tendency.
Also evaluation of clustering tendency can be done by Nbclust package.
   >data(iris) > iris$Species = NULL. NbClust method is applied but
prioir the seed function is applied to any value so that the result is reproducible.

4 Clustering Methods and Implementation in R

4.1 K-Means Algorithm

K-means is an unsupervised learning algorithm. It results in a distinct number of
clusters, k of the input data set that. The value of k is fixed prior. The objective of
the algorithm is to minimize squared error function defined in Eq. 1 as

$$J = \sum_{j=1}^{k} \sum_{i=1}^{k} \left\| x_i^{(j)} - c_j \right\|^2 \tag{1}$$

where $\left\| x_i^{(j)} - c_j \right\|^2$ is the distance measure selected between a data point $x_i^{(j)}$ and the
cluster centre c_j.

Implementation
k-means algorithm is implemented on iris dataset in R language. The seed value is
taken 101 and the result is plotted on the two-dimensional space.

Result
See (Fig.1).

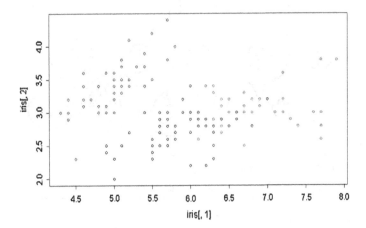

Fig. 1 K-means clustering result

4.2 Self Organising Map Algorithm

Self Organising Map is a neural network method of clustering. The basic SOM algorithm uses Euclidean distance in clustering. Clustering is performed in the learning phase of the algorithm. [2, 4, 11].

Implementation

```
> set.seed(101)
> train.obs <- sample(nrow(iris), 50)
> train.set <- scale(iris[train.obs,][,-5])
> test.set <- scale(iris[-train.obs, ][-5],
+ center = attr(train.set, "scaled:center"),
+ scale = attr(train.set, "scaled:scale"))
> som.iris <- som(train.set, grid = somgrid(5, 5,
"hexagonal"))
> plot(som.iris)
```

Result
See (Fig.2).

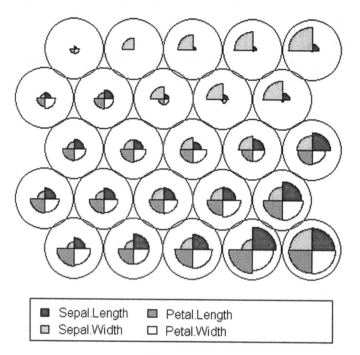

Fig. 2 Cluster of iris dataset using SOM

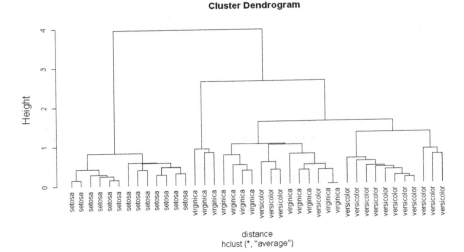

Fig. 3 Dendrogram of hierarchical clustering

4.3 Hierarchical Clustering

Hierarchical clustering algorithms result in a tree of clusters of data instances. The output of nested clusterings is a dendrogram. A dendrogram is an inverted tree like structure [4, 12].

Implementation
Hierarchical clustering of iris dataset is performed in R with seed value of 101. The distance measure used is Euclidean distance measure and method is Average Link clustering [13]. A dendrogram is plotted to display the result.

Result
See (Fig.3).

References

1. Li, Y., Zhong, N.:Web Mining Model and Its Application on Information gathering. Knowledge Based Systems, vol 17, pp. 207–217 (2004).
2. Narang, Tulika., Tewari, R.R.: Multilevel Approach to Ontology Driven Clustering of Web Documents. In: International Conference on Information and Knowledge Engineering, pp. 21–25. USA (2012).
3. Kosala, Raymond., Blockeel, Hendrick.: Web Mining Research: A Survey. ACM SIGKDD, vol 2(1) (2000).
4. Kaufmann, L., Rousseeuw, P.J.: Finding Groups in Data: An Introduction to Cluster Analysis. John Wiley and Sons, USA (1999).
5. Yates, Baeza, R,.Neto, Ribeiro, B.: Modern Information Retrieval, Addison Wesley (1999).

6. Nigro, Oscar, Hector., Cisaro, Gonzalez, Sandra., Xodo, Hugo, Daniel.: Data Mining with Ontologies-Implementations, Findings, and Frameworks. IGI Global (2008).
7. Han, Jiawei., Kamber, Micheline.: Data Mining Concepts and Techniques. Morgan Kaufman, USA (2012).
8. Dunham, H, Margaret., Sridhar, S.: Data Mining–Introductory and Advanced Topics. Pearson Education, India (2006).
9. Roiger, J, Richard.,Geatz, W, Michael.: Data Mining: A Tutotrial-based Primer, Addison-Wesley (2005).
10. Gardener, Mark.: Beginning R: The Statistical Programming Language. Wrox (2013).
11. Narang, Tulika., Tewari, R.R.: SOM Based Clustering of Web documents using an ontology. International Journal of Engineering Research and Science and Technology vol2, pp. 167–174 (2013).
12. Narang, Tulika.: Hierarchical clustering of Web documents. International Journal of Innovations & Advancement in Computer science vol.4, pp. 154–159 (2015).
13. Chang, K., Liu, B.: Editorial: Special issue on web content mining, SIGKDD Explorations 6(2) (2004).

A Quality-Concordance Metric Based Contour Detection by Utilizing Composite-Cue Information and Particle Swarm Optimisation

Sandipan Choudhuri, Nibaran Das and Mita Nasipuri

Abstract Contour detection forms a significant module of computer vision frameworks, and is still an active area of research. This paper presents a feature-based edge detection strategy on color images, where the likeliness of a pixel to lie on a border separating two distinct regions is estimated by utilizing joint information obtained from two different visual cues. The first cue draws special attention to regions with presence of discontinuities and is constructed by exploiting standard deviation, busyness and entropy measures on the input image and its intrinsic map. The second cue diminishes the chances of broken edge generation by utilizing a population-based global optimisation heuristic (Particle Swarm Optimization) to detect the final edges from highlighted regions of the former cue. The result achieves noteworthy performance that is orders of magnitude better than most of the competing standard approaches, while attaining promising detection results on BSDS300 dataset.

Keywords Edge detection · Image segmentation · Busyness · Entropy · Particle swarm optimization · BSDS300

S. Choudhuri (✉) · N. Das · M. Nasipuri
Department of Computer Science & Engineering, Jadavpur University,
Kolkata 700032, West Bengal, India
e-mail: sandipanchoudhuri90@gmail.com
URL: http://www.jaduniv.edu.in

N. Das
e-mail: nibaranju@gmail.com

M. Nasipuri
e-mail: mitanasipuri@gmail.com

© Springer Nature Singapore Pte Ltd. 2017
S.C. Satapathy et al. (eds.), *Proceedings of the 5th International Conference on Frontiers in Intelligent Computing: Theory and Applications*, Advances in Intelligent Systems and Computing 515, DOI 10.1007/978-981-10-3153-3_64

641

1 Introduction

The potential to identify and localize object boundaries forms one of the rudimentary operations of image processing and computer vision domain [1]. The fact that edges can represent an image by capturing most of the shape information (thereby achieving a compact region-based interpretation of the image scene) more compactly than the image itself makes edge detection an active field of research [2].

Several contour detection strategies have been proposed till date, and are available in the literature, where each technique caters to certain categories of images. Classical methods of edge pixel identification [3] perform inefficiently when operating on noisy images, as both edge and noise possess contents of high frequency. In order to circumvent the issue of incorporating noisy pixels when exploiting neighborhood information, Gaussian-based techniques [4] have been proposed. These approaches establish a trade-off between removing noise and obtaining precision in localization of detected edge pixels, thereby compromising with accuracy. These techniques also disappoint when it comes to detection of corners. Issues like refraction of light and reduced focus steer the detection process to select plausible but false edges which do not correspond to region boundaries. Hence, issues such as elimination of fake edges, missing of true edges, noisy context, inordinate time complexity, etc. still persist even after years of dynamic research, making it justified to further explore this crucial low-level image processing task.

The outline of this paper is as follows: Sect. 2 briefly summarizes information relating to the works that inspired the presented approach. The proposed work is presented in Sect. 3, followed by results and discussions in Sect. 4. Concluding remarks and future scope are discussed in Sect. 5.

2 Related Work

Recently, analysing of color images has been in the forefront of image processing domain [5]. The benefits of color images can be easily validated citing the fact that those boundaries existing between two distinct regions of different colors but with same intensity values can be easily detected by exploiting color information. Simple yet effective legacy color edge detection method was utilized by [6] where gradient maps are computed for each color channel, followed by conglomerating information from all the gradient maps to obtain the final gradient image. Inspired by these prospects of using color images, we use the RGB data of images in our edge detection algorithm.

The edge detection discipline has witnessed a number of approaches exploiting artificial feature information from images. Ganguly et al. [7] introduced a technique where the features (*standard deviation, entropy* and *busyness*) were extracted by performing statistical calculations on image intensity values. This work was carried forward by Neupane et al. [8], where the authors proposed an iterative *k*-means clustering technique over four artificial features—*entropy, variance, busyness,* and *gradient.* While comparing the performance of their algorithms with the standard legacy edge detectors, it was noted that not only did the feature-based edge detectors perform effectively in determining edges, they were far more resistant to noise. Inspired by the approaches of exploiting artificial features, presented in [7, 8], we tackle the initial problem of highlighting regions with edges by making use of a statistical measure-based discontinuity determination technique formed with some of the mentioned features.

Many noise removal edge detection procedures do not perform satisfactorily and often end up with broken edges. Edge-linking procedures such as sequential edge linking [9] and hough transform [10] are not effective as they suffer from the absence global information of edge structures. Various evolutionary algorithms like bacteria-foraging algorithm, ant-colony optimization technique have also been put to use which yield satisfactory results, but at the cost of inordinate time complexity. In order to overcome these issues, Setayesh et al. [11] proposed a new Particle Swarm Optimization(PSO) technique modelled over a new encoding scheme and fitness function, such that the edges are realized by the best fitting collection of pixels. Inspired by a social model, PSO is used for solving optimization problems. The fact that it has a high rate of convergence and is easy to implement [12] makes it a notable candidate among the class of evolutionary algorithms. Motivated by this intuition of utilizing PSO-based heuristic to locate edges, we incorporate the method proposed in [11] in our edge detection framework.

3 Proposed Approach

We begin with a feature-based approach to identify regions with discontinuities. The initial criteria for a pixel to pose as a potential edge pixel is established by utilizing three artificial features obtained from two visual maps—the input RGB image I and the *intrinsic map IM* (a color variant image of I). The following subsections provide a detailed summary of the necessary steps involved in our approach.

3.1 Noise Reduction

Prior to identifying discontinuities, each channel in I is subjected to a noise reduction procedure by *median filtering* [13] (a nonlinear filtering technique effective in eliminating salt-and-pepper noise) followed by *Gaussian smoothing* [14] with a kernel of size 3×3 to yield a modified image I'.

3.2 Generation of Intrinsic Map

The intrinsic map *IM* is a ratio image produced by applying statistical operations over the noise-reduced image I'. Each channel of *IM* emphasizes on the disparity that each channel pair of I' harbors with respect to the combined disparities between all the channel pairs. This can be represented mathematically as: $P^{IM_1} = \frac{\left|P'_1 - P'_2\right| \times 255}{\left|P'_1 - P'_2\right| + \left|P'_2 - P'_3\right| + \left|P'_3 - P'_1\right|}$, $P^{IM_2} = \frac{\left|P'_2 - P'_3\right| \times 255}{\left|P'_1 - P'_2\right| + \left|P'_2 - P'_3\right| + \left|P'_3 - P'_1\right|}$, and $P^{IM_3} = \frac{\left|P'_3 - P'_1\right| \times 255}{\left|P'_1 - P'_2\right| + \left|P'_2 - P'_3\right| + \left|P'_3 - P'_1\right|}$, where P^{IM_1}, P^{IM_2}, and P^{IM_3} represent pixels from the first, second, and third channels of *IM*, respectively. P'_1, P'_2 and P'_3 denote pixels from the red, green, and blue channel components of I', respectively. Operations pertaining to identifying regions of interest are given in the following subsection.

3.3 Highlighting Regions of Interest

The characteristic information that a pixel represents draws association with its neighboring pixels. This connection is utilized by extracting artificial features, namely, *Standard Deviation*, *Entropy* and *Busyness* using a 3×3 kernel. These measures harbor notable properties to perceive image discontinuities.

Standard deviation (*sd*) is a typical estimate of how far a set of data is spread out from the mean, i.e., increase in standard deviation values in a region increases the likeliness of edge presence. As stated above, *sd* is calculated by utilizing a 3×3 kernel, and is computed as:

$$sd(P_c^{x,y}) = \sqrt{variance(P_c^{x,y})}, \tag{1}$$

where

$$variance(P_c^{x,y}) = \frac{1}{3^2} \left(\sum_{i,j=-1}^{1} \left(P_c^{x+i,y+j} - \mu(P_c^{x,y}) \right)^2 \right). \tag{2}$$

Here, $variance(P_c^{x,y})$ and $\mu(P_c^{x,y})$ represent the variance and mean of the pixels $P_c^{i,j}$ covered by the 3×3 kernel centered at pixel $P_c^{x,y}$ of the *c*th channel in image I', respectively, [c = {1, 2, 3}].

Entropy (*ent*) is customarily discerned as a measure for disorderedness, with a big local entropy valued pixel which has more chances to lie on an edge [15]. In our approach, the entropy of pixel values covered by 3×3 neighborhood matrix centered at pixel $P_c^{x,y}$ of the *c*th channel in image I' is evaluated as:

$$ent(P_c^{x,y}) = - \sum_{i=x-1}^{x+1} \sum_{j=y-1}^{y+1} p_{ij}^c \log p_{ij}^c, \tag{3}$$

where

$$p_{x,y}^c = \frac{P_c^{x,y}}{\sum_{i=x-1}^{x+1} \sum_{j=y-1}^{y+1} P_c^{i,j}}. \tag{4}$$

Busyness metric (*bn*) was coined in the work [16] where it was introduced as a measure of spacial dispersion. This measure can be introduced to perceive how a pixel diverges locally from its neighboring pixels. In our approach, we compute busyness by measuring Minimum Total Variations (MVT) of the gray-level differences by making use of a 3×3 kernel centered at pixel $P_c^{x,y}$, as given below

$$bn(P_c^{x,y}) = \frac{1}{12} \left(\sum_{i=x-1}^{x+1} \left(\left| P_c^{i,y-1} - P_c^{i,y} \right| + \left| P_c^{i,y+1} - P_c^{i,y} \right| \right) \right.$$

$$\left. + \sum_{j=y-1}^{y+1} \left(\left| P_c^{x-1,j} - P_c^{x,j} \right| + \left| P_c^{x+1,j} - P_c^{x,j} \right| \right) \right) \tag{5}$$

Adjacent-Pixel-Concordance: (APC) is a measure that represents the *concordance* or closeness of the gray value of a pixel *P* with one of its adjacent pixels in the direction of maximum gradient, in a 8-connected neighborhood. It lies between [0, 1], with a lower value implying presence of discontinuities. For a pixel $P_c^{x,y}$, the *APC* is computed by employing the following equation:

$$APC(P_c^{x,y}) = \frac{255 - \max_{\substack{-1 \leq i \leq 1 \\ -1 \leq j \leq 1 \\ i \neq 0, j \neq 0}} \left| P_c^{x,y} - P_c^{x+i,y+j} \right|}{255} \tag{6}$$

As *APC* is inversely proportional to the intensity differences between two adjacent pixels. Therefore, *probability of presence of discontinuity at* $P_c^{x,y} \propto \frac{1}{APC(P_c^{x,y})}$.

Adjacent-Pixel-Concordance Threshold: As it can be conjectured from the presented theory in this section, *probability of presence of discontinuity at* $P_c^{x,y} \propto$ $sd(P_c^{x,y})$, $ent(P_c^{x,y})$, and $bn(P_c^{x,y})$. The *adjacent-pixel-concordance threshold* (*APCT*) for a pixel $P_c^{x,y}$ of the *c*th channel in image I' estimates its likeliness to become an edge pixel, and therefore is modelled in agreement with this probability. This is framed as

$$APCT(P_c^{x,y}) = k \times \frac{sd(P_c^{x,y}) \times ent(P_c^{x,y}) \times bn(P_c^{x,y})}{\max_{\substack{0 \leq i \leq I.rows \\ 0 \leq j \leq I.cols}} \left\{ sd(P_c^{i,j}) \times ent(P_c^{i,j}) \times bn(P_c^{i,j}) \right\}} \tag{7}$$

In the equation above, the denominator represents maximum value of the product of three features in channel I_c. The constant *k* is used as a tuning parameter and ranges between [0, 255].

In our approach, this measure is exploited along with APC to figure out edge locations. The underlying intuition behind this technique originates from the fact that an increase in the intensity difference between two adjacent pixels will decrease the APC value, while increasing the $APCT$ measure. When represented graphically, the two lines representing these measures will intersect. The parameter k is employed to ensure that these two lines intersect at points resulting in maximization of edge pixels and minimization of outlier pixels.

3.4 Determination of Regions of Interest from Two Visual Maps

The regions of interest map is constructed by convolving each channel of a color image with Gaussian directional derivatives in eight orientations ranging between $[0°, 180°)$. Only those pixels P where the inequality $APC(P) < APCT(P)$ holds are assigned edge strengths given by the maximum response over eight orientations. Extracting derivatives in this form disregards the smooth variations that might lead to spurious edges. In order to circumvent the difficulty of estimating the precise value of k for an image, we repeat the process for multiple values of k followed by averaging the intermediate maps, thus generating a fuzzy contour map.

Motivated by the work presented by Ren [17], in order to perceive both fine and coarse details we consider a multiscale variant of our algorithm at three scales [original (σ), half $(\frac{\sigma}{2})$ and double resolution (2σ) version] for each channel of the two visual maps (I' and IM), followed by resizing them to original dimensions (σ) and then averaging the results to form two region of interest maps $RM^{I'}$ and RM^{IM} from images I' and IM, respectively. A non-maximum suppression step is conducted prior to producing the two maps to create thin contours.

In order to incorporate both groups of salient information represented by $RM^{I'}$ and RM^{IM}, we propose a linear weighted combination of the two to produce our first visual cue, i.e., the final regions of interest map RM, where

$$RM = \alpha \, RM^{I'} + \beta \, RM^{IM}. \tag{8}$$

The weights α and β are estimated by employing gradient ascent technique on the F-measure using training images of BSDS300 dataset.

3.5 Edge Detection Using Particle Swarm Optimization

One of the most rudimentary objectives associated with edge detection is that of identifying subtle structured similarities between groups of edge pixels, thereby reducing the side effects of broken and jagged edges. In order to mitigate such

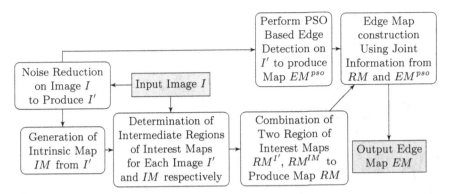

Fig. 1 Schematic flow diagram of the proposed framework

effects, we follow the work presented by Setayesh et al. [11], where a new encoding scheme for a particle is proposed as a collection of pixel orientations, for PSO-based algorithm. This technique operates on a fitness function that ensures to obtain the best fitting curve as a collection of pixels with their curvature minimized. We apply this PSO-based algorithm to the noise suppressed image I', producing our second visual cue EM^{pso}.

3.6 Generation of Edge Map Using Joint Information

To circumvent the task of eliminating spurious edges from both cues RM and EM^{pso}, the final output (EM) of our algorithm is produced by exploiting common information from two cues (RM and EM^{pso}). This is represented as

$$EM = RM \cap EM^{pso} \tag{9}$$

Figure 1 exhibits the flowchart detailing of our contour detection methodology.

4 Results

To test the performance of our contour detection algorithm, we make use of the popular Berkeley Segmentation Dataset and Benchmark (BSDS300) [18] and compare our edge detector against competing methods.

The experiment was performed for five values of k [5, 25, 40, 70, 120]. For PSO-based contour detection module, the population size was set to 50 with maximum of 200 iterations. Based on the initial set of experiments conducted on training images, the minimum contour length was set to 23. In each run of the algorithm, 4×4 pixels

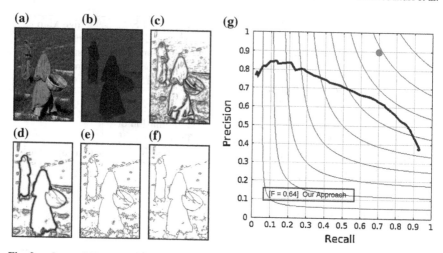

Fig. 2 **a** Image I (from BSDS500 dataset), **b** intrinsic map IM, **c, d** represent regions of interest maps $RM^{I'}$ and RM^{IM}, respectively, before non-maximal supression, **e** regions of interest map RM based on joint information, **f** final edge map produced by our algorithm, **g** precision-recall curve for our edge detector of BSDS300

Table 1 Results of eight different methods on BSDS300

Method	Average precision (AP)
gPb [19]	0.68
Our approach	0.64
Mean shift [21]	0.54
Normalized cuts [22]	0.43
EDContours [20]	0.62
Felz-Hutt [23]	0.53
Canny	0.58
Setayesh et al. [11]	0.61

were evaluated. The weight factors to calculate the inter-set distance (between two regions), intra-set distance (within a region), and curvature cost were set to 85, 36, and 36, respectively. The user-defined threshold to check the penalized fitness value (penalisation based on the number of pixels on the curve that should be marked as edges) was set to 0.6.

Figure 2g plots the overall precision-recall curve for our edge detector. As an evaluation measure, we use the average precision(AP) (equivalent to area under the precision-recall curve) in our experiment. Table 1 shows the AP scores for eight different techniques. Although our approach bags the second position with an AP score of 0.64, following gPb [19], it is computationally more efficient ($\frac{1}{45}$ FPS) as compared to the gPb contour detector ($\frac{1}{240}$ FPS). Figure 3 compares our edge detection results with that of gPb [19] and EDContours [20].

Fig. 3 Edge detection results on the BSDS300 dataset—*column 1* original images, *column 2* ground truth, *column 3* thresholded gPb edge maps [19], *column 4* results for EDContours [20], *column 5* contours generated by our edge detector

5 Conclusion

Here, we present a contour detection algorithm that efficiently captures object boundaries and reduces broken edges. It operates on different scale-space implementations of an image and exploits information from channels of two different images, jointly with that obtained from an evolutionary algorithm-based edge detection technique. The results presented are preliminary, but adequate enough to emphasize on the importance of multiple cue information when detecting edges. As with the future work, we plan to incorporate two different features in our proposed framework: a texture channel providing local textural regularity-based information for gathering coherent statistics within objects, followed by representing region information in terms of correlated superpixels to capture image redundancy, thereby yielding more semantic edge maps.

References

1. Appia, Vikram, and Anthony Yezzi. "Active geodesics: Region-based active contour segmentation with a global edge-based constraint." In Computer Vision (ICCV), 2011 IEEE International Conference On, pp. 1975–1980. IEEE, 2011.
2. Shaikh, Masood. "A Comparative Study of Color Edge Detection Techniques." Image 10: 5.
3. Gonzalez, Rafael C. Digital image processing. Pearson Education India, 2009.
4. Basu, Mitra. "Gaussian-based edge-detection methods-a survey." IEEE Transactions on Systems, Man, and Cybernetics, Part C 32, no. 3 (2002): 252–260.
5. Evans, Carolyn J., Stephen J. Sangwine, and Todd A. Ell. "Colour-sensitive edge detection using hypercomplex filters." In Signal Processing Conference, 2000 10th European, pp. 1–4. IEEE, 2000.
6. Fan, Jianping, Walid G. Aref, Mohand-Said Hacid, and Ahmed K. Elmagarmid. "An improved automatic isotropic color edge detection technique." Pattern Recognition Letters 22, no. 13 (2001): 1419–1429.
7. Ganguly, Debashis, Swarnendu Mukherjee, Kheyali Mitra, and Partha Mukherjee. "A novel approach for edge detection of images." In Computer and Automation Engineering, 2009. ICCAE'09. International Conference on, pp. 49–53. IEEE, 2009.
8. Neupane, Bijay, Zeyar Aung, and Wei Lee Woon. "A new image edge detection method using quality-based clustering." In Proceedings of the 10th IASTED International Conference on Visualization, Imaging, and Image Processing, pp. 20–26. 2012.
9. Farag, Aly A., and Edward J. Delp. "Edge linking by sequential search." Pattern Recognition 28, no. 5 (1995): 611–633.
10. Parker, Jim R. Algorithms for image processing and computer vision. John Wiley & Sons, 2010.
11. Setayesh, Mahdi, Mengjie Zhang, and Mark Johnston. "A novel particle swarm optimisation approach to detecting continuous, thin and smooth edges in noisy images." Information Sciences 246 (2013): 28–51.
12. AlRashidi, Mohammed R., and Mohamed E. El-Hawary. "A survey of particle swarm optimization applications in electric power systems." Evolutionary Computation, IEEE Transactions on 13, no. 4 (2009): 913–918.
13. Brownrigg, D. R. K. "The weighted median filter." Communications of the ACM 27, no. 8 (1984): 807–818.
14. Lee, Jong-Sen. "Digital image smoothing and the sigma filter." Computer Vision, Graphics, and Image Processing 24, no. 2 (1983): 255–269.
15. Dai, Wenzhan, and Kangtai Wang. "An image edge detection algorithm based on local entropy." In Integration Technology, 2007. ICIT'07. IEEE International Conference on, pp. 418–420. IEEE, 2007.
16. Dondes, P. A. and Rosenfeld, A. (1982). Pixel classification based on gray level and local 'busyness'. Pattern Analysis and Machine Intelligence, IEEE Transactions on, (1):79–84.
17. Ren, Xiaofeng. "Multi-scale improves boundary detection in natural images." In Computer Vision–ECCV 2008, pp. 533–545. Springer Berlin Heidelberg, 2008.
18. Martin, David, Charless Fowlkes, Doron Tal, and Jitendra Malik. "A database of human segmented natural images and its application to evaluating segmentation algorithms and measuring ecological statistics." In Computer Vision, 2001. ICCV 2001. Proceedings. Eighth IEEE International Conference on, vol. 2, pp. 416–423. IEEE, 2001.
19. Arbelaez, Pablo, Michael Maire, Charless Fowlkes, and Jitendra Malik. "Contour detection and hierarchical image segmentation." Pattern Analysis and Machine Intelligence, IEEE Transactions on 33, no. 5 (2011): 898–916.
20. Akinlar, Cuneyt, and Cihan Topal. "EDContours: High-speed parameter-free contour detector using EDPF." In Multimedia (ISM), 2012 IEEE International Symposium on, pp. 153–156. IEEE, 2012.

21. Comaniciu, Dorin, and Peter Meer. "Mean shift: A robust approach toward feature space analysis." Pattern Analysis and Machine Intelligence, IEEE Transactions on 24, no. 5 (2002): 603–619.
22. T. Cour, F. Benezit, and J. Shi, Spectral segmentation with multiscale graph decomposition. CVPR, 2005.
23. Felzenszwalb, Pedro F., and Daniel P. Huttenlocher. "Efficient graph-based image segmentation." International Journal of Computer Vision 59, no. 2 (2004): 167–181.

Analysis of Pancreas Histological Images for Glucose Intolerance Identification Using Wavelet Decomposition

Tathagata Bandyopadhyay, Sreetama Mitra, Shyamali Mitra, Luis Miguel Rato and Nibaran Das

Abstract Subtle structural differences can be observed in the islets of Langerhans region of microscopic image of pancreas cell of the rats having normal glucose tolerance and the rats having pre-diabetic (glucose intolerant) situations. This paper proposes a way to automatically segment the islets of Langerhans region from the histological image of rat's pancreas cell and on the basis of some morphological feature extracted from the segmented region the images are classified as normal and pre-diabetic. The experiment is done on a set of 134 images of which 56 are of normal type and the rests 78 are of pre-diabetic type. The work has two stages: primarily, segmentation of the region of interest (roi), i.e., islets of Langerhans from the pancreatic cell and secondly, the extraction of the morphological features from the region of interest for classification. Wavelet analysis and connected component analysis method have been used for automatic segmentation of the images. A few classifiers like OneRule, Naïve Bayes, MLP, J48 Tree, SVM, etc, are used for evaluation among which MLP performed the best.

T. Bandyopadhyay (✉) · S. Mitra
School of Computer Engineering, KIIT University, Bhubaneswar, India
e-mail: tathagatabanerjee15@rocketmail.com

S. Mitra
e-mail: msreetama10@gmail.com

S. Mitra
Department of Electronics and Telecommunication Engineering, Jadavpur University, Kolkata, India
e-mail: shyamali.mitraa@gmail.com

L.M. Rato
University of Evora, Évora, Portugal
e-mail: lmr@di.uevora.pt

N. Das
Department of Computer Science and Engineering, Jadavpur University, Kolkata, India
e-mail: nibaran@gmail.com

© Springer Nature Singapore Pte Ltd. 2017
S.C. Satapathy et al. (eds.), *Proceedings of the 5th International Conference on Frontiers in Intelligent Computing: Theory and Applications*, Advances in Intelligent Systems and Computing 515, DOI 10.1007/978-981-10-3153-3_65

653

Keywords Automatic segmentation · Pancreas cell · Morphological feature · Wavelet analysis · Connected component · Feature extraction · Classification

1 Introduction

The challenge related to conventional clinical wisdom and expertise level has paved the way to develop automatic computer-aided diagnostic system. The broad goal of automatic computerized diagnostic system is to improve patient care by moving from an experienced based form of clinical practice to one informed by computer-aided systematic application of medical knowledge. Moreover, it involves the risk of the pathologist to be infected with some contagious diseases while testing the samples. In the present interest, it is being observed that diabetic prone rats develop a cell mediated auto immune destruction of β-cells in pancreas resulting in an abrupt onset of diabetes Maleates. The symptoms are generally observed between 60 and 120 days of age in both males and females. The onset of disease is characterized by glycosuria, hyper glycaemia, weight loss, ketosis, etc. Detailed analysis shows that DP rats develop insulin that destroys only the pancreatic β-cells resulting in an islet devoid of insulin positive cells.

To address the challenges as mentioned above, an approach of automatic segmentation of images of pancreatic cells of Wister rats using wavelet analysis has been developed in the present paper. The process is followed by connected component analysis to automatically segment the islets of Langerhans region from the histological images of rats' pancreas cell in order to calculate some morphological features to classify the images as glucose tolerant (normal) or glucose intolerant (pre-diabetic) [1]. The segmentation of the image is done through 2-D-DWT decomposition using "*haar*" wavelet. The cooccurrence matrices are calculated for the original subimage block under three detailed coefficients are obtained for each subimage block.

This paper is categorically divided into six sections. In Sect. 2, the usage of wavelet transform strictly in relation to the work is briefly discussed in a nutshell. Section 3 gives a brief description of the color enhancement technique used in the present work. Connected component analysis techniques along with the masking out the original image are briefly mentioned in the Sect. 4. Section 5 covers the feature extraction and classification of the images. Finally in Sect. 6, results and corresponding detailed analysis are given in tabular form. It is worthy to mention here that a real data set of 134 images (56 from rats with normal glucose tolerance and 78 from pre-diabetic ones) used in [1], are also used here. Few samples of the dataset are shown in Fig. 1.

A convenient flow chart of developed methodology is given in Fig. 2 for better understanding of the system.

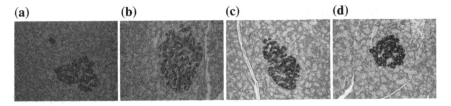

(a)　　　　**(b)**　　　　**(c)**　　　　**(d)**

Fig. 1 Histological sample images of **a–b** Normal **c–d** Pre-diabetic pancreas cell. The brown color regions denotes the beta cells

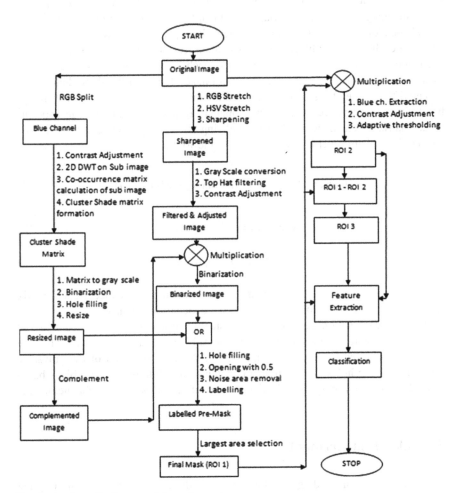

Fig. 2 A schematic diagram of the entire system

2   Wavelet Analysis

As mentioned before that wavelet analysis is applied first to segment out the β-cell region (the region with reddish brown shade). To do that first, the input images are split into RGB channels. The blue channel is used for processing the image due to having priority of blue colors. Consecutive 32×32 subimage blocks [2] of the extracted blue channel, starting from the top left corner are used for single level 2-D-DWT (discrete wavelet transform) decomposition using *"haar"* [3] wavelet. We have used the *"haar"* wavelet due to its simplicity, low computational cost and having power to identify the potential differences between two adjacent pixels. Each sub image blocks are overlapping but differs by one row or column in the spatial location from its adjacent subimage block [2]. By 2-D wavelet decomposition on each subimage block one approximate (cA) and three detail (cH, cV, cD) coefficients are obtained corresponding to each block. Now, the cooccurrence matrices are calculated for the original sub image block and the three detail coefficients [2] using eight gray level approximation. That means for each sub image block four cooccurrence matrices are obtained, from which one average cooccurrence matrix is calculated. After normalizing the average cooccurrence matrix, cluster shade feature [2] is calculated using the formula:

$$\text{Cluster shade} = \sum_{i,j=1}^{N} (i - Mx + j - My)^3 C(i,j) \tag{1}$$

$$\text{where } M_x = \sum_{i,j=1}^{N} iC(i,j) \tag{2}$$

$$\text{and } M_y = \sum_{i,j=1}^{N} jC(i,j) \tag{3}$$

for this experiment $N = 8$ (8 gray level approximation for cooccurrence matrix calculation) and C is the normalized average cooccurrence matrix.

Thus for each subimage block, one cluster shade feature value is obtained and storing all the feature values for all the subimage blocks a matrix is formed. Then the matrix is converted to grayscale image. After binary conversion of the grayscale image and filling holes the β-cell area is obtained as white region on a black background. This binary image is resized to the size of the extracted blue channel image.

3   Color Enhancement

As shown in the block diagram (Fig. 1), RGB stretching is done on the original image followed by HSV stretching to enhance the contrast of color. Then the obtained image is sharpened. These operations are done to make the Islets of Langerhans area surrounding the β-cell region more conspicuous. After converting the sharpened colored image to a grayscale image top hat filtering [4] is done to subtract an approximate background from the image. After contrast adjustment the filtered image is masked with the complement of the resized binary image obtained in wavelet analysis. Now the resulting image is converted to binary. Then OR operation is performed on this binarized image with theresized image obtained in the wavelet analysis stage. Now hole-filling is done on the output image. Next the filled image is opened with square of width 5 to disconnect the thin connections and regions having area less than 50 pixels are considered as noise and are removed. The final output image obtained in this stage is named as "pre_mask."

4 Connected Component Analysis

All the connected components in the "pre_mask" edimage are labeled and area of each region is calculated. Then, the connected region having maximum area is kept and others are discarded. Image, thus obtained, is named as "final_mask." This is also termed as roi1.

4.1 Masking Original Image

Original image is masked with the "final_mask" to segment out the β-cell region along with the surrounding islets of Langerhans region. Now blue channel is extracted from this masked image and then the contrast adjustment is done. Then on resulting image adaptive thresholding [5] (median method with 32 window size) is performed to separate out only the β-cell region from the mixture of β-cell and islets of Langerhans. Resulting image is named as "roi2." Another region, "roi3" is calculated by subtracting "roi2" from "roi1".

5 Feature Extraction and Classification

Here, we have used three morphological features for classification of the image, (1) the ratio between the area of the β-cells and the area of islets of Langerhans, (2) ratio of the area of "roi2" to "roi1", (3) the perimeter per area ratio of islets of Langerhans. First one is calculated by taking the ratio of area of "roi3" and "roi1." Second one is calculated by taking the ratio of area of "roi2" and "roi1." To calculate the third feature first, the boundary points of "roi1" is obtained. Then perimeter is calculated by taking the sum of the distances of two successive boundary points. Now this perimeter value is divided by the area of the same region to obtain the perimeter per area (PPA) ratio. Thus, three features are calculated for all the images. It is worthy to mention here that all the image processing modules have been developed using MATLAB.

The obtained extracted features are then used for training and testing using different standard classifiers. To do that, Weka 3.6, a well-known open source software is used. Five different classifiers such as (a) OneRule, (b) Naïve Bayes, (c) J48Tree, (d) Multilayer Perceptron (MLP), and (e) SVM [6] are used here to evaluate the performance with 10-fold cross validation.

6 Results

As mentioned before, the entire work is divided into parts: (1) automatic segmentation of Islets of Langerhans and (2) classification of image as normal or pre-diabetic image. It is worthy to mention here, in the paper [1], Ratio et al. used the manual segmentation to achieve their results. A picture of automatic segmentation has been shown in Figs. 3 and 4. From the images, it can be said that the

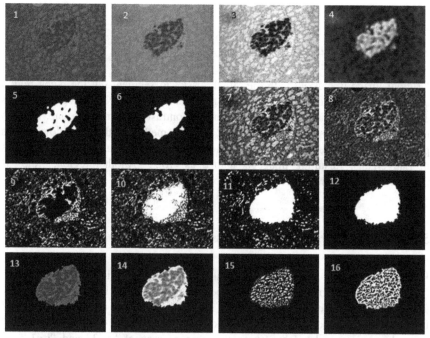

1.Original image, 2. Blue channel, 3. Contrast adjustment of 2, 4. Gray scale of Cluster Shade matrix, 5. Binarization of 4, 6. Hole filling of 5, 7. RGB & HSV stretched sharpened image of 1, 8. Top hat filtered and contrast adjusted image of gray scale of 7, 9. B-cell portion masked out, 10. After OR operation between 6 and 9, 11. Labelled pre-mask (Hole filling, opening and labelling of 10) , 12. Final Mask (ROI1), 13. Masked original image, 14. Contrast adjusted blue channel of 13, 15. ROI2, 16. ROI3

Fig. 3 Intermediate images

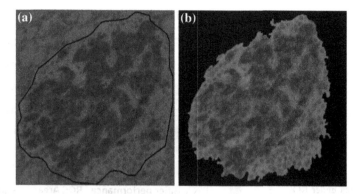

Fig. 4 a Manual segmentation of Islets of Langerhans by the method [1] **b** Automatic segmentation of Islets of Langerhans from the same image using our proposed method

Table 1 Performance of different classifiers using 10 fold cross validation techniques

Classifier	Correctly classified (in percent %)	TP rate	FP rate	Precession	Recall	F-measure	ROC area
OneR	70.8955	0.709	0.305	0.71	0.709	0.709	0.702
J48 tree	78.3582	0.784	0.211	0.79	0.784	0.785	0.827
SVM	82.0896	0.821	0.169	0.829	0.821	0.822	0.826
MLP	84.3284	0.843	0.133	0.861	0.843	0.844	0.866
Naive Bayes	71.6418	0.716	0.319	0.714	0.716	0.712	0.837

results are quite satisfactory when compared to the manual segmentation. Results of classification are shown in Table 1. From Table 1, it is observed that maximum recognition accuracy observed for is 84.3284% using MLP classifiers. It can also be observed that all the classifiers provide recognition accuracy above 70%. Furthermore, most of the classifiers used here (except One Rule) give ROC area higher than 0.8. The results indicate that the proposed automatic segmentation and classification algorithm can diagnose diabetic condition accurately in most of the cases. However, accuracy can vary depending on the classifiers and their specification selected. Comparative chart of using different classifiers is shown in Figs. 5 and 6.

Fig. 5 Accuracy (%) of different classifiers

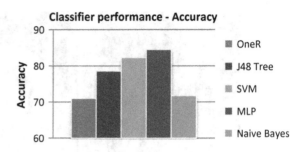

Fig. 6 Area under ROC curve for different classifiers

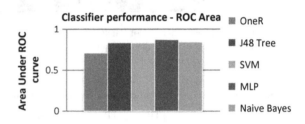

7 Conclusion

The proposed segmentation model is capable enough to segment out the β-cell regions along with the surrounding Islets of Langerhans (which is more challenging to segment) without human intervention. The classification results can also be considered satisfactory. Moreover, all the classifiers (except "one rule") used here gives ROC area higher than 0.8. That means the proposed automatic segmentation and classification algorithm can diagnose diabetic condition accurately in most of the cases. However, accuracy can vary depending on the classifiers and their specification selected. Further improvement in the results can be made by considering more robust features. Our algorithm cannot segment out the smaller β-cell areas in the images having multiple β-cell regions. Moreover, the proposed algorithm is slow as it uses wavelet analysis in a particular sub region. So, for real-time implementation of this method some improvement should be done in the algorithm to achieve faster processing and to address multiple β-cell regions.

Acknowledgements The authors thank Professor Fernando Capela e Silva, from the Department of Biology and Ana R. Costa and Célia M. Antunes, from the Department of Chemistry, University of Évora, Portugal, for the data set used in this article.

References

1. L. M. Rato, F. C. e Silva, A. R. Costa, and C. M. Antunes, "Analysis of pancreas histological images for glucose intolerance identification using imagej-preliminary results," in *4th Eccomas Thematic Conference on Computational Vision and Medical Image Processing (VipIMAGE)*, 2013, pp. 319–322.
2. S. Arivazhagan and L. Ganesan, "Texture segmentation using wavelet transform," *Pattern Recognition Letters*, vol. 24, no. 16, pp. 3197–3203, 2003.
3. A. Gavlasov´a, A. and Proch´azka, and M. Mudrov, "Wavelet based image segmentation," in *Proceedings of the 14th Annual Conference Techincal Computing*, 2006, pp. 1–7.
4. "http://in.mathworks.com", (last accessed 15th June 2016).
5. "http://homepages.inf.ed.ac.uk/rbf/HIPR2/adpthrsh.htm.", (last accessed 15th June 2016).
6. B. Nunes, L. Rato, F. Silva, A. Rafael, and A. Cabrita, "Processing and classification of biological images," in *Technology and Medical Sciences*, CRC Press, 2011, pp. 233–237.

A Systematic Review on Materialized View Selection

Anjana Gosain and Kavita Sachdeva

Abstract The purpose of materialized view selection is to minimize the cost of answering queries and fast query response time for timely access to information and decision support. Besides various research issues related to data warehouse evolution, materialized view selection is one of the most challenging ones. Various authors have given different methodologies, strategies and followed algorithms to solve this problem in an efficient manner. The main motivation behind this systematic review is to provide a path for future research scope in materialized view selection. Various techniques presented in the papers are identified, evaluated, and compared in terms of memory storage space, cost, and query processing time to find if any particular approach is superior to others. By means of a review of the available literature, the authors have drawn several conclusions about the status quo of materialized view selection and a future outlook is predicted on bridging the large gaps that were found in the existing methods.

Keywords Data warehouse · Data warehouse evolution · View materialization · View selection · Business intelligence · Query optimization · Caching

1 Introduction

Materialized view selection aims at speeding up execution of queries, ranging from traditional query processing, online analytical processing, data mining, business intelligence, decision support, and web database caching [1–3]. It is a crucial step in data warehouse evolution [4]. Therefore, great attention is paid toward materialized view selection by precomputing and selecting a set of materialized views under

A. Gosain · K. Sachdeva (✉)
University School of Information and Communication Technology,
Guru Gobind Singh Indraprastha University, New Delhi 110078, India
e-mail: kavitasachdeva4@gmail.com

A. Gosain
e-mail: anjana_gosain@hotmail.com

© Springer Nature Singapore Pte Ltd. 2017
S.C. Satapathy et al. (eds.), *Proceedings of the 5th International Conference on Frontiers in Intelligent Computing: Theory and Applications*, Advances in Intelligent Systems and Computing 515, DOI 10.1007/978-981-10-3153-3_66

specific resource constraints, such as disk space and maintenance time, so as to minimize the total query processing cost. A few of these techniques have been surveyed, evaluated, and compared by many researchers [5–18] but, a generalized conclusion has not been drawn by any of them.

During the course of research, preliminary literature survey indicated that, no systematic review has been published so far in this field. This paper presents a systematic literature review of all the existing materialized view selection techniques presented till date, along with their classification and comparison, based on some common parameters. It is an attempt to review the amount of efforts already been put into the field of view materialization and to come up with a base for the advancement of future work. To achieve the same, 890 papers were located, out of which a qualitative analysis of 58 [10–12, 15, 18–71] relevant papers was performed by comparing them with respect to the various measures like classification technique used, subclassification method used, framework used, constraints addressed, etc. Research methodology adopted for this review study is described by defining the research questions, data collection, its evaluation, summarization, and concluding results on the basis of observations made. This has been discussed in detail in further sections and subsections.

2 Related Work

Many authors in [5–7, 13, 14, 16] have given survey and review papers on materialized view selection. However, no study has been done so far to present a systematic literature review on materialized view selection. In [5], a critical survey of the methodologies to select materialized view is presented. According to the authors, advanced solutions to solve the view selection problem focused on evolutionary optimization methods. In [6], researchers discussed the approaches for selection of views in static (where the query workload is known in advance) as well as dynamic setting (where the query workload changes over time) and have classified view selection techniques based on configuration evolution, resource constraints, optimization goals, candidate views, and candidate interrelationships. Authors in [7] have categorized the problem of answering queries using views into cost-based rewriting (i.e., query optimization and maintenance of physical data independence) and logical rewritings (i.e., in the context of data integration). A comparison of traditional query optimizer with one that exploits materialized views is best explained in this work, with the help of real-life examples. A comprehensive and comparative study of different approaches based upon various parameters is presented in a tabular manner, in [13]. Furthermore, a very critical survey on Materialized View Selection was presented by [16]. It reported a detailed summary of the current state of art and identified the main dimensions that are required for the classification of view selection methods in relational databases and for distributed setting. It was not a systematic literature review. Therefore, various classification methods used, subclassification methods used, framework/data structure used for representation and constraints addressed, etc., have been reported, analyzed, and reviewed in our work.

3 Review Questions

This article reviews various techniques for materialized view selection and tries to answer the following research questions:

RQ1. What are the empirical evidences to support the usage of methods for selection of materialized views?

RQ2. Frameworks used for representing views and the assessment of tradeoffs among the constraints defined for materialized view selection.

RQ3. Summarized effects of the comparison with respect to the approach used and publication trends.

4 Classification of Materialized View Selection Techniques

To better study the progress of research in materialized view selection, we have classified the techniques into three broad categories, as defined in Fig. 1. This classification strategy is inspired from the previous work by Mami and Bellahsene [16, 63]. It is further subclassified, as per the approach followed for view selection, by various researchers.

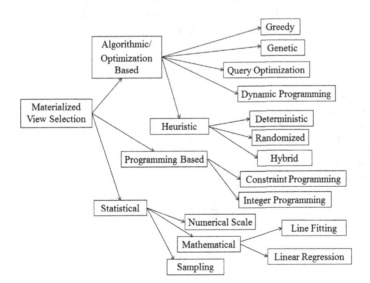

Fig. 1 Classification of materialized view selection

5 Observations and Results

After studying various articles for materialized view selection in our review work, we have made various observations regarding the methods and techniques used. These observations are represented graphically using bar graphs and pie chart as discussed below. It tries to answer the questions defined in Sect. 3.

6 Conclusion

A systematic review on materialized view selection is presented in this study which interprets and evaluates all the research work related to the area. It presents a brief summary of the best available evidences. This research study identified over 58 articles, since 1996 and observed that the optimization and algorithmic based, methods reserve highest level of contribution in materialized view selection (Fig. 2). This is further justified by the use of the majority of heuristic and hybrid techniques as observed by Fig. 3. Though at most only 10 techniques were found and compared in this study, the results obtained provided useful insights into the MVS (Materialized View selection) field. Heuristic techniques were found to be better than the other techniques (Fig. 3). Despite the rich body of research and practice of using various frameworks and data structures for selecting materialized views, Graph or lattice-based framework can be used to address majority of the issues (Fig. 4). This review work answers the research problems identified and provides an informational evidence, by the publication trends observed, that it is an open area for research, by which researchers can extend the research on materialized view selection further to bring about a new method and better solution to solve the materialized view selection problem (Figs. 5, 6 and 7).

Fig. 2 Number of articles reviewed and categorized as per classification technique; OB-Optimization based, AB-Algorithmic based, PB-Programming based, S-Statistical, O-Others

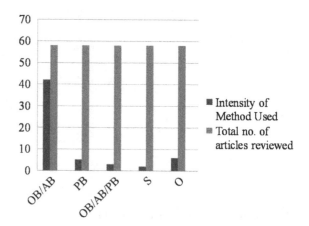

Fig. 3 Percentage of Articles reviewed according to subclass of methods

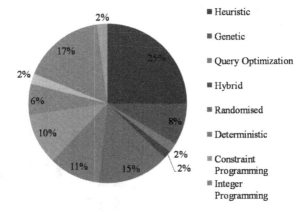

Fig. 4 Percentage of articles using a particular framework

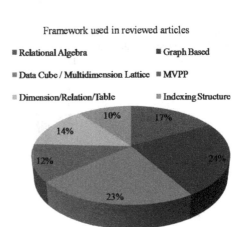

Fig. 5 Number of articles addressing the respective constraints

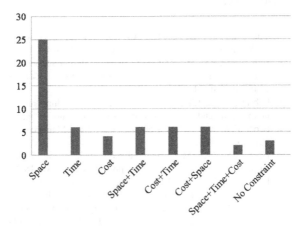

Fig. 6 Publication Trends of articles reviewed according to year of publication

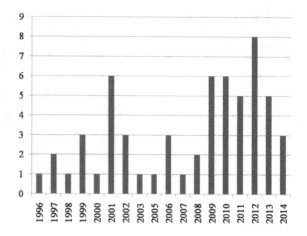

Fig. 7 Publication Trends of articles reviewed according to source

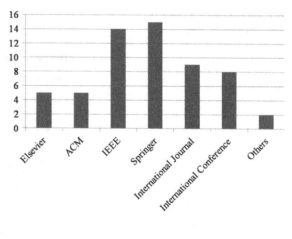

References

1. Chaudhuri, S., Dayal, U.: An overview of data warehousing and OLAP technology, ACM SIGMOD Record, vol. 26, no. 1, pp. 65–74, Março (1997).
2. Lim, E.P., Chen, H., Chen, G.: Business intelligence and analytics: Research directions. ACM Transactions on Management Information System 3(4), Article 17 DOI:http://dx.doi.org/10. 1145/2407740.2407741 (2013).
3. Labrinidis, A., Luo, Q., Xu, J., Xue, W.: Caching and materialization for web databases, Foundations and Trends in Databases, vol. 2, no. 3, pp. 169–266 (2009).
4. Ferri, Fernando, Datawarehouse Evolution Management, IRPPS-CNR, Italy. fernando.-ferri@irpps.cnr.it.
5. Dhote, C.A., Ali, M.S.: Materialized view selection in Data warehouse: A survey, Journal of Applied Sciences, Vol 9, No. 3, pp. 401–414, ISSN 1812-5654 (2009).
6. Li, Xiang: Materialized View Selection: A Survey, Informatik 5, RWTH Aachen University, doi:10.4018/978-1-60566-816-1.ch005. lixiang@dbis.rwth-aachen.de.
7. Halevy, A.Y.: Answering queries using views: A survey, The VLDB Journal 10: 270–294, 2001/ Digital Object Identifier (DOI) 10.1007/s007780100054 (2001).

8. Gupta, H.: Selections of Views to materialize in a Data warehouse, International Conference On Database Theory, Delphi, Greece, Pg. 98–112 (1997).

9. Ashadevi, B., Navaneetham, P., Balasubramanian, R.: A Framework for the View Selection Problem in Data Warehousing Environment, International Journal on Computer Science and Engineering, Vol. 02, No. 09, pp. 2820–2826 (2010).

10. Mistry, H., Roy, P., Sudarshan, S., Ramamritham. K.: Materialized view selection and maintenance using multi-query optimization, In: SIGMOD Conference, pp. 307–318, (2001).

11. Yu, J.X., Yao, X., Choi, C.H., Gou, G.: Materialized view selection as constrained evolutionary optimization, IEEE Transactions on Systems, Man, and Cybernetics, Part C 33 (4), 458–467 (2003).

12. Aouiche, K., Darmont, J.: Data mining-based materialized view and index selection in data warehouses, Journal of Intelligent Information Systems, Springer, (2009).

13. Jain, H., Gosain, A.: A Comprehensive Study of View Maintenance Approaches in Data Warehousing Evolution, ACM SIGSOFT Software Engineering Notes, Vol. 37, No. 5 (2012).

14. Zi-Yu, L., Dong-Qing, Y., Teng-Jiao, W., Guo-Jie, S.: Research on Materialized View Selection, Journal of software, vol. 20, pp. 193–213 (2009).

15. Mami, I., Bellahsene, Z., Coletta, R.: A Declarative Approach to View Selection Modeling, Springer-Verlag Berlin Heidelberg (2013).

16. Mami, I., Bellahsene, Z.: A Survey of View Selection Methods, SIGMOD Record, Vol. 41, No. 1 (2012).

17. Agrawal, S., Chaudhuri, S., Narasayya, V.: Automated Selection of Materialized Views and Indexes for SQL Databases, Proceedings of the 26th International Conference on Very Large Databases, Cairo, Egypt (2000).

18. Kalnis, P., Mamoulis, N., Papadias, D.: View selection using randomized search, Data & Knowledge Engineering, Elsevier (2002).

19. Hylock, R., Currim, F.,: A maintenance centric approach to the view selection problem, Information Sciences, Elsevier (2013).

20. Bhan, M., Kumar, S., Rajanikanth, K.: Materialized view size estimation using sampling, IEEE International Conference on Computational Intelligence and Computing Research (2013).

21. Harinarayan, V., Rajaraman, A., Ullman, J.: Implementing data cubes efficiently, ACM SIGMOD, pp:205–216 (1996).

22. Yang, J.H., Chung, I.J.: ASVMRT: Materialized View Selection Algorithm in Data Warehouse, International Journal of Information Processing Systems, Vol. 2, No. 2, pp. 67–75 (2006).

23. Theodorates, D., Sellis, T.: Datawarehouse Configuration, Proceedings of the 23rd International Conference on very Large Databases, Morgan Kaufmann Publishers Inc., San Francisco, CA,USA, pp. 126–135 (1997).

24. Shukla, A., Deshpande, J.F., Naughton, J.F.,: Materialized View Selection for Multidimensional Datasets, Proceedings of the 24th Very Large Databases Conference, (1998).

25. Theodorates, D., Sellis T.,: Designing Datawarehouses, Data and Knowledge Engineering, 31: 279–301 (1999).

26. Agrawal, S., Chaudhari, S., Narasayya, V.: Automated selection of materialized views and indexes for SQL databases, In Proceedings of 26th International Conference on Very Large Databases (2000).

27. Theodorates, D., Dalamagas, T., Simitsis, A., Stavropoulos, M.: A randomized approach for the incremental design of an evolving data warehouse. Proceedings of the Twentieth International conference on Conceptual Modeling, pp:325–338 (2001).

28. Goldstein, J., Larson, P.A.: Optimizing queries using materialized views: A practical, scalable solution, ACM SIGMOD, pp:328–339 (2001).

29. Liang, W., Wang, H., Orlowska, M.E.,: Materialized View Selection under the maintenance time constraint, Data and Knowledge Engineering, Elsevier (2001).

30. Lee, M., Hammer, J.,: Speeding up materialized view selection in data warehouses using a randomized algorithm, Int. Journal of Cooperative Information Systems 10(3), 327–353 (2001).
31. Valluri, S.R., Vadapalli, S., Karlapalam, K.,: View relevance Driven Materialized view selection in data warehousing Environment, IEEE Computational Society, USA, pp:187–196 (2002).
32. Nadeua, T.P., Teorey, T.J.: Achieving Scalability in OLAP Materialized View Selection, Proceedings of DOLAP'02, pp. 28–34, ACM (2002).
33. Yousri, N.A.R., Ahmed, K.M., EI-Makky, N.M.,: Algorithms for selection Materialized views in a Data Warehouse, IEEE (2005).
34. Shah, B., Ramachandran, K.,: A Hybrid Approach for Data Warehouse View Selection, International Journal of Datawarehouse and Datamining (2006).
35. Gou, G., Yu, J.X., Lu, H.,: A* Search: An Efficient and Flexible Approach to Materialized View Selection, IEEE Transactions on Systems, Man and Cybernetics Part C, Vol 36, No. 3 (2006).
36. Lawrence, M., Chaplin, A.R.,: Dynamic View Selecting for OLAP, DaWak 2006, LNCS 4081, pp. 33–44, Springer (2006).
37. Hung, M.C., Huang, M.L., Yang, D.L., Hsueh, N.L.,: Efficient approaches for materialized view selection, Information Sciences, Elsevier (2007).
38. Ashadevi, B., Balasubramanian, R.,: Cost effective approach for materialized views selection in data warehouse environment, In International Journal of Computer Science and Network Security, vol. 8 (2008).
39. Mahboubi, H., Aouiche, K., Darmont, J.,: Materialized View Selection by Query Clustering in XML Data Warehouses, arXiv.org (2008).
40. Lijuan, Z., Xuebin, Ge., Linshuang, W., Qian, S.,: Efficient Materialized View Selection Dynamic Improvement Algorithm, Sixth International Conference on Fuzzy Systems and Knowledge Discovery (2009).
41. Zhang, Q., Sun, X., Wang, Z.,: An Efficient Ma-Based Materialized Views Selection Algorithm, IITA, International Conference On Control, Automation And Systems Engineering (2009).
42. Baralis, E., Cerquitelli, T., Chiusano, S.,: Imine: Index Support For Item Set Mining, IEEE Transactions On Knowledge And Data Engineering, Vol. 21, No. 4 (2009).
43. Kumar., T.V., Ghoshal: A Greedy Selection of Materialized Views, Int. Journal of Computer and Communication Technology Vol-1., No.-1, pp. −156–172 (2009).
44. Talebi, Z.A., Chirkova, R., Fathi, Y.,: Exact and inexact methods for solving the problem of view selection for aggregate queries, Int. Journal of Business Intelligence and DataMining (2009).
45. Zhou, L., Xu, M., Shi, Q., Hao, Z.,: Research on Materialized Views Technology in DataWarehouse, Beijing Educational Committee science and technology development plan project (2010).
46. Karde, P., Thakare, V.,: Selection of materialized views using query optimization in database management: An efficient methodology, International Journal of Management Systems, vol. 2 (2010).
47. Li, X., Qian, X., Jiang, J., Wang, Z.,: Shuffled Frog Leaping Algorithm for Materialized Views Selection, Second International Workshop on Education Technology and Computer Science, IEEE (2010).
48. Nalini, T., Kumaravel, A., Rangarajan, K.,: An Efficient I-Mine Algorithm For Materialized Views In A Data Warehouse Environment, IJCSI, International Journal Of Computer Science Issues, Vol. 8, Issue 5, No.1 (2011).
49. Huang, X., Chen, Q.,: A Maintainable Model of Materialized View Based on Data Warehouse, International Conference on Mechatronic Science, Electric Engineering and Computer, IEEE (2011).
50. Drias, H.,: Generating Materialized Views Using Ant Based Approaches And Information Retrieval Technologies, IEEE Symposium Series on Computational Intelligence (2011).

51. Chen, W., Bo, X.,: A dynamic materialized view Selection in a Cloud-based Data Warehouse, IEEE (2011).
52. Rashid, N.M.B., Islam, M.S., Latiful Hoque, A.S.M.,: Dynamic Materialized View Selection Approach for Improving Query Performance, Computer Networks and Information Technologies Communications in Computer and Information Science, Volume 142, pp 202–211 Springer (2011).
53. Chaudhari, M.S., Dhote, C.,: Dynamic Materialized View Selection Algorithm: A Clustering Approach, LNCS 6411, pp. 57–66 2012, Springer (2012).
54. Zhu, C., Zhu, Q., Zuzarte, C., Ma, W.,: DMVI: A Dynamic Materialized View Index for Efficiently Discovering Usable Views for Progressive Queries, Conference of the Center for Advanced Studies on Collaborative Research (CASCON 2012) pp 42–56, IBM, Canada, (2012).
55. Katsifodimos, A., Manolescu, I., Vassalos, V.,: Materialized View Selection for XQuery Workloads, ACM SIGMOD (2012).
56. Huang, R., Chirkova, R., Fathi, Y.,: Deterministic View Selection for Data Analysis Queries: Properties and Algorithms, ADBIS, Advances in Databases and Information Systems, LNCS 7503, pp. 195–208, Springer (2012).
57. Goswami, R., Bhattacharyya, D.K., Dutta, M.,: Multiobjective Differential Evolution Algorithm Using Binary Encoded Data in Selecting Views for Materializing in Data Warehouse, Swarm, Evolutionary, and Memetic Computing, Lecture Notes in Computer Science, Part II, LNCS 8298, pp. 95–106, Springer (2013).
58. Sohn, J.S., Yang, J.H., Chung, I.J.,: Improved View Selection Algorithm in Data Warehouse, IT Convergence and Security, Lecture Notes in Electrical Engineering 215, DOI:10.1007/978-94-007-5860-5_111, Springer (2013).
59. Perez, L.L., Jermaine, C.M.,: History-aware Query Optimization with Materialized Intermediate Views, IEEE International Conference on Data Engineering, (2014).
60. Mbaiossoum, B., Bellatrechel, L., Jean, S.,: Materialized View Selection Considering the Diversity of Semantic Web Databases, Springer, ADBIS, Advances in Databases and Information Systems, LNCS 8716, pp. 163–176 (2014).
61. Kehua, Y., Diasse, A.: A dynamic materialized view Selection in a Cloud-based Data Warehouse, IJCSI International Journal of Computer Science Issues, Vol. 11, Issue 2, No.1 (2014).
62. Horng, J.T., Chang. Y.J., Lin. B.J., Kao, C.Y.: Materialized view selection using genetic algorithms in a data warehouse system, Proceedings of the Congress on Evolutionary Computation, pp:22–27 (1999).
63. Zhang, C., Yang, J.,: Genetic Algorithm for Materialized View Selection in Data Warehouse Environments, Springer (1999).
64. Zhang, C., Yao, X., Yang, J.: An evolutionary Approach to Materialized View Selection in a Data Warehouse Environment, IEEE Transactions on Systems, Man and Cybernetics, Part C vol. 31, no.3, pp. 282– 293 (2001).
65. Bellahsene, Z., Cart, M., Kadi, N.: A Cooperative Approach to View Selection and Placement in P2P Systems, Springer (2010).
66. Kumar, T.V.V., Haider, M.: A Query Answering Greedy Algorithm for Selecting Materialized Views, Springer (2010).
67. Song, X., Lin, G.: An Ant Colony based algorithm for optimal selection of Materialized view, International Conference on Intelligent Computing and Integrated Systems (2010).
68. Talebi, Z.A.: An integer programming approach for the view and index selection problem, Data & Knowledge Engineering, 111–125, Elsevier (2012).
69. Li, J., Li, X., Juntao, L.V.: Selecting Materialized Views Based on Top-k Query Algorithm for Lineage Tracing, Third Global Congress on Intelligent Systems, IEEE (2012).
70. Mami, I., Bellahsene, Z., Coletta, R.: View Selection under Multiple Resource Constraints in a Distributed Context, Springer (2012).
71. Kumar, T.V.V., Kumar, S.: Materialized View Selection Using Simulated Annealing, Springer (2012).

SQLI Attacks: Current State and Mitigation in SDLC

Daljit Kaur and Parminder Kaur

Abstract The SQL injection is a predominant type of attack and threat to web applications. This attack attempts to subvert the relationship between a webpage and its supporting database. Due to widespread availability of valuable data and automated tools on web, attackers are motivated to launch high profile attacks on targeted websites. This paper is an effort to know the current state of SQL injection attacks. Different Researchers have proposed various solutions to address SQL injection problems. In this research work, those countermeasures are identified and applied to a vulnerable application and database system, then result are illustrated.

Keywords Attack · Security · Sql injection · Secure development · Web applications

1 Introduction

SQL injection (SQLI) continues to be one of the most predominant web application threat as it has compromised large number of websites including those of some high profile companies. It allows attackers to obtain unauthorized access to the backend database to change the intended application-generated SQL queries. This type of attack exploits vulnerabilities existing in web applications or stored procedures in the backend database server [1, 2]. It allows attackers to inject crafted malicious SQL query segment to change the intended effect, so that attacker can view, edit or make the data unavailable to other users, or even corrupt the database server. When an application becomes susceptible to SQLI Attack (SQLIA), attacker can get total control and access to database [3]. A successful SQLIA can read sensitive data from

D. Kaur (✉)
Lyallpur Khalsa College, Jalandhar, India
e-mail: jeetudaljit@gmail.com

P. Kaur
Guru Nanak Dev University, Amritsar, India
e-mail: parminder.kaur@yahoo.com

© Springer Nature Singapore Pte Ltd. 2017
S.C. Satapathy et al. (eds.), *Proceedings of the 5th International Conference on Frontiers in Intelligent Computing: Theory and Applications*, Advances in Intelligent Systems and Computing 515, DOI 10.1007/978-981-10-3153-3_67

673

database, modify database data (insert/update/delete), execute administration operations on database (such as shut down DBMS and make it unavailable), recover the content of given file present on DBMS file system and in some cases can also issue commands to operating system [1]. This research paper examines the current state of SQLIAs by following various related news in recent years and analyzing previous years attacks data and scanning few websites with automated vulnerability scanner in Sect. 2. Section 3 discusses the developed vulnerable application and analyses the SQLI vulnerability status by implementation of known countermeasures from different Researchers and security organizations. The scan result before and after are shown and illustrated. Section 4 concludes the paper and provides the future directions.

2   Current State of SQLI Attacks

SQL injection attacks and its prevention has become one of the most active topics of research in industry and academia. There have been significant progress in the field and number of models have been proposed and developed to prevent and counter SQLIAs, but lots of web applications are still suffering from SQLI vulnerability. This attack is still popular among hackers. To know the truth of the current state of SQLIAs, data has been collected from various news and security sites like thehackernews.com, reddit.com, and hackmageddon.com. Also a test bed is prepared with various offline and online web applications, and those websites have been scanned using automotive web scanners to detect vulnerabilities and result are shown in Table 1.

Table 1 OWASP-ZAP scan result

Web App	Total alerts	No. of vulnerability categories	Alerts				SQLi alerts
			High	Medium	Low	Info	
Giftshop	305	10	3	2	5	0	12
Jewelry	115	11	1	4	6	0	02
Shoestore	339	9	1	3	5	0	08
Sportskart	660	10	0	4	6	0	–
Caterer	203	11	1	2	8	0	04
Socialnet	42	5	1	0	4	0	01
Careerguidance	120	9	0	4	5	0	–
Webshop	11	4	0	1	3	0	–
Testwebsite1	175	7	1	2	4	0	04
Testwebsite2	75	7	0	1	6	0	–
Lemoncrow.com	130	6	2	1	3	0	–
Eball.net	15	4	0	1	3	0	–
	2190		10	25	58		31

2.1 SQLI Attacks in News

Some major hacks regarding SQLI in recent years are:

- In April 2016, SQLI vulnerability was found at Panama papers firm Mossack Fonseca.
- In November 2015, the personal information of almost 5 million parents and more than 200,000 kids was exposed who bought products sold by VTech.
- In October 2015, SQLIA on British telecommunication company Talk-Talk's servers revealed personal details including credit card numbers and passwords of 4 million customers.
- In 2014, a group of Russian hackers stole more than one billion passwords from almost 400,000 sites including big and small.
- In 2013, a hacker group "RedHack" breached Istanbul administration site and erased people's debts to water, electricity, internet, and telephone companies. Also they published username and password for other citizens to login and clear their debt.
- In July 2012, yahoo confirms hacking of 4 million accounts.
- In our recent research [4], data has been collected from 2012 to 2015 w.r.t. attacks on web applications, and found that SQLI is the leading attack among through all these years.

2.2 Web Vulnerability Scanner Result

There are different automatic tools available for testing the security of a web applications. To know the state of SQLI Vulnerability existence in today's web applications, set of few web applications from some developers has been collected and test bed is created using XAMPP server and Kali Linux. XAMPP stands for Cross-Platform (X), Apache (A), MariaDB (M), PHP (P), and Perl (P). It is an open source, simple, lightweight Apache distribution to create a local web server for testing purposes [5]. Kali Linux is a powerful and yet very commonly used operating system for testing purposes which provides more than 300 penetration testing tools [6]. For testing our set of web applications, we have used OWASP-ZAP and Vega Vulnerability scanner. OWASP-ZAP (Open web application Security Project —Zed Application Proxy) is an open source tool for finding vulnerabilities in web applications. It allows to see all of the requests and responses of a web application. It is very simple, easy to configure, spider based, stable and fast tool which gives trustable result [7]. The results of the OWASP-ZAP scan are shown in Table 1.

Scan result reveals that almost every web application is insecure with less or more serious vulnerability and more than 58% of web applications are seriously vulnerable indicating high alerts. Among those seriously vulnerable applications, 60% high alerts are only because of SQLI vulnerability and in total 50% of applications are vulnerable to SQLI. One interesting thing about SQLI vulnerability is that almost every web application infected with this vulnerability has multiple SQLI alerts, i.e., multiple entries are infected.

2.3 Why It Still Exists?

Many SQLI preventive techniques are known and discovered by researchers, also OWASP provides clear, simple, actionable guidance for preventing SQL Injection flaws in web applications [1]. But still this attack exists and is one of amongst the most impactful attacks. The reasons behind this are:

- Very sensitive/lucrative data in the database attracts the attackers
- It is an Easy and simple attack that does not require much skills to start (also automating tools like SQLmap and Havij require almost nil knowledge to exploit)
- Insecure development architecture
- Trusting Input (poorly filtered strings and incorrect type handling)
- Code samples outdated
- Database Access Rights not properly assigned
- Budget shortfalls
- Lack of awareness

3 Solution (Practical Implementation)

Prevention is better than cure and to prevent from SQLIA, there are various preventive measures offered by different researchers. Here we have classified those preventive measures/activities in SDLC phases (Design, Coding, Testing, and Implementation). Also to know the effectiveness of the prevention activities, we have started with the completely SQLI vulnerable web application in PHP and implemented preventive measures identified in each phase one by one in different versions of the web application and scanned each version using OWASP-ZAP and Vega vulnerability scanner available in Kali Linux. Classification of SQLI preventive measures in different SDLC phases is as follows.

Application and Database Design Activities

A1. Use minimum text boxes and try radio buttons/drop down list/check boxes instead [8].
A2. Use principles of least privileges and disable default accounts and passwords [1, 9–13]. Also use Read only views for SQL statements that do not require any modification.

A3. Choose names for tables and fields that are not easy to guess [11].

A4. Identify the list of SQL statements that will be used by application and only allow those [11].

Coding Activities

A5. Sanitize/Validate Input by ensuring data is properly typed and does not contain escaped code [1, 9–15].

A6. Validate inputs with Data Type, Data Length and Data Format at both client and server side. [3, 10, 11, 16].

A7. Encode string in such a way that all meta-characters are interpreted by the database as normal characters [3, 14–16].

A8. Use Stored procedure with static SQL wherever possible [1, 3, 10–12, 15].

A9. Use parameterized queries instead of dynamic queries [11].

A10. Use prepared statements in programming languages like Perl, Java [1, 3, 11–13].

A11. Use POST method instead of GET method for form submission [8].

A12. Ensure that Error Messages do not disclose any internal database structure, table names, or account names. Use proper error handling mechanism (Custom errors) also keep error messages and usable [3, 9, 11, 13, 15].

Testing Activities

A13. Conduct penetration tests against applications, servers and perimeter security [9].

Configuration & Implementation Activities

A14. Install the database on different machine than Web server or Application server [11].

A15. Update and Patch production servers (including operating system and application) [7, 9, 11, 13].

A16. Disable potentially harmful SQL stored procedure calls [9, 11].

A17. Delete system stored procedures [10].

A18. Delete/Disable unnecessary stored procedures/prepared statements [9].

For the purpose of testing the effectiveness of SQLI preventive activities, a simple web application is developed, in which a page displays the detail of the user (or it can be something else like product information) depending on the ID entered by the user. For the very first version, not any security requirement is considered and it is a complete vulnerable application. This vulnerable version shows SQLI alerts in both OWASP-ZAP and Vega scanner. To make this application free from SQLI vulnerabilities, SQLI prevention activities (A1 to A12) are implemented in version to version till it shows zero SQLI vulnerability in scan result.

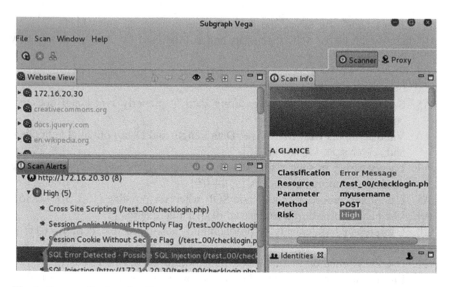

Fig. 1 Scan result of version 2

In next version of the same web application, Design phase prevention activities, i.e., activities A1, A2, A3, and A4 are implemented. For this text box used in the design page is changed to drop down menu and in database a new user is created who can only execute select command on the specified table of the database. It may has reduced the risk but scan result of both Vega and OWASP-ZAP still shows the existence of SQLI Vulnerability as shown in Fig. 1 and even same number of security alerts as in the previous version with Vega scanner.

In the final version, w.r.t requirement of the application, Design and Coding countermeasures i.e. activities A1 through A5, A6, and A9 through A11 are implemented. In this version, input is sanitized properly and validated at both client and server side. Queries are parameterized, POST method is used for form submission and errors are handled properly. Finally this version was scanned with both Vega and OWASP-ZAP scanner as shown in Fig. 2, and it displayed zero SQLI vulnerability.

Also penetration testing was conducted on the web application(not considering the server configuration and other details) and found that final version is secure from SQLI attack. We cannot say that it is complete secure page as it contains some server side flaws in configuration and also 100% security is not possible. But yes, it is secure from SQLI attack just by following few steps in design and coding phase of the development.

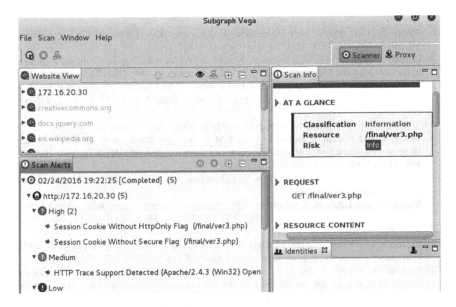

Fig. 2 Scan result of version 3 with Vega

4 Conclusion and Future Works

In this paper, we have concentrated on the most popular SQL injection vulnerability. According to OWASP's ten most critical Web Application Security Vulnerabilities [1], many SQL injection-related issues are among the most harmful threats to the web applications. Since we have covered only SQLI aspects, we would like to suggest that further studies should be made on other security attacks and vulnerabilities. Also organizations should consider the importance of application layer security and developers should understand and implement the security guidelines during development process, which could be used in developing more secure web applications.

References

1. SQL Inject Prevention cheat sheet, https://www.owasp.org/index.php/SQL_Injection_Prevention_Cheat_Sheet.
2. Protecting Websites from advanced and automated SQL injection, http://www.imperva.com/docs/WP_SQL_Injection20.pdf.
3. Torgby. W.K, Asabere, N.Y. Structured Query Language Injection (SQLI) Attacks: Detection and Prevention Techniques in Web Application Technologies. International Journal of Computer applications Vol. 71-No.11. 29–40.ISSN: 0975-8887. (2013).
4. Kaur D, Kaur P. Empirical Analysis of Web Attacks. In Procedia of Computer Science. Elsevier Publications. DOI:10.1016/j.procs.2016.02.057 (2016).

5. XAMPP, https://en.wikipedia.org/wiki/XAMPP.

6. Muniz J., Lakhani A. Web Penetration Testing with Kali Linux. PACKT Publishing. MUMBAI-India. (2013).

7. Gandhi M. and Baria J. SQL Injection Attacks in Web Application. International Journal of Soft computing and Engineering (IJSCE), Vol 2, Issue 6. 189–191. ISSN:2231-2307. (2013.).

8. Parmar.G. and Mathur K. Proposed Preventive measures and strategies Against SQL injection Attacks. Indian Journal of Applied Research, Vol. 5, Issue 5. 664–671. ISSN- 2249555X. (2015).

9. SQL Injection, https://www.us-cert.gov/sites/default/files/publications/sql200901.pdf.

10. Madan.S. and Madan S. Bulwark Against SQL Injection attack – An Unified Approach. International Journal of Computer Science and Network Security(IJCSNS), Vol. 10 No.5. 305–313. (2010).

11. Steps to Protect your Websites from SQL Injection attacks, https://www.whitehatsec.com/resource/whitepapers/SQL.html.

12. Mahapatra and Khan. S. A Survey of SQL Injection Countermeasures, International Journal of Computer science & engineering (IJCSES) Vol. 3, No.3. **55–74**. DOI:10.5121/ijcses.2012. 3305 55. (2012).

13. Kalaria S. and Vivekanandan. M. Dark Side of SQL Injection. In the proceedings of ASAR International Conference, Bangalore. 67–72. ISBN: 978-81-927147-0-7. (2013).

14. Helford. W, Viegas. J. and Orso. A. 2006. A Classification of SQL Injection attacks and countermeasures. In the proceedings of the International symposium on secure software Engineering. Washington, USA (2006).

15. Gollmann. D. Securing Web Applications. **Article** *in* ELSEVIER Information Security Technical Report Volume 13 Issue1. Elsevier Advanced Technology Publications Oxford, UK. 1–9.DOI:10.1016/j.istr.2008.02.002.

16. Aggarwal. U, Saxena. M. and Rana. K.S. A Survey of SQL Injection attacks. International Journal of Advanced Research in Computer Science and Software Engineering (IJARCSSE), vol. 5, Issue 3. 286–289. ISSN:2277128X. (2015).

Theoretical Validation of Object-Oriented Metrics for Data Warehouse Multidimensional Model

Anjana Gosain and Rakhi Gupta

Abstract Metrics are commonly used to guide the designers to build quality data warehouse models. Recently, researchers have defined various object-oriented metrics for data warehouse conceptual model to access their quality. These metrics require theoretical and empirical validation to confirm their applicability in real time. Empirical validation of object-oriented metrics has already been carried out but theoretical validation has not been taken into account. In this paper, theoretical validation for object-oriented metrics using Zuse's framework is presented to show that these metrics may be considered as strong measures for evaluating quality of object-oriented conceptual models of data warehouse.

Keywords Object-oriented metrics · Object-oriented data warehouse · Theoretical validation · Zuse's framework

1 Introduction

Data warehouses are complex systems that support managers in strategic decisions. They are designed primarily for better and faster decision making for top level management. Because of their consideration in decision making, it becomes essential for organizations to assure quality of data warehouse. The quality of data warehouse depends on data models (physical, logical, and conceptual) [1]. Hence, the choice of a relevant data model is one of the important tasks in the data warehouse system development process. Piattini et al. have proposed various object-oriented metrics [1] for accessing the quality of data warehouse conceptual model. Metrics helps in measuring quality aspect of data warehouse in a stable and

A. Gosain · R. Gupta (✉)
University School of Information and Communication Technology, Guru Gobind Singh Indraprastha University, Dwarka, New Delhi 110078, India
e-mail: rakhigupta751@yahoo.com

A. Gosain
e-mail: anjana_gosain@yahoo.com

© Springer Nature Singapore Pte Ltd. 2017
S.C. Satapathy et al. (eds.), *Proceedings of the 5th International Conference on Frontiers in Intelligent Computing: Theory and Applications*, Advances in Intelligent Systems and Computing 515, DOI 10.1007/978-981-10-3153-3_68

681

factual manner. Theoretical and empirical validation of metrics is vital to prove the practicality of the metrics. Though empirical validation has been performed through controlled experiments [1–4] these metrics lack theoretical validation. Theoretical validation helps to confirm systematically that the metrics are proper mathematical interpretation of the measured attribute (adhering to a set of systematic properties) and to interpret if the metric actually measures that it is assumed to measure [5, 6]. There are two major ways of theoretical validation in software measurement work: (i) framework builds on measurement theory principles like distance framework [7] and (ii) framework builds on axioms like Briand's framework and Zuse's framework [8, 9]. We have used Zuse's framework to validate the object-oriented metrics as proposed in [1]. The paper is structured as follows: Sect. 2 discusses the various object-oriented metrics as proposed by Piattini et al. Section 3 describes the Zuse's framework and Sect. 4 presents theoretical validation of the OO metrics using Zuse's framework and in the last section, conclusion, and future work is discussed.

2 Object-Oriented Metrics for Data Warehouse Multidimensional Model

This section briefly presents the object-oriented metrics as defined by Piattini et al. [1] for data warehouse multidimensional model.

1. NFC: Number of fact classes.
2. NDC: Number of dimension classes.
3. NBC: Number of base classes.
4. NC: Total number of classes, i.e., NC = NFC + NDC + NBC.
5. RBC: Ratio of base classes, i.e., Number of base classes per dimensional class.
6. NSDC: Number of dimensional classes shared by more than one star.
7. NAFC: Number of FA attributes of the fact classes.
8. NADC: Number of D and DA attributes of the dimensional Tables.
9. NASDC: Number of D and DA attributes of the shared dimensional classes.
10. NA: Number of FA, D, and DA attributes.
11. NH: Number of hierarchies.
12. DHP: Maximum depth of the hierarchical relationships.
13. RDC: Ratio of dimensional classes. Number of dimensional classes per fact class.
14. RSA: Ratio of attributes. Number of FA attributes divided by the number of D and DA attributes.

3   Zuse's Framework

This is the framework based on measurement theory. Its objective is to ascertain to which scale (ordinal scale, ratio scale, or above ordinal scale) the metric pertains. This framework functions for three main mathematical structures:

- Modified extensive structure
- Independence condition
- Modified relation of belief

Details about these mathematical structures and the complete formal framework can be found in [10]. Any given metric should fulfill one of the three structures to be validated. If a metric comply with modified extensive structure, it will essentially comply with independence conditions and metric pertains to ratio scale [8]. When metric does not comply with modified extensive structure but only with independence conditions, then it pertains to ordinal scale. When metric only satisfies modified relation of belief and not the other two structures then that metric pertains to above ordinal scale. For a metric validation, the axioms and the conditions as given in Table 1 should be satisfied.

4   Theoretical Validation of OO Metric Using Zuse's Framework

In this section, we theoretically validate the object-oriented metrics as presented in Section II using Zuse's framework.

4.1   NA (Number of Attributes)

Let there be three classes A, B, C, having attributes as (a, b), (b, c), (d, e), and methods M1, M2, M3, respectively.

NA as extensive modified structure:

Axiom 1: If NA (A) $\geq$ NA (B) and NA (B) $\geq$ NA (C), then NA (A) $\geq$ NA (C). Thus, NA accomplishes axiom 1.

Axiom 2: When attributes in class A are concatenated with attributes of the other class B, then it will always be greater than or equal to number of attributes in class A alone. Hence NA satisfies this axiom.

Axiom 3: Since concatenation operation is associative, this axiom is satisfied.

Axiom 4: Since concatenation operation is commutative, axioms 4 is satisfied.

Axiom 5: If NA (A) is greater or equal to NA (B), then after concatenation with attributes of class C the value may or may not be greater or equal; In Fig. 1 NA

Table 1 Summary of Zuse's formal framework [11]

Modified extensive structure
Axiom1: $(X, \bullet >=)$ (weak order)
Axiom2: X1 o X2 $\bullet$ > = X1 (Axiom of weak positivity)
Axiom3: X1 o (X2 o X3) ~ (X1 o X2) o X3 (Axiom of weak associativity)
Axiom4: X1 o X2 ~ X2 o X1 (Axiom of weak commutativity)
Axiom5: X1 $\bullet$ > = X2 => X1 o X $\bullet$ > = X2 o X (Axiom of weak monotonicity)
Axiom6: If X3 $\bullet$ > X4 then for any X1, X2, then there exists a natural number n. such that X 1 o nX3 $\bullet$ > X2 o nX4 (Archimedean axiom)
Binary relation $\bullet$ > = is weak order if it is transitive and complete: X1 $\bullet$ > = X2, and X2 $\bullet$ > = X3 => X1 $\bullet$ > = X3 X1 $\bullet$ > = X2 or X2 $\bullet$ > = X1

Independence conditions
C l: X1 ~ X2 => X1 o X ~ X2 o X and X1 ~ X2 => X o X1 ~ X o X2
C2: X1 ~ X2 <=> X1 O X ~ X2 o X and X1 ~ X2 ⇔ X o X1 ~ X o X2
C3: X1 $\bullet$ > = X2 => X1 o X $\bullet$ > = X2 o X and X1 $\bullet$ > = X2 => X o X1 $\bullet$ > = X o X2
C4: X1 $\bullet$ > = X2 ⇔ X1 o X $\bullet$ > = X2 o X and X1 $\bullet$ > = X2 ⇔ X o X1 $\bullet$ > = X o X2
Where X1 ~ X2 if and only if X1 $\bullet$ > = X2 and X2 $\bullet$ > = X1. and
X 1 $\bullet$ > X2 if and only if X 1 $\bullet$ > = X2 and not (X2 $\bullet$ > = X 1).

Modified relation of belief
MRB l: For all X, Y ∈ ℑ: X $\bullet$ > = Y or Y $\bullet$ > = X (completeness)
MRB2: For all X, Y, Z ∈ ℑ: X $\bullet$ > = Y and Y $\bullet$ > = Z => X $\bullet$ > = Z (transitivity)
MRB3: For all X ⊇ Y => X $\bullet$ > = Y (dominance axiom)
MRB4: For all (X ⊃ Y and X ∩ Z = Φ) => (X $\bullet$ > = Y => X U Z $\bullet$ > Y U Z) (partial monotonicity)
MRB5: For all X ℑ: X $\bullet$ > = 0 (positivity)

(A) $\bullet$ > = NA (B), but after concatenation operation, NA (A o C) < NA (B o C). Hence, NA need not always satisfy this axiom. Thus, NA does not satisfy this axiom.

Axiom 6: Archimedian axiom

A concatenation rule is idempotent, if for all objects a ∈ A holds

$$a \ o \ a = a.$$

Whenever we have an idempotent concatenation operation, then archimedian axiom can never be fulfilled [10]. Thus, we cannot have extensive structure. In NA (A), if class A is concatenated with itself, the NA (A) remains same (Concatenation

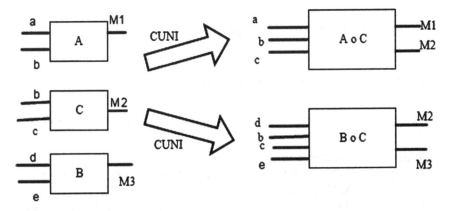

Fig. 1 Axiom 5 of extensive modified structure and C1 for Independence condition for NA

removes duplicate attributes). Thus, NA is idempotent and hence does not satisfy this axiom.

Thus, the metric NA fails to satisfy all the modified extensive structure axioms, and hence NA is not an extensive modified structure.

NA and independence conditions:

C1: If NA (A) = NA (B), then, if we combine these classes with class C, it is not necessary that we obtain that NA (A o C) = NA (B o C). In Fig. 1, NA (A) = NA (B) = 2. Now, When these classes are concatenated with class C, NA (A o B) < NA (B o C). Thus first condition is not satisfied.

C2: If a metric does not satisfy first condition, then it cannot accomplish second condition. Hence NA does not satisfy this condition.

C3: NA does not satisfy fifth axiom of extensive modified structure, hence it cannot accomplish this third condition.

C4: If C3 is not satisfied, C4 cannot be established. Hence, NA does not satisfy this condition also.

Hence, independence conditions are not satisfied by NA.

NA and the modified structure of belief:
When weak order property is accomplished by a metric, then *MRB1* and *MRB2* are also satisfied.

MRB3: Consider class B as a subset of class A, then NA (A) ≥ NA (B). Thus, the metric accomplishes MRB3.

MRB4: Consider metric satisfies MRB3 and the two classes A and C do not share any common attribute, then NA (A o C) > NA (B o C). Thus, the metric accomplishes MRB4.

MRB5: It takes into account positivity, which is always true; since NA can never have a value less than zero, thus always positive.

Thus, NA only satisfies modified relation of belief and not the other two structures, and hence is a metric that pertains to above ordinal scale.

4.2 NFC (Number of Fact Classes)

Let there be three schemas X, Y, Z. In Fig. 2, solid rectangles depict fact classes and others are dimension classes.

NFC as extensive modified structure:
Axiom 1: If NFC (X) ≥ NFC (Y) and NFC (Y) ≥ NFC (Z), then NFC (X) ≥ NFC (Z). Thus, NFC satisfies first axiom.
Axiom 2: When fact classes in schema X are concatenated with fact classes in schema Y, then the total fact classes in the resultant schema will always be greater than or equal to NFC (X) alone. Hence NFC satisfies this axiom.
Axiom 3: Since concatenation operation is associative, axioms 3 is satisfied.
Axiom 4: Since concatenation operation is commutative, axioms 4 is satisfied.
Axiom 5: If NFC(X) > = NFC(Y), then after concatenation with fact class in schema Z, the value may or may not be greater or equal; In Fig. 2 NFC (X) ≥ NFC (Y), after concatenation operation, NFC (X o Z) < NFC (Y o Z). Thus, for NFC, axiom 5 is not satisfied.
Axiom 6: Archimedian axiom

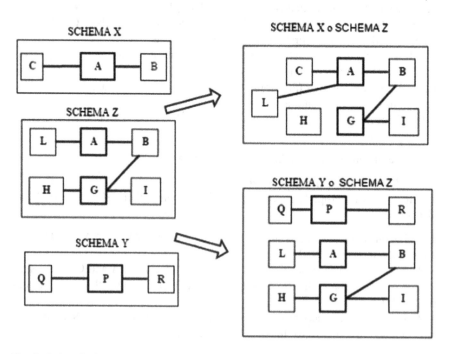

Fig. 2 Axiom 5 of extensive modified structure and C1 of Independence condition for NFC

Whenever we have an idempotent concatenation operation, then Archimedean axiom can never be fulfilled.

When a fact class in a schema is concatenated with itself, in the same schema, then we get the same fact class. Thus, NFC is idempotent and hence does not satisfy this axiom.

Thus, the metric NFC fails to fulfill all the modified extensive structure axioms and hence NFC is not an extensive modified structure.

NFC and independence conditions:

C1: If NFC (X) = NFC (Y), then, if we combine the fact classes with fact classes in schema Z, it is not necessary that we obtain that NFC (X o Z) = NFC (Y o Z). In Fig. 2, NFC (X) = NFC (Y) = 1. Now, when these fact classes are concatenated with fact classes in schema Z, NFC (X o Z) < NFC (Y o Z). Thus, first condition is not satisfied.

C2: If a metric does not satisfy first condition, then it cannot accomplish second condition. Hence NFC does not satisfy this condition.

C3: NFC does not satisfy fifth axiom of extensive modified structure, hence it cannot accomplish this third condition.

C4: If C3 is not satisfied, C4 cannot be established. Hence, NFC does not satisfy this condition also.

Hence, independence conditions are not satisfied by NFC.

NFC and the modified structure of belief
When weak order property is accomplished by a metric, then *MRB1* and *MRB2* are also satisfied.

MRB3: When schema Y is subset of schema X, then NFC (X) > = NFC (Y) Thus, the metric accomplishes MRB3.

MRB4: Consider metric satisfies MRB3 and the two schemas X and Z do not share any common fact table, then NFC (X o Z) > NFC (Y o Z). Thus, the metric accomplishes MRB4.

MRB5: It takes into account positivity, which is always true; since NFC can never have a value less than zero, thus always positive

Thus, NFC only satisfies modified relation of belief and not the other two structures and hence is a metric that pertains to above ordinal scale.

Similarly, NDC (number of dimension classes), NBC (number of base classes), and NC (NFC + NDC + NBC) can also be validated and can be found to be the measure that is above ordinal scale.

4.3 NH (Number of Hierarchies)

Let there be three schemas X, Y, Z having F as fact class and P, Q, R, S as dimension classes.

NH as extensive modified structure:

Axiom 1: If NH (X) $\geq$ NH (Y) and NH (Y) $\geq$ NH (Z), then NH (X) $\geq$ NH (Z). Thus, NH satisfies first axiom.

Axiom 2: When number of hierarchies in schema X is concatenated with number of hierarchies in schema Y, then it will always be greater than equal to NH (X) alone. Hence NH satisfies this axiom.

Axiom 3: As concatenation operation is associative, axioms 3 is satisfied.

Axiom 4: As concatenation operation is commutative, axioms 4 is satisfied.

Axiom 5: If number of hierarchies in schema X is greater or equal to number of hierarchies in schema Y, then after concatenation with hierarchies in schema Z, the value may or may not be greater or equal; In Fig. 3, NH (X) $> = NH$ (Y), after concatenation operation, NH (X o Z) $< NH$ (Y o Z). Thus, for NH, axiom 5 is not satisfied.

Axiom 6: Archimedean axiom

When hierarchies in a schema are concatenated with themselves, then we get the same hierarchies. Thus, NH is idempotent and hence does not satisfy this axiom.

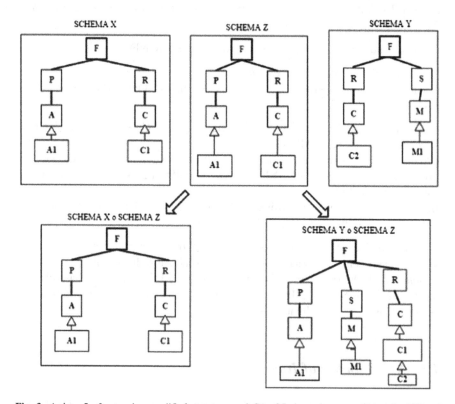

Fig. 3 Axiom 5 of extensive modified structure and C1 of Independence condition for NH and DHP

Thus the metric NH fails to fulfill all the modified extensive structure axioms, and hence NH is not an extensive modified structure.

NH and independence conditions:
Let there be three schemas X, Y, Z

C1: If *NH* (X) = *NH* (Y), then, if we combine the respective schemas with schema Z, it's not necessary that we obtain NH(X o Z) = *NH* (Y o Z). This happen when schema z has a hierarchy common either to schema X or schema Y. In Fig. 3, NH (X) = NH(Y) = 2, but NH(X o Z) < NH(Y o Z). Thus, first condition is not satisfied.
C2: If a metric does not satisfy first condition, then it cannot accomplish second condition. Hence *NH* does not satisfy this condition.
C3: *NH* does not satisfy fifth axiom of extensive modified structure, hence it cannot accomplish this third condition.
C4: If C3 is not satisfied, C4 cannot be established. Hence, *NH* doesn't satisfy this condition also.

Hence, independence conditions are not satisfied by NH.

NH and the modified structure of belief
When weak order property is accomplished by a metric, then *MRB1* and *MRB2* are also satisfied.

MRB3: Consider schema Y is subset of schema X, then *NH* (X) > = *NH* (Y) Thus, the metric accomplishes MRB3.
MRB4: Consider metric satisfies MRB3 and schema X and schema Z don't share any common hierarchy, then, *NH* (X o Z) > *NH* (Y o Z), thus satisfying MRB4.
MRB5: It takes into account positivity, which is always true; since NH can never have a value less than zero, thus always positive.

Thus, *NH* (number of hierarchies) only satisfies modified relation of belief and not the other two structures and hence is a metric that pertains to above ordinal scale level.

4.4 DHP (Maximum Depth of the Hierarchical Relationships)

Let there be three schemas X, Y, Z having F as fact class and P, Q, R, S as dimension classes.

DHP as extensive modified structure:

Axiom 1: If *DHP* (X) $\geq$ *DHP* (Y), and *DHP* (Y) $\geq$ *DHP* (Z), then *DHP* (X) *DHP* (Z). Thus, DHP satisfies first axiom.
Axiom 2: This axiom is also satisfied by *DHP* as when schema X is concatenated with schema Y, maximum depth of hierarchy in the new schema (X o Y) is always greater than or equal to *DHP* (X) alone. Hence DHP satisfies this axiom.

Axiom 3: As concatenation operation is associative, axioms 3 is satisfied.

Axiom 4: As concatenation operation is commutative, axioms 4 is satisfied.

Axiom 5: If DHP in schema X is greater or equal to DHP in schema Y, then after concatenation with schema Z, the value may or may not be greater or equal; In Fig. 3 *DHP* (X) > = *DHP* (Y), after concatenation, *DHP* (X o Z) < *DHP* (Y o Z). Thus, for metric *DHP*, this axiom is not satisfied.

Axiom 6: Archimedian axiom

When a schema is concatenated with itself, then we get the same schema. Thus, *DHP* is idempotent and hence does not satisfy this axiom.

Thus the metric DHP fails to all the modified extensive structure axioms, and hence DHP is not an extensive modified structure.

DHP and independence conditions:

C1: If *DHP* (X) = *DHP* (Y), then, if we combine the respective schemas with schema Z, it is not necessary that we obtain *DHP* (X o Z) = *DHP* (Y o Z). In Fig. 3, *DHP* (X) = *DHP* (Y) = 2. Now, when these schemas are concatenated with schema Z, *DHP* (X o Z) < *DHP* (Y o Z).Thus, first condition is not satisfied.

C2: If a metric does not satisfy first condition, then it cannot accomplish second condition. Hence *DHP* does not satisfy this condition.

C3: *DHP* does not satisfy fifth axiom of extensive modified structure, hence it cannot accomplish this third condition.

C4: If C3 is not satisfied, C4 cannot be established. Hence, *DHP* does not satisfy this condition also.

Hence, independence conditions are not satisfied by DHP.

DHP and the modified structure of belief

When metric accomplishes weak order property, then *MRB1* and *MRB2* are satisfied.

MRB3: Consider schema Y as a subset of schema X, then *DHP* (X) > = *DHP*(Y). Thus the metric accomplishes MRB3.

MRB4: Consider metric satisfies MRB3 and schema X and schema Z don't share any common hierarchy, then, *DHP* (X o Z) > *DHP* (Y o Z), thus satisfying MRB4.

MRB5: It takes into account positivity, which is always true; since DHP can never have a value less than zero, thus always positive.

Thus, DHP only satisfies modified relation of belief and not the other two structures and hence is a metric that pertains to above ordinal scale.

5 Conclusion and Future Work

In this paper, we theoretically validate ten object-oriented metrics namely NA, NAFC, NADC, NABC, NC, NFC, NDC, NBC, NH, and DHP using Zuse's framework. We conclude that these object-oriented metrics are the measures which are above the ordinal scale that confirms that these metrics are formally valid software metrics and validates their practical utility in real time. The interpretation of metrics that lies above the ordinal scale level is important since not much can be done with ordinal numbers and majority of statistical tests that are applied on metric pertaining to ratio scale can be applied to metrics pertaining to above ordinal scale. Though many empirical validations have been carried out for all these object-oriented metrics, but there is need to replicate the empirical validation by considering increased number of schemas with real time industrial data to verify their significance in terms of understandability, efficiency and effectiveness.

References

1. Serrano M, Calero C, Trujillo J., Lujan S., Piattini M.: Empirical validation of metrics for conceptual models of data warehouse. In: Persson, A., Stima, J., Advanced Information Systems Engineering, vol. 3084, pp. 506–520. Springer, Heidelberg (2004).
2. Kumar M, Gosain A, Singh Y.: Empirical validation of structural metrics for predicting understandability of conceptual schemas for data warehouses. In: International Journal of System Assurance Engineering and Management, vol 5, pp. 291–306. Springer, India (2014).
3. Gosain A., Mann S.: Empirical validation of metrics for object oriented multidimensional model for data warehouse. In: International Journal of System Assurance Engineering and Management, vol 5, pp. 262–275. Springer, India (2014).
4. Serrano M., Trujillo J., Calero C., Piattini M.: Metrics for data Warehouse conceptual models understandability. In: Information and Software Technology (INFSOF). Elsevier 49(8) pp. 851–870 (2007).
5. Gosain A., Nagpal S, Sabharwal S.: Validating dimension hierarchy metrics for the understandability of multidimensional models for data warehouse. In: IET software, vol 7, pp. 93–103 (2013).
6. Fenton N,: Software measurement: a necessary scientific basis. In: IEEE Transactions on Software Engineering, vol 20, pp. 199–206 (1994).
7. Poels G., Dedene, G.: Distance: a framework for software measure construction. In: Research Report DTEW9937, Dept Applies Economics Katholieke Universiteit Lueven, Belgium, (1999).
8. Briand L., Morasca S., and Basili V.: Property based software engineering measurement. In: IEEE transactions on software Engineering, vol 22, pp. 68–86 (1996).
9. Briand, L.C., Morasca, S., Basili, V.R.: Response to: Comments on "Property-Based Software Engineering Measurement: Refining the Additivity Properties. In: IEEE Transactions on Software Engineering, vol.23, no. 3, pp. 196–197 (1997).
10. Zuse H.: Properties of Object-Oriented Software Measures. In: Proceedings of the Annual Oregon Workshop on Software Metrics. (1995).
11. Calero C., Piattini M., Pascual, C. and Serrano M: Towards Data Warehouse Quality Metrics. In: 3rd International Workshop on Design and Management of Data Warehouses (DMDW'01) Interlaken Switzerland, pp 1–10 (2001).

Paving the Future of Vehicle Maintenance, Breakdown Assistance and Weather Prediction Using IoT in Automobiles

B.J. Sowmya, Chetan, D. Pradeep Kumar and K.G. Srinivasa

Abstract There is an immense potential to solve many of the challenging and persistent traffic and accident related problems by implementing IoT in vehicles. Various sensors come in-built with most of the vehicles, simplifying our task. By monitoring engine parameters, we can warn the vehicle owners of potential breakdowns and also notify the nearest service centre. In case of a breakdown, help is immediately dispatched to the vehicle. This is particularly useful for heavy vehicles and public transport which causes major traffic jams. The ubiquity of vehicles on the roads also makes possible the use of sensors for weather detection by using onboard sensors for collecting real-time data.

Keywords Smart vehicles · Cloud computing · Iot in vehicles · Engine health monitoring · Smart service · Accurate weather prediction

1 Introduction

An automobile, as seen from a layman's view is just a means of transportation, to go from place A to place B. However, from the automobile's perspective, it is a source that experiences a lot of real time factors. The automobile gets stuck in traffic, traverses long distances through varying weather conditions and air quality, gets involved in collisions. In the modern age of the automobile, it is equipped with

B.J. Sowmya (✉) · Chetan · D. Pradeep Kumar · K.G. Srinivasa
M S Ramaiah Institute of Technology, Bangalore, India
e-mail: sowmyabj@msrit.edu

Chetan
e-mail: chetanshetty@msrit.edu

D. Pradeep Kumar
e-mail: pradeepkumard@msrit.edu

K.G. Srinivasa
e-mail: kgsrinivas@msrit.edu

© Springer Nature Singapore Pte Ltd. 2017
S.C. Satapathy et al. (eds.), *Proceedings of the 5th International Conference on Frontiers in Intelligent Computing: Theory and Applications*, Advances in Intelligent Systems and Computing 515, DOI 10.1007/978-981-10-3153-3_69

top- notch sensors- a rain sensor, a light sensor, a temperature sensor, a humidity sensor, an impact sensor, sensors for various functional units, along with embedded systems that log all this data as shown in Fig. [1]. This is just the data a single automobile possesses [1]. Imagine now, thousands of millions of vehicles, travelling every part of the world can gather humungous volumes of data. And, such volumes of information is no longer a problem, but an unmisable opportunity to gather data in real-time, which can be in turn help in addressing many of the traffic and accident associated problems as well as making lives easier. What we do with this data is only limited to where our imagination can take us (Figs. 1, 2, 3 and 4).

An issue of major concern is the breakdown of public transport or other huge vehicles in the middle of the road. A long time is taken to attend to these vehicles, thus leading to severe traffic jams and inconvenience to a lot of commuters. This problem can be effectively addressed by (1) warn the vehicle owners well in advance of the deteriorating health of their vehicle (2) alerting the nearest service centre immediately so that they can dispatch help in case of a breakdown. The meteorological departments predict weather, especially rains, using the data from cloud cover (obtained via satellite images) and also based on their previous predictions. However, these predictions deviate a lot to real time conditions and the accuracy is off the mark by quite a bit. Considering, the ubiquity of vehicles in a region, and the fact that most of them are already equipped with humidity sensors, we can provide more real-time rain data to these institutions. This data can supplement the satellite data to make more accurate predictions.

1.1 Objectives of the System

Breakdown of a vehicle on Indian roads creates a bottleneck and affects the traffic. This system will essentially help owners of vehicle and other drivers on the roads reduce the unpleasantness of traffic problems, by giving them a fair warning about

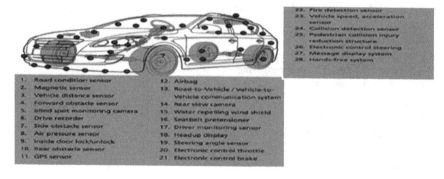

Fig. 1 Sensors in vehicle

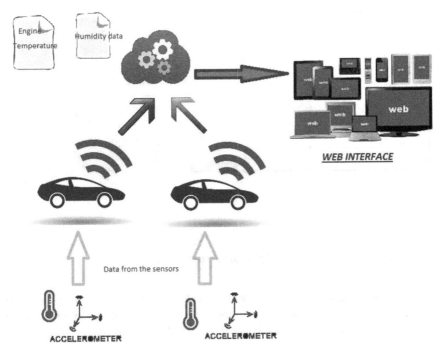

Fig. 2 Architecture

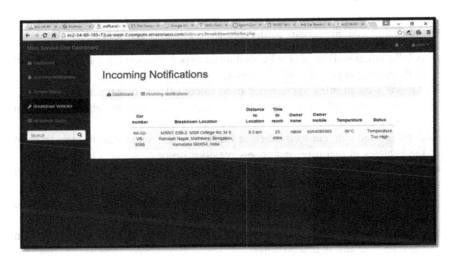

Fig. 3 Incoming notifications about broke down vehicles and even for warned vehicles

Fig. 4 Automatic warning email and SMS to the vehicle owner

the potential breakdown of their vehicles and also help reduce the time required to service them. The system also aims to use the ubiquity of vehicles on the roads to help improve weather prediction by meteorological organizations.

1.2 System Deliverables

The deployed system consists of the following products and services:

1. A fully functional Raspberry Pi Module equipped with sensors (temperature and humidity sensors) and a Wi-Fi module.
2. An AWS cloud infrastructure deployed to collect and gather data from the Pi, perform computations and to notify the service centre and vehicle owners in case of breakdowns.
3. A robust and efficient web interface for the service centre staff to keep track of the vehicles in real time and access previous history.
4. A user-friendly web interface for the vehicle owners to view the engine health and previous repair history of their vehicle.
5. An automatic SMS generation system which alerts the vehicle owner when the engine health is deteriorating.
6. An automatic email generation system which notifies the vehicle owner when the engine health is deteriorating.

2 Literature Survey

The Internet of Things, regarded as the "the next big technological revolution", is estimated to generate about US $746 billion revenue opportunity worldwide in the manufacturing industry (including automotive) [2]. The IoT is an intelligent network which facilitates exchange of information and communication through the information sensing devices by connecting all things to the Internet. It achieves the goal of intelligent identifying, locating, tracking, monitoring, and managing things [3]. It is an extension and expansion of Internet-based network and expands the communication from human and human to human and things or things and things. In the IoT paradigm, many objects surrounding us (in our case, vehicles) will be connected into networks in one form or another. Traditional practices and processes in the automobile industry are being changed forever with the advent of IoT. It has enabled automotive industry to have direct contact with the customers by embracing the latest technology and making vehicle servicing a smooth task. In Predictive Maintenance, the vital vehicle information and warning alerts provided through telematics capabilities are combined with other data points to enable advanced real-time analytics to identify trends, predict machine health and more [4]. Vong et al. [5] have applied IoT in vehicles to notify car owners and governmental authorities of the emission information of the car so they can get their engines fixed before the mandatory vehicle emission test is carried out. Alam et al. [6] have described the Internet of Vehicles (IoV) as one of the key members of IoT. Using this technology, vehicles can easily exchange safety, efficiency, infotainment and comfort related information with other vehicle and infrastructures using vehicular ad hoc networks (VANETs). The true potential of IoT is unleashed when powered by the cloud infrastructure. The vast volumes of data collected every few seconds from billions of vehicles can be stored and processed seamlessly on the cloud. The definition by National Institute of Standards and Technology (NIST) for "Cloud Computing" is a model for enabling convenient, on-demand network access to a shared pool of configurable computing resources (e.g., networks, servers, storage, applications and services) that can be rapidly provisioned and released with minimal management effort or service provider interaction [7]. Cloud framework has striking traits such as pay-as-you-go model, multitenancy. Cloud computing has become a research hotspot among modern technologies, researchers pay more attentions to its applications. The system on chip (SoC) used in this system is Raspberry Pi Model 2, because of it is brilliant configuration, having a quad core processor running at 700 MHz, 512 MB DRAM and a Broadcom GPU [8].

3 Design

The design of the system includes multiple sensors like the temperature sensors for keeping track of the engine temperature and the humidity sensors for weather related data.

The deployed system is implemented using the following modules as illustrated in Fig. [3].

The different processes that are involved are

1. *Vehicle condition reporting*

This function is to report to the driver, the condition of the vehicle. The vehicle is given this information from the server based on the data it has about other vehicles. If in case there is imminent failure, it is reported to the driver as well as the service station.

Gather data from the sensors

The SoC runs its function to continuously sense the parameters from the sensors every 5 s. Once the data is gathered, it readies it for sending to the server.

Send data to server

The SoC then calls upon another function to make a HTTP post request to the server with the attributes as post parameters. If the request fails, it stores the data and tries again after 5 s.

Receive data and display to user

Once the response is obtained from the server, the SoC runs a function to check the returned values and what it means. If it is a warning of failure, it displays a red LED to the driver.

2. *Service centre notification*

This function runs on the server continuously, finding vehicles which have reported warnings. The service centre representative can then see the vehicles which are reporting warnings and can prepare the parts required to repair the vehicle, if the driver confirms the repair.

Identify vehicles needing repair

As the vehicle keeps reporting its data, the server checks with available data sets to find if its parameters indicate potential failure. If it is the case, it sends a warning to the driver and also stores the pertinent data on the database at the back end. It sends a notification to nearest few service centres automatically.

Obtain confirmation

This function is used to automatically obtain a confirmation from the driver if he plans to repair the vehicle at the suggested service centre, by replying to an SMS that is sent to him. This SMS service is automat.

Prepare for repair

Once the confirmation is obtained, the service representative should search for the particular vehicle problem details on the web interface. For the respective problem, the representative should then prepare the parts for the repair once the vehicle comes in.

3. Automated breakdown assistance

This function is used by the server to automatically identify vehicles that have broken down. It sends notifications to the nearest service centre, about the break down location and the parameters at the time of the breakdown. The service centre can then dispatch help immediately. All through this, the driver is kept informed about the activities through automated SMS.

Breakdown identification

This function would identify the vehicles which are broken down. The vehicle when it reports its parameters sends a status update whether it is running or not. If the car life status is of failure, the server identifies it automatically, obtains its location and the parameters recorded when it failed and sends it to the nearest service centre.

Service representative action

The service representative through the web interface will access a separate tab for breakdown vehicles. The representative can then find details about its failure through the parameters, and immediately dispatch help.

Report updates

This function from the server is responsible to keep the driver informed about the actions taken by the service centre. It is an automated message service, as the service representative follows through on the notification.

When a vehicle is operated at a temperature above a safe threshold temperature, say 35°, there is a mail and a SMS automatically sent to the vehicle owner about the engine temperature, along with instructions regarding the nearest service centre and the time taken to reach it from the current location of the vehicle. The owner can confirm his arrival by responding to a link in the mail. In case of a vehicle breakdown, the service centre id immediately notified of the vehicle location and time needed to reach this vehicle. As far as the weather detection is concerned, this system collects the data of atmospheric humidity from the ever increasing presence of vehicles on the roads and directs it into the hands of the meteorological department to help them with future prediction and validation of their predictions.

4 Implementation

Our system consists of the following modules:

Sensing Modules:

These Modules will obtain the required data and send it to the monitoring module. The sensing modules used in our system are as follows:

Sensing module for engine breakdown reporting
This module will consist of temperature sensors that will be used to report the engine temperatures. This can later be used to predict the break down temperature of engines of the same model and warn the vehicle owners if the vehicle is in danger of breaking down. These details, along with the geographical co-ordinates of the location will be sent to the monitoring module.

Sensing module for weather detection by using humidity sensor
This module consists of a humidity sensor that is used to collect data of humidity in the air. This data can be relayed to meteorological department personnel to help validate their predictions and also to help with better and more accurate rain prediction.

Monitoring Module:
The monitoring module will consist of Raspberry pi, which collect all the reports and data from the various modules, that is, Engine breakdown reporting module, Humidity sensing module and sends it to the cloud application module for analytics. Each vehicle consists of a monitoring module.

Cloud Application Module:
This module receives data from many monitoring modules of different vehicles, performs analytics and sends related information/results to the corresponding web interface modules.

Engine Breakdown Application Module:
Based on the engine temperature data received from the monitoring module of vehicles (focusing on public transport buses), the average breakdown temperature of the engine for a particular model is computed from vehicles which have broken down. The service centre is notified about the location of the broken down vehicles. The warning temperature is set corresponding to the breakdown temperature and vehicles nearing this temperature are notified using a mobile app.

Weather Detection Application Module
The real data obtained from the monitoring modules of numerous vehicles combined with the satellite data can be used to predict weather conditions more accurately. This data can help the meteorological departments.

Web Interface modules:
This module provides the users the required information in a clean and user-friendly format. Our system consists of the following web interfaces:

Web Interface for Service Centre:
The service centre staffs are informed about the breakdown vehicles so that they can immediately send help to the vehicle. They can also obtain the general history of the particular vehicle, which will further help in diagnosing the problem with the vehicle well in advance.

Mobile Interface for the Vehicle (Bus) Drivers:

The vehicle (Bus) drivers receive warnings/alerts to get the vehicle serviced when the engine temperature of their vehicle reaches the warning temperature. This way, they can avoid their vehicle breaking down in the middle of the roads, amidst heavy traffic.

5 Results

The system could successfully warn the user of deteriorating engine health and notify the service centre of broken down vehicles. Also, the system can intelligently send notifications to the nearest service centre.

6 Scope and Future Work

This concept of applying IoT to vehicles can solve many more traffic and accident related problems by further expanding it in the following ways:

Pothole Detection and Reporting

There are IoT of accidents reported on a daily basis in news channels and newspapers, many of which can be attributed to potholes on the road. Either the driver tries to avoid a pothole thereby colliding with adjacent vehicles or loses control over the vehicle and falls/injures himself. The civic authority's attempts to cover up the potholes are hindered by the lack of prioritization as they are unaware of which potholes require immediate attention and as a result small potholes are covered up and serious ones which may cause severe accidents are left unattended. Currently, civic authorities rely on manual registration of complaints by discomfited commuters to take action. It is possible to automate reporting of potholes by sending the location of the potholes and also their priority, which is determined by the number of vehicles reporting the pothole, to the authorities. Also the traffic density will help in prioritizing which potholes need to be attended to first, because the more cars report a pothole in a particular area, the quicker the potholes of that area need to be attended to.

Traffic Jam Detection and Alternate Route Suggestions

Based on the GPS data regarding the location of the vehicles, we can detect whether the vehicles are stuck in a traffic jam taking into account the number of vehicles and speed of the vehicles. We can use this data to inform vehicles moving towards this area or via this route to take alternative routes to help ease this traffic jam.

Collision Evidence for Claiming Insurance

The extension of Internet of things to cars also helps us use data from vehicle collisions to prove the presence and involvement of entities so that when the authorities or insurance agents require proof they can access and use this data for legal proceedings.

7 Conclusion

In this paper, we explore the Internet of things technology by applying it to vehicles in a real-time situation to solve civic problems, such as traffic and accidents. The solution is feasible because most of the modern vehicles are already equipped with a range of high end sensors. The breakdown of vehicles, which affects not only the owner but also the surrounding vehicles and traffic density across the city, this suggests that warning the driver of when his vehicle may breakdown and speeding up the repair process by reporting breakdowns to service centre will alleviate this problem.

References

1. http://tf.nist.gov/seminars/WSTS/PDFs/10_Cisco_FBonomi_ConnectVehiclespdf: The Smart and Connected Vehicle and the Internet of Things.
2. "Worldwide Internet of Things Spending by Vertical Market 2014–2018 Forecast," IDC, June 2014.
3. J. A. Stankovic, "Research directions for the Internet of Things," *IEEE Internet Things J.*, vol. 1, no. 1, pp. 3–9, Feb. 2014.
4. https://www.sap.com/bin/sapcom/fr_ca/downloadasset.2014-10-oct-31-20.ceo-perspective-the-interrnet-of-things-for-the-automotive-industry-pdf.html: CEO Perspective: The Internet of Things for the Automotive Industry.
5. Chi-Man Vong, Pak-Kin Wong, Weng-Fai Ip,"Framework of Vehicle Emission Inspection and Control through RFID and Traffic Lights", Proceedings of 2011 International Conference on System Science and Engineering, Macau, China - June 2011, pp 597–561.
6. Kazi Masudul Alam, Mukesh Sainiy, and Abdulmotaleb El Saddik, "Towards Social Internet of Vehicles: Concept, Architecture and Applications", IEEE 2015.
7. Mell, P., & Grance, T. (2009, 7 10). The NIST Definition of Cloud Computing, http://www.nist.gov/itl/cloud/upload/cloud-def-v15.pdf.
8. http://www.mouser.com/applications/galileo-2-raspberry-pi-2/: Intel Galileo2 vs. Raspberry Pi2.

Cognitive Architectural Model for Solving Clustering Problem

Meenakhi Sahu and Hima Bindu Maringanti

Abstract Human in an environment sees and then perceives objects of interest before he/she tries to find the correlation and association between the various objects in the region of their interest (ROI). By doing so, the agent here, the human develops an understanding of the environment, may it be static and certain or dynamic and uncertain. This paper simulates such an ability of humans, vital to his/her understanding, after being exposed to a visual stimulus. Filtration or selective attention happens then followed by clustering based on identified associations. These clusters form the basis of understanding and stored as Concept maps inside the long-term memory. In order to simulate this feature, various techniques and clustering algorithms exist. This work is a cognitive architectural approach to tackle the clustering problem, as it is a more natural and intuitive approach followed by humans. The ACT-R architecture has been chosen for the task.

Keywords Language comprehension · Filtration · Selective attention · ACT-R · Clustering · Concept map

1 Background

Natural language is one of the important means of communication, with a fixed grammatical structure and usage rules. Language understanding and generation involve cognitive tasks, hidden, and subtle, but decipherable. Human, when exposed to the environment tries to understand the objects that belong to his/her own region of interest. Thus, the human agent initially perceives and then observes the correlation and association among the objects seen. Grouping a set of different objects into a close proximity with another is referred as clustering. The objects of one

M. Sahu (✉) · H.B. Maringanti
North Orissa University, Baripada, Odisha, India
e-mail: meenakhii@yahoo.com

H.B. Maringanti
e-mail: mhimabindu@yahoo.com

© Springer Nature Singapore Pte Ltd. 2017
S.C. Satapathy et al. (eds.), *Proceedings of the 5th International Conference on Frontiers in Intelligent Computing: Theory and Applications*, Advances in Intelligent Systems and Computing 515, DOI 10.1007/978-981-10-3153-3_70

group, with similarities, are called as one cluster. The objects of one cluster are more similar to each other, whereas these are different from objects of other clusters. In order to simulate this, various techniques and algorithms exist, called as clustering algorithms, in the domain of data mining or in unsupervised learning techniques.

The different techniques to classify the clustering are hierarchical clustering and partitioned clustering based on the properties of clusters generated [1, 2]. The different clustering algorithms are based on distance and similarity measures, hierarchical clustering, vector quantization, pdf estimation via mixture densities, graph theory-based clustering, combinatorial search techniques-based clustering, fuzzy clustering, neural network-based clustering, kernel based, and genetic algorithm-based clustering [3].

ACT-R is a cognitive architecture, implemented using LISP programming language. This architecture can be used for language and communication, Problem solving and decision making, Emotion and cognitive development of learning and memorization, Human computer interaction, etc. It is a programming environment, where the model consists of different modules. The modules are intentional, declarative, visual, and manual. The modules are associated with buffers. They are goal, retrieval, visual, and motor, respectively [4]. Each module implements buffers. Declarative memory is that part of human memory that can store items and procedural memory is the long-term memory of skills and procedures. Declarative module contains facts or intermediate results of information processing. It uses a typed data structure called CHUNK. The CHUNK has a unique name, defined as the slot and value pairs. The visual buffer act as an input buffer, percept from the environment, and manual or motor buffer acts as an output buffer (Fig. 1).

The authors adopt [5] a memory-based approach to develop the functional and cognitive model of the brain of a human. According to functions, the brain can be

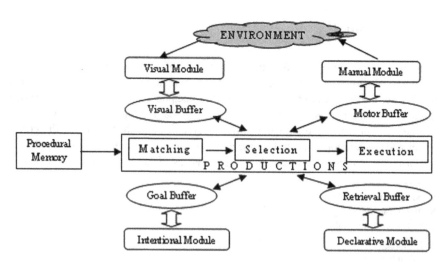

Fig. 1 Overview of ACT-R Architecture (*Source* from [4])

divided, into four different types of components. Those components are the sensory buffer memory (SBM), the short-term memory (STM), the long-term memory (LTM), and the action buffer memory (ABM). The SBM is an input buffer; it receives the inputs from the environment using sensors like eye, ear, nose, tongue, etc., and pass to STM for further processing. In STM, it will store for few minutes and the collected processed information then transferred to the LTM and ABM. LTM for permanent storage and future use, send to ABM to perform some actions like write, walk, etc.

The human brain is broadly classified into four lobes. The lobes that are involved in the different language processing functions are as follows:

i. **Frontal Lobe**: Broca's area and motor cortex are situated in this lobe [6–8]. Mainly involves in attention and conscious thinking, voluntary movement for process execution. Functions like cognition, language processing, comprehension, problem solving, planning, etc., are executed here.

ii. **Parietal Lobe**: It integrates the information from different sensory organs and creates a coherent structure [7, 9]. The dorsal and ventral streams of the occipital lobe also integrate here. Important functions involved are like spatial attention, visuospatial processing, and taste, touch.

iii. **Temporal Lobe**: Wernicke's and Hippocampus area [6–8] are situated in this lobe attributed with functions of recognition, perception, language comprehension, learning and memorizing. This lobe also takes part in audio, visual, verbal information processing, and used as long-term memory (LTM) for memorizing.

iv. **Occipital Lobe**: This is the primary visual area. Human perceives the visual stream through thalamus for processing color, orientation, and motion information of the perceived object [7, 9, 10]. Two streams originate from this lobe, one dorsal pathway toward parietal and other ventral which reaches temporal lobe, respectively.

Korbian Broadmann was the first person who provided a cytoarchitectonic [10, 11] description of the human brain cortex. According to functional localization, human brain is numbered in the range *from* 1 *to* 52 referred to as Brodmann Areas (BAs), *BA n* and further subdivided into *na, nb,* etc. Though the process of complete meaning understanding involves other sensory inputs by focusing only on one aspect of sense viz. visual, the authors [12] highlight a cognitive model for language comprehension. The process of language comprehension has been hypothesized to constitute of the following processes, which are mostly in a sequence: perception, filtration, analysis, token-meaning extraction, association, overall meaning extraction, matching, error correction, concept updation and action.

The various functional areas of the brain, called as Brodmann areas (BAs) and their functionalities are studied and the model of sentence understanding has been developed for the visual sensory input. The input signal is constrained to be a written sentence, albeit, a heard one. Hence the areas covered by such signal propagation would be different from other input signal types.

According to the authors [10], the BAs of brain that are involved in visual sentence comprehension; for the above-mentioned processes, according to the different lobes are as follows: the Occipital Lobe regions (BA 17, BA 18, BA 19), which perceive and forward to Dorsal and Ventral side, the Frontal Lobe areas involved are (BA 44, BA 45), Broca's area and motor cortex, the regions of Parietal Lobe (BA 4, BA 6) which integrate the Occipital lobe streams and the Wernicke's and Hippocampus and of the Temporal Lobe (BA 22) involved in language comprehension (Figs. 2 and 3).

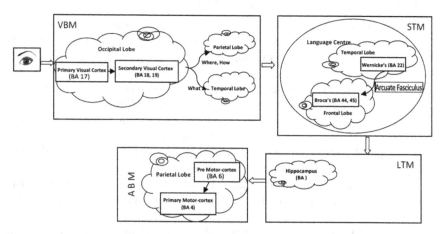

Fig. 2 Brodmann Areas (BAs) involved in the proposed model of Language Comprehension (*Source* from [10])

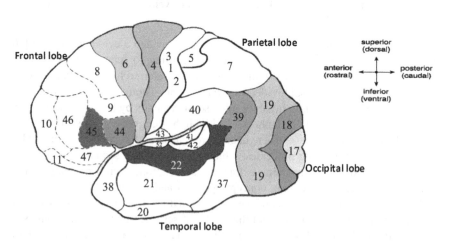

Fig. 3 BAs indicating the areas involved in Language Comprehension (*Source* from [10])

2　Contributory Work

The natural and intuitive approach to clustering problem is simulated using ACT-R cognitive architecture. The various modules of the ACT-R architecture are intentional, declarative, visual, and manual that functions as the goal or intention, working, or short-term memory (STM), visual perception and action memory, respectively. This cognitive architecture has been used to model an instance for visual perception and clustering according to the attributes of the various objects. Whatever is perceived from the environment becomes the input to this model and considered for clustering; it is not domain specific. The model is to be equipped with detailed knowledge about all objects in the environment; more the number, better classification into the appropriate cluster. The perceived input is fed to the ACT-R architecture using LISP, an AI programming language; prior to which declarative memory is filled with environmental knowledge. Then the association and clustering occurs in the STM memory and clusters are output.

The objects are displayed one after the other, on the visual window for input as shown in screen shot Fig. 4. The visual buffer (VBM) is used to perceive these. The objects are perceived and internal processing occurs.

Depending upon their similarities in attributes, they are grouped into different clusters. And finally those clusters are displayed as the result shown in screen shot

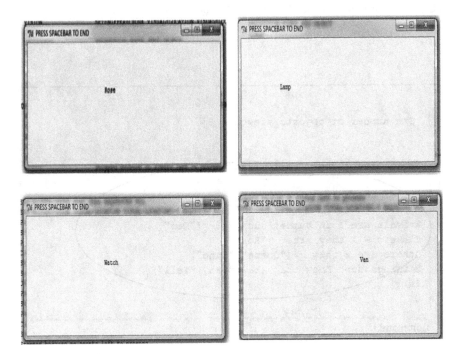

Fig. 4 Screen shots of objects displayed randomly one after other

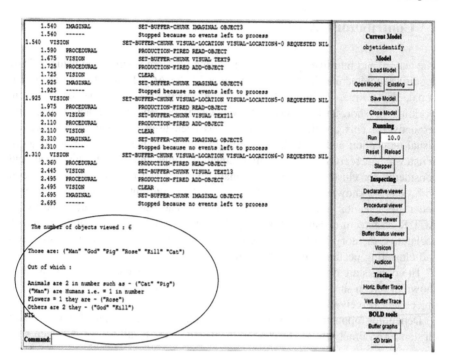

Fig. 5 Clustering results of the above perceived 7 objects

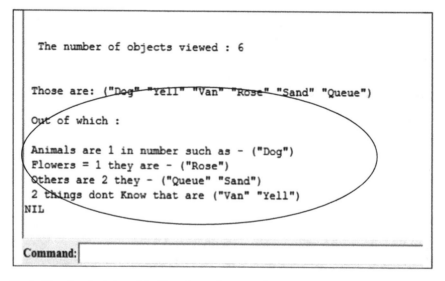

Fig. 6 Showing clustering of 6 objects in another execution

Fig. 5. In response to a query about the nature of objects/veracity of the objects, the structurally different object groups are displayed as shown in the screen shot Fig. 6. The same is true with any example set of environmental objects read into the VBM module of the architecture, which are stored in the LTM.

3 Conclusion

An intuitive and natural cognitive model for clustering has been designed using the ACT-R cognitive architecture. An instance of clustering functionality, which is very vital for understanding, has been simulated/implemented in LISP. The results obtained were found to be as per expectation.

The existing methods of clustering are simulatory, using learning models which need to be designed in terms of number of neurons, hidden layers, threshold value, activation function, data normalization, etc., yielding only approximate solutions. While this intuitive and natural model proposed, reproduces and/or recalls exactly whatever is declared inside its memory, always yielding 100 % result and hence the best.

4 Future Work

Whatever is perceived from the environment becomes the input to this model and considered for clustering; it is not domain specific. The model is to be equipped with detailed knowledge about all objects in the environment; more the number, better classification into the appropriate cluster. This task of improving the performance of this model is continuous. However, the perception modality could be varied. In addition to clustering, other complex cognitive functions could also be attempted, to completely exploit the functionality of the ACT-R architecture.

References

1. B. Evertt, S. Landau, M. Leese: Cluster analysis (2001)
2. A. Jain, M Murty and P Flynn: Data Clustering: A review, Vol. 31, No. 3, pp. 264–323 ACM Comput. Surv. (1999)
3. Rui Xu: Survey of Clustering algorithms: Transactions on Neural Networks, IEEE Vol.16, No 3, pp. 645–678, doi:10.1109/TNN.2005.845141 (2005)
4. Meenakhi Sahu, Hima Bindu M.: Cognitive Model of Sentence Understanding: ICRTCSE, IEEE Conference, pp. 307–312, ISBN - 978-81-8487-391-7, Narosa, CUB, Patna (2014)
5. Yingxu Wang, Ying Wang: Cognitive Models of the Brain: ICCI'02, International Conference on Cognitive Informatics, IEEE, ISSN – 0-7695-1724-2/02 (2002)
6. http://en.wikipedia.org

7. http://www.btbuddies.org.uk/about-high-grade-brain-tumours/areas-of-the-brain-and-their-functions.html
8. http://www.skiltopo.com/brodmann.htm
9. http://en.wikipedia.org/wiki/Visual_cortex-26
10. http://www.korbinian-brodmann.de/english/brodmann.html-20
11. http://www.trans-cranial.com/local/manuals/cortical_functions_ref_v1_0_pdf.pdf-24
12. Meenakhi Sahu, Hima Bindu Maringanti: Brain Functional Model of Natural Language Understanding: IJAER, Vol. 10, Number 13, pp. 33283–33286, ISSN - 0973-4562 (2015)
13. Meenakhi Sahu, M Hima Bindu: Multi-Modal Perception Model of Sentence Comprehension: NCCTC, pp. 82–86, ISBN - 978-3-642-24819-6 (2014)
14. Nina F. Dronkers, David P. Wilkins, Robert D. Van Valin Jr., Brenda B. Redfern, Jeri J. Jaeger: Lesion Analysis of the Brain Areas Involved in Language Comprehension: Cognition 92, 145–177, doi:10.1016/j.cognition.2003.11.002, Elsevier (2004)
15. And U. Turken, Nina F. Dronkers: The Neural Architecture of the Language Comprehension Network: Converging Evidence from Lesion and Connectivity Analyses: Frontiers in Systems Neuroscience, doi:10.3389/fnsys.2011.00001 (2011)
16. Mark D'Esposito: From Cognitive to Neural Models of Working Memory: Phil. Trans. R. Soc. B 2007 362, doi:10.1098/rstb.2007.2086 (2007)
17. Angela D. Friederici: The Brain Basis of Language Processing from Structure to Function: doi:10.1152/physrev.00006.2011
18. Mazahir T. Hasan, Samuel Hernández-González, Godwin Dogbevia, Mario Treviño, Ilaria Bertocchi, Agnès Gruart, José M. Delgado-García: Role of Motor Cortex NMDA Receptors in Learning-Dependent Synaptic Plasticity of Behaving Mice: Nature Communications, doi:10.1038/ncomms3258 (2013)
19. Alan Baddeley, Graham J. Hitch: Working Memory: doi:10.4249, Scholarpedia, 3015 (2010)
20. http://www.sciencedaily.com, ACT-R Environment Manual Working Draft Dan Bothell; http://act-r.psy.cmu.edu, ACT-R 6.0 Reference Manual Working Draft Dan Bothell; http://act-r.psy.cmu.edu

Action Classification Based on Mutual Difference Score

Shamama Anwar and G. Rajamohan

Abstract Human action recognition refers to the classification of human action from video clips automatically. Images extracted from the video clips at regular time interval are processed to identify the action contained in them. This is done by comparing these images with images taken from appropriate standard action databases. Thus, human action recognition becomes the task of verifying the similarity between two images. This paper proposes *mutual difference score* as a measure of similarity between two images. The proposed measure has been validated using the Weizmann and KTH datasets.

Keywords Entropy · Human action recognition · Joint entropy · Mutual difference score · Mutual information

1 Introduction

Human action recognition has applications in various fields like surveillance, sports, action prediction, traffic analysis, etc. It refers to the process of extraction of *frames* (*snapshots* or *images*) from the video clips at specified time interval and classifying them into apposite actions. These frames form the basis for human action recognition by matching them with template images or standard datasets. The Weizmann dataset consisting of 10 action classes and 90 videos [4], KTH dataset consisting of six action classes and 600 videos (192 for training, 192 for validating and 216 for testing) [12], UCF Sports dataset consisting of 182 videos for six action

S. Anwar
Department of Computer Science and Engineering, Birla Institute of Technology, Ranchi, India
e-mail: shamama@bitmesra.ac.in

G. Rajamohan (✉)
Department of Manufacturing Engineering, National Institute of Foundry & Forge Technology, Ranchi, India
e-mail: grajamohan.nifft@gov.in

© Springer Nature Singapore Pte Ltd. 2017
S.C. Satapathy et al. (eds.), *Proceedings of the 5th International Conference on Frontiers in Intelligent Computing: Theory and Applications*, Advances in Intelligent Systems and Computing 515, DOI 10.1007/978-981-10-3153-3_71

711

classes [11, 14], etc., are some of the datasets widely used for human action recognition. The Weizmann and KTH datasets are used in the present work due to their popularity.

2 Related Work

Shannon generalized the *thermodynamics* concept of *entropy* to make it applicable to *information theory* [13]. He used it as a measure of information content in a message in terms of its uncertainty. A logarithmic of the probability distribution of events was used for calculating the entropy. Guisado et al. [5] used Shannon's entropy to classify the emergent behaviour of laser system dynamics during simulation. Marvizadeh and Choobineh highlighted the application of entropy for dispatching of automated guided vehicles, ranking and selection of simulated systems based on their mean performance and comparison between random variables using cumulative probability distributions [8]. Thum proposed one of the first applications of entropy in image processing. He established the connection between the histogram and entropy of an image and also suggested a method for quick and efficient estimation of entropy [15]. Kapur et al. proposed a Shannon entropy based algorithm for thresholding an image using its histogram. They used the global property of histograms for segmentation of images [7]. Pal and Pal mooted the concept of *probabilistic entropy* based on the exponential behaviour of information gain and emphasized its use in image segmentation. They also developed two algorithms based on global entropy and two more algorithms based on local and conditional entropies [10]. Finlayson et al. proposed an entropy minimization based method for finding the invariant direction on intrinsic, shadow-free images [2]. Yanai and Barnard used the probabilistic region selection for images and computed an entropy measure called the *visualness*, and suggested it as a measure for automatic image annotation [16]. Huang et al. used Shannon entropy and image resolution to construct a logical predicting model for image similarity [6]. Mistry et al. used joint histograms to find the image similarity in medical images [9].

Some of the works related to the use of mutual information (MI) are as follows. Fish et al. proposed a mutual information based feature selection method for human action recognition. The action recognition was done using a binary decision tree based on Naïve Bayes classifier [3]. Russakoff et al. used the mutual information computed on (a) pixel-by-pixel basis and (b) regional basis (considering the neighbouring pixels as well) for measuring the similarity between images [12]. Yuan et al. proposed a branch and bound algorithm for 3-D bounding box cropping by maximizing the mutual information of the features and actions and validated it using the KTH dataset [18]. A method of obtaining the similarity between different action images has been proposed by using a distance measure [17]. In this paper, mutual information has been used to propose *mutual difference score* as a new similarity measure between two images.

3 Mutual Difference Score (MDS)

Mutual difference score is based on the mutual information, which in turn is based on individual entropies of two images and their joint entropy. The Shannon entropy of an image A can be defined as

$$H(A) = - \sum_{i=1}^{G} p(i) \cdot \log(p(i)). \tag{1}$$

where G is the number of grey levels of image's histogram ranging between 0 and 255 and $p(i)$ is the normalized frequency of occurrence of each grey level. The $p(i)$ value can be calculated using Eq. (2), where $f_k(i)$ represents the frequencies in the grey level k and N represents the number of states $(=2^n$, where n is the number of bits in pixels).

$$p_k = \frac{\sum_{i=1}^{G} f_k(i)}{N}. \tag{2}$$

Joint entropy is a measure of uncertainty associated with a set of variables (here a set of images) [9]. The joint entropy $H(A, B)$ of a pair of images (discrete variables) A and B with a joint distribution $p(i, j)$ is defined as

$$H(A, B) = - \sum_{i \in A} \sum_{j \in B} p(i,j) \cdot \log(p(i,j)). \tag{3}$$

Joint entropy in grey level images is computed using joint histograms. Each point pair (i, j) in a joint histogram specifies the number of intensity values having intensity i in the first image and intensity j in the second image. Zero entries are then removed from the joint histogram so as to calculate the logarithm of joint probability for use in Eq. (3). The resulting joint histogram is finally normalized. As joint entropy finds the uncertainty between two images, a lower uncertainty value means higher similarity between them. The *mutual information* is a commonly used image similarity measure that makes use of both individual and joint entropies [10, 12]. Taking two images A and B, their mutual information (MI) can be calculated as

$$MI = H(A) + H(B) - H(A, B). \tag{4}$$

where $H(A)$ and $H(B)$ are the individual entropies of the images and $H(A, B)$ is their joint entropy. It is evident from Eq. (4) that minimizing the joint entropy increases the MI value. Both the joint entropy and MI are computed for the overlapping parts of the images and these measures are dependent on the size and extent of the overlap. The MI works better than joint entropy even in cases where background subtraction is not done due to unavailability of the background image.

Assume that there are k images in an action class. On comparing the input image with each image of the action class, k numbers of MI values can be calculated using Eq. (4). It may be more appropriate to assign a single value to the entire action class. The *mutual difference score*, expressed as in Eq. (5), is introduced for this purpose.

$$MDS_{action\_class} = \frac{\sum_{i=1}^{k} MI_i}{\max(MI) - \min(MI)}. \tag{5}$$

The mutual difference score will be low for similar action classes, while it will be high for dissimilar action classes. The algorithm for computing the mutual difference score is shown in Fig. 1.

Input   Given image A and various *action classes*
Output   Mutual Difference Score
Steps
    For each *action class* C
        For each *agent* P of an *action class* C
            For each *image* B of the *agent* P
                • Calculate the Individual Entropies $H(A)$ and $H(B)$ of Images A and B using Eq. (1).
                • Compute the Joint Histogram of Images A and B.
                • Remove all Zero Entries in the Joint Histogram.
                • Calculate the Joint Probability (p).
                • Calculate the Joint Entropy of Images A and B using Eq. (3).
                • Compute the Mutual Information (MI) between the Images A and B using Eq. (4).
            End
            • Compute the MI for the agent of the action class using,

$$MI_{C,P} = \frac{\sum_{i=1}^{k} MI_{C,P}(i)}{k}$$, where k is the number of images (frames) in the

action class C performed by agent P.
        End
        • Compute the Mutual Difference Score for the action class using,

$$MDS_C = \frac{\sum_{i=1}^{n} MI_C(n)}{\max(MI_C) - \min(MI_C)}$$, where n is the number of agents.

    End

Fig. 1 Algorithm for calculating the mutual difference score

4 Results and Discussion

As first step in human action recognition, *frames* (*images* or *snapshots*) are extracted from the desired video clips at some specified time interval by running the video clip on a media player and executing a small piece of code at the command prompt. The VLC Media Player is used in the present work.

The Weizmann and KTH datasets have been used to validate the proposed human action recognition method. The Weizmann dataset consists of 10 action classes, viz. walk, run, bend, side gallop, jump, wave by one hand, wave by two hands, jump in place, jumping jack and skip, performed by different agents. The agents are different people performing the same action, but in their own style. A total of 126 images have been taken (first 7 action classes, performed by 3 agents and 6 images per agent) from this dataset. Figure 2 shows the partial list of these images. The KTH dataset consists of 6 action classes performed by different agents, viz. walk, run, jog, hand waving, hand clapping and boxing. A total of 72 images (3 agents and 4 images per agent per action class) have been taken from this dataset. Considering different agents performing the same action adds some variability in the manner in which the action is performed. The use of such variability will increase the efficiency of the action recognition method.

The noise-free input image, containing the action to be classified, is first resized to match the size of images in the standard dataset. Noisy input images can be improved using some image enhancement method, such as improved decision median filter [1], after resizing. The input image is then background subtracted to separate the object of interest from its background. After the background subtraction, individual and joint entropies of input image and images from dataset are

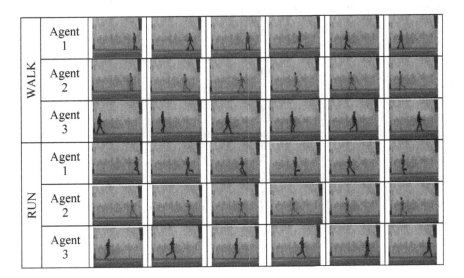

Fig. 2 Partial list of images extracted from Weizmann dataset videos

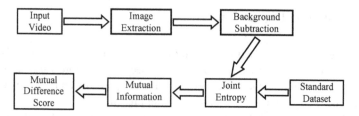

Fig. 3 Steps involved in computing the mutual difference score

calculated. This joint entropy is used to calculate the MI, which in turn is used to calculate the MDS (Fig. 3). The MI values for different images in the action classes considered for the input image taken from *run action class* of Weizmann dataset are shown in Table 1 and the MDS values for all the action classes are shown in Table 2.

Table 1 Mutual information values for different images from the Weizmann dataset and the input image from *run class* (*MI value in last column is independent of agent*)

Action class	MI values for each image			MI values for whole action class			
	Agent 1	Agent 2	Agent 3	Agent 1	Agent 2	Agent 3	—
Walk1	0.6029	0.5949	0.6116	0.595	0.591	0.636	0.607
Walk2	0.6121	0.6142	0.6708				
Walk3	0.5578	0.5456	0.6017				
Walk4	0.6301	0.6336	0.6740				
Walk5	0.6019	0.5888	0.6439				
Walk6	0.5703	0.5707	0.6139				
Run1	0.5964	0.5906	0.5074	0.593	0.591	0.506	0.563
Run2	0.6162	0.6158	0.5157				
Run3	0.5546	0.5556	0.5017				
Run4	0.6316	0.6215	0.4971				
Run5	0.5817	0.5944	0.5023				
Run6	0.5799	0.5675	0.5116				
Bend1	0.6284	0.6013	0.6274	0.606	0.601	0.603	0.603
Bend2	0.6247	0.6259	0.6218				
Bend3	0.5679	0.5613	0.5655				
Bend4	0.6343	0.6278	0.6296				
Bend5	0.6028	0.5977	0.5977				
Bend6	0.5819	0.5831	0.5760				
Jump1	0.6018	0.6023	0.6044	0.604	0.603	0.608	0.605
Jump2	0.6302	0.6255	0.6235				
Jump3	0.5695	0.5687	0.5681				
Jump4	0.6365	0.6341	0.5994				
Jump5	0.6052	0.6049	0.6381				
Jump6	0.5834	0.5825	0.6172				
Side1	0.5892	0.5910	0.6046	0.589	0.587	0.622	0.599
Side2	0.6152	0.6089	0.6590				
Side3	0.5502	0.5534	0.5901				
Side4	0.6250	0.6167	0.6777				
Side5	0.5861	0.5869	0.6037				
Side6	0.5696	0.5659	0.5995				

(continued)

Table 1 (continued)

Action class	MI values for each image			MI values for whole action class			
	Agent 1	Agent 2	Agent 3	Agent 1	Agent 2	Agent 3	—
Wave-11	0.5975	0.5889	0.6115	0.595	0.608	0.624	0.609
Wave-12	0.6171	0.6116	0.6787				
Wave-13	0.5617	0.6096	0.5676				
Wave-14	0.6256	0.6883	0.6634				
Wave-15	0.5935	0.5863	0.6248				
Wave-16	0.5751	0.5639	0.5975				
Wave-21	0.5908	0.5695	0.6266	0.592	0.593	0.631	0.605
Wave-22	0.6154	0.6065	0.6514				
Wave-23	0.5586	0.5502	0.6206				
Wave-24	0.6228	0.6510	0.6642				
Wave-25	0.5922	0.6265	0.6223				
Wave-26	0.5712	0.5590	0.6048				

It may be seen from Table 2 that the *run* class has the lowest MDS. Therefore, it may be concluded that this action has been classified correctly. Another key observation is that the MDS is very high for actions that are very different. The MDS between *bend* and *run* (*input image*) classes is the highest as they are very different in terms of body part movements while the MDS between *walk* and *run* (*input image*) classes is less as they are more similar actions. The MDS is also independent of the agent performing the action and the direction in which the action is performed. Similar computations have been performed for KTH dataset also

Table 2 MDS for Weizmann dataset

Action	MDS
Walk	13.48
Run	6.47
Bend	201
Jump	121
Side Gallop	18.15
Wave1	21
Wave2	15.51

Table 3 MDS for KTH dataset

Action	MDS
Walk	14.25
Run	23.91
Jog	19.00
Wave	306.0
Clap	15.89
Boxing	10.42

Table 4 Confusion matrix for Weizmann dataset

%	Walk	Run	Bend	Jump	Side gallop	Wave1	Wave2
Walk	100	0	0	0	0	0	0
Run	5	95	0	0	0	0	0
Bend	0	0	100	0	0	0	0
Jump	0	0	0	100	0	0	0
Side gallop	0	0	0	0	100	0	0
Wave1	0	0	0	0	0	100	0
Wave2	0	0	0	0	0	0	100

Table 5 Confusion matrix for KTH dataset

%	Walk	Run	Jog	Hand waving	Hand clapping	Boxing
Walk	90	5	5	0	0	0
Run	0	85	15	0	0	0
Jog	0	5	95	0	0	0
Hand waving	0	0	0	100	0	0
Hand clapping	0	0	0	0	100	0
Boxing	0	0	0	0	0	100

taking an image from *boxing* class. The table of mutual information values for this dataset is not shown considering the length of paper. However, its MDS are shown in Table 3. It may be seen that the *boxing* class has the lowest MDS and hence the action is taken to be classified properly.

The classification accuracy is represented using the confusion matrix (Table 4 for the Weizmann dataset and Table 5 for KTH dataset), created by testing of 20 images per action class. The values in the confusion matrix are basically the ratio of number of correct classifications to the total number classifications tried. The classification accuracy has been found to be about 99.3% for Weizmann dataset and 95% for KTH dataset. These values are higher than or comparable to one of the reported accuracy of 96% [3] and higher than the 93.3% accuracy reported for KTH dataset [17].

5 Conclusion

Mutual difference score (MDS) has been proposed in this paper as a measure to verify the similarity between two images for human action recognition. Performance of the proposed measure has been verified using the images of different action classes, taken from Weizmann and KTH datasets. The classification accuracy has been found to be about 99% for Weizmann and 95% for KTH datasets. The proposed measure is also independent of the agent performing the action. It may be computationally faster than some existing action recognition methods as it does not require classifiers or training.

References

1. Bhateja, V., Malhotra, C., Rastogi, K. Verma, A.: Improved Decision Median Filter for Video Sequences Corrupted by Impulse Noise, Int. Conference on Signal Processing and Integrated Networks, 716–721 (2014)
2. Finlayson, G. D., Drew, M. S., Lu, C.: Intrinsic Images by Entropy Minimization. 8th European Conference on Computer Vision, LNCS, 582–595 (2004)
3. Fish, B., Khan, A., Chehade, N. H., Chien, C., Pottie, G.: Feature Selection based on Mutual Information for Human Activity Recognition, IEEE International Conference on Acoustics, Speech and Signal Processing, 1729–1732 (2012)
4. Gorelick, L., Blank, M., Shechtam, E., Irani, M., Basri, R.: Actions as Space - Time Shapes. Tenth IEEE International Conference on Computer Vision - ICCV (2005)
5. Guisado, J. L., Jimenez-Morales, F., Guerra, J. M.: Application of Shannon's entropy to Classify Emergent Behaviors in a Simulation of Laser Dynamics. Computational Methods in Sciences and Engineering, 213–216 (2003)
6. Huang, Q. M., Tong, X. J., Zeng, S., Wang, W. K.: Digital Image Resolution and Entropy. IEEE International Conference on Machine Learning and Cybernetics, 3, 1574–1577 (2007)
7. Kapur, J. N., Sahoo, P. K., Wong, A. K.: New Method for Gray-Level Picture Thresholding using Entropy of the Histogram. Computer Vision, Graphics and Image Processing, 29, 273–285 (1985)
8. Marvizadeh, S. Z., Choobineh, F. F.: Entropy based Dispatching for Automatic Guided Vehicles. International Journal of Production Research, 52(11), 3303–3316 (2014)
9. Mistry, D., Banerjee, A., Tatu, A.: Image Similarity based on Joint Entropy (Joint Histogram). International Conference on Advances in Engineering and Technology (2013)
10. Pal, N. R., Pal, S. K.: Entropy: A New Definition and its Application. IEEE Transactions on Systems, Man and Cybernetics, 21(5), 1260–1270 (1991)
11. Rodriguez, M. D., Ahmed, J., Shah, M.: Action MACH: A Spatio-temporal Maximum Average Correlation Height Filter for Action Recognition. Computer Vision and Pattern Recognition (2008)
12. Russakoff, D. B., Tomsai, C., Rohlfing, T., Maurer Jr., C. R.: Image Similarity Using Mutual Information of Regions. 8th European Conf. on Computer Vision, 596–607 (2004)
13. Shannon, C. E.: A Mathematical Theory of Communication. The Bell System Technical Journal, 27, 379–423 (1948)
14. Soomro, K., Zamir, R. A.: Action Recognition in Realistic Sports Videos. In Computer Vision in Sports. Springer International Publishing (2014).
15. Thum, C.: Measurement of the Entropy of an Image with Application to Image Focusing. Optica Acta: International Journal of Optics, 31(2) (1984)
16. Yanai, K., Barnard, K.: Image Region Entropy: A Measure of Visualness of Web Images Associated with One Concept. 13th ACM Int. Conference on Multimedia, 419–422 (2005)
17. Yao, B., Khosla, A., Li, F. F.: Classifying Actions and Measuring Action Similarity by Modeling the Mutual Context of Objects and Human Poses. 28th International Conference on Machine Learning, Bellevue, USA (2011)
18. Yuan, J., Liu, Z., Wu, Y.: Discriminative Subvolume Search for Efficient Action Detection. IEEE Conference on Computer Vision and Pattern Recognition (CVPR), 2442–2449 (2009)

Preprocessing and Feature Selection Approach for Efficient Sentiment Analysis on Product Reviews

Monalisa Ghosh and Gautam Sanyal

Abstract In the recent years opinion mining plays an important role by business analyst before launching a product. Opinion mining mainly concerns about detecting and extracting the feature from various opinion rich resources like review sites, discussion forum, blogs and news corpora so on. The data obtained from those are highly unstructured in nature and very large in volume, therefore data preprocessing plays an essential role in sentiment analysis. Researchers are trying to develop newer algorithm. This research paper attempts to develop a better opinion mining algorithm and the performance has been worked out.

Keywords Information retrieval · Web data analysis · Preprocessing · Opinion mining · Feature selection · N-gram model

1 Introduction

Sentiment analysis also known as opinion mining is the process of determining the emotional tones behind a series of words, in recent years it has been receiving a lot of attention from researchers. This field has many interrelated sub problems rather than a single problem to solve, which makes this field more challenging. Sentiment classification task is performed for several reasons like to indicate the ups and downs of overall attitude of a brand or product, to pull out customer feedback on some topic or brand, to compare the attitude of one product with another. Sentiment analysis performed not for marketing purpose only, it has been useful in other areas also politics; law/policy making; social media monitoring etc. Many online sites including epinions.com, amazon.com, ebay.com, are highly depends on customers

M. Ghosh (✉) · G. Sanyal
Department of Computer Science and Engineering, National Institute
of Technology, Durgapur, West Bengal, India
e-mail: monalisa_05mca@yahoo.com

G. Sanyal
e-mail: nitgsanyal@gmail.com

© Springer Nature Singapore Pte Ltd. 2017
S.C. Satapathy et al. (eds.), *Proceedings of the 5th International Conference on Frontiers in Intelligent Computing: Theory and Applications*, Advances in Intelligent Systems and Computing 515, DOI 10.1007/978-981-10-3153-3_72

feedback for product evaluation. Properly identified reviews present a baseline of information that indicates ideal levels and supports the business intelligence [1].

This research paper discuss the pre processing [2, 3] technique which applied on structured as well as unstructured data to perform sentiment classification task in a significant way. Our work aims to explain how to get preprocess the online reviews in order to identify the sentiment and determine the sentiment polarity [4, 5] whether it is positive or negative [5–7].

The rest of the paper is structured as follows. Section 2 presents the existing works which can relate with our approach. Then Sect. 3 describes the various pre-processing techniques which required to process huge volume of user generated content. Section 4 explains in detail the feature selection techniques that are found to be suitable. Sections 5 and 6 deals with the classification technique. Section 7 presents details about experimental setup and elaborate the results also. Section 8 concludes the proposed method by providing a summary of the work.

2 Related Work

Sentiment Analysis is the field where many studies have been carried out on Sentiment-based Classification. Sentiment classification technique can help researchers to identify first whether a text is subjective or objective and then to determine whether the subjective text contain positive or negative sentiments. There are mainly two approaches considered from previous sentiment classification studies, *machine learning approach* and *lexicon based approach* or *semantic orientation approach* [6, 8, 9]. Both approaches have their own advantages and disadvantages and here we discuss some related works with these methods as well as their combined approaches.

The task of automatic word classification according to their polarity was may be the first major attempt by Hatzivassiloglou and Mc Keown (1997); the Wall Street Journal corpus was used instead of internet to determine whether a word was positive or negative [1].

Jiang et al. [10] work on a novel approach by using WordNet. All features for polarity mining extracted and stored in seed list, then checked with extracted opinion word from reviews. With the help of WordNet, whenever a synonym is matched it stored in seed list with same polarity while with antonym it is stored with opposite polarity and consistently seed list keeps on increasing and updated.

It means that one can estimate the polarity of a word according to its connected conjunction and adjective whose polarity is known by using WordNet or Senti-WordNet [11]. But now, no lexicon could cover the whole words and its semantic orientation, because different words have different semantic orientation in different contexts.

Therefore, machine learning approach was a great deal for sentiment classification problem. Machine learning method of sentiment classification can build more features and effectively change the dataset [12–14].

Pang Bo et al. [12] applied machine learning technique with statistical feature selection methods for the data of various field like product review and movie review. They claimed that Machine learning algorithm performed very well on sentiment classification. In Zha et al. [15] categorized the positive and negative opinions on the aspects from the pros and cons of the reviews. To determine the customer opinion in the context of free text reviews, they trained the classifier by pros and cons reviews. They evaluated the performances of three supervised classifiers SVM, NB, ME using the evaluation matrices F1-Measure, which defined as the harmonic mean of precision and recall. SentiView tool [3] used as an interactive visualization system and it focuses on analysis of public sentiments for popular topics on the Internet. Uncertainty modeling and model-driven adjustment is combined in SentiView, it mines and models the changes of the sentiment on public topics, by searching and correlating frequent words in text data.

Furthermore, the statistical method TF-IDF has been successfully applied in text classification. It can evaluate how the word is important for a file set. Kim et al. [7] proposed a novel approach to use only the term frequency part of TF-IDF, as an unsupervised weighting scheme which assigned an adjusted score to each term. The comparison is done against traditional TF-IDF weighting scheme on multiple benchmark and the proposed method gives better results.

Govindarajan [13] found that there are the disadvantages for the SVM classifier that performed not up to the mark for small dataset. Then proposed to combine both classifier SVM and NB which known as Hybrid approach to perform sentiment classification tasks, with the aim of efficiently integrating the advantages of the NB and SVM.

Dang et al. [9] proposed a lexicon enhanced method for sentiment classification by incorporating both approaches Machine learning and semantic orientation into single framework. Specifically, they combined sentiment features with content- free and content-specific features used in the existing machine-learning approach.

Therefore, our previous work [1] was based on unsupervised linguistic method for classifying sentiment of online product reviews at sentence level. SentiWordNet used to calculate the overall sentiment score of each sentence, and after summing up all opinion score the review can be classified either positive review or negative review.

A large number of research papers [16–18] published in the field of sentiment analysis with novel ideas as well as new techniques. As we discussed about previous research work it's clear that most of the work mainly focuses on identifying the sentiment orientation of the text, where very few consider data pre-processing and feature selection as to improve the accuracy. This work proposed an efficient data pre-processing and feature selection technique to get better accuracy.

3 Methodology

The reviews of a specific product collected from online sources usually those online data contains a lot of noisy text as well as uninformative data. Then we consider the dataset to prepare for sentiment classification. The whole process can be summarized into few steps: online text cleaning, white space removing, tokenization, removal of stop words, stemming, negation handling, noun phrase selection and then feature selection. This section presents the architecture and functional details of our proposed study. Figure 1 show the architecture of our proposed system, which consists of different functional components.

3.1 Preprocessing Phase

3.1.1 Tokenization or Segmentation

In the pre-processing phase, tokenization can be done by splitting documents (crawled reviews) into a list of words. Reviews are scanned to extract tokens consisting of words as well as numbers and after that documents are ready to be used for further processing.

Fig. 1 Architecture of proposed framework for sentiment classification

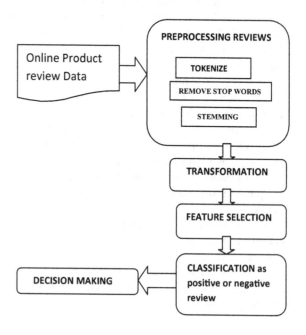

3.1.2 Removal of Stop Words

Stop words are very common and high frequency words that must be filtered to enhance performance of feature selection algorithm. The stop words removal method reduces dimensionality of the data sets, then the rest of the key words in the review corpus can be easily identified by the automatic feature extraction techniques. Some of the high frequency stop words (prepositions, irrelevant words) like "a", "of", "the", "I", "he", "she", "at", "it", "about", "and" etc. These words don't carry any sentiment information are generally known as 'functional words'. In our experiment we remove stop words for reducing file index size without effecting of user's accuracy level.

3.1.3 Stemming

Stemming is one of the essential parts of preprocessing phase during feature extraction. It is the process converts all the words of the text into their stem, or root form. Stemming is a fast and simple approach which makes the feature extraction process easier. The basic stemming process works like 'automatic,' 'automate,' and 'automation' are each converted into the stem 'automat'. The Porter's stemmer is a popular stemming algorithm for English language. The basic Stemming process can transform the words in following way (Table 1).

4 N-Grams

An n-gram language model becomes popular in researchers because of its simplicity and scalability.

N-gram model mainly works to find a set of n-gram words from a review document. Models those are generally used are 1-gram sequence or unigrams where n = 1, 2-gram sequence or bigrams where n = 2 and 3-gram sequence or trigrams where n = 3 and the sequence can be extended. The following example can define n-gram model in better way

Example— Text Data: "Something is better than nothing."

(n = 1) Unigrams: "something", "is", "better", "than", "nothing".

(n = 2) Bigrams: "something is", "is better", "better than", "than nothing".

(n = 3) Trigrams: "something is better", "is better than", "better than nothing".

Table 1 Transforming word into base form	List of Words	Stem form
	Played, plays, playing	Play
	Argue, argued, argues, arguing.	argu

5   Feature Selection

Feature selection is a process where we run through the corpus before the classifier has been trained and remove any features that seem unnecessary.

To perform sentiment classification, at the starting we considered a large no of words, terms or phrases as features that may express opinion. But very few of them actually express positive or negative opinion so we used several methods to filter those features very efficiently to improve accuracy. We applied feature weighting method with certain threshold for the targeted features. The different features weights for a feature set discussed below:

5.1   Feature Presence (FP)

Feature Presence is nothing but to check whether the feature appears in the text or not. Multiple occurrence of the same feature are ignored. We get a vector of binary values like 1 for each feature that presence in the document otherwise the value becomes 0. Feature Presence used by many researches for sentiment classification and Pang et al. [12] were first to use this method.

5.2   Feature Frequency (FF)

This method used in sentiment classification is one of the simplest methods to represent a document with a vector. Feature value is the number of times that feature occurs in the document. For an example, if the word "zoom" appeared in a document 14 times, the associated feature would have a value of 14. Sometimes many high-frequency features are very weak to distinguish the document according to low frequency features.

5.3   Term Frequency Inverse Document Frequency (TFIDF)

The TF-IDF weight is a statistical measure used to evaluate how important a word is to a document in a collection. TF-IDF value mainly consists of two scores one is term frequency and another one is inverse document frequency. Term frequency actually counts how frequent the term has appeared in a given document and inverse document frequency is calculated by dividing the total number of documents by the number of documents that a given term occurs in. Tf-IDF for each feature is calculated by using the following formulae.

$$TF - IDF = f/n.(-\log_2 N_f/D) \tag{1}$$

Where f defines the frequency count of the term in the review document of size n. N_f is the number of review documents contains the term, and D is the total number of documents in the database.

6 Classification

Supervised learning methods are commonly applied for Sentiment analysis or text classification problem to classify the opinion as positive or negative. Sentiment classification problem is generally of two types one is binary sentiment classification with positive and negative classes and another one is multi-class sentiment classification. We work on binary sentiment classification task with the use of the Naive Bayes classifier and Support Vector Machines for classifying the review documents.

6.1 Naive Bayes (NB)

Naive Bayes classifier one of the simple probabilistic classifier which is based on the applying Bayes theorem. The Naive Bayes classifier uses a feature vector matrix to determine whether a document is under positive classes or negative classes. The probability of a class c given document d is estimated using bayes' rule as follows.

$$P(c|d) = \frac{P(c)P(d|c)}{P(d)} \tag{2}$$

where P(c) is the prior probability of any document in class c, P(d|c) is the probability of a document d given it is in class c, and P(d) is the probability of the document d.

6.2 Support Vector Machine (SVM)

Support vector machines classifier is a standard non probabilistic binary classifier; it can classify data as either linear or nonlinear. These classifier includes some property like high dimensional feature space, sparse instance vector etc. SVM classifier finds a maximum margin hyperplane which used to separate the d-dimensional data perfectly into its two classes. SVM performs more efficiently to compare with others Naïve Bayes; Maximum Entropy classifier for almost all combination of features.

7 Experiments and Results

In order to evaluate our proposed method for preprocessing dataset and feature selection to perform sentiment analysis. All the experiments are performed based on the following dataset prepared by collecting online reviews of digital camera.

7.1 Dataset Preparation

In this section the online customer reviews are crawled from target review site (amajon.com, ebay.com, epinion.com) and store locally after filtering markup language tags. We extract the review of different types of digital cameras like Nikon, Sony, Fuji Film, Canon etc. To evaluate our proposed work we also used some publicly available corpus for product review polarity dataset [19] and a popular corpus on Amazon Product Review Data (more than 5.8 million reviews). We present the whole database it shows the no of items, no of customer reviews, no of sentences per review [1] (Table 2).

7.2 Performance Evaluation

We used precision, recall, F1 measures as evaluation matrices [1] to evaluate the experiment. **Precision is** the ratio of true positives among all retrieved instances; **recall is** the ratio of true positives among all positive instances and **F1 measure** is the combination of precision (π) and Recall (ρ).

$$\textbf{Precision } (\pi): \quad \frac{TP}{TP + FP} \tag{3}$$

$$\textbf{Recall } (\rho): \quad \frac{TP}{TP + FN} \tag{4}$$

Table 2 Initial review dataset on camera

Product	Reviews	Average line/review	Unique feature
Canon EOS40D	269	9	112
Nikon coolpix 4300	129	4	42
Nikon D3SLR	117	7	56

Table 3 Performance results obtained for feature selection techniques

	TF-IDF		FF		FP	
	$R^{(bp)}$	$R^{(ap)}$	$R^{(bp)}$	$R^{(ap)}$	$R^{(bp)}$	$R^{(ap)}$
Accuracy	63.23	78.49	71.12	78.32	76.337	81.5
Precision	61.29	76.34	68.86	76.66	75.33	80.21
Recall	73.21	78.21	72.21	79.31	79.21	81.66
F-Measure	72.43	79.92	71.77	77.96	77.86	82.72

$$\text{F1:} \quad \frac{2TP}{2TP + FP + FN} \tag{5}$$

The above table presents the classification accuracies on both not pre-processed and preprocessed data for each of the features matrices (TF-IDF, FF, FP). The column $R^{(bp)}$ refers to before preprocessing and $R^{(ap)}$ refers to the result after applying preprocessing method (Table 3).

8 Conclusion

Sentiment analysis is one of the most challenging fields which involves with natural language processing. It has a wide range of applications like marketing; politics; news analytics etc. and all these areas are benefited from the result of sentiment analysis. In our research we focused on to applying different preprocessing techniques to remove the noise and irrelevant features from the text for reducing feature space, while at the same time trying to improve the accuracy in sentiment orientation task. We considered two popular classifiers Support Vector Machine and Naïve Bayes in the context of sentiment classification where as the performance of SVM classifier is much better to compare with Naïve Bayes classifier for almost all combination of features, although previous research had already identified the same. The proposed method in this research paper is just an initial step towards the improvement in the techniques for sentiment classification. In future we aim to include more feature selection technique like Information Gain and Chi Square method to find optimal feature subset and the result will make the classifiers more efficient and accurate.

References

1. Ghosh, M., Kar, A.: Unsupervised Linguistic Approach for Sentiment Classification from Online Reviews Using SentiWordNet 3.0. International Journal of Engineering Research and Technology (IJERT). Vol. 2, no. 9 (2013) September (2013).

2. Haddi, E., Liu, X., Shi, Y.: The Role of Text Pre-processing in Sentiment Analysis. Procedia Computer Science. Vol. 17, 26–32 (2013).
3. Wang, C., Xiao, Z., Liu, Y., Xu, Y., Zhou, A., Zhang, K.: SentiView: Sentiment Analysis and Visualization for Internet Popular Topics. Human-Machine Systems. IEEE Transactions. Vol. 43, no. 6 (2013).
4. Hatzivassiloglou, V., McKeown, K.R.: Predicting the semantic orientation of adjectives. Proceedings of the 8th conference on European chapter of the Association for Computational Linguistics. 174–181 (1997).
5. Subrahmanian, V.S., Reforgiato, D.: AVA: Adjective-verb-adverb combinations for sentiment analysis. Intelligent Systems. IEEE 23(4), 43–50 (2008).
6. Turney, P.D.: Thumbs up or thumbs down?: Semantic orientation applied to unsupervised classification of reviews. Proceedings of the 40th Annual Meeting on Association for Computational Linguistics.417–424 (2002).
7. Kim, Y., Zhang, O.: Credibility Adjusted Term Frequency: A Supervised Term Weighting Scheme for Sentiment Analysis and Text Classification. Sentiment and Social Media Analysis. Proc. Of the 5th Workshop on Computational Approaches to Subjectivity. 79–83. Baltimore, Maryland, USA (2014).
8. Neviarouskaya, A., Prendinger, H., Ishizuka, M.: SentiFul: A Lexicon for Sentiment Analysis. IEEE Transactions on Affective Computing. Vol. 2, no.1, 22–36 (2011).
9. Dang, Y., Zhang, Y., Chen, H.: Lexicon Enhanced Method for Sentiment Classification: An Experiment on Online Product Reviews. Intelligent Systems. IEEE vol. 25, no. 4, 46–53 (2010).
10. Jiang, P., et al.: An approach based on tree kernels for opinion mining of online product reviews. In Data Mining (ICDM), IEEE 10th International Conference, 256–265 (2010).
11. Esuli, A., Sebastiani, F.: SentiWordNet: A Publicly Available Lexical Resource for Opinion Mining. Proceedings from International Conference on Language Resources and Evaluation (LREC). Genoa (2006).
12. Pang, B., Lee, L., Vaithyanathan, S.: Thumbs up? Sentiment Classification Using Machine Learning Techniques. ACL Press. In. Proceedings of the Conference on Empirical Methods in Natural Language Processing ACL Press. 79–86 (2002).
13. Govindarajan, M.: Sentiment Classification of Movie Reviews Using Hybrid Method. International Journal of Advances in Science Engineering and Technology, Vol. 1, no. 3 (2014).
14. Bollegala, D., Weir. D., Carroll, J.: Cross Domain Sentiment Classification using a Sentiment Sensitive Thesaurus. IEEE Transactions on Knowledge and Data Engineering. Vol. 25, no. 8, 1719–1731 (2012).
15. Zha, Z.J., Yu, J., Tang, J., Wang, M., Chua, T.S.: Product aspect ranking and its applications. IEEE Transaction of Knowledge and Data Engineering. Vol. 26, no. 5 (2014).
16. Fang, X., Zhan, F.: Sentiment Analysis Of Product Review Data. Journal Of Big Data. a Springer Open Journal (2015).
17. Arshad, S., Yaqub, N., Inayat, M.: Sentiment Classification Of Product Reviews At Different Level: A Survey. Proceedings of the 2nd International Conference on Engineering & Emerging Technologies (ICEET). 26–27. Lahore (2015).
18. Tan, S., Li, Y., Sun, H., Guan, Z., Yan, X., Bu, J.: Interpreting the Public Sentiment Variations on Twitter. IEEE Transactions on Knowledge and Data Engineering. vol. 26, no. 5 (2014).
19. http://www.cs.cornell.edu/people/pabo/product-review-data.

BDN: Biodegradable Node, Novel Approach for Routing in Delay Tolerant Network

Afreen Fatimah and Rahul Johari

Abstract In the recent past, Delay tolerant network has gradually evolved as a viable solution to various needs arising in intermittently connected wireless networks. In this paper, a border security architecture, has been proposed where DTN has been used to securely transfer messages within the army units of a particular region that are deployed in harsh and in adaptable terrains, thereby, ensuring integrity and security of the message. In the proposed approach, a special property of the nodes is used which makes them self-disposable in nature after their defined TTL gets over. So, the message as well as the node would be always safe from getting into unauthorized hands.

Keywords DTN · Routing · Security · Biodegradable node · Architecture

1 Introduction

Delay tolerant networks fall in the category of wireless networks which are characterized by ad hoc connectivity with frequent disconnections when the transfer of message takes place. Such disruptions may vary depending upon the geographical terrains, size of message et al. In a DTN, each node has the ability to store and then forward the data packets using opportunistic contact. But security of the message in such vulnerable networks is a challenging area. In the below presented work, a border security scenario has been showcased, where the communication among army men is guarded from any unauthorized access. For the security of message, Caesar Cipher technique is used. But in such cases, data security as well as security

A. Fatimah (✉) · R. Johari (✉)
USICT, GGS Indraprastha University, New Delhi, India
e-mail: fatimahafreen@yahoo.co.in

R. Johari
e-mail: rahul.johari.in@ieee.org

© Springer Nature Singapore Pte Ltd. 2017
S.C. Satapathy et al. (eds.), *Proceedings of the 5th International Conference on Frontiers in Intelligent Computing: Theory and Applications*, Advances in Intelligent Systems and Computing 515, DOI 10.1007/978-981-10-3153-3_73

of the DTN node, both are mandatory. To address this issue, it is proposed to use DTN nodes made up of self dissolvable biodegradable materials. Rest of the paper is organized as follows: Sect. 2 below shows the problem scenario prevailing in the border areas followed by Sect. 3 which presents the motivation of doing this work. Sections 4, 5 and 6 show the related work and our approach to this problem, followed by the conclusion and future work.

2 Problem Scenario

Today cross-border terrorism is a growing threat in the world. In this multifaceted problem, territories are trying to intrude into their own neighbor's land in order to get hold of sensitive information or resources. This has already resulted in various wars between the neighboring countries across the globe. Such wars have made the army men of respective countries to be deployed on the border region to protect their territories from getting invaded. But there have been multiple instances reported by the intelligence agencies of the countries regarding information getting leaked or the dialogues being intercepted when message is transferred among the deployed army units. This might have resulted as an effect of the harsh terrains existing in the geographical regions. So in order to plug in these loop holes, a message transfer architecture for securing both data and node to avoid data pilferage has been proposed.

Although there are various security architectures that have been applied on the border region, but no effort has been made to protect the node carrying the message. This is crucial because if enemy gets hold of the node, he can easily extract the message inside it and this may lead to compromising the security of a message.

3 Motivation of the Present Work

Since there have been multiple problems encountered in border security, there is a need to implement a security architecture that not only protects the message but also makes the message carrying node resistant to attacks. This motivated us to work in this direction by proposing a property in the DTN node that make them self disposable or self destructing. As stated in [1], the minimum hop count has increasing effect on the vulnerability of the node, so if the minimum hop count increases, it means the message is more prone to attacks. In the below presented approach, the message carrying node is dropped at the particular battalion location by the army helicopter/airplane thereby reducing the hop count to just one.

The DTN nodes used are made up of a completely biodegradable material with non toxic battery so that it does not harm the environment in due course of time. Such wireless sensor nodes can be procured in bulk to reduce the overall expenditure on such nodes and consequently reducing the budget of the security in harsh terrain areas.

4 Related Work

There has been a lot of research work done on the security in DTNs, various attacks, and security considerations [2, 3] have been studied. A hierarchical identity-based encryption technique has been proposed in [4] which deals with fine grained revocation and access method of the messages within the DTN infrastructure. A more technique for message protection has been proposed in [5] called iTrust which finds out any abnormality in the node's behavior that may take place while data transfer. Also in [6], UN peacekeeping scenario has been studied to analyze the security requirements in similar areas. Considerable research work on routing [7, 8] has also been carried out, but there have not been any significant attempts to safeguard the data and node both during message transmission.

5 Our Approach

Border defence system includes a scenario where delay tolerant network is used to send and receive messages between the army units deployed on the border of a country, say country 'A'. While in the battle field, there are many messages or short commands that need to be transferred from the commanding officer to the jawans in a confidential manner. So instead of deploying full fledged wireless networks, a Delay Tolerant Network would be used, as it is economical and also message would be encrypted to ensure that they are always safe. Figure 1 shows the architecture of this system. It depicts two countries that are in war mode. The commanding officer (CO) wishes to send a secret message to his battalions. So as per the encryption scheme, he prepares a message and encodes it in a DTN node. The helicopters then carry the nodes and drop them near their respective battalions. In the meanwhile, sensing some activity over the border, enemy team may also try to retrieve the message. But due to highly sensitive nature of the node, any type of tampering or wrong credentials may lead to its decomposition and thus making the message safe from unauthorized access.

Our aim is to deploy DTN in the border areas or war zones such that the message security, which includes authenticity, integrity and confidentiality of the message is maintained. For this, we use an encryption scheme to encrypt the message and also a special type of DTN node which is self destructive or biodegradable [9] in nature.

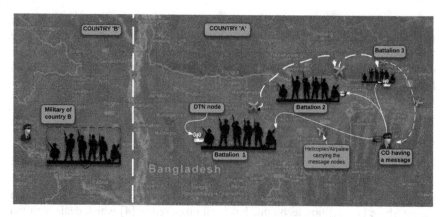

Fig. 1 Architecture of the proposed usage of biodegradable nodes in border defence system [15]

This would ensure that the DTN nodes sprinkled all over the border region do not eventually cause environment pollution of any sort after their message is read by the authorized personnel. But the major challenge here lies in the security and reduction of vulnerability of the message inside the node.

5.1 Structure of DTN Node

The proposed structure of DTN node is similar to the imote2 [10] developed by Crossbow Technology Inc. The node weighing around 10–12 g is a tiny networking element which has the capability of storing short messages up to 256 kb in size. This size may vary as per the type of node that we choose to use but here we are using nodes with only 256 kb size. These nodes are powered with a 256 kb RAM, a transceiver and a processor of a particular make, depending on the manufacturer. The power consumption of these nodes is 31 mA when the radio is off and 44 mA when it is on. The data rate is 250 kb/s.

5.2 Message to Be Encoded in the Node

As stated above, the main aim of the DTN node is to convey the message and after this is done, self destroy itself or start degrading. Therefore, it is foremost that the message should be comprehensive enough to guide the receiver in this direction. The message header consists of all the information regarding the Battalion that it is meant for.

Table 1 Message header

Latitude (512 B)	Longitude (512 B)	Drop_Time (512 B)
No. of nodes (200 B)		Load_Time (512 B)
Node ID (512 B)		Size (300 B)
TTLn (262 B)	Crack_Time (262 B)	CO-ID (512 B)
Key(1 KB)		
Encryption (1 KB)		

5.3 Description of Fields in the Message Header

1. Latitude: This gives the latitude value of the battalion present on the border.
2. Longitude: This gives the longitude value of battalion present on the border.
3. TTL: It is the TTL of the node. It is one of the most crucial fields in the header. This time is set such that if no one comes to read the message from the node, it will get destroyed after this much amount of time. It is the maximum time to live of the node.
4. Number of nodes: It is the total number of nodes present in the network.
5. Time to load: Time at which the node is loaded on the helicopter/airplane.
6. Time to drop: Time at which the node is dropped at the prescribed latitude and longitude value.
7. CO-ID: Unique ID of the commanding officer. This clearly tells about the sender of the message.
8. Node-ID: Unique ID of the node.
9. Size: Size of the message inside the node.
10. Key: This is the unique alphanumeric code of the node which is matched with the corresponding code of the receiver to unlock the node and retrieve the message.
11. Encryption: It gives the encryption technique used.
12. Time to crack: This is the time required to crack the message of the node when key is not known.

Total size of the message header is 6 kb and therefore the maximum size of the message that can be stored in the node is 250 kb. So, the whole message can be retrieved in just 1 s owing to the data rate of 250 kb/s. Table 1 shows the header format of the message.

6 Algorithms to Implement the Border Defence System

The below mentioned algorithms describe the full procedure of implementing the border defence system. Figure 2 shows a detailed flow chart of the same.

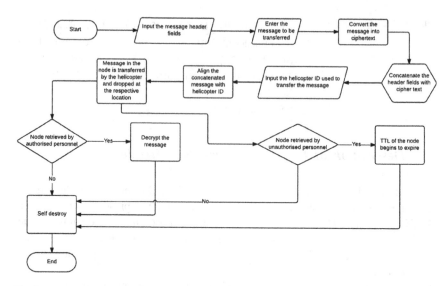

Fig. 2 Flow chart showing the proposed implementation of the border defence system

---

6.1 Algorithm 1: Populating the node

Notation

H : header of the message

PT: plain text

Trigger: Storage of data into the node

1. Input message header fields [input(H)]

    Let l': Latitude value of the Battalion

                l'': Longitude value of the Battalion

                t: TTL of the node

                n: Number of nodes in the network

                n_i: i^{th} node in the network

                t_c: Time to crack the message in the node, where $t_c > t$

                l_i: Load time of the node

                d_i: Drop time of the node

                c: CO-ID

                N: node ID

                m: size of the message to be sent

                k: key used to retrieve the message

                e: encryption technique used

2. Store header fields in the node

    $H \leftarrow \{ l', l'', t, n, t_c, l_i, d_i, c, N, m, k, e \}$

3. Enter the message in the node [enter(PT)]

    $PT \leftarrow \{$message to be sent$\}$

6.2 Algorithm 2: Obtaining the Ciphertext

Notation
CT: Ciphertext
Trigger: Encrypting the plain text message
1. Convert PT into ciphertext [encrypt(PT)]
    $CT \leftarrow$ encrypt(PT)

6.3 Algorithm 3: Concatenation

Notation
X: Prepared message ready to be sent to the node
Trigger: Preparation of the message
1. Concatenate H with CT [prepare_message (CT, H)]
2. $X \leftarrow$ {prepare_message (CT, H)}

6.4 Algorithm 4: Sending the message

Notation
Y: Prepared node ready to be transferred
hid: helicopter ID
Trigger: Transfer of the message
1. Align the message with the respective helicopter ID
2. $Y \leftarrow$ align(hid, X)
3. Transfer(Y)

6.5 Algorithm 5: Self destruction of the node

Trigger: After the node is dropped
if ($n_i \leftarrow$ retrieved by authorized personnel)
    decrypt (CT, k)
        self_destroy (n_i)
elseif ($n_i \leftarrow$ retreived by unauthorized personnel)
    if ($t_c < t$)
        $t++$
    else
        self_destroy (n_i)
else
    self_destroy (n_i)

7 Conclusion

To the best of our knowledge, this is the first, of its kind, approach for providing data security in DTN in such a manner. This proposal is bug free as no breach or pilferage of the data takes place. It is so because even if the message does not reach the authorized personnel, it will still not go into the hands of any unauthenticated person.

After the implementation of such a defence system on the border areas, the messages, and sensitive information can be kept safe from all types of hackers, crackers, and unintended personnel.

8 Future Work

As a part of the future work to our existing work, we intend to implement the above algorithms using ONE [11] simulator. The messages would be encrypted using Caesar cipher at the source and the encrypted message would then be processed as in Algorithms 3 and 4 above. For routing of the data packets, as presented in [12], we would use any of the routing algorithms like MaxProp [13] or Epidemic [14] to transfer such messages. This architecture can be deployed by defence establishments in order to safeguard their data without disturbing their defence budget.

References

1. Choo, Fai Cheong, Mun Choon Chan, and Ee-Chien Chang. "Robustness of DTN against routing attacks." *Communication Systems and Networks (COMSNETS), 2010 Second International Conference on*. IEEE, 2010.
2. Farrell, Stephen, and Vinny Cahill. "Security considerations in space and delay tolerant networks." *Space Mission Challenges for Information Technology, 2006. SMC-IT 2006. Second IEEE International Conference on*. IEEE, 2006.
3. Ahuja, Sachin, Rahul Johari, and Chetna Khokhar. "EAST: exploitation of attacks and system threats in network." In *Information Systems Design and Intelligent Applications*, pp. 601–611. Springer India, (2015).
4. Patra, Rabin, Sonesh Surana, and Sergiu Nedevschi. "Hierarchical identity based cryptography for end-to-end security in DTNs."*Intelligent Computer Communication and Processing, 2008. ICCP 2008. 4th International Conference on*. IEEE, 2008.
5. Punitha, K., et al. "A Secure I-Trust Scheme Towards Periodic Trust Establishment in Delay-Tolerant Networks." (2015).
6. Bhutta, N., et al. "Security analysis for delay/disruption tolerant satellite and sensor networks." (2009): 385–389.
7. Johari, Rahul, and Dhari A. Mahmood. "GA-LORD: Genetic Algorithm and LTPCL-Oriented Routing Protocol in Delay Tolerant Network." In *Wireless Communications, Networking and Applications*, pp. 141–154. Springer India, (2016).

8. Johari, Rahul, and Sakshi Dhama. "Routing Protocols in Delay Tolerant Networks: Application-Oriented Survey." In *Wireless Communications, Networking and Applications*, pp. 1255–1267. Springer India, (2016).
9. Anderson, Noel Wayne. "Non-toxic, biodegradable sensor nodes for use with a wireless network." U.S. Patent No. 8,063,774. 22 Nov. 2011.
10. http://www.xbow.jp/imote2.pdf.
11. Keränen, Ari, Jörg Ott, and Teemu Kärkkäinen. "The ONE simulator for DTN protocol evaluation." *Proceedings of the 2nd international conference on simulation tools and techniques. ICST*, 2009.
12. Afreen Fatimah, Rahul Johari, "PART: Performance analysis of routing techniques in delay tolerant networks", *International conference of internet of things and cloud computing, Cambridge, UK (ICC'16)*, ACM, 2016.
13. El Mastapha Sammou and Abdelmounaim Abdali, "Routing in Delay Tolerant Networks (DTNs), Improved Routing with MaxProp and the Model of "Transfer by Delegation" (Custody Transfer)", Int. J. Comm, Network and System Sciences, 2011, 4, 53–58.
14. A. Vahdat and D. Becker, "Epidemic Routing for Partially Connected Ad Hoc Networks," Technical Report CS-200006, Duke University, Durham, April 2000.
15. http://www.news18.com/news/india/5-2-earthquake-jolts-india-bangladesh-border-593941. html.

Minimization of Energy Consumption Using X-Layer Network Transformation Model for IEEE 802.15.4-Based MWSNs

Md. Khaja Mohiddin and V.B.S. Srilatha Indra Dutt

Abstract This analysis of MWSNs illustrates that IEEE 802.15.4 has very wide applications in the province of mobile wireless sensor networks as per the related research in this extent is concerned. One of the major investigation in the consideration of MWSNs sustains from the problem of system throughput and end-to-end delay along with the issue of energy consumption. This idle illustrates a X-layer network (cross-layer network) transformation model that can decline the problem of energy consumption and along with end-to-end delay in these networks. In this paper, the moderate model restrains four layers in the network transformation: (a) application layer, which has been utilized to update the node locality and information; (b) network layer, which has been accomplished to recognize the routing of the internetwork through links; (c) MAC (medium access control) layer, which has been centralized on the effectiveness of the networks, and (d) physical layer has also been recommended keeping the intentional view on transmission power from sensor node to the sink node. The place/location and position/status of the mobile node is interlinked in the routing transformation immediately as soon as the route finding process is successfully terminated and then it is been employed by controlling the transmission power of the MAC layer to rectify the range of the transmission with respect to the route. As per the future expectations of practical characteristics is concerned, adjacent NB(neighbor) list discovery broadcast will be engaged only for active nodes. But this paper is accomplished an modern technique, i.e., mobility aware protocol, which recognizes the mobility (velocity/speed) of the nodes so that only those respective nodes will be upgraded with its adjacent NB-list broadcasting, resulting in minimum power utilized by the network interface and also in the degradation of the energy consumption of the node's. In spite of the above concern, one more additional cause approaching in this model is the issue of bottleneck problem, which has been established due to multiple sources resulting in huge packet loss. This

Md.K. Mohiddin (✉) · V.B.S.S.I. Dutt
Department of ECE, GITAM University, Visakhapatnam, India
e-mail: khwaja7388@gmail.com

V.B.S.S.I. Dutt
e-mail: srilatha06.vemuri@gmail.com

© Springer Nature Singapore Pte Ltd. 2017
S.C. Satapathy et al. (eds.), *Proceedings of the 5th International Conference on Frontiers in Intelligent Computing: Theory and Applications*, Advances in Intelligent Systems and Computing 515, DOI 10.1007/978-981-10-3153-3_74

issue has been solved by pipe-lining those packets of the sensor which are nearer to the sink node, which results in reduction of end-to-end delay and energy utilization resulting in high system throughput. Through NS-II simulation, the results of energy consumption, system throughput, end-to-end delay, etc., has been shown.

Keywords X-layer network transformation model · IEEE 802.15.4 protocol · Energy efficiency · End-to-end delay · System throughput · Mobile WSNs

1 Introduction

As per the WSNs issues is concerned, mobility plays a vital role with various merits and demerits of the network issues. The scope of this paper is estimate the mobility of the sensor nodes by calculating the distance between the nodes which be done through the application layer with modern mechanisms [1]. A fret system of mobile sensors and stable sensors opens modern frontiers in a diversity of marine and troops applications and in some expert disciplines. Here the excitable sensor can reallocate recognition, networking, and estimating contrivance to foresee the essential coverage and indicated recognizing fidelity, to gather data from nearby stable sensors with higher energy efficiency, to employ the stable/static sensors, to retrieve and sustain the meshwork, while the wireless network and stable sensors stipulates the environmental recognition and communication [2].

In this paper, a mobility aware algorithm is been intended for resolving a specific circumstances of the problem in which bonding accessibility probabilities has been provided with highest priority. In this case, the algorithm utilized the roundness of the enhancement to exhibit an competent diversified explanation that together improves transmission power control. Here, the common/universal issue cannot be transform into a convex issue and the duality difference between best solutions to the dual and chief problems. Since the usual enhancement is non-convex, so a modern distributed heuristic algorithm has been proposed for its resolution [3].

Here, a modern scheme has been came into existence, i.e., XL detection and allocation (XLDA), which is proposed to solve the issue caused due to hidden terminal problem in IEEE 802.15.4. XLDA involves its mechanism in both physical as well as MAC layers in which the hidden devices are been dual verified. Depending upon the interference of the signal, the location and posture (position) of the device can be identified from the signals that been clashed in the PHY layer to recognize the hidden devices. The addresses that are been confirmed after verification will be updated in the hidden device address list (HDA-List) which is pipelined to operate the medium through specific time slots in various subplots according to the number of pairs of hidden devices as well as the HDA-List to suppress the possibility of causing the HDP [4].

2 Motivation for IEEE 802.15.4 Using X-Layer Network Transformation Protocol Model

In this paper, a X-layer approach has been proposed for trustworthy and efficient energy data gathering in MWSNs based on IEEE 802.15.4 disciplines. This approach focuses on energy sensitive adaptation model, which seizes the applications reliability needs as well as independently monitors the MAC layer supported by the network analysis and authentic traffic circumstances. Practically, the adaptive access parameter tuning (ADAPT) mechanism has been implemented as it is very simple and effective with less complication in which utilizes the local database to the sensor nodes depending upon the divisive approach of the protocol standards. It can be easily merged into MSWNs with any modification in the standards. This mechanism consumes less energy and its exploitation is very optimum for multiple operating circumstances [5].

The requirement to supervise the cluster of mobile entities has very vast application in various fields, which intercepts the contemplation of living things. In this paper, WSNs devices is been applied to the current problem that contains hitherto accepted insignificant consideration. It has been observed that 3 points of the solution space exists: At one end, group membership database is proactively and unitedly preserved by the individual node in the cluster, At the other end, the updates of the cluster membership is triggered reactively depending upon the least-level adjacent discovery policy, At the center, the solution adopts the ideas from the remaining two ends [6].

3 Related Work

Here, the problem of scheduling is been investigated in MWSNs. The main aim is to contribute meshwork distant enhanced TDMA scheduling that can accomplish large power efficiency, collision less and limited end-to-end delay. To execute the above objective, we prior need to construct a X-layer model surrounding the network, MAC and physical layer that focuses to reduce the energy utilized in MWSNs. This problem is been solved by converting the above approach in two subproblems: One is based on the meshwork wide distribution flow estimated from the existing model and the other is the transmitter power on individual link. After completion of the above issue an modern algorithm is been implemented for the scheduling of the TDMA utilizing reuse agenda to obtain least frame length [7].

Keeping the prospective view on low-power, low-cost, and low data rate WPANs the Zigbee protocol has been influenced by the IEEE 802.15.4 standards. This protocol is mainly utilized to originate the WSN for transmission applications. But however, due to mobility of the nodes the transmission failure occurs often in the meshwork topology which is been compensated by taking correct measures in developing the route reestablishing process but it exhausts large amount of resources, so proper scheduling is also been preferred for successful and efficient packet delivery. Therefore to enhance the PDR as well to minimize the consequences occurring due to the mobility of sensor node, a Zigbee protocol node deployment as well as tree construction framework has been introduced. As due to above mentioned enhancement the frequently existing of nodes mobility will be scheduled properly. It also involves the overhearing concept which rectifies the PDR to achieve good results [8].

While implementation of real-time network protocols, there are few challenges related to the bandwidth and end-to-end delay in MWSNs. Therefore, endowing such parameters results in the awareness of energy and QoS in various layers. In several cases, the network traffic is high due to delay might be one of the reliable parameter. Henceforth, the QoS becomes one the key parameter in real-time applications. In this paper, an EQSR protocol has been introduced to improve the lifetime of the network by using energy balancing algorithm within multiple nodes. It utilized the residual energy, buffer size and SNR to foretell the utmost next hop through the links. Depending upon the conception of service differentiation, the EQSR serves a queuing mechanism for both real and non-real-time applications [9].

In this paper, the IEEE 802.15.4 protocol is used to contour the features of PHY and MAC sublayer in the WSNs. The main aim of the MAC protocol is to navigate the sensor nodes accessibility in the unguided medium. In spite having several advantages, IEEE 802.15.4 MAC sustains from many problems that results in regressing its exploitation. Practically, IEEE 802.15.4 supported meshworks frequently deployed in the neighborhood of the other WSNs that also works in the same ISM Band that results in match with interference from the neighbor networks [10]. In this paper, rolling gray model is been forecast instead of GM(1,1) Model for reducing the energy consumption of WSNs. This proposed model delivers efficient accuracy by deducing the no. of data transmission packets which uses less memory. In this model, only recent values are claimed equal to the window size. Apart from the above, cluster-based routing protocol along with the residual energy & distance mechanisms are implemented to minimize the packet overheads while cluster functioning. In this, both first node dead and half node dead performances are been evaluated and proving the proposed model is better than GM (1,1) [11]. Figure 1 illustrates the various IEEE 802.15.4 MAC approaches to obtain efficient performance.

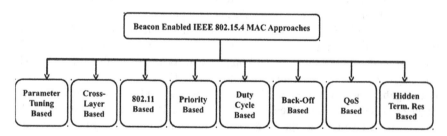

Fig. 1 Various IEEE 802.15.4 MAC approaches

4   Proposed Model

As soon as the initialization stage has been accomplished, append the data to the node present in the network, which has data of interest resulting in locality message of the mobile node. The GPS model established along the node provide the information about location of the node. After the location of the node is known, it originates transmitting the RREQ packets toward the sink node from the origin to establish the link in the medium. In this, the AODV routing protocol utilizes a cyclic adjacent sustenance information which is called as "Hello Packets," which constitutes the choice of using the NB-List from the N/W layer present in the D-Link layer, which suppresses the necessity of sending the NB discovery message which has to be sent to the MAC. After the fortunate transmission of the RREQ message from the sensor node to the sink node, then an declaration message in the form of acknowledgment is been sent back from that corresponding sink node in the form of RREP message. After the establishment of the sink node location, the next hop node will receives this information in the opposite route direction. Thereafter, that particular node estimates the distance between self and sink node intercepting which the data is being shifted to the data-link layer. Based on the distance information the MAC protocol calculates the power that been needed for the successful transmission of the Hello Packets. In this concept, two-ray ground propagation model is been utilized to compute the transmission power also its range relying upon the distance estimated by the nodes. The suitable/conformable distance has been possessed as the Euclidian distance between the corresponding nodes. Figure 2 represents the operation structure model of the wireless sensor network (WSN). To deduce the end-to-end delay and also to minimize the energy consumption, the nodes are directed only in the routes, which are active to broadcast the Hello Packets to the corresponding NB-List. Active route is defined as the link that is been created between the sensor node and the sink node after the completion of route discovery process.

4.1   Energy Model of the Efficient X-Layer Network Transformation Model

At the commencing of the starting stage of the network projection, the node originates to broadcast the Hello Packets to constitute the neighbor tables. The overall energy in the meshwork is nothing but the energy used by the individual node after successful transreception of the hello packets. After which, to detect the link, it transmits the Hello Packets to save the RREQ packets between the nodes. The sum of the power used to transmit the Hello Packets and broadcasting the RREP packets results in the total power utilized. The suggested X-layer network transformation model restricts the cyclic transmission of hello packets toward the sink nodes which are present the active link establishment. Considering all of the above, the transmitted power that is a function of distance and time taken to complete the broadcasting

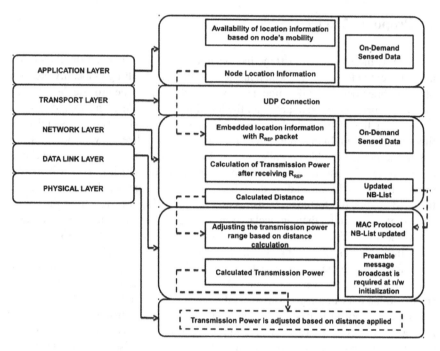

Fig. 2 Operation structure model

has been estimated depending upon the distance determined. The data packets that has been transmitted includes the power consumption which in turn delimited by the route lifetime. It is determined in the characteristics of the routing protocol.

4.2 Model Evaluation Environment

The network animation (NAM) for seven sensor nodes with single sink nodes is shown in Fig. 3. Figure 3 describes the broadcasting of the hello packets within the active routes by diminishing the demerit of the bottleneck issue problem. The suggested two-ray ground propagation model accomplished in the evaluation process in which the line of sight exists and obstacles are absent between the nodes. It has been frequently utilized in evaluating MWSNs operation. The energy model contains of few states as follows: transmission, reception, sleep, idle/listening, and transition states. All the simulation parameters has been considered as per the requirement of the results. The remaining details are being shown in Fig. 4.

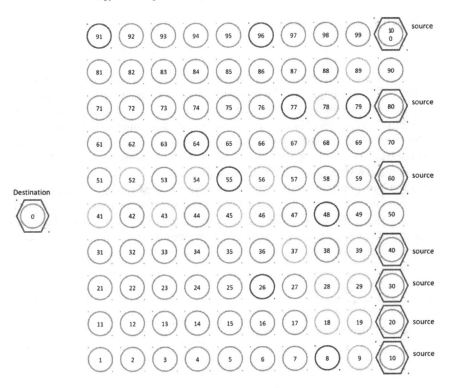

Fig. 3 Network animation process

Simulation Parameters	Values
No. of nodes	10, 20, 30, 40, 60, 80, 100
Initial Energy (Joules)	1000
Mobility	1m/s – 3m/s
Radio propagation model	Two Ray Ground Model
Transmission Range	40 meters
Simulation Time	500 seconds
Mobility Pause Period	50 seconds
Routing Protocol	AODV
MAC Protocol	IEEE 802.15.4
No. of Sources	7 nodes
Transport Protocol	UDP
Application	CBR
Packet Size	100 bytes
Queue Length	150 packets

Fig. 4 Simulation parameters

5 Results and Discussions

To compute the performance parameters of the X-layer network transformation model advances, comprehensive simulations has been calculated and performed with the help of NS-II simulation. The suggested model also implement the X-layer network approach but the only difference when compared to the existing model is that, a mobility aware protocol has been implemented for determining the mobility of the nodes from which the fast traveling nodes can be observed through which the NB-List can be updated only to those fast mobile nodes so as to energy consumption can be minimized resulting in efficient network lifetime.

Energy Consumption Results: In the network topology, energy consumption in terms of joules is shown in Fig. 5. The energy consumption per packet is also shown is Fig. 6. As per the output results are examined, the energy consumption per network as well as per data packet with view to modern model is more efficient when compared to the conventional model.

Fig. 5 Energy consumed per network

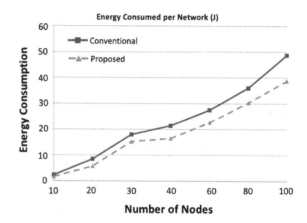

Fig. 6 Energy consumed per package

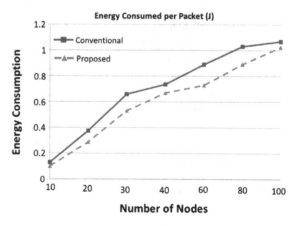

System Throughput: The system throughput is high in the proposed model as due to the low energy consumption as well as control packet minimization is also high when compared to the conventional model. Figure 7 shows the system throughput of the IEEE 802.15.4 with respect to the X-layer model. The problem of packet overhead occurs in the route between source nodes and destination node, which is due to the more number of data packets at the bottleneck of the destination node which results in the bottleneck problem.

End-to-nd Delay: The end-to-end delay is minimum in the proposed model due the less time taken for the fast mobility nodes to be updated with the NB-List. Figure 8 shows the end-to-end delay which is better in the proposed system.

Fig. 7 System throughput

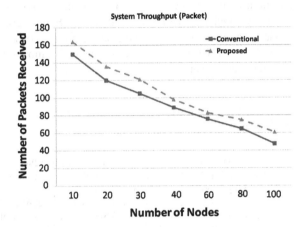

Fig. 8 End-to-end delay

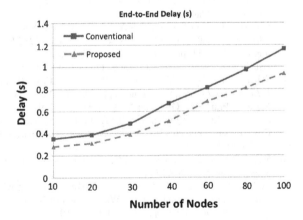

6 Conclusion and Future Scope

The proposed design standard regulates two important clockworks: Primarily, is the navigating and controlling the number of control packets being broadcasting in the meshwork to serve the assistance for the medium between the nodes. The mechanism of deduction of the control packets has an immediate effect on the broadcast packets as well as on the NB discovery packets at the routing layer. Second, is the transmitted power control mechanism that is proportionate to the place and locality of the nodes. This mechanism is valid only after the discovery of the route process. Combination of the above two important clockworks and also updating the NB-list only to those fast mobility nodes resembles in efficient energy consumption, high system throughput and minimum end-to-end delay. The bottleneck issue has been rectified to certain extent by scheduling the fast mobility nodes. Future scope for the efficient suggested model is to minimize end-to-end delay as much as possible as well as the bottleneck issue. This can be achieved by implementing few more efficient scheduling algorithm to obtain better results.

References

1. Marwan Al-Jemeli, Fawnizu A. Hussin: "An Energy Efficient Cross-Layer Network Operation Model for IEEE 802.15.4-Based Mobile Wireless Sensor Networks," *IEEE Sensor Journal*, vol. 15, no.2, pp. 684–692, February 2015.
2. Tao Yang, Tetsuya Oda, Leonard Barolli, Jiro Iwashige, Arjan Durresi and Fatos Xhafa: "Investigation of Packet Loss in Mobile WSNs for AODV Protocol and Different Radio Models," *26th IEEE International Conference on Advanced Information Networking and Applications*, pp. 709–715, March 2012.
3. Shibo He, Jiming Chen, David K.Y. Yau and Youxian Sun: "Cross-Layer Optimization of Correlated Data Gathering in Wireless Sensor Networks," *IEEE Transactions on Mobile Computing*, vol. 11, no.11, pp. 1678–1691, November 2012.
4. Hsueh-Wen Tseng, Shan-Chi Yang, Ping-Cheng Yeh and Ai-Chun Pang: "A Cross-Layer Scheme for Solving Hidden Device Problem in IEEE 802.15.4 Wireless Sensor Networks," *IEEE Sensor Journal*, vol. 11, no.2, pp. 493–504, February 2011.
5. Mario Di Francesco, Giuseppe Anastasi, Marco Conti, Sajal K. Das and Vincenzo Neri: "Reliability and Energy-Efficiency in IEEE 802.15.4/ZigBee Sensor Networks: An Adaptive and Cross-Layer Approach," *IEEE Journal in Communications*, vol. 29, no.8, pp. 1508–1524, September 2011.
6. Marco Cattani, S tefan Guna and Gian Pietro Picco: "Group Monitoring in Mobile Wireless Sensor Networks," *IEEE Transactions*.
7. Liqi Shi and Abraham O. Fapojuwo: "TDMA Scheduling with Optimized Energy Efficiency and Minimum Delay in Clustered Wireless Sensor Networks," *IEEE Transactions on Mobile Computing*, vol. 9, no.7, pp. 927–940, July 2010.
8. Yuan-Yao Shih, Wei-Ho Chung, Pi-Cheng Hsiu and Ai-Chun Pang: "A Mobility-Aware Node Deployment and Tree Construction Framework for ZigBee Wireless Networks," *IEEE Transactions on Vehicular Technology*, vol. 62, no.6, pp. 2763–2779, July 2013.
9. Bashir Yahya and Jalel Ben-Othman: "An Energy Efficient and QoS Aware Multipath Routing Protocol for Wireless Sensor Networks," *IEEE 34th Conference on Local Computer Networks*, pp. 93–100, October 2009.

10. Mounib Khanafer, Mouhcine Guennoun, and Hussein T. Mouftah: "A Survey of Beacon-Enabled IEEE 802.15.4 MAC Protocols in Wireless Sensor Networks," *IEEE Communications Surveys & Tutorials*, vol. 16, no.2, pp. 856–876, 2014.

11. Dhirendra Pratap Singh, Vikrant Bhateja & Surender Kumar Soni: "Energy Optimization in WSNs employing Rolling Grey Model," *Proc. (IEEE) International Conference on Signal Processing and Integrated Netwroks (SPIN-2014)*, pp. 801–808, Feb. 2014.

An Automated Approach to Prevent Suicide in Metro Stations

Anindya Mukherjee and Bavrabi Ghosh

Abstract Every year, we find a lot of people committing suicide in metro. To avoid such incidents, we have designed a system that would take the captured video of preinstalled CCTV cameras in metro stations and analyze them. On breaking down the captured video into frames, the region of interest for each will be calculated and their histogram found. Processing the histogram values, if danger is detected then an alert message will be fired. In future with this triggering of the message a physical barrier, installed beforehand at the edge of the platform, will come up thereby preventing the victim from making the suicide attempt, saving his/her life and hours of harassment for others.

Keywords Histogram · Trigger message · Suicide attempt · CCTV footage

1 Introduction

Since its inception, the metro rail service has served as a viable, cost-effective, and convenient mode of transportation. Yet, there is an issue that diminishes the certain mentioned glow. That issue is the rampant rate of suicide. According to the journal by Roychowdhury et al. [1], a total of 58 suicide cases have been reported since 1987 till 2008. Suicide not only takes away a person's precious life but also results in hours of delay leading to unnecessary harassment for those who rely on the uninterrupted service. Now, the death in some cases is not even caused by physical injury inflicted by the moving train, even after applying brakes the person may get electrocuted due to the high current in the third rail.

A. Mukherjee (✉) · B. Ghosh
Department of Computer Science and Engineering,
Institute of Engineering & Management, Kolkata 700091, India
e-mail: anindya.pg@gmail.com

B. Ghosh
e-mail: bavrabi@gmail.com

© Springer Nature Singapore Pte Ltd. 2017
S.C. Satapathy et al. (eds.), *Proceedings of the 5th International Conference on Frontiers in Intelligent Computing: Theory and Applications*, Advances in Intelligent Systems and Computing 515, DOI 10.1007/978-981-10-3153-3_75

Statistics show that most of the suicide attempts take place in the busy office hours, disrupting the day to day life of passengers, causing a lot of trouble. The journal by Roychowdhury et al. [1] reported an average of 15 deaths between 1 pm and 3 pm followed by an average of 13 deaths between 3 pm and 5 pm, while the death toll was comparatively less at other times of the day.

A measure to prevent all these unfortunate deaths can be taken by adopting our system. We plan to capture the live feed of existing CCTV (Closed-circuit television) cameras in metro rail stations and apply our algorithm to that.

We split the CCTV camera feed (in the form of video) into a series of frames. We choose the region of interest from each frame and do further processing. During further processing, we calculate the histogram of consecutive frames and find their difference. If the difference lies in a particular range, we identify it to be a danger situation, during which someone is trying to commit suicide. In this situation, we trigger the alert message. If the histogram difference does not lie in the specified range, then the system goes on without any interruption.

The different sections of this paper are organized as follows: the "Literature Review" section contains the summary of research works related to our topic or from where we have got hints about how to proceed with our work and what the possible ways of obtaining a certain result are. It also has discussions about how our designed system is different from others and why we cannot implement the discussed methods in our systems because of presence of a few constraints. The "Proposed Method" section contains a detailed description of how the system has been implemented, with pictorial representation. Following this section, we have the "Result and Discussion" part where the ultimate result obtained by the system is discussed in details with an overview of what the ultimate goal of the project is. The "Future Scope" section has further proposals about the hardware implementation of the project, followed by the "Conclusion" section that gives an insight of how our project can benefit society. The "References" section holds record of all the materials we referred to, for our work.

2 Literature Review

We have come across a few research works in the same field comprising various other methods to detect unusual behavior or prevent suicide. The outcome of our analysis of these systems is given below:

1. Zhong [2] developed an unsupervised technique for detecting unusual events in a large video set. In this technique, the goal was to detect unusual activity in certain scenarios, like patient monitoring in a hospital or cheating in a poker game. The researcher has made a few prototypes of the probable case that can take place and then matches them with the video he/she is capturing of the event taking place in front. For example, in case of a poker game the different types of moves are recorded and stored as a prototype. Now, when the actual game is

going on, the whole match is recorded through a camera. If it is a clean game, then there would be huge amount of similarity between the actual video and the prototype of a fair game. If any dissimilarity is found, then unusual activity is detected and in case of a poker game, it has to be some foul move for cheating. Same applies for patient monitoring in a hospital.

2. Shim et al. [3] proposed a model of suicide prevention along one side of a bridge which has been based on region of interest. Through the video footage captured by CCTV cameras, the region of interest (target danger zone) is identified and marked beforehand. Now, the images (frames) are converted to binary form and blob analysis of the frames is done. Here, the background image, that is, the river on which the bridge is situated, is constant. So, whenever anyone tries to commit suicide by climbing the railing of the bridge, a percentage of the person's binary image's blob is calculated to be present in the danger zone. If the percentage is higher than allowed threshold value, then an alarm is raised.

3. Rodriguez et al. [4] designed a crowd detection technique where optimization of joint energy function, combining crowd density function, is used to track the number of people in a crowd.

4. Ali Dağlı et al. [5] took into consideration the presence of security cameras on metro stations and that the staff cannot monitor all camera feeds simultaneously. The detection of unusual events and generation of warnings to the staff is their main goal. This paper was published in April 2013, yet, there have been multiple incidents of suicide even after using this approach.

Studying the above-mentioned works and models, we have come across a few components which are not applicable in our project. These include:

1. Like the model by Zhong et al. [2] to detect unusual activity, maintaining a prototype for this project is next to impossible as it will result in huge amount of memory usage and file usage. Also matching a live video with prototype will need some time, which will not make the project efficient and here, we need utmost efficiency as being late by a fraction of a second, can cause a life to end.

2. In our project, the maximum number of suicides occurs when the train is entering the station. So the background is not constant. Therefore, doing a blob analysis in a changing background is not a wise decision for our project as our region of interest is the tunnel through which the train is traveling. This is in contrast with the model proposed by Shim et al. [3].

3. We cannot follow the crowd density function used in the technique by Rodriguez et al. [4] as we plan to take into account the edge of the platform and the position of the camera which would not capture major part of the crowd. Therefore, following this method would not help much in our system.

4. In our paper, the method that we are following is simpler and easy to implement compared to the approach by Ali Dağlı et al. [5]. We are not specifically detecting the train when it is entering the platform, neither does the system have to detect any yellow line intrusion. Thus, computational complexity reduces and

it is better when one is going to implement it in real time because fast processing is of high importance in life and death situations.

3 Proposed Method

There are roughly four major steps in which we have carried out the work to prevent suicide in metro. First step is to connect our system to the CCTV cameras installed in the metro rail platforms and read the video stream. This video stream is read at the rate of 25 frames per second, in the video format 'RGB24'.

Following this, in the second step, the frames from the video are extracted. This is done in such a way that, at any time, we have the current frame and can also cross-reference with a subset of the elapsed frames. We have stored the current frame and its preceding frame in two separate variables. These variables contain information about the spatial distribution of pixel intensity values in raster graphics format, for consecutive frames. We retrieve these frames using a loop that ends with the stream (Figs. 1, 2 and 3).

In the third step, we perform different image processing techniques on the extracted frames (Figs. 1 and 4), within the body of the loop. Since the CCTV cameras in metro stations have fixed fields of view, the location of the tunnel does not change with each frame. Hence, select the tunnel from the entire scene by specifying coordinates to represent the boundary of the tunnel within the scene. Now, using the coordinates, we apply polygonal region of interest by calling the "roipoly" function and thus extract the region of interest (Figs. 2 and 5) from each

Fig. 1 Original preceding frame

Fig. 2 Extracting region of interest of preceding frame

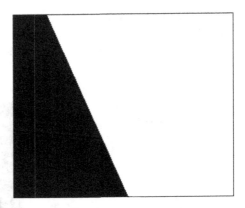

Fig. 3 Result after extraction (preceding frame)

Fig. 4 Original next frame

frame, storing the results of consecutive frames (Figs. 3 and 6) in two different variables. The tunnel is taken as the region of interest for better detection while reducing time complexity (Figs. 7, 8 and 9).

Fig. 5 Extracting region of interest of next frame

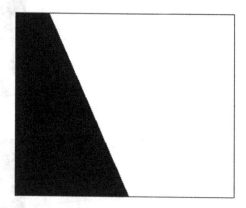

Fig. 6 Result after extraction (next frame)

Fig. 7 Region of interest of preceding frame

In the fourth step, using the previously stored frames, following the extraction of region of interest, we calculate histogram of consecutive frames. These histogram values (Figs. 8 and 10) are now stored for further comparison.

In the fifth and final step, we calculate the difference in histogram values for the consecutive frames. When there is no train in the tunnel then its histogram does not change much. Also, when the train enters the platform, the changes in the histogram take place at a certain rate. In normal situations, when no one is trying to commit suicide or no foreign object enters the tunnel, then the histogram value either lies below 0.0037 or above 0.0040. However, in both situations, if a foreign body enters the frame, the histogram change is noticeable. When a person tries to jump in front

Fig. 8 Histogram of the preceding frame

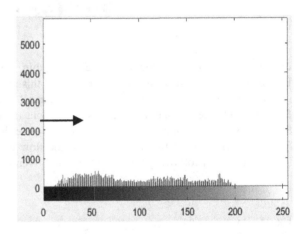

Fig. 9 Region of interest of next frame

Fig. 10 Histogram of the
next frame

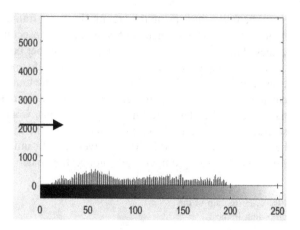

of the train, then the histogram value lies within the range of 0.0037–0.0040. Hence, when the histogram difference lies in this range, we detect an unusual scenario in the tunnel. On detection of the unusual scenario, an alert message is fired (Fig. 11) thereby alerting the authority about the intrusion in metro service. Seeing this message, the security personnel can save the person.

The algorithm used by our system and the flow of events through which our system passes on following this algorithm, are given on the next page.

Fig. 11 This is the alert
message generated when a
suicide attempt by any person
is detected

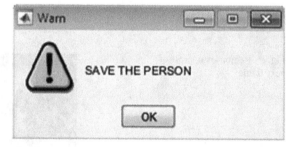

3.1 Algorithm

1. Obj ← Read video file
2. For i ← 140 to 200
3. u ← read current frame
4. a ← read next frame
5. Assign 4 coordinates values to 'r' and 'c' in order to denote the edges of the region of interest.
6. BW ← roiploy(u,r,c)
7. Calculate 'R' and 'C' of BW
8. For i ← 1 to R , j ← 1 to C
9. If BW = true
10. Display u(I,j)
11. m(i,j) ← u(i,j)
12. Repeat step 5 to 8 for 'a'
13. p(i,j) ← a(i,j)
14. [c,n] ← Histogram of 'm'
15. [c2,n2] ← Histogram of 'p'
16. d ← Difference of c and c2
17. if d>0.0037 and d<0.0040
18. Display Alert message
19. break
20. Else goto step 3

3.2   *Flow of Events*

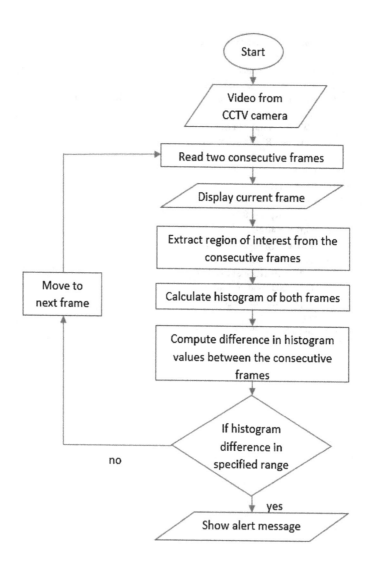

4   Result and Discussion

During execution of our system, we have used as input, recorded CCTV footage from a metro station in the form of a video file where it is seen that a person commits suicide. Our system can take video footage from CCTV cameras without needing any enhancement and it also accepts video files in Audio Video Interleaved format. If the test video is not in this format, preprocessing is needed to convert it. We have performed our experiments on an Intel® Core™ i5-3230 M machine with 4 GB of RAM, running Windows 10 64-bit.

For the result, we have designed a GUI (Graphical User Interface) window where we have made two panels. In one panel we run the video without applying the algorithm. The train passes over the person and she dies (Fig. 12). We find the person was able to commit suicide without any intervention when the algorithm is not applied.

Upon applying the algorithm, we find that as soon as the person's leg enters the region of interest (the tunnel) the system detects a danger situation based on histogram difference and an alert message is fired (Fig. 13).There is very little chance that someone will stoop to confirm the arrival of the train as the sound is enough. Hence, the possibility of generation of false alarm is very feeble. Also, if this system is installed then passengers will be told beforehand about the rules and regulations to be followed on the platform when the train has not yet arrived.

The number of experiments performed by us is 30, including both situations where suicide does not occur and also situations where suicide occurs. We found that the alert is fired only in the latter case. Thus, the success rate is 100%. Time

Fig. 12 Result without algorithm applied

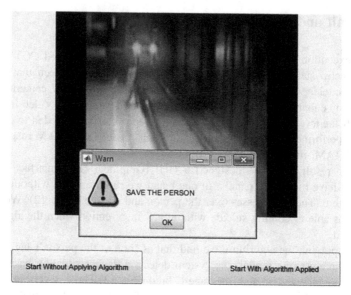

Fig. 13 Result with algorithm applied

taken to simply display 60 frames was found to be 4.21 s and time taken after processing using our algorithm was found to be 7.95 s. Hence the lag, per frame was found to be 0.062 s using the formula (1).

$$\text{lag} = (\text{time with algorithm} - \text{time without algorithm}) / \text{frames}. \quad (1)$$

The time required to generate the graphical user interface on the screen was found to be 0.027 s on the first run, and the time needed to display the message upon detection of suicide activity was faster than what could be measured.

5 Future Work

Till now, the system that we have developed can display an alert message on detection of a person trying to commit suicide in metro. That alone would not help much. In future, we plan to implement a physical barrier (Fig. 14) that will stop any person from falling or jumping on the track, using thin fiber sheets. These sheets will be installed at platform edge with sponge padding to prevent any injury. This will further automate the system by directly saving life the moment a suicide attempt is detected.

Fig. 14 Physical barrier comes up preventing the person from jumping onto the track

6 Conclusion

Suicide is the extreme form of decision a depressed mind takes when he/she feels there is no more hope left for them. Our system aims to save the lives of all these people, saving their families from grievance and also ensuring that co-passengers do not face any inconvenience. Hence, this system can be beneficial to many people.

References

1. U.B. Roychowdhury, M. Pal, B. Sukul: Suicides in Kolkata metro railway. In: Journal of Indian Academy of Forensic Medicine, 2009. vol.: 31, Issue: 2, pp. 118–121.
2. Hua Zhong: Detecting Unusual Activity in Video. In: Computer Vision and Pattern Recognition, 2004. CVPR 2004. Proceedings of the 2004 IEEE Computer Society Conference, Page(s):II-819 - II-826 Vol.2, DOI:10.1109/CVPR.2004.1315249.
3. Young-Bin Shim, Hwa-Jin Park, Yong-Ik Yoon.: A Study on Surveillance System of Object's Abnormal Behavior by Blob Composition Analysis. In: International Journal of Security and Its Applications Vol.8, No.2 (2014), pp. 333–340.
4. Mikel Rodriguez, Ivan Laptev, Josef Sivic, Jean-Yves Audibert.: Density-aware person detection and tracking in crowds by Blob Composition Analysis. In: 2011 International Conference on Computer Vision, pp. 2423–2430, DOI:10.1109/ICCV.2011.6126526.
5. Mehmet Ali Dağlı, ÇiğdemEroğluErdem: Auto-detection of unusual events in metro stations via security cameras. In: Signal Processing and Communications Applications Conference (SIU), 2013 21st, pp. 1–4,DOI:10.1109/SIU.2013.6531414.

B-Secure: A Dynamic Reputation System for Identifying Anomalous BGP Paths

A.U. Prem Sankar, Prabaharan Poornachandran, Aravind Ashok,
R.K. Manu and P. Hrudya

Abstract BGP (Border Gateway Protocol) is one of the core internet backbone protocols, which were designed to address the large-scale routing among the ASes (Autonomous System) in order to ensure the reachability among them. However, an attacker can inject update messages into the BGP communication from the peering BGP routers and those routing information will be propagated across the global BGP routers. This could cause disruptions in the normal routing behavior. Specially crafted BGP messages can reroute the traffic path from a source ASN to a specific destination ASN via another path and this attack is termed as AS Path Hijacking. This research work is focused on the detection of suspicious deviation in the AS path between a source and destination ASNs, by analyzing the BGP update messages that are collected by passive peering to the BGP routers. The research mainly focuses on identifying the AS Path Hijacking by quantifying: (1). How far the deviation occurred for a given AS Path and (2). How much credible is the deviated AS path. We propose a novel approach to calculate the deviation occurred by employing weighted edit distance algorithm. A probability score using n-gram frequency is used to determine credibility of the path. Both the scores are correlated together to determine whether a given AS Path is suspicious or not. The experi-

A.U. Prem Sankar (✉) · P. Poornachandran · A. Ashok · R.K. Manu · P. Hrudya
Amrita Center for Cyber Security Systems and Networks,
Amrita Vishwa Vidyapeetham, Amritapuri Campus, Kollam, India
e-mail: premsankar@am.amrita.edu

P. Poornachandran
e-mail: praba@am.amrita.edu

A. Ashok
e-mail: aravindashok@am.amrita.edu

R.K. Manu
e-mail: manurk@am.amrita.edu

P. Hrudya
e-mail: hrudyap@am.amrita.edu

© Springer Nature Singapore Pte Ltd. 2017
S.C. Satapathy et al. (eds.), *Proceedings of the 5th International Conference on Frontiers in Intelligent Computing: Theory and Applications*, Advances in Intelligent Systems and Computing 515, DOI 10.1007/978-981-10-3153-3_76

767

mental results show that our approach is capable of identifying AS path hijacks with low false positives.

Keywords Border gateway protocol · AS Path Hijacking · Weighted edit distance · N-Gram probability · Anomaly detection

1 Introduction

Internet can be decomposed into connected subnetworks that are under separate administrative authorities [1]. Routers act as the intermediate nodes between the networks. Routing protocol operates at the control plane in which the control flow is in the opposite direction of the data flow. Routing protocol exchanges reachability information among the routers, and finally helps the router to select the optimum path to reach a given destination. IP routing prefix is a logical subdivision of network and is expressed as CIDR notation [2]. For example, 10.30.8.0/24 is a prefix notation that indicates that 2^24 IPV4 addresses can be contained by this subnet starting from the IP address 10.30.8.0. A prefix or a collection of prefixes under the administrative control of a single organization is termed as an autonomous system (AS) [3]. The routing process between the different ASes is termed as the interdomain routing. Global scale interdomain routing on the internet is enabling by the border gateway protocol (BGP). Each AS will announce their respective organization prefixes to the other ASes as BGP update Messages via the BGP control plane. A unique ASN (autonomous system number) is assigned to each AS, which is used for the BGP routing. When an AS announces a prefix it reaches the immediate neighbor ASes. The neighbor ASes append their own ASN to the AS Path in update message and again forwards the update message to their neighbors and this goes on. Finally, every router participating in the BGP routing process learns the AS Path to reach the AS who has announced the prefix. There are 53829 ASes participating in the BGP routing system and 339754 CIDR aggregated prefixes are present in BGP routing table as of 2016 April 29 CIDR report [4].

BGP is designed with less notion of security approach in the BGP communication. BGP works on the trust agreed between different ASes. BGP has no mechanism to validate the attributes of the BGP update messages which are received from the peering ASes. Attacker can announce the BGP messages with specifically crafted attributes from his malicious AS and this will propagate across the global BGP routers. Eventually the best path selection algorithm evaluating usual means (prefix length, shorter as path vs several origin points, etc.) will choose the defined path by the attacker [5] between two ASNs. This type of man-in-middle attack performed against BGP is termed as AS Path Hijack as the attack is performed by hijacking the path and the traffic still reaches the actual destination, the attacker keeps on eavesdropping on the traffic.

Prefix Hijack is well-known type of attack faced by BGP in which the attacker will announce a prefix or a more specific sub prefix or a publically unused address

space, which belongs to the victim AS. Eventually, some portion of the traffic destined to the original owner AS of this particular prefix will go to the malicious AS. There the traffic will be dropped and this will be acting as a black hole effect for some portion of internet trying to reach this particular prefix [6]. Instead of black holing the traffic, the attacker may still forward the traffic to the actual destination AS. This is also a type of BGP MITM attack where the traffic is deviated.

Dell secure works uncovered multiple man-in-the-middle BGP attacks used to steal bitcoins. The thief earned about $83,000 in profits in more than four months, compromising 51 networks from 19 different ISPs in the year 2014 [7]. Thousands of prefixes were mis-originated and were rerouted to Indonesia and Syria. Many Internet users were routed over the path via Syria for about 1500 prefixes in the year 2014 [7]. Icelandic Traffic Diversion, Belarusian Traffic Diversion reported by Renesys in the year 2014 [8] shows some real-world cases in which the traffic was supposed be in the limit of south America or nearby neighboring routers, instead deviated to Russia or Iceland and comes back to the destination ASN.

We infer the connectivity topology of ASNs by analyzing the historic BGP data using probabilistic graphical models. The contributions of this research are as follows:

- Developing a framework capable of analyzing BGP updates and alerting AS path hijacks happening worldwide.
- The framework focuses on analyzing the AS paths between a source and destination ASN to quantify the amount of deviation present and the credibility of the deviated path.
- We propose a novel approach for calculate the deviation occurred in a given path by calculating the weighted Edit Distance from known reputed historic paths.

2   Literature Survey

Zheng et al. [9] proposed a work on BGP prefix hijack detection in realtime. The approach is mainly focused on stability of hop count as well as the path similarity. This path similarity is obtained by using hamming distance. Jian et al. [10] introduced a system, AS_CRED that focus on detecting anomalous BGP updates. After detecting the anomalous traffic, reputation checker will check the credibility of detection. Wavelet-based BGP anomaly detection is explained in [11]. Origin and subprefix hijack detection and prevention is proposed in ROVER [12] developed by Gersch and Joseph E. This approach exploits the reverse DNS concept. Shue et al. in [13] explores the maliciousness of ASes by checking the ISPs ASes in 10 popular blacklists. They have come up with a conclusion that some of the detected ASes has over 80% of IP addresses as blacklisted. Through this study they have explained the lack of security in the internet. Shivani et al. in [14] used Edit Distance algorithm as one of the features for finding AS path instability. However, the normal edit distance algorithm fails to provide the accurate measurement of the deviation occurred in the path. In this research we introduce weighted edit distance, a novel approach which is an enhancement of the Damerau–Levenshtein edit distance algorithm.

3 System Architecture

A software-based BGP router is configured with a private ASN (private ASN range is 64,512–65,534) and is peered with the University BGP router (ASN 58703). All the routing updates are forwarded to this private ASN. Proper rules are enforced at the data collector to make sure no internal prefixes are announced from the collector to the external BGP routers. Figure 1 shows deployment of the private ASN, which is peering with the University BGP router to collect the BGP update messages. University BGP router is peered with two different ISP BGP routers and participates in the BGP routing process. Eventually, at the collector we will be able to get all the AS Path information from ASN 58703 to reach all the publically announced

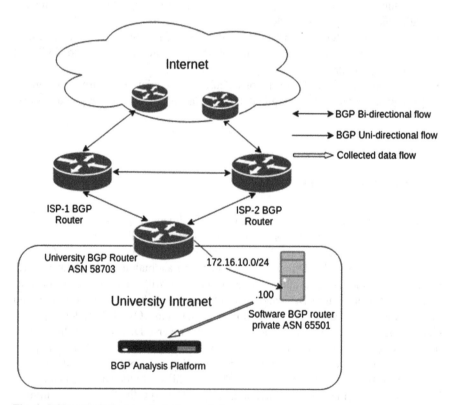

Fig. 1 BGP analysis framework deployment diagram

prefixes by rest of the global ASes. The collected BGP messages will be temporarily stored as files in the collector itself and will be pushed to our BGP Analysis Platform periodically (during every 15 min).

4 Implementation Details

B-Secure framework mainly focuses on identifying a suspicious path between a Source and Destination AS by analyzing two aspects from the path:

a. **Path Deviation**: The purpose of this analysis is to quantify the deviation in the observing path when compared with the known benign paths. Using Damerau–Levenshtein edit distance algorithm we can find the distance between two AS path sequence. Considering the following example showcasing two paths between ASN 65501 and ASN 42010:

New Path = 42010 **6939 1299** 6453 4755 45820 58703 65501
Historic Path = 42010 **174** 6453 4755 45820 58703 65501

In this case the edit distance score will be 2. Instead of ASN 174 in the historic path ASN 6939 and ASN 1299 are present in the new path. We propose a slight modification in the Damerau–Levenshtein edit distance algorithm, which we term it as weighted D-L edit distance approach that is explained with the help of Fig. 2 and example given below.

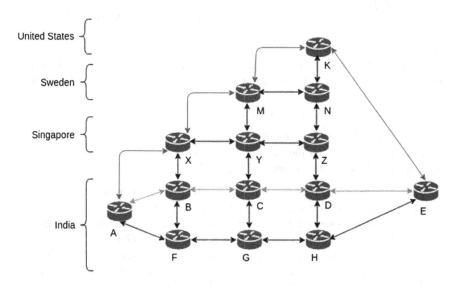

Fig. 2 Example for demonstrating weighted D-L algorithm

In Fig. 2, A is the source ASN and E is the destination ASN, belonging to India.

- Let Path 1, A-BCD-E be the usual AS Path seen between A and E. The intermediate nodes B, C, and D belong to India.
- Consider new path 2, A-FGH-E is seen with the intermediate nodes F, G and H belongs to India. This can be a less suspicious deviation as the routing happens within the same country.
- Consider new path 3, A-XMK-E. Here the intermediate node X belongs to Singapore, node M belongs to Sweden and node K belongs to US. This could be highly suspicious deviation.

In the above example, if we go by the standard D-L edit distance algorithm:

- Edit distance score from A-BCD-E to A-FGH-E is 3.
- Edit distance score from A-BCD-E to A-XMK-E is 3.

From these score, a lot of information is lost about the deviation occurred in the path.

To address this problem, we assign a weight factor to each transformation operation in the algorithm. As in Fig. 2, considering the paths, A-BCD-E and A-XMK-E, Damerau–Levenshtein transform score from D to K is 1. Instead, we add a weight factor—the number of hops in the shortest path connecting the nodes D and K.

$$D - L_{Distance}(P_1, P_2) = \min[\#Sub(i) + \#Del(i) + \#Ins(i) + \#Trns(i)] \quad (1)$$

$$W\_D - L_{Distance}(P_1, P_2) = \min \begin{bmatrix} \#Sub(i)^* wf + \#Del(i)^* wf + \\ \#Ins(i)^* wf + \#Trns(i)^* wf \end{bmatrix} \quad (2)$$

where

#Sub, #Del, and #Ins are the number of substitutions, deletions, insertions, and transpositions required in the session i and wf is the positive numbered weighting factor.

As in Fig. 2, the shortest path from:

- D to K is D-E-K and the number of hops is 2.
- C to M is C-Y-M and the number of hops is 2.
- B to X is B-X and the number of hops is 1.

Adding all the hops count 2 + 2 + 1 gives the weighted edit distance score as 5. Weighted edit distance score from A-BCD-E to A-FGH-E is still the same as 3.

Thus, weighted approach provides a higher degree of suspiciousness to the AS path A-XMK-E (score 5) compared to the AS path A-FGH-E (score 3) which reduces the information loss and helps to increase the efficiency of the system.

If the deviation of the path is more than the sum of mean and standard deviation of all the historic paths, we consider the path as suspicious.

b. **Path Credibility**: To identify the path credibility we first calculate the bigram and trigram frequencies of the intermediary nodes in the historic paths between a source and destination ASN. Let a path P = {O, X, Y, Z, D} where O and D are the origin and destination ASNs and X, Y, and Z are the intermediate ASNs. We calculate the overall probability of the path by

$$P(Path\_P) = p(X|O)^{*}p(Y|O,X)^{*}p(Z|X,Y)^{*}p(D|Y,Z) \tag{3}$$

If the probability of the path is less than a threshold value, we consider the path as suspicious.

5 Experimental Results

15 GB of data is being collected over a period of 18 months (July 2014 to March 2016) in which we were able to get 189.421638 million BGP control messages, 48564 unique ASNs and their neighbors details, 510,743 unique prefixes and their AS paths to different destination ASes. Initial 12 months data is used to train the system in order to build the graph connectivity among the ASes and to learn the communication behavior between the different ASes. Remaining 6 months of data we gave as the input to the system for the analysis.

Our system was able to identify 109 suspicious AS Paths for the given 6 months of input dataset. Listed are two case studies based on our experimental results:

Case 1:

Traffic from India to US deviates via Japan. Figure 3 shows the historic legitimate ASN path from 58703 (IN) to 747 (US). Figure 4 shows the identified suspicious ASN path from 58703 to 747. There is deviation in the path via Japan (725, 17676).

Case 2:

Traffic inside India deviates via US, Sweden, and UK. Figure 5 shows the historic legitimate ASN path from 58703 (IN) to 9583 (IN). Figure 6 shows the identified suspicious ASN path from 58703 to 9583. There is deviation in the path via US, Sweden, and UK (6453, 1299 and 1273 respectively).

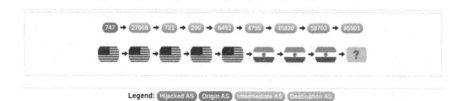

Fig. 3 Normal path between AS 58703 and AS 747

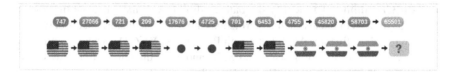

Fig. 4 Detected suspicious path between AS 58703 and AS 747

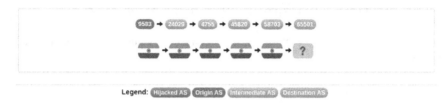

Fig. 5 Normal path between AS 58703 and AS 9583

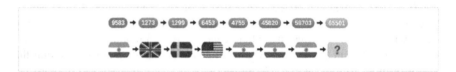

Fig. 6 Detected suspicious path between AS 58703 and AS 9583

We injected some of the well-known AS Path Hijack incidents to the system along with the normal input. 2013 Renesys reported path hijack, 2014 Syrian telecom path hijack were some of them. Our system successfully identified both these events as highly suspicious Path Hijack attacks.

6 Conclusion

In this research, we developed a framework, which could detect AS Path deviation by analyzing the BGP update messages. Currently the framework is focused on the weighted Damerau–Lavenshtein edit distance algorithm and the path probability calculation to determine the deviation and the credibility of an AS Path, respectively. There are more attributes like AS Path length, prefix announcement behavior by an AS, number of different countries where the path is spanning, integrating with publically available knowledge bases like ASN reputation, etc., which can be a part of the path reputation decision. We are looking forward to work on those aspects and leverage them using complex machine learning algorithms to increase the efficiency of the framework.

References

1. Faloutsos, Michalis, Petros Faloutsos, and Christos Faloutsos. "On power-law relationships of the internet topology." ACM SIGCOMM computer communication review. Vol. 29. No. 4. ACM, 1999.
2. Fuller, V., and T. Li. "IETF RFC 4632-Classless Inter-domain Routing (CIDR): The Internet Address Assignment and Aggregation Plan. Online document. Updated in August 2006. Cited on 7.8. 2010."
3. Butler, Kevin RB, et al. "A Survey of BGP Security Issues and Solutions. "Proceedings of the IEEE 98.1 (2010): 100–122.
4. Bates, Tony, Philip Smith, and Geoff Huston. "CIDR Report." CIDR Report. Web. 30 Apr. 2016. <http://www.cidr-report.org/as2.0/>.
5. Pilosov, Alex, and Tony Kapela. "Stealing the Internet: An Internet-scale man in the middle attack." NANOG-44, Los Angeles, October (2008): 12–15.
6. Turk, D. "Configuring BGP to block Denial-of-Service attacks." (2004).
7. Toonk, Andree. "BGP Routing Incidents in 2014, Malicious or Not?" BGPmon. 17 Feb. 2015. Web. 30 Apr. 2016. <http://www.bgpmon.net/bgp-routing-incidents-in-2014-malicious-or-not/>.
8. Cowie, Jim. "The New Threat: Targeted Internet Traffic Misdirection - Dyn Research." Dyn Research. 19 Nov. 2013. Web. 30 Apr. 2016. <http://research.dyn.com/2013/11/mitm-internet-hijacking/>.
9. Zheng, Changxi, et al. "A light-weight distributed scheme for detecting IP prefix hijacks in real-time." ACM SIGCOMM Computer Communication Review. Vol. 37. No. 4. ACM, 2007.
10. Jian Chang, Krishna K. Venkatasubramanian, Andrew G.West, Sampath Kannan, Insup Lee, Boon Thau Loo and Oleg Sokolsky, "AS-CRED: Reputation and Alert Service for Interdomain Routing".
11. Zhang, Jian, Jennifer Rexford, and Joan Feigenbaum. "Learning-based anomaly detection in BGP updates." Proceedings of the 2005 ACM SIGCOMM workshop on Mining network data. ACM, 2005.
12. Gersch, Joseph E. ROVER: A DNS-based method to detect and prevent IP hijacks. Diss. Colorado State University, 2013.
13. Shue, Craig A., Andrew J. Kalafut, and Minaxi Gupta. "Abnormally malicious autonomous systems and their internet connectivity." IEEE/ACM Transactions on Networking (TON) 20.1 (2012): 220–230.
14. Deshpande, Shivani, et al. "An online mechanism for BGP instability detection and analysis." Computers, IEEE Transactions on 58.11 (2009): 1470–1484.

Active Steganalysis on SVD-Based Embedding Algorithm

P.P. Amritha, Rithu P. Ravi and M. Sethumadhavan

Abstract Steganography is an art of hiding of secret information in an innocuous medium like an image. Most of the current steganographic algorithms hide data in the spatial or transform domain. In this paper, we perform attacks on three singular value decomposition-based spatial steganographic algorithms, by applying image processing operations. By performing these attacks, we were able to destroy the stego content while maintaining the perceptual quality of the source image. Experimental results showed that stego content can be suppressed at least by 40%. PSNR value was found to be above 30 dB and SSIM obtained was 0.61. Markov feature and BER are used to calculate the percentage of stego removed.

Keywords Radiometric operations · Geometric operations · Singular value decomposition · Markov feature · PSNR · SSIM

1 Introduction

Steganography is an act of covert communication, which means that only the authorized parties are aware of secret communication. To achieve this, the secret is hidden within an innocent-looking media such as text, image, audio, and video files. Some common steganographic methods [1] are implemented in the spatial and frequency domain of the multimedia signal. Currently, one of the large categories of techniques against digital steganography is steganalysis [2], which is a kind of

P.P. Amritha (✉) · R.P. Ravi · M. Sethumadhavan
TIFAC-CORE in Cyber Security, Amrita School of Engineering, Amrita Vishwa Vidyapeetham, Amrita University, Coimbatore, India
e-mail: pp_amritha@cb.amrita.edu

R.P. Ravi
e-mail: rithupravi@gmail.com

M. Sethumadhavan
e-mail: m_sethu@cb.amrita.edu

© Springer Nature Singapore Pte Ltd. 2017
S.C. Satapathy et al. (eds.), *Proceedings of the 5th International Conference on Frontiers in Intelligent Computing: Theory and Applications*, Advances in Intelligent Systems and Computing 515, DOI 10.1007/978-981-10-3153-3_77

method trying to detect and estimate the hidden information. Steganalysis can be broadly divided into two categories: detecting hidden information (passive attack) and disabling hidden information (active attack). Passive attack steganalysis can be classified into targeted steganalysis and universal steganalysis. Targeted steganalysis is effective only for certain kind of image steganographic method because attacks are designed based on the unique properties of their steganographic methods. Therefore, an important weakness for such steganalysis is that the type of the steganographic method has to be known in advance. So these methods are useful only for analyzing a certain group of suspicious targets rather than monitoring the general Internet multimedia streams. On the other hand universal steganalysis is claimed to be adaptive to multiple kinds of digital steganographic methods. However, these methods are based on the pattern classification and impossible to finish the learning process because of the lack of training information. Obviously, the above two methods are not able to classify and detect image streams on the real time. In all, the limitation of the steganalysis method is that, it is built on a passive way. An active steganalysis differs from a passive steganalyst because if the existence of a secret is found, the active warden would modify the work such that the integrity of the message is broken.

This paper focuses on performing and analyzing the active attacks on singular value decomposition (SVD)-based steganographic algorithms by applying image processing operations on the stego image. In particular, we mainly focused on radiometric and geometric operations, which when applied on the stego image will destroy the stego content to some extent. SVD-based steganographic algorithm resist most of the targeted attack like chi-square and RS steganalysis. These algorithms also resist noise attack, compression and cropping attack. So, only method to attack SVD-based algorithm is through active attack using some of the image processing operations which was successfully done as described in this paper. After performing image processing operations on the stego image, stego content was removed but at the same time these operations were able to maintain the perceptual quality of the source image. The performance of our method is measured using PSNR and structured similarity index measure (SSIM).

The remainder of the paper is organized as follows. Section 2 briefly reviews the SVD-based embedding algorithms used. The proposed active steganalysis method along with some metrics used to assess the performance of our system is described in Sect. 3. Section 4 presents the experimental results. The paper is concluded with some closing remarks in Sect. 5

2 SVD-Based Embedding Methods Used for Analysis

There are different variants of SVD-based steganographic algorithms which are discussed in [3, 4]. Normally SVD-based embedding algorithms are categorized into three. In the first category, embedding is done only in diagonal values (eigen values of the original matrix), in second embedding in any one of the orthogonal matrix and in third, embedding is performed using two orthogonal matrices and diagonal matrix. We have focused on the analysis of three SVD-based embedding algorithms mentioned below.

Method 1 is introduced by Chang et al. [5] in 2007. Here an image A is partitioned into small blocks, of size 4×4. Singular value decomposition is utilized to each block to get three component matrices $U,\ S,\ V$ in which U and V are orthogonal and S is the diagonal matrix. The secret message is embedded in the singular diagonal matrix.

Method 2 is developed by Chung et al. [6] in 2007. In this method, the secret message is embedded in any one of the orthogonal matrices.

Method 3 is a slight modification to the one introduced by Chanu et al. [7] in 2012. Instead of embedding secret in the low frequency sub bands of an image [7], in our work we have directly embedded the secret message in the left singular, right singular, and the singular values of the SVD of the image itself.

3 Proposed Active Steganalysis Attack Method

The main objective of active steganalysis is to destroy the hidden information while maintaining minimum distortion of source image. The active steganalysis done on stego image obtained after embedding using SVD-based algorithm, can be expressed as Eqs. (1) and (2)

$$I_d(x, y) = I(x, y) \otimes h(x, y) \tag{1}$$

$I_d\ (x,y)$ is the filtered image which is stego free
$I\ (x,y)$ is the stego image
$h\ (x,y)$ is the filters like Gaussian, Weiner, Average of window size 3, 5, 7

$$I_r(x, y) = R_\theta(I(x, y) \quad where \quad \theta \in \{\ 1,\ 5,\ 20\ \} \tag{2}$$

$I_r(x, y)$ is the rotated image
$R_\theta(I(x, y))$ is the operator that rotates stego image I by θ°

The stego image created using above-mentioned SVD-based stego algorithms (Sect. 2) was subjected to active steganalysis by applying image processing operations. 12 different operations were applied on the stego image. This process can suppress stego content in different proportions depending on the filters used. The radiometric operations used are Median, Gaussian and Wiener filters of different window sizes (3, 5 and 7), and the geometric operations include rotation through various angles (1, 5 and 20°). So a total of 12 different filters were applied on stego images to remove the stego content.

After performing active steganalysis we have extracted the secret from the filtered stego image. Then the extracted secret image is compared with the original secret image to check how much percentage of bits have been destroyed by operations. The quality of the source image is calculated using PSNR and SSIM. BER and Markov features are used to calculate the percentage of stego removed.

3.1 Effectiveness of Our Proposed Attack

It is necessary to assess the performance of attacking techniques in two different aspects to get an accurate comparison

3.1.1 Image Quality Metrics

PSNR [8] and SSIM [9] are used to assess the visual quality of the output image.

3.1.2 Metric for Calculating the Quantity of Stego Data Removed

We use two metrics to calculate the quantity of stego removed at different payload and hence to evaluate performance of our proposed system.

- Bit Error Rate: This metric is designed to capture bit by bit error in the secret message extracted after the showering process is applied on stego image with the original secret image.
- First Order Markov Feature: This proposed Markov feature set gives percentage of stego content removed. The feature calculation starts by forming the matrix along four directions: horizontal, vertical, diagonal, and minor diagonal (further denoted as $H\,(i,j)$, $V\,(i,j)$, $D\,(i,j)$, $MD\,(i,j)$, respectively).

Let the secret image be $I_1 = [x_{ij}]$ where $1 \leq i \leq M, 1 \leq j \leq N$ and the secret image extracted from restored image be $I_2 = [y_{ij}]$ where $1 \leq i \leq M, 1 \leq j \leq N$.

Four matrices are constructed as follows:

$$H(i,j) = [\partial(x_{ij}, y_{ij}) \,\&\&\, \partial(x_{ij+1}, y_{ij+1})] \; where \quad 1 \leq i \leq M, 1 \leq j \leq N-1$$
$$V(i,j) = [\partial(x_{ij}, y_{ij}) \,\&\&\, \partial(x_{i+1j}, y_{i+1,j})] \; where \quad 1 \leq i \leq M-1, 1 \leq j \leq N$$
$$D(i,j) = [\partial(x_{ij}, y_{ij}) \,\&\&\, \partial(x_{i+1j+1}, y_{i+1j+1})] \; where \quad 1 \leq i \leq M-1, 1 \leq j \leq N-1$$
$$MD(i,j) = [\partial(x_{i+1j}, y_{i+1j}) \,\&\&\, \partial(x_{ij+1}, y_{ij+1})] \; where \quad 1 \leq i \leq M-1, 1 \leq j \leq N-1$$

where $\&\&$ is the logical operator AND. We define delta function as

$$\partial(x,y) = 0 \begin{cases} if & x=y \\ else & 1 \end{cases}$$

The above matrices will represent the count giving the number of non-equal adjacent pixels from which we can calculate the percentage of stego content removed. This method is found to be more accurate than BER.

4 Experimental Results

Our database contains of textured and nontextured images of size 256 × 256 and secret message of size 64 × 64. These images were subjected to three SVD-based stego algorithms described in Sect. 2. 12 different filters were applied on stego images created from these algorithms.

4.1 Results on the Analysis of SVD-Based Embedding Methods

Tables 1, 2 and 3 shows the effect of threshold on image quality metric PSNR, Mean Square Error (MSE), SSIM, Entropy, normalized correlation (NC) in methods 1, 2, 3. We have found that when threshold value increases all the metric values were gradually decreasing and was not able to extract the secret back from the stego image. By experiments we fix the threshold value to 0.3, 0.6 and 0.50, 0.50, 0.30 for method 1, 2 and 3 respectively for effective retrieval of secret.

Table 1 Effect of threshold on different image quality metric in method 1

Threshold (T)	MSE	PSNR	Entropy	NC
0.30	**24.70**	**34.23**	**7.13**	**0.62**
0.60	33.16	32.95	7.14	0.56
0.85	36.18	32.57	7.14	0.50

Table 2 Effect of threshold on different image quality metric in method 2

Threshold (T)	MSE	PSNR	Entropy	NC
0.30	22.07	34.72	7.02	0.63
0.60	**23.56**	**34.44**	**6.83**	**0.62**
0.85	30.56	30.44	6.00	0.59

Table 3 Effect of threshold on different image quality metric in method 3

Threshold (T_1, T_2, T_3)	MSE	PSNR	Entropy	NC
0.60, 0.60, 0.30	53.89	30.84	6.72	0.53
0.50, 0.50, 0.30	**53.22**	**30.90**	**6.83**	**0.55**
0.50, 0.50, 0.50	54.52	30.79	6.83	0.46

4.2 Results on the Analysis of Attacks on SVD-Based Embedding Methods

From the results, we concluded that we can destroy the stego content to some extent in such a way that secret extracted from the filtered image is not enough to identify the original secret. Median filter of window sizes 3, 5, and 7 were able to destroy the secret completely when applied to all three methods. Wiener filter of window size 5 and 7 were able to destroy the secret completely when applied to all three methods but method 1 was resisting the attack by wiener filter of window size 3. Gaussian filter of window size 3 and 5 is not a good choice of attack because method 1 could resist this attack. By applying rotation 1 and 5 on method 3 secret cannot be completely destroyed hence rotation of degree 1 and 5 is not a good choice. When 20° rotation was applied on the stego images it was able to destroy the secret to large extent. All these results are summarized in Table 4.

Table 4 Comparison of our proposed attacks on different methods

Attacks	Method 1	Method 2	Method 3	
Median Filter	**Secret cannot be**	**Secret cannot be**	**Secret cannot**	be
(Window size 3)	**extracted**	**extracted**	**extracted**	
Median Filter	**Secret cannot be**	**Secret cannot be**	**Secret cannot be**	
(Window size 5)	**extracted**	**extracted**	**Extracted**	
Median Filter	**Secret cannot be**	**Secret cannot be**	**Secret cannot be**	
(Window size 7)	**extracted**	**extracted**	**Extracted**	
Gaussian Filter	Secret can be	Secret cannot be	Secret can be	
(Window size 3)	Extracted	extracted	Extracted	
Gaussian Filter	Secret can be	Secret cannot be	Secret can be	
(Window size 5)	extracted	extracted	Extracted	
Gaussian Filter	Secret can be	Secret cannot be	Secret cannot be	
(Window size 7)	extracted	extracted	Extracted	
Wiener Filter	Secret can be	Secret cannot be	Secret cannot be	
(Window size 3)	extracted	extracted	Extracted	
Wiener Filter	**Secret cannot be**	**Secret cannot be**	**Secret cannot**	be
(Window size 5)	**extracted**	**extracted**	**extracted**	
Wiener Filter	**Secret cannot be**	**Secret cannot be**	**Secret cannot be**	
(Window size 7)	**extracted**	**extracted**	**Extracted**	
Rotation by 1	Secret can be	Secret cannot be	Secret cannot be	
degree	extracted	Extracted	Extracted	
Rotation by 5	Secret can be	Secret cannot be	Secret cannot be	
degree	extracted	extracted	Extracted	
Rotation by 20	**Secret cannot be**	**Secret cannot be**	**Secret cannot be**	
degree	**extracted**	**extracted**	**Extracted**	

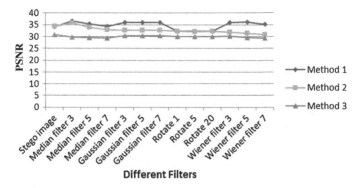

Fig. 1 PSNR value compared for three methods before and after attacks

From Figs. 1 and 2, we can observe that by performing our attack on stego algorithms we was able to maintain the PSNR and SSIM value above 30 dB and 0.60, respectively, on method 1 and 2. For method 3 PSNR and SSIM was 29 dB and 0.52, respectively. Figure 3 shows percentage of secret removed from the stego image after performing attacks.

Pictorial representation results of secret image extracted after performing attacks is shown in above Fig. 4. Median filter of size 3 (Fig. 4c), Weiner filter of size 5 Fig. 4d, Gaussian filter of size 3 Fig. 4e and Rotate 20° Fig. 4f, on the stegoed image Fig. 4a.

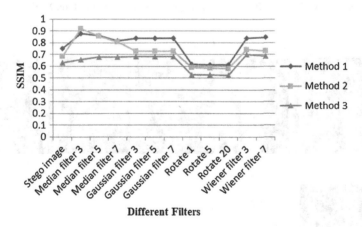

Fig. 2 SSIM value compared for three methods before and after attack

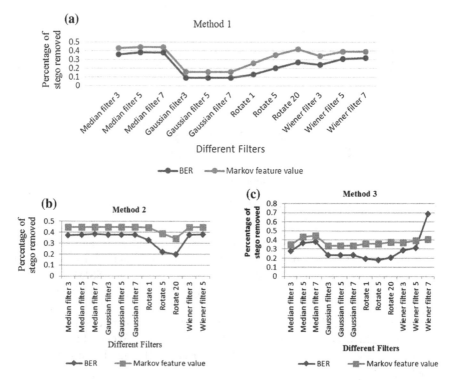

Fig. 3 **a** Percentage of stego removed by our attack on Method 1. **b** Percentage of stego removed by our attack on Method 2. **c** Percentage of stego removed by our attack on Method 3

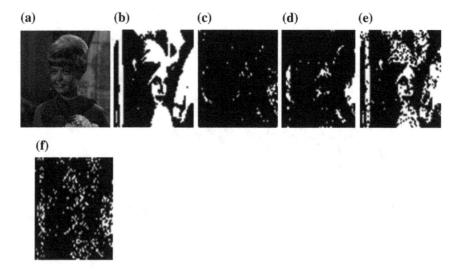

Fig. 4 **a** Stego image, **b** secret image, **c** secret extracted from median filtered image, **d** secret extracted Weiner filtered image, **e** secret extracted from Gaussian filtered image, **f** secret extracted from 20° rotated image

5   Conclusions

The proposed steganalysis provides an effective attack on hidden data, embedded using SVD-based embedding method, which was resisting compression, Impulse noise and cropping attack. In our proposed system by applying different filters, we found that varying percentage of stego was removed still maintaining the perceptual quality. We concluded that Median filter, Weiner filter of size 5 and 7 and Rotation of degree 20 is best in removing secret from the stego images embedded using all the three SVD methods. First order Markov feature gave better results than BER. By maintaining the perceptual quality, the above-mentioned image processing operations were able to destroy the stego content at least by 40%.

References

1. Amritha, P. P., & Gireesh, T. K. A Survey on Digital Image Steganographic Methods. In Cyber Security, Cyber Crime and Cyber Forensics: Applications and Perspectives, pp. 250–258. (2010).
2. R. Chandramouli., M. Kharrazi., N. Memon. Image steganography and steganalysis: Concepts and practice. In: T. Kalker, Y.M. Ro, and I. Cox, editors, Digital Watermarking, vol. 2939, pp. 35–49. Springer, Heidelberg (2004).
3. Y. J. Chanu., K. M. Singh., and T. Themrichon. A robust steganographic method based on singular value decomposition. Int. J. Inf. Comput. Technol 4 (7), pp. 717–726. (2014).
4. C. Bergman., and D. Jennifer. Unitary embedding for data hiding with the SVD. In Electronic Imaging, International Society for Optics and Photonics, pp. 619–630. (2005).
5. C. C. Chang, C.C. Lin and Y.S. Hu, An SVD oriented watermark embedding scheme with high qualities for the restored images, IJICIC, vol. 3, no. 3, pp. 609–620, (2007).
6. K.L. Chung, W.N. Yang, Y.H. Huang, S. T. Wu and Y. C. Hsu, On SVD- based watermarking algorithm, Elsevier Science Direct Applied Mathematics and Computation, vol. 188 pp. 54–57. (2007).
7. Chanu, Yambem Jina, Kh Manglem Singh, and Themrichon Tuithung. Steganography Technique based on SVD. International Journal of Research in Engineering and Technology (IJRET) vol. 6, pp. 293–297. (2012).
8. Gupta, P., Srivastava, P., Bhardwaj, S., & Bhateja, V. A modified PSNR metric based on HVS for quality assessment of color images. In Communication and Industrial Application (ICCIA), IEEE, pp. 1–4. (2011).
9. Bhateja, Vikrant, Aastha Srivastava, and Aseem Kalsi. Fast SSIM index for color images employing reduced-reference evaluation. International Conference on Frontiers of Intelligent Computing: Theory and Applications. Springer, vol. 247, pp. 451– 458. (2014).

Handwritten Mixed-Script Recognition System: A Comprehensive Approach

Pawan Kumar Singh, Supratim Das, Ram Sarkar and Mita Nasipuri

Abstract Most of the researchers around the world focus on developing mono-lingual Optical Character Recognition (OCR) systems. But in a multilingual country like India, it is quite common that a single document page includes text words written in more than one script. Therefore, OCRing such documents need a script identification module as a prerequisite. This paper reports a complete script recognition system for handwritten mixed-script documents. The document pages are first segmented into their corresponding text-lines and words. Then, the script recognition is done at word-level using texture-based features. The present technique is applied on 100 mixed-script document pages written in *Bangla* or *Devanagari* text mixed with *English* words. Encouraging outcomes would motivate more researchers to work on multilingual handwriting recognition domain.

Keywords Handwritten script recognition · Mixed-script document · Text-line segmentation · Word segmentation · Laws' texture energy measures · Tamura texture features

P.K. Singh (✉) · S. Das · R. Sarkar · M. Nasipuri
Department of Computer Science and Engineering, Jadavpur University, Kolkata, India
e-mail: pawansingh.ju@gmail.com

S. Das
e-mail: supratimdas21@gmail.com

R. Sarkar
e-mail: raamsarkar@gmail.com

M. Nasipuri
e-mail: mitanasipuri@gmail.com

© Springer Nature Singapore Pte Ltd. 2017   787
S.C. Satapathy et al. (eds.), *Proceedings of the 5th International Conference on Frontiers in Intelligent Computing: Theory and Applications*, Advances in Intelligent Systems and Computing 515, DOI 10.1007/978-981-10-3153-3_78

1   Introduction

There is an enormous diversity of the scripts used in the world. Almost every country has its own script(s), which can be distinguished from the others in different facades. Due to the colonial past, *English*, written in *Roman* script, is extensively used among the larger populace of India and it has also become the binding language due to the diversity of languages and scripts used in India. For this reason, during writing a text document in their native language, people in the Indian subcontinent spontaneously make use of *English* words. A document containing text information in more than one script is called a mixed-script document. In the scientific articles/text-books written in regional languages, it is really very common to have popular terms written in *English* along with the native language. Therefore, to convert such documents through OCR system, a script identification system to recognize the script of the documents is of pressing need as recognizing text documents written in different scripts by a common recognizer is not feasible.

Script identification can be done at any of the three levels: (a) Page-level, (b) Text-line level, and (c) Word-level. Among these, word-level script recognition is much more challenging task because the information gathered from a few characters present in a word image may not be enough, at times, for the identification of the scripts. When we deal with mixed-script documents, it is generally observed that script variation happens mostly at word-level rather than at text-line level. Hence, script identification at word-level remains an eventual choice for mixed-script documents.

The problem of script identification has received reasonable attention since last two decades. A detailed state-of-the-art on *Indic* script identification described by P. K. Singh et al. [1] reveals that a significant amount of work has been done at page-level [2–4], text-line level [5, 6], and word-level [7–10]. But one of the major drawbacks of the preceding techniques is that the researchers have tested their scheme on either individual document pages or text-lines or words taken directly as the input. In addition, they hardly deal with the mixed-script documents. So, a complete model for mixed-script documents is really missing till date. This has motivated us to design an automatic handwritten mixed-script recognition system for *Indic* documents, which inputs a bi-script document page and extracts its corresponding text-lines, words and finally, the scripts are identified at word-level using some conventional texture-based features.

2   Proposed Methodology

The proposed mixed-script recognition system consists of three basic modules: (1) text-line segmentation, (2) word segmentation, and (3) script identification. These modules are described below in detail.

2.1 Text-Line Segmentation

The extraction of text-lines from the document images have been performed using the method as described in [11]. Initially, the Gray-level Distance Transform (GDT) is computed on the gray-scale images. This is also known as modified geodesic distances which is the distance between any two pixels p and q and is defined as the minimum lengths of the paths $[p = p_0, p_1, \ldots, p_n = q]$. The length of the path, $l(p)$, can be written as

$$l(p) = \sum_{i=0}^{n-1} f(d(p_i, p_{i+1}), I(p_i), I(p_{i+1})) \tag{1}$$

where f is a distance and $d(p_i, p_{i+1})$ is the slope between any two consecutive pixels. *Seams* are then calculated by producing an energy map (or *seam map*) which finds the minimum cost of the optimal paths using dynamic programming [12]. The *seam map* is computed as follows:

$$\mathrm{MAP}[i,j] = 2GDT(i,j) + \min_{l=-1}^{1}(w_l * MAP[i+1, j-1]) \tag{2}$$

where $w_0 = -1$ and $(w_1 = w_{-1} = 1/\sqrt{2})$. Equation (2) is used to decide the optimal path containing the minimum cost. This is computed bidirectionally: (a) from left-to-right and (b) from right-to-left. These two resulting seam maps are then bilinearly interpolated into the final *seam map*.

This *seam map* produces seams of two types: *medial seams* and *separating seams*. A sequence of pixels belonging to the text region of a particular text-line is called a *medial seam*. A sequence of pixels lying below or above the text region is called a *separating seam*. As a result, an entire text-line region is determined by the *medial seam* whereas both the upper and lower boundaries of the corresponding text-line defines its *separating seams*. These two types of *seams* for each of the text-lines are considered in a given document image depending on the *seam maps*, which is calculated on the constructed GDT. The segmentation of text-lines for a document image written in *Bangla–Roman* mixed-script is shown in Fig. 1.

2.2 Word Segmentation

The word segmentation is performed by using one of our earlier technique described in [13]. The extraction of text words follows a two-stage approach. In the first stage, the key points are estimated using Harris corner point detection algorithm [14], whereas, in the second stage, the estimated key points are grouped into clusters using Density-Based Spatial Clustering of Applications with Noise (DBSCAN) clustering algorithm [15]. Finally, a convex hull is formed considering the corner points belonging to a particular cluster, which ultimately forms the required word

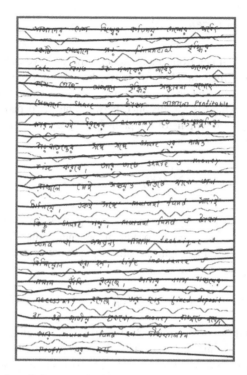

Fig. 1 Generation of *medial seams* (in *blue*) and *separating seams* (in *red*) for text-line segmentation on a sample handwritten *Bangla–Roman* mixed-script document image

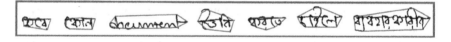

Fig. 2 Estimation of word boundaries from a text-line written in *Bangla–Roman* mixed-script

boundaries. Figure 2 displays an example of word formation from a sample text-line written in *Bangla* script mixed with *Roman* script words.

2.3 Extraction of Texture-Based Features

This module explores some well-known texture-based features for the identification of the script in which the words are written. These features are extracted from the word images by using local filter masks to detect the value of coarseness. The repetitive nature of the local regions, in turn, helps to identify similar script words. Laws [16] developed a texture energy approach that measures the amount of variation of local pixel distribution within a fixed-sized window. This method applies a set of 5 × 5 filter masks and then the texture energy is computed using these masks. This energy can be represented in the form of a nine-element vector for each pixel of the word image being analyzed. The image characteristics are examined from the following vectors:

$$L5(\text{Level}) = \begin{bmatrix} 1 & 4 & 6 & 4 & 1 \end{bmatrix}$$

$$E5(\text{Edge}) = \begin{bmatrix} -1 & -2 & 0 & 2 & 1 \end{bmatrix}$$

$$S5(\text{Spot}) = \begin{bmatrix} -1 & 0 & 2 & 0 & -1 \end{bmatrix}$$

$$R5(\text{Ripple}) = \begin{bmatrix} 1 & -4 & 6 & -4 & 1 \end{bmatrix}$$

The $L5$ vector gives a center-weighted local average. The $E5$, $S5$ and $R5$ vectors detect edges, spots, and ripples, respectively. At first, local averaging is done by moving a sliding window across the segmented word image. This averaged value is then subtracted from each pixel. This pre-processing is done in order to make the average intensity of neighboring pixels close to zero. In this paper, we have chosen the window of size 15 × 15. The 2-D convolution masks are formed by the product of each vector. A list of 16 possible 5 × 5 masks is applied to the pre-processed image, producing 16 filtered images. From these 16 energy maps, we have selected a set of 9 final maps whose sum of filter elements is zero. Let $F_k(i,j)$ be the result of filtering with the kth mask at pixel (i,j). Then the texture energy map E_k for filter k is defined by

$$E_k(r,c) = \sum_{i=r-7}^{r+7} \sum_{j=c-7}^{c+7} F_k(i,j) \tag{3}$$

The nine resultant energy maps used in the present work are as follows: L5E5/E5L5, L5S5/S5L5, L5R5/R5L5, E5E5, E5S5/S5E5, E5R5/R5E5, S5S5, S5R5/R5S5, and R5R5. Finally, the five standard texture features *namely*, energy, entropy, contrast, correlation, and homogeneity [17] are computed from these maps which generate a 45-element (F1-F45) feature vector.

For quantitative analysis of texture images, Tamura [18] proposed 6 different texture-based features: coarseness, contrast, directionality, linelikeness, regularity, and roughness. But it is perceived from the literature analysis that the first three features are highly informative since they strongly correlate with human perception. So, for the present work, we have hired the first 3 features (F46-F48) using Tamura texture-based features. Hence, a set of 48 texture features (i.e., 45 + 3) are extracted from each of the word images for performing script recognition task. The results, described in the next section, validate that the combination of these two texture-based features provide complementary information to the classifier.

3   Results and Analysis

A database consisting of 100 handwritten document pages has been collected for the experimentation. Among them, 50 pages are written in *Bangla* mixed with *Roman* script words and the other 50 pages are written in *Devanagari* mixed with *Roman* script words. Scanning of these document pages is performed at 300 dpi resolution using a flat-bed HP scanner and images are then stored as gray-tone images. The well-known Gaussian filter [17] is applied for removing noisy pixels from the images. The accuracies of text-line and word segmentation methods are detailed in Tables 1 and 2 respectively. For evaluating the performances of these methods, both the over-segmentation and under-segmentation errors are considered. If a single text-line (or word) is erroneously broken down into two/more parts, then it is considered as an over-segmentation error (O), whereas if multiple text-lines (or words) are identified as a single text-line (or word), then we call it as an under-segmentation error (U). Denoting the number of actual text-lines (or words) present in the document page as T, the success rate (SR) of the said methods can be calculated as follows:

$$SR = \left[\frac{(T - (O + U)) * 100}{T} \right] \tag{4}$$

After performing the text-line and word segmentation algorithms, the well-known Multi Layer Perceptron (MLP) classifier has been used for the script recognition purpose. For designing the recognition model using MLP-based classifiers, back propagation-based learning algorithm is used iteratively. Here, the values of two parameters *namely*, learning rate (η) and momentum term (α) are taken as 0.6 and 0.7 respectively. The performance is also measured by varying the number of neurons present in its hidden layer. For bi-script identification of *Bangla/ Devanagari* with *Roman* script words, we have considered 48 neurons in its input layer, 30 neurons in its hidden layer and 2 neurons in its output layer, i.e., the configuration becomes 48-30-2. Though it is very uncommon, still we have also evaluated the proposed script identification scheme for tri-script scenario which proves the robustness of the said technique. For tri-script identification, the configuration is taken as 48-30-3. Each model is trained for 1000 epochs and the testing is done using 3-fold cross-validation scheme. Individual performances of both the

Table 1 Performance evaluation of the text-line segmentation on mixed-script documents

Database	Number of text-lines present	Number of text-lines found experimentally	SR (%)
Bangla–Roman	1177	1067	**90.65**
Devanagari– Roman	997	924	**92.68**

Table 2 Performance evaluation of the word segmentation on mixed-script documents

	Database	
	Bangla– Roman	Devanagari– Roman
Number of *Bangla/Devanagari* words present	5817	5830
Number of *Roman* words present	2013	1750
Total number of words present	7830	7580
Number of *Bangla/Devanagari* words found experimentally	5388	5625
Number of *Roman* words found experimentally	1896	1565
Total number of words found experimentally	7284	7190
SR (%)	**93.02**	**94.85**

texture-based features along with their combination are listed in Table 3. The accuracies of three different runs of script identification scheme on the datasets of *Bangla–Roman, Devanagari–Roman,* and *Bangla–Devanagari-Roman* scripts are found to be 93.17, 95.11, and 94.23%, respectively.

Table 4 provides a statistical performance analysis with respect to the six parameters for both bi-script and tri-script scenarios. As the number of samples of *Roman* script words is relatively less in comparison to *Bangla/Devanagari* words, the model probably became biased towards (i.e., overfitting) these words, which, in turn, causes the misclassification errors. Due to cursive handwriting styles of the individuals, sometimes the Matra, a prominent feature in *Bangla/Devanagari* script, is absent in the writing. This causes the misclassification too. Again, certain alphabets of *Bangla* and *Devanagari* scripts are misclassified between them due to high structural similarities.

Table 3 Recognition accuracies of the present script identification technique using threefold cross-validation scheme (shaded cells indicate best accuracies)

Scenario	Script	Number of words trained	Number of words tested	Recognition Accuracies (%)		
				F1– F45	F46– F48	F1– F48
Biscript	*Bangla*	3878	1939	**89.51**	**55.61**	*93.17*
	Roman	1342	671			
	Total	**5220**	**2610**			
	Devanagari	3887	1943	**92.93**	**60.45**	*95.11*
	Roman	1167	583			
	Total	**5054**	**2526**			
Triscript	*Bangla*	3878	1939	**88.63**	**52.7**	*94.23*
	Devanagari	3887	1943			
	Roman	2509	1254			
	Total	**10274**	**5136**			

Table 4 Statistical performance analysis achieved by the proposed technique

Database	Script	TPR	FPR	Precision	Recall	F-measure	AUC
Bangla-Roman	*Bangla*	0.953	0.089	0.914	0.953	0.933	0.958
	Roman	0.911	0.047	0.951	0.911	0.930	0.966
Devanagari-Roman	*Devanagari*	0.957	0.055	0.946	0.957	0.951	0.980
	Roman	0.945	0.043	0.956	0.945	0.951	0.983
Bangla–Devanagari-Roman	*Bangla*	0.919	0.029	0.941	0.919	0.930	0.990
	Devanagari	0.959	0.019	0.963	0.959	0.961	0.997
	Roman	0.949	0.039	0.924	0.949	0.936	0.992

4 Conclusion

In this paper, we design a complete system to recognize the scripts from the mixed-script document images following the text-line and word segmentations. The words of the corresponding scripts are identified on the basis of texture-based features. MLP classifier has been used for classifying the scripts and an overall accuracy of 93.17, 95.11, and 94.23% have been attained for bi-script (*Bangla–Roman, Devanagari–Roman*) and tri-script scenarios, respectively. The aim of this paper is to explore the mixed-script handwritten OCR system and script-based retrieval of offline handwritten documents. In future, we will incorporate some post-processing technique for improving the error cases found during the segmentation of text-lines and words. Again, some shape-based features, complement to texture-based features, would be explored to enhance the script recognition accuracies. Another plan will be to develop handwritten mixed-script document databases for other regional languages too.

References

1. Singh, P.K., Sarkar, R., Nasipuri, M.: Offline Script Identification from Multilingual Indic-script Documents: A state-of-the-art. Computer Science Review (Elsevier). 15–16, 1–28 (2015).
2. Obaidullah, S.M., Kundu, S.K., Roy, K.: A System for Handwritten Script Identification from Indian Document. Journal of Pattern Recognition Research. 8, 1–12 (2013).
3. Padma, M.C., Vijaya, P.A.: Global Approach for Script Identification using Wavelet Packet Based Features. International Journal of Signal Processing, Image Processing and Pattern Recogntion. 20, 29–40 (2010).
4. Hiremath, P.S., Shivshankar, S., Pujari, J.D., Mouneswara, V.: Script identification in a handwritten document image using texture features. In: IEEE 2nd International Conference on Advance Computing. pp. 110–114 (2010).
5. Hangarge, M., Dhandra, B. V: Offline Handwritten Script Identification in Document Images. International Journal of Computer Applications (IJCA). 4, (2010).
6. Singh, P.K., Sarkar, R., Nasipuri, M.: Line-level Script Identification for six handwritten scripts using texture based features. In: 2nd Information Systems Design and In-telligent Applications, AISC. pp. 285–293 (2015).

7. Sarkar, R., Das, N., Basu, S., Kundu, M., Nasipuri, M., Basu, D.K.: Word level script Identification from Bangla and Devnagari Handwritten texts mixed with Roman scripts. Journal of Computing. 2, 103–108 (2010).

8. Pati, P.B., Ramakrishnan, A.G.: Word level multi-script identification. Pattern Rec-ognition Letters. 29, 1218–1229 (2008).

9. Singh, P.K., Sarkar, R., Das, N., Basu, S., Nasipuri, M.: Identification of Devnagari and Roman script from Multiscript Handwritten documents. In: 5th International Conference on PReMI, LNCS 8251. pp. 509–514 (2013).

10. Singh, P.K., Mondal, A., Bhowmik, S., Sarkar, R., Nasipuri, M.: Word-level Script Identification from Multi-script Handwritten Documents. In: 3rd International Conference on Frontiers in Intelligent Computing Theory and Applications (FICTA). pp. 551–558 (2014).

11. Saabni, R., Asi, A., El-Sana, J.: Text line extraction for historical document images. Pattern Recognition Letters. 35, 23–33 (2014).

12. Saabni, R., El-Sana, J.: Language-independent text lines extraction using seam carving. In: IEEE International Conference on Document Analysis and Recognition. pp. 563–568 (2011).

13. Singh, P.K., Chowdhury, S.P., Sinha, S., Eum, S., Sarkar, R.: Page-to-Word Extraction from Unconstrained Handwritten Document Images. In: 1st International Conference on Intelligent Computing and Communication(ICIC2) (2016).

14. Harris, C., Stephens, M.: A combined corner and edge detector. Alvey vision Conference. 15, (1988).

15. Ester, M., Kriegel, H.P., Sander, J., Xu, X.: A density-based algorithm for discovering clusters in large spatial databases with noise. In: 2nd International Conference on Knowledge Discovery and Data Mining. pp. 226–231 (1996).

16. Laws, K.: Rapid Texture Identification. Image Processing for Missile Guidance. SPIE. 238, 376–380 (1980).

17. Gonzalez, R.C., Woods, R.E.: Digital Image Processing. vol. 1, Prentice-hall, (1992).

18. Tamura, H., Mori, S., Yamawaki, T.: Textural Features Corresponding to Visual Perception. IEEE Transactions on Systems, Man, and Cybernetics. 8, 460–473 (1978).

A Rule-Based Approach to Identify Stop Words for Gujarati Language

Rajnish M. Rakholia and Jatinderkumar R. Saini

Abstract Stop words removal is an important step in many natural language processing (NLP) tasks. Till now, there is no standardized, exhaustive, and dynamic stop word list created for documents written in Indian Gujarati language which is spoken by nearly 66 million people worldwide. Most of the existing stop words removal approaches are file or dictionary based, wherein a hard-coded static, nonstandardized, and individually created list of stop words is used. The existing approaches are time consuming and complex owing to file or dictionary preparation by collecting possible stop words from a large vocabulary, complex framework and a morphologically variant Gujarati document. Even the other proposed approaches in the literature are also very restricted due to their dependence on word-length, word-frequency, and/or training data set. For the first time in scientific community worldwide, this paper proposes a dynamic approach independent of all factors namely usage of file or dictionary, word-length, word-frequency, and training dataset. An 11 rule-based approach is presented focusing on automatic and dynamic identification of a complete list of Gujarati stop words. Extensive empirical evidence has been presented through deployment of proposed algorithm on nearly 600 Gujarati documents, categorized into routine and domain-specific categories. The respective results with 98.10 and 94.08% average accuracy show that the proposed approach is effective and promising enough for implementation in NLP tasks involving Gujarati written documents.

Keywords Gujarati · Natural Language Processing (NLP) · Rule-based approach · Stop word

R.M. Rakholia (✉)
School of Computer Science, R K University, Rajkot, Gujarat, India
e-mail: rajnish.rakholia@gmail.com

J.R. Saini
Narmada College of Computer Application, Bharuch, Gujarat, India
e-mail: saini_expert@yahoo.com

© Springer Nature Singapore Pte Ltd. 2017   797
S.C. Satapathy et al. (eds.), *Proceedings of the 5th International Conference on Frontiers in Intelligent Computing: Theory and Applications*, Advances in Intelligent Systems and Computing 515, DOI 10.1007/978-981-10-3153-3_79

1 Introduction

Natural language processing is a field of computational linguistic and the goal of NLP is to analyze, understand, and generate human understandable language. But this goal is not easy to reach, because different language has own grammatical structure. To understand dependency among words and sentences, ambiguity of word, and how to link those concepts together in a meaningful say it is challenging task in NLP [1].

1.1 Stop Words

Stop word is a word which has less significant meaning than other tokens. Identification of stop words and its removal process is a basic preprocessing phase in NLP and data mining applications. For any NLP tool there is no single universal list of stop words used for a specific language, because stop words list is generally domain specific [2].

1.2 Diacritics

Diacritic is a mark that is used to change the sound value of the character. Diacritic mark could be identifying by unique UTF-8 value. And by using with any consonant in Gujarati language, it is possible to produce multiple meaning. A list of diacritic marks presented in Table 1 and is further elaborated based on wide and rare usage of the concerned diacritic [3].

Table 1 Diacritics for Gujarati document

Widely Used Diacritics in Gujarati Document				
◌ા (U+0ABE)	◌િ (U+0ABF)	◌ી (U+0AC0)	◌ુ (U+0AC1)	◌ૂ (U+0AC2)
◌ે (U+0AC7)	◌ૈ (U+0AC8)	◌ો (U+0ACB)	◌ૌ (U+0ACC)	◌ં (U+0A82)
◌ૃ (U+0AC3)	◌ૅ (U+0AC5)	◌ૉ (U+0AC9)	◌્ (U+0ACD)	◌ઁ (U+0A81)
Rarely Used Diacritics in Gujarati Document				
◌ઃ (U+0A83)	◌ૄ (U+0AC4)	◌઼ (U+0ABC)	◌ૢ (U+0AE2)	◌ૣ (U+0AE3)

1.3 Gujarati Language

Gujarati is an official and regional language of Gujarat state in India. It is 23rd most widely spoken language in the world today, which is spoken by more than 46 million people. Approximately 45.5 million people speak Gujarati language in India and half million speakers are from outside of India that includes Tanzania, Uganda, Pakistan, Kenya and Zambia. Gujarati language is belongs to Indo-Aryan language of Indo-European language family and it is also closely related to Indian Hindi language [4].

1.4 Unicode Transformation Format (UTF)

Unicode Transformation Format (UTF) is a character set [5] which is used to display the character written in Indian languages. We have used 8-bit encoding system to process Gujarati written document which is not possible to display each character using American Standard Code for Information Interchange (ASCII). There are many representations of UTF including utf8, utf16 and utf32 in which UTF-8 is widely used in web technology and mobile application for Indian languages.

2 Related Works and Existing Approaches

Pandey and Siddiqui [6] prepared a list of stop words for Hindi language based on its frequency and some manually operations. For experiment they used EMILLE corpus dataset, precision, and recall was used for evaluation. By removing stop words from raw content, it is possible to improve the accuracy of retrieval [6].

Kaur and Saini [7] they presented natural language processing approach to identify stop words in Panjabi literature in which they concentrates on poetry and other news articles for data collection. They identify 256 stop words from selected category and released for public use [7]. Kaur and Saini [8] described pre-processing phases for Punjabi language, in which, they have manually analyzed the data set (Punjabi text documents) and identified 1,500 stop words. High-frequency terms occurring in document, they have also considered stop word [8]. Kaur and Saini [9] they have provided enhanced understanding of stop words in Panjabi language based on Part-of-speech tagging. They constructed data set from different five categories of Panjabi literature: natures, romantic, religious, patriotic and philosophical, are manually populated with 250 poems. They prepared 256 stop words manually, due to unavailability of Punjabi stop words in public domain [9].

Thangarasu and Manavalan [10] developed stemmer for Tamil language; stemming algorithm pay important role to create stop words list. They created a list of tokens which is available in text corpus. After shorting that list and based on token frequency they prepared stop words list and other words to be discarded [10].

Yao and Zen-wen [11] created list of 1289 Chinese-English stop words by combining domain-specific stop words with list of classical stop words [11]. For Mongolian language, [12] used entropy calculation to create stop words list. They calculate entry for each word that is available in initial created stop words list. To prepare final stop words list, they combine this result with Mongolian part-of-speech [12]. Alajmi et al. [13] have used statistical approach to generate stop words list for Arabic language [13].

Chauhan et al. [14] presented stemmer for Gujarati language by using rule-based approach to improve Information Retrieval System. They used Gujarati news paper corpus for experiment purpose and created list of 280 stop words based on a word which is frequently occurring and it is less importance in document [14]. Joshi et al. [15], presented stop word elimination approach for information retrieval (IR) of Indian Gujarati language to improve mean average precision (MAP). They have collected data from FIRE corpus, based on their experiment, they constructed list of 400 words which is less importance and extensively used in Gujarati language. They created 282 stop words list from constructed list by analyzing and manually inspection by linguistic expert [15]. Rakholia and Saini [16] they study and analyzed different stemmer algorithms and pre-processing approaches are available for Gujarati language to process Gujarati written document. Through of their literature they found that, stop words removal is important pre-processing step in natural language processing application [16].

Based on this detailed literature review of the most relevant research works found in research community, our analysis based on stop words identifying process for Gujarati written document, it has been found by us that most of the researchers have obtained average accuracy for training and testing phase for Gujarati stop word identification at 85 and 67%, respectively. This motivated us for the presented research work as there is no effective stop word identification method or approach developed for Gujarati written document, which can yield a performance enough to make it practically acceptable in real world.

3   Our Approach

We have used rule-based approach to identify stop words from Gujarati written document. We have not considered the length of the word to identify the stop word because Gujarati document can be written using consonants, vowels and diacritics signs as well. It is noteworthy to mention here that the length of the stop word found by methods used by other researchers hence is dependent on and influenced by the usage of diacritics as well. To design and implement a length independent approach, we have deployed the usage of the fact that each diacritic mark in written

Gujarati document considers a single character. Also, from linguistic computational perspective, each diacritic mark has a unique UTF-8 hexadecimal value.

Following rules are applied to identify stop words appearing in Gujarati document

Rule 1: All single consonant or vowel words, with or without diacritics, were considered stop word and eliminated, except only {મા, ચા, બા, દી, પી **and** ગૌ}.

For instance: With diacritics: {તે, જે, જી, તો, છે, મે, એ, કે etc. }
    Without diacritics: {ન, પ, સ, ક, વ, દ, છ, ઈ, ઉ etc.}

Rule 2: A word that contains three regular Gujarati characters other then diacritic sign, if a word is terminated with "થી" and if a middle character has "ાા" diacritic sign and first character has either "ાા" or "ાા" sign, then it was considered stop word and eliminated.

For instance: {તારાથી, કેનાથી, જેનાથી, તેનાથી, આનાથી, કોનાથી, એનાથી, હોવાથી etc.}

If a word that contains two regular Gujarati characters other then diacritic sign and if word is terminated with "થી", then it was considered stop word and eliminated.
For instance: {આથી, તેથી, જેથી, નથી etc.}

Rule 3: A word that contains only two regular Gujarati characters other then diacritics sign and if word is terminated with "રૂ" or "વો", then it was considered stop word and eliminated.

For instance: {આવૂ, કેવુ, જેવુ, લેવો, કેવો, તેવુ, etc.}

Rule 4: A word that contains only two regular Gujarati characters other then diacritics sign and the word is terminated with "ની" and word does not start by using this three diacritics sign {ાા, ાિ, ાી} then it was considered stop word and eliminated. Because in most cases, these three diacritic signs {ાા, ાિ, ાી} are used to make proper nouns (e.g., name of girls) in Gujarati language.

For instance: {તેની, એની, જેની, દેની, લેની, કોની etc.}

Rule 5: A word that contains only two regular Gujarati characters other then diacritics sign and if word is terminated by "ણે" with at least one diacritic sign and does not start with "થા", then it was considered stop word and eliminated.

For instance: {કોણે, એણે, તેણે, જેણે, ગણે, etc.}

Rule 6: A word that contains only two regular Gujarati characters other then diacritics sign and if word is terminated by "નુ" and starting character has only "ાે" diacritic sign, then it was considered stop word and eliminated.

For instance: {કેનુ, જેનુ, તેનુ, દેનુ, એનુ, સેનુ, etc.}

Rule 7: A word that contains only two regular Gujarati characters other then diacritics sign and if word is terminated by "ઈ" and first character either does not contain diacritic sign or have only "ૌ" diacritic sign, then it was considered stop word and eliminated.

For instance: {કઈ, ગઈ, લઈ, જઈ, કોઈ, etc.}

Rule 8: A word that contains only two regular Gujarati characters {હ, ત } other then diacritic sign and last character has at least one diacritic sign when first character has "ૌ" or "ૌ" sign, then the word under consideration was treated as a stop word and eliminated. Using this rule, it was also possible to identify past tense sentences written in Gujarati language.

For instance: {હતુ, હતી, હોત, હતો etc.}

Rule 9: A word that contains only two regular Gujarati characters other then diacritics sign and if word is terminated with "ને", then it was considered stop word and eliminated.

For instance: {કોને, તને, એને, કેને, અને, જેને, શાને, ઓને, રેને, તેને etc.}

Rule 10: A word that contains two regular Gujarati characters other then diacritic sign and if word is terminated with "મ" and first character contained at least one diacritic sign except "ા", then it was considered stop word and eliminated.

For instance: {જેમ, કેમ, તેમ, એમ, etc.}

Rule 11: A word that contains two or three regular Gujarati characters other then diacritic sign and if word is terminated with "તો" or "તે", then it was considered stop word and eliminated.

For instance: {થતે, એતો, જતો, કેતો, વખતે, આવતે, જતે etc.}

4 Comparison with Other Approaches

Almost researchers have created stop words list for Indian Gujarati language by manually inspection of linguistic expert and based on words frequency. A list of existing approaches that are used for Indian language to identify stop words is presented in Table 2.

Other than these approaches, statistical approach is also used to generate stop words list. In almost all existing approaches, first step is frequency calculation for each word. But in many cases a word that has high frequency with significant meaning in document, but it cannot be consider as stop word. Second, many researchers have used statistical approach for English language and they achieved good accuracy, because many stop words in English language does not have multiple form, for instance: "any", "is," "a," "the," "an." But for the Indian Gujarati

Table 2 Existing approaches

S. No.	References	Dataset/corpus	Existing approach	Language and no. of stop words
1	Pandey and Siddiqui [6]	EMILLE	Frequency-based and manual operations	Hindi (Not Provided)
2	Kaur and Saini [7]	Dataset from Poetry and news article	Manual inspection	Panjabi (256)
3	Kaur and Sharma [8]	Panjabi text document	Frequency based	Panjabi (1500)
4	Thangarasu and Manavalan [10]	Text corpus (Not Provided which corpus used)	Frequency based	Tamil (Not Provided)
5	Chauhan, Patel and Joshi [14]	Gujarati news paper corpus	Frequency based	Gujarati (280)
6	Joshi H et al. [15]	FIRE Corpus	Manual inspection	Gujarati (282)
7	Proposed Approach	Routine Gujarati and domain specific	Rule-based	Gujarati (Dynamic)

language statistical approach will lead to the loss of accuracy because single stop words has multiple form, for instance: "હજુ", "હતી", "હોત", "હતા".

4.1 Precise Benefits of Proposed Approach Over Existing Approaches

The research works found in the related literature are based on training dataset and/or the length of the word. The proposed approached is free from the length of the word as well as the requirement of the training data set. It is noteworthy to mention that deploying a training dataset often leads to biased training of the system, more so in absence of availability of a standard text corpus for a resource scarce language like Gujarati. The proposed approached is hence free from machine learning based techniques. The proposed approach is also, hence, free from the risk of getting obsolete with time.

4.2 Known Limitations of Proposed Approach

The proposed work, in its present state, "will not perform well" only in case of stop words that contain more than three characters. It will also "not perform well" with specific words that belong to a peculiar domain. Still, two points are worth mentioning here. Firstly, the phrase "will not perform well" here should be taken with a pinch of salt as the only detrimental thing from the system will be a slight reduction in the accuracy. Second, the probability of peculiar domain stop word identification is very less, more so during the usual text processing and natural language processing tasks for any language, again much more so for resource scarce language Gujarati. In neither case, the proposed rules prove to be injurious enough preventing the system from wide implementation and its acceptability with good reputation in the scientific community.

5 Empirical Setup and Results

Indeed there is no a priori definition of stop words and their handling is governed by the domain and application area they are used for. Still, the NLP tasks like machine translation (MT), POS-tagging, and classification make use of general stop-word removal phase. To say "general stop-word" removal emphasizes on the fact that there are words with high frequency and their removal helps in faster processing as well as also helps in dimension reduction in terms of space requirement. This paper does neither intend to highlight the domain or application area in which stop words should be removed, nor does it focus on the number of stop words to be removed. The scientific literature of natural language processing has many instances of stop-word removal. This is true for Gujarati language, other Indo-Aryan languages as well as various International languages. This paper emphasizes on the fact that if the stop words have to be removed for Gujarati documents, there is no need to implement word frequency-based approach, word-length based approach, or manual inspection. Exploiting the morphological structure and symmetry of Gujarati stop words, this paper proposes a rule-based approach for stop word removal from Gujarati documents. This approach could be used anywhere where general (i.e., non-application and non-domain specific) removal of stop words is required. Even for cases where application and domain-specific removal of stop words is required for extrinsic evaluation of any system, the proposed "generic" rules could be applied before implementing domain and application specificities. As the proposed rules could be applied anywhere where removal of stop words is required, we term them 'generic'.

This section described the source of data collection for empirical implementation of the proposed rules. The system was implemented using Java Server Pages (JSP) technology and the results follow.

5.1 Data Sets

The data was collected randomly from multiple free Gujarati websites, to avoid the bias of a single website on the proposed work. For experimental purpose,

373 documents were prepared for routine Gujarati document and each document contained more than 400 words. We also prepared 224 documents for domain-specific (medical and engineering) categories and each document in these categories contained more than 275 words.

5.2 Results

In Gujarati language, there is no automated tool readily available to calculate the accuracy. Hence, we had to manually go through each document and evaluate the performance of the system. The obtained results on accuracy were recorded side by side. The average accuracy of routine Gujarati written documents was obtained at 98.10%. Similarly, for domain-specific medical and engineering categories, the obtained average accuracy was 94.08%. We also pondered on the reasons of getting less accuracy for routine Gujarati written documents and found that the reason is the presence of stop words containing more than three characters. Similarly, the non-availability of 100% accuracy in case of domain-specific categories owes to the presence of peculiar domain biased words. The average accuracy of routine Gujarati written documents is greater than the average accuracy of specific domain category documents by 4.02% because of presence of many domain-specific words, in such documents, which were not identify by any rules.

6 Conclusion and Future Work

We have presented an effective approach to accurately identify and eliminate a high percentage of the stop words in the Gujarati written documents. The proposed work used rule-based approach to identify stop words dynamically. The average accuracy for routine Gujarati written documents was obtained at 98.10% and for the specific domain (Medical and Engineering), we got 94.08% accuracy. We advocate that these results are reproducible on other large corpuses of routine Gujarati written documents as well. We propose and strongly claim that this approach is more efficient than any other existing approaches, which are available for identification of stop words from Gujarati written documents. The approach to finding stop words that presented here is currently limited in its applicability only for the word that contains more than three characters and for the word that belongs to a specific domain. This is our focus for future work. The proposed approach can be well applied as a preprocessing step for many NLP tasks including text classification, information retrieval, as well as document clustering, to name a few.

References

1. Microsoft Research, Natural Language Processing [online] available: http://research. microsoft.com/en-us/groups/nlp/ [Feb 10 2016].
2. Wikipedia, Stop Words Basic [online] available: https://en.wikipedia.org/wiki/Stop_words [Feb 5, 2016].
3. Rakholia R and Saini J, "The Design and Implementation of Diacritic Extraction Technique for Gujarati Written Script Using Unicode Transformation Format", Proceeding of ICECCT, IEEE, 2015, pp. 654–659.
4. UCLC, Gujarati Language [online]: http://www.lmp.ucla.edu/Profile.aspx?LangID=85&menu= 004 [Feb 10 2016].
5. The Unicode Consortium, USA; The Unicode Standard [Online]. Available: http://www. unicode.org/standard/standard.html [December 15, 2015].
6. Pandey A and Siddiqui T, "Evaluating Effect of Stemming and Stop-word Removal on Hindi Text Retrieval", Proceedings of the First International Conference on Intelligent Human Computer Interaction, Springer, 2009, pp. 316–326.
7. Kaur J and Saini J, "POS Word Class based Categorization of Gurmukhi Language Stemmed Stop Words", accepted for publication in the proceedings of International Conference on ICT for Intelligent Systems (ICTIS-2015), supported by ACM, CSI and Information Security Research Association and held during November 28–29, 2015, Ahmedabad.
8. Kaur R and Sharma S, "Pre-processing of Domain Ontology Graph Generation System in Punjabi", International Journal of Engineering Trends and Technology, Volume 17 Number 3 – Nov 2014, pp. 141–146.
9. Kaur J and Saini J, "A Natural Language Processing Approach for Identification of Stop Words in Punjabi Language", published in International Journal of Data Mining and Emerging Technologies; ISSN: 2249-3212 (eISSN: 2249-3220); Indian Journals, New Delhi, India; vol. 5, issue 2, November 2015; pages 114–120.
10. Thangarasu M and Manavalan R, "Design and Development of Stemmer for Tamil Language: Cluster Analysis", International Journal of Advanced Research in Computer Science and Software Engineering, Volume 3, Issue 7, pp. 812–818, July 2013.
11. Yao Z and Ze-wen C, "Research on the construction and filter method of stop-word list in text Preprocessing", Fourth International Conference on Intelligent Computation Technology and Automation, 2011.
12. Zheng G and Gaowa G, "The Selection of Mongolian Stop Words", IEEE International Conference on Intelligent Computing and Intelligent Systems (ICIS), 2010.
13. Alajmi A. et al., "Toward an ARABIC Stop-Words List Generation", International Journal of Computer Applications, Volume 46– No. 8, May 2012.
14. Chauhan K, Patel R and Joshi H "Towards Improvement in Gujarati Text Information Retrieval by using Effective Gujarati Stemmer" Journal of Information, Knowledge and Research in Computer Engineering, Nov 12 TO Oct 13, Volume – 02, Issue – 02, Page 218.
15. Joshi H. et al., "To stop or not to stop — Experiments on stopword elimination for information retrieval of Gujarati text documents" Engineering (NUiCONE), 2012 Nirma University International Conference on, 6–8 Dec. 2012, Page 1–4, IEEE.
16. Rakholia R and Saini J, "A Study and Comparative Analysis of Different Stemmer and Character Recognition Algorithms for Indian Gujarati Script", published in International Journal of Computer Application (IJCA); Digital Library ISSN: 0975-8887; ISBN: 973-93-80883-64-4; Foundation of Computer Science, USA; vol. 106, issue 2; November 2014; pages 45–50; DOI: 10.5120/18496-9558

Author Index

© Springer Nature Singapore Pte Ltd. 2017
S.C. Satapathy et al. (eds.), *Proceedings of the 5th International Conference on Frontiers in Intelligent Computing: Theory and Applications*, Advances in Intelligent Systems and Computing 515, DOI 10.1007/978-981-10-3153-3